T0310426

VACUUM NANOELECTRONIC DEVICES

VACUUM NANOELECTRONIC DEVICES

NOVEL ELECTRON SOURCES AND APPLICATIONS

Anatoliy Evtukh

National Academy of Sciences of Ukraine, Kyiv, Ukraine

Hans Hartnagel

Technische Universität Darmstadt, Germany

Oktay Yilmazoglu

Technische Universität Darmstadt, Germany

Hidenori Mimura

Shizuoka University, Hamamatsu, Japan

Dimitris Pavlidis

Boston University, USA

This edition first published 2015

Registered office
John Wiley & Sons Ltd, The Atrium, Southern Gate, Chichester, West Sussex, PO19 8SQ, United Kingdom

For details of our global editorial offices, for customer services and for information about how to apply for permission to reuse the copyright material in this book please see our website at www.wiley.com.

Library of Congress Cataloging-in-Publication Data Applied for.

A catalogue record for this book is available from the British Library.

ISBN: 9781119037958

Set in 10/12pt TimesLTStd by SPi Global, Chennai, India
Printed and bound in Singapore by Markono Print Media Pte Ltd

1 2015

Contents

Preface

Vacuum micro- and nanoelectronics resulted from the interaction of vacuum electronics with the well-developed micro- and nanotechnologies. Solid-state technologies allowed miniaturization of the structures and consequently the devices. On the other hand, electrons in a vacuum can travel much faster with less energy dissipation than in any semiconductor where electron scattering plays an important role. These properties enable faster electron modulation and higher electron energies than with semiconductor structures. Therefore vacuum micro-nanoelectronic devices can operate at higher frequencies, higher power, and wider temperature range, as well as in high radiation environments.

Vacuum micro-nanoelectronics can find in the near future a variety of potential applications, such as miniaturized microwave power amplifier tubes, electron sources for microscopes in nanovision systems, miniaturized x-ray tubes, electron beam lithography systems, flat panel field emission displays, ultra-bright light sources, multiple sensors, and so on. One of the most attractive applications is the high frequency electronics. The growth in wireless and optical communications systems has closely followed the intensive growth in microelectronics. The need for high-frequency processing as well as transfer of large packets of electronic data via Internet, wireless systems, and telephony increases the demands on the bandwidth of these systems. Hardware used in these systems must be able to operate at higher frequencies and output power levels. In the case of solid-state semiconductor electronics the highest frequencies were obtained with GaAs, InP, and GaN based materials. They have inherently higher electron mobility compared to silicon. Usually high-frequency solid-state electronic devices have limited output power and difficulties in harsh environment. The alternative to them for high-frequency, high-power, and harsh environmental operation are vacuum nanoelectronics devices.

The device dimensions decreased to the range where quantum mechanical effects start and even become dominant. These quantum mechanical effects are important in current and future devices based on solid-state and vacuum nanoelectronics. The device performance can be adapted to create novel devices based on new physical principles with important functionalities.

The most important physical process in vacuum nanoelectronics is electron field emission, that is, quantum mechanical tunneling of electrons through the energy barriers. The continuous development of vacuum nanoelectronic devices is associated with research and development of novel efficient field emission cathodes.

For the realization of desirable electron beams in vacuum the nanostructured cathodes with quantum size phenomena are intensively researched and developed. New quantum mechanical effects in electron emitting cathodes and electron transport cause new properties of electron beam and improved parameters of vacuum nanoelectronic devices. Many applications aim at a tight electron beam with small energy distribution, high current density, and high-frequency electron density modulation.

Electron emission cathodes with quantum mechanical effects in cathodes or during electron transport are so-called quantum cathodes or quantum electron sources.

The purpose of this book is to describe the physical processes in nanocathodes with quantum phenomena as well as the electron transport through nanostructure into vacuum. It will also present and analyze the results on research and development of novel quantum electron sources that will determine the future development of vacuum nanoelectronics. The influence of quantum mechanical effects on high frequency vacuum nanoelectronic devices is also considered and analyzed.

The book contains nine chapters in total. It is divided into two parts. Part I (Chapters 1–5) describes the physical backgrounds of quantum electron sources, whereas Part II considers and analyses the novel electron sources with quantum effects and their applications.

All chapters of Part I are connected by their consideration of the peculiarities and influence of quantum effects on electron transport in nanocathodes and emission into vacuum. The presented consideration in many cases can be equally applied to solid-state nanostructures and electron field emitters.

In Chapter 1 the electron transport through energy barriers and wells is considered. Based on the transfer matrix technique the tunneling probability through different nanostructures, quantum barriers, and quantum wells is described. Tunneling through a triangular barrier at electron field emission is considered as a special case. The effects of charge trapping in barrier as well as temperature effects are also analyzed. The great attention is paid to resonant tunneling of electrons and time parameters of this process.

The electron supply function as an important part of electron field emission description is analyzed in Chapter 2 for three-, two-, one-, and zero dimension cases. The density of electron states and Fermi distribution function is considered for nanostructures. As an example of such analyses the two-dimensional electron gas in GaN/AlGaN heterojunction and quantum sized semiconductor films are presented.

Other important parameters during the analysis of electron field emission from semiconductors are band bending and work function. They are considered in Chapter 3. The surface space-charge region, quantization of electron energy spectra at semiconductor surface, image charge potential are analyzed in detail. The work function as the most important parameter at electron field emission and its changing at cathode coating with thin films, under electric field, temperature, and surface adatoms are considered here.

Chapter 4 is devoted to the consideration of current transport through the nanodimensional barriers at electron field emission. It combines with the three previous chapters in obtaining one of the most important electron field emission parameters, namely emission current. The currents through one barrier at electron field emission from semiconductors are considered. Special attention is paid to current transport through a double barrier resonant tunneling structure. Coherent, sequential, and single electron tunneling is analyzed thoroughly. The theoretical consideration of novel current transport mechanisms at electron field emission, namely resonant tunneling, two-step electron tunneling in nanoparticle and single-electron field emission through nanoparticles is presented.

For many applications it is important to have electron beam with narrow energy distribution of electrons. The electron energy distribution of emitted electrons is considered in Chapter 5. A simple theory and an experimental set up are presented. The peculiarities of the electron energy distribution spectra at emission from semiconductors and Spindt-type metal microtips are considered. Special attention is paid to electron energy distribution spectra of emitted electrons from silicon and nanocrystalline silicon due to their importance for perspective integration of solid-state and vacuum nanoelectronics devices.

In Chapter 6 the novel electron sources based on silicon with quantum effects are considered. Silicon electron sources are important first of all due to highly developed Si-based technologies and perspectives of integration of solid-state and vacuum nanoelectronics devices. The peculiarities of electron field emission from porous silicon, silicon tips with multilayer coating and laser formed silicon tips with thin dielectric layer are described. In all cases the peaks on emission current-voltage characteristics have been revealed and explained in frame of resonant tunneling theory. Electron field emission from $SiO_x(Si)$ and $SiO_2(Si)$ films containing Si nanocrystals is considered in detail. Special attention is paid to Metal-Insulator-Metal and Metal-Insulator-Semiconductor emitters. They are integrated in solid state and have plain design. The peculiarities of electron transport mechanism and role of nanoparticles are considered in detail and the importance of further development of Si-based electron field emitters is shown.

The wide bandgap semiconductors GaN and related materials (AlGaN, AlN) are promising for many applications in electronics and photonics due to their unique properties. The electron emission properties of GaN based cathodes are considered in Chapter 7. The electron sources based on wide bandgap semiconductors, namely AlGaN electron source, solid-state controlled emitter, emission from nanocrystalline GaN films, graded electron affinity source are described. The polarization field emission enhancement model is analyzed in detail. The resonant tunneling at electron field emission from nanostructured cathodes is also considered. Special attention is paid to electron field emission from GaN nanorods and nanowires. The photo-assisted field emission from GaN nanorods is also an important part of this chapter.

The carbon based cathodes are considered in Chapter 8. Different modifications of carbon materials are promising for application as electron sources. In this chapter the peculiarities of electron field emission from diamond, diamond-like carbon films, carbon nanotubes, graphene, and nanocarbon are considered and analyzed. Some features of material properties are taken into account in the proposed models of electron transport and electron field emission, such as negative electron affinity in the case of diamond, electrically nanostructured heterogeneous in the case of diamond-like carbon films, high electric field enhancement coefficient in the case of carbon nanotubes, graphene, and nanocarbons. The perspectives of carbon based materials for efficient electron field emission sources are emphasized.

In Chapter 9 some of the most important applications of quantum electron sources such as high frequency vacuum nanoelectronics devices are considered. The importance of resonant tunneling for realization of microwave devices is considered and function principles and parameters of field emission resonant tunneling diode are analyzed. Some proposals for generation of THz signals as in solid state and in field emission vacuum devices are presented. The possibility of Gunn effect integration at electron field emission is described. The theory of field emission microwave sources including gate-modulated current density is considered in detail. The possibility of microwave sources based on CNT FEAs (carbon nanotube field emission arrays) and their advantages are also analyzed.

This book may be a useful source of potential information for students of electrical engineering and physics, and also engineers and physics who research, develop, and apply vacuum nanoelectronic components and devices in a variety of technological fields.

The book was reviewed by highly qualified experts, who devoted hours of free time to improve the book with their critical comments and valuable suggestions. We would like to thank all of them.

We want to acknowledge our colleagues from V. Lashkaryov Institute of Semiconductor Physics, Ukraine (Prof. V. Litovchenko, Dr N. Goncharuk, Dr M. Semenenko), Technical University of Darmstadt, Germany (Dr -Ing. K. Mutamba, Dr -Ing. J.-P. Biethan, Prof. J. J. Schneider, Dr R. Joshi), and Shizuoka University, Japan ((Prof. Y. Neo), Hachinohe Institute of technology Japan (Prof. H. Shimawaki), Tohoku University, Japan (Prof. K. Yokoo)) for many years of fruitful collaborations in electron field emission research.

Special thanks to the Research Institute of Electronics, Shizuoka University (Japan) for financial support.

We also appreciate the permissions granted to us from the respective journals and authors to reproduce their original figures cited in this book.

The authors hope that this book will stimulate further interest to researches and developments in such important fields of solid-state and vacuum nanoelectronics and the readers of this book will find it useful.

May, 2014

A. Evtukh, (Kyiv, Ukraine)
H. L. Hartnagel (Darmstadt, Germany)
O. Yilmazoglu (Darmstadt, Germany)
H. Mimura, (Hamamatsu, Japan)
D. Pavlidis (Boston, USA)

Part One

Theoretical Backgrounds of Quantum Electron Sources

1

Transport through the Energy Barriers: Transition Probability

In this chapter electron transport through energy barriers and wells is considered. Based on transfer matrix technique, tunneling probability through different nanostructures, quantum barriers, and quantum wells is described. Tunneling through triangular barrier at electron field emission is considered as a special case. The effects of charge trapping in barrier and temperature effect are also analyzed. Great attention is paid to resonant tunneling of electrons and time parameters of this process.

1.1 Transfer Matrix Technique

In order to describe the electron transport through structure containing energy barriers and wells the matrix method is commonly used [1–3]. The matrix method is based on the continuity of the wave function and its first derivative at any heterostructures (Figure 1.1). It allows determining the incidence energy dependence of transmission probability. Using the envelope wave function under effective mass approximation the wave function of particle with the incident and reflected waves amplitudes of A_n and B_n at any segment n is:

$$\Psi_n(x) = A_n e^{ik_n x} + B_n e^{-ik_n x}. \tag{1.1}$$

with wave vector, k_n

$$k_n = \sqrt{2m'^*_n(E - U_n)/\hbar^2} \tag{1.2}$$

where E is the incident electron energy and U_n is the potential related to the reference n segment (Figure 1.1).

The following matrix equation can be written:

$$\begin{pmatrix} A_{n+1} \\ B_{n+1} \end{pmatrix} = \prod_{p=1}^{n} M_p \begin{pmatrix} A_1 \\ B_1 \end{pmatrix} \tag{1.3}$$

Vacuum Nanoelectronic Devices: Novel Electron Sources and Applications, First Edition.
Anatoliy Evtukh, Hans Hartnagel, Oktay Yilmazoglu, Hidenori Mimura and Dimitris Pavlidis.

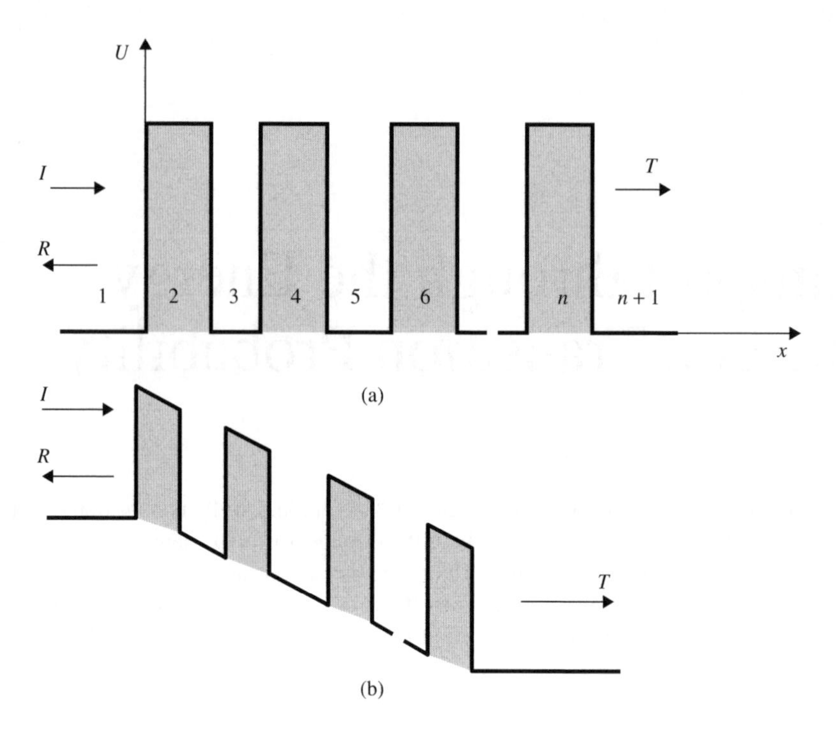

Figure 1.1 Multilayer structure with barriers and wells at (a) zero and (b) applied bias

The matrix M_p is generated by invoking the continuity of the wave function $\Psi(x)$ and its first derivative by properly accounting for the effective mass,

$$\frac{1}{m_n^*}\frac{d\Psi_n}{dx} \tag{1.4}$$

at the interface n. The transmission probability at any energy $T(E)$ is given as

$$T(E) = \frac{m_1^*}{m_{n+1}^*}\frac{k_{n+1}}{k_1}\left|\frac{A_{n+1}}{A_1}\right|^2. \tag{1.5}$$

If we assume $A_1 = 1$

$$T(E) = \frac{m_1^*}{m_{n+1}^*}\frac{k_{n+1}}{k_1}|A_{n+1}|^2. \tag{1.6}$$

To clarify the idea of transfer matrix technique let's consider one obstacle (potential barrier border) (Figure 1.2). Equation (1.3) for one barrier can be rewritten as

$$\begin{pmatrix} A_2 \\ B_2 \end{pmatrix} = M_1^{(21)} \begin{pmatrix} A_1 \\ B_1 \end{pmatrix} = \begin{pmatrix} M_{11}^{(21)}, M_{12}^{(21)} \\ M_{21}^{(21)}, M_{22}^{(21)} \end{pmatrix} \begin{pmatrix} A_1 \\ B_1 \end{pmatrix} \tag{1.7}$$

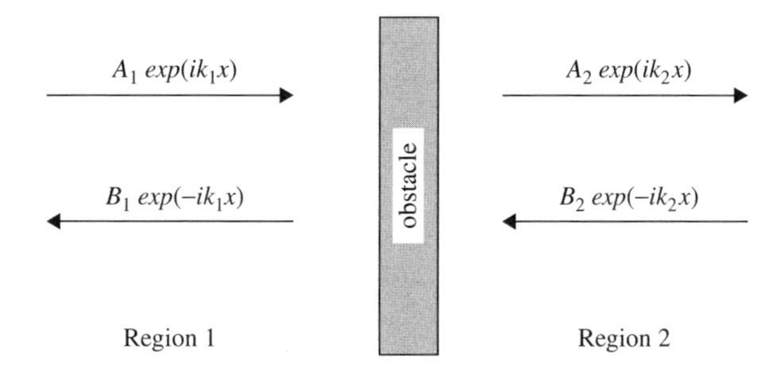

Figure 1.2 Scattering of quantum particle on one obstacle

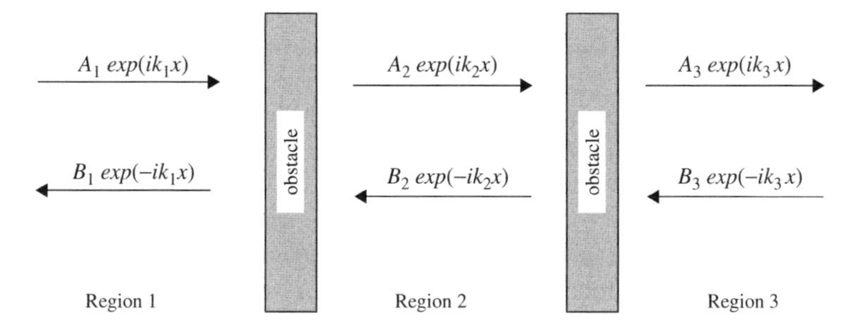

Figure 1.3 Scattering of quantum particle on two obstacles

There are no advantages of transfer matrix technique for scattering process on one barrier. But if we consider a more complicated process of subsequent scattering of particles on two barriers (Figure 1.3), the transfer matrix technique has significant advantages. The amplitudes of particle waves that move from region 1 into region 2 are given by wave amplitudes in region 1 and transfer matrix $M^{(21)}$. The wave amplitudes in region 3, in turn, are connected with wave amplitudes in region 2 by matrix $M^{(32)}$. Accordingly

$$\begin{pmatrix} A_2 \\ B_2 \end{pmatrix} = M_1^{(21)} \begin{pmatrix} A_1 \\ B_1 \end{pmatrix}, \quad \begin{pmatrix} A_3 \\ B_3 \end{pmatrix} = M_2^{(32)} \begin{pmatrix} A_2 \\ B_2 \end{pmatrix} \tag{1.8}$$

Then it is easy to connect wave amplitudes in region 3 with wave amplitudes in region 1:

$$\begin{pmatrix} A_3 \\ B_3 \end{pmatrix} = M_2^{(32)} M_1^{(21)} \begin{pmatrix} A_1 \\ B_1 \end{pmatrix} \equiv M^{(31)} \begin{pmatrix} A_1 \\ B_1 \end{pmatrix}. \tag{1.9}$$

Now it is easy to generalize the method of calculation of transmission coefficient of the quantum particle moving through the multilayer structure. Particle movement in the structure containing n barriers with known transmission coefficient for each of them is shown in Figure 1.4. Sequent consideration of scattering process on each barrier, as in the case of two barrier structure, allows us to write

$$\begin{pmatrix} A_n \\ B_n \end{pmatrix} = M^{(n,n-1)} M^{(n-1,n-2)} \ldots M^{(32)} M^{(21)} \begin{pmatrix} A_1 \\ B_1 \end{pmatrix} \equiv M^{(n1)} \begin{pmatrix} A_1 \\ B_1 \end{pmatrix}. \tag{1.10}$$

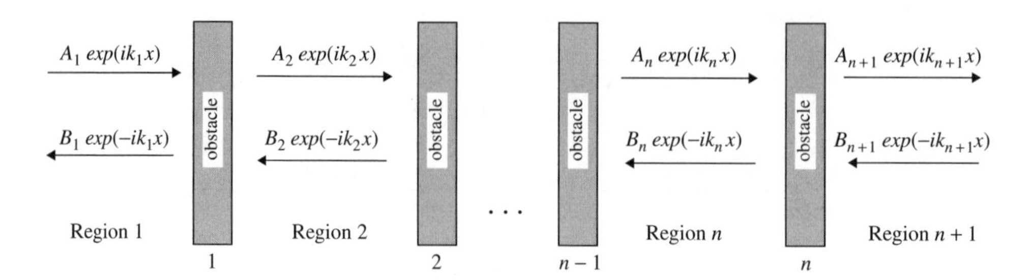

Figure 1.4 Scattering of quantum particle on n obstacles

Thus, to find the amplitudes of waves with n time scattering process, it is necessary simply to find corresponding transfer matrix, which is the product of n matrices for each scattering process.

In this way we obtain the very important result that is the base of transfer matrix technique, namely: transition coefficient in case of n barrier structure is the product of transition matrices of each barrier.

Sometimes instead of $M^{(n,n-1)}$ matrix that connects the wave function amplitudes of the n region from the wave function amplitudes of the $n-1$ region it is useful to use the $M^{(n-1,n)}$ matrix that connects the wave function amplitudes of the $n-1$ region from the wave function amplitudes of the n region. In that case Equation (1.3) can be written:

$$\begin{pmatrix} A_1 \\ B_1 \end{pmatrix} = \prod_{l=1}^{n} M_l^{/} \begin{pmatrix} A_{n+1} \\ B_{n+1} \end{pmatrix} \tag{1.11}$$

$$\begin{pmatrix} A_1 \\ B_1 \end{pmatrix} = M_1^{/} M_2^{/} \ldots M_n^{/} \begin{pmatrix} A_{n+1} \\ B_{n+1} \end{pmatrix}. \tag{1.12}$$

The reverse matrix for wave amplitude can be obtained by changing the matrices of wave vectors $k_1, k_2, k_3, \ldots k_n, k_{n+1}$ into $k_{n+1}, k_n, \ldots k_3, k_2, k_1$, respectively.

The full set of matrices includes the transition through barrier and well regions and borders between regions. In addition to matrices in Equations (1.10) and (1.12) at description of wave function transmission through the heterostructure it is necessary to use the additional matrices which characterize the wave transition inside the barriers and wells. Because of the wave function amplitude of the particle changes only at transition of the barrier border (obstacle) the moving of the particle inside the barrier or well regions causes only the wave function phase shift. The incident wave in point 0 of the barrier has view $A_2exp(ik_2x)$, and in point d the wave function is $A_2exp(ik_2x)\ exp(ik_2d)$. This can be represented by diagonal matrix

$$M_1^{(22)} = \begin{pmatrix} M_{11}^{(22)}, M_{12}^{(22)} \\ M_{21}^{(22)}, M_{22}^{(22)} \end{pmatrix} = \begin{pmatrix} e^{ik2d}, 0 \\ 0, e^{-ik2d} \end{pmatrix}, \tag{1.13}$$

and for reverse matrix

$$M_1^{/(22)} = \begin{pmatrix} e^{-ik2d}, 0 \\ 0, e^{ik2d} \end{pmatrix}. \tag{1.14}$$

To demonstrate the transfer matrix method let's consider some simple cases that are the basis for the creation of more complicated multilayer structures. In the following description of the electron transport through barriers and wells we will use reverse matrices.

1.2 Tunneling through the Barriers and Wells

The quantum description of the particle movement through barriers and wells includes the incident wave package, which represents the electron going from the left. This package will go to the barrier and some of them will be reflected, and some will be transmitted. The reflected part of the wave package will give the reflection probability of the electron, and the transmitted part will be the probability of passing on. The package is assumed to be wide, that the incident wave can be represented approximately by the wave function $A_1 exp(ik_1x)$, where $k_1 = \sqrt{2m_1^* E/\hbar^2} = \frac{2\pi}{\lambda_1}$.

Then the incident wave will give a constant in time density of probability at which the steady flow of electrons will be moving to the right. The average value of the flux density of probability will be $j_0 = (\hbar k_1/m) \times |A_1|^2$. So, despite the presence of flow, to maintain the constant density of probability there must be continuous addition of electrons from the left.

The integral of the normal component of the flux vector on a surface represents the probability that a particle crosses a specified surface in unit time. The flux densities of the incident, reflected, and transmitted particles can be written respectively as

$$j_0 = \frac{\hbar k_1}{m_1^*}|A_1|^2, \; j_r = \frac{\hbar k_1}{m_1^*}|B_1|^2, \; j_t = \frac{\hbar k_2}{m_2^*}|A_2|^2. \tag{1.15}$$

where A_1, B_1, A_2 are the amplitudes of incident, reflected, and transmitted waves respectively, k_1, k_2 are the wave-vectors in regions 1 and 2; and m_1^*, m_2^* are the effective masses of electrons in regions 1 and 2.

For simplicity we assumed $m_1^* = m_2^* = m^*$.

1.2.1 *The Particle Moves on the Potential Step*

A particle moving toward a finite potential step U_2 at $x = 0$ illustrates the reflection and tunneling effects which are basic features of nanophysics. Suppose $U = 0$ for $x < 0$ and $U = U_2$ for $x > 0$ (Figure 1.5).

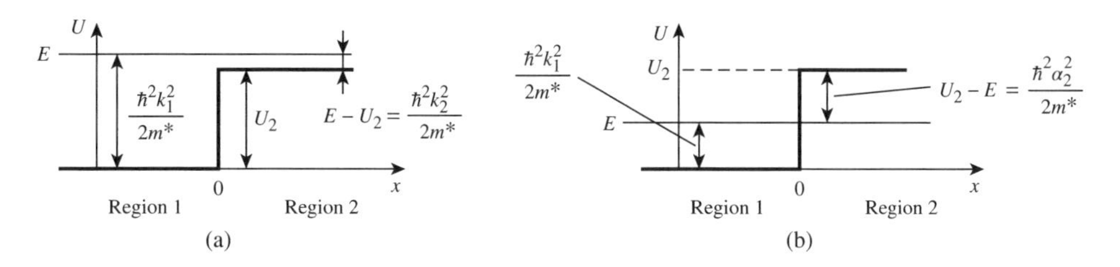

Figure 1.5 The particle moves on the potential step: $E > U_2$ (a) and $E < U_2$ (b)

Let's write a one-dimensional stationary Schrodinger equation for both regions.

For region 1 ($x < 0$)

$$\frac{d^2\Psi_1}{dx^2} + \frac{8\pi^2 m}{h^2} E\Psi_1 = 0, \tag{1.16}$$

for region 2 ($x > 0$)

$$\frac{d^2\Psi_2}{dx^2} + \frac{8\pi^2 m}{h^2} (E - U_2)\Psi_2 = 0, \tag{1.17}$$

where E is the total energy of the electron.

Then wave vectors of the particle moving in region 1 and region 2 are correspondingly

$$k_1 = \sqrt{2m_1^* E/\hbar^2} = \frac{2\pi}{\lambda_1}, \tag{1.18}$$

$$k_2 = \sqrt{2m_2^* (E - U_2)/\hbar^2} = \frac{2\pi}{\lambda_2}, \tag{1.19}$$

where λ_1 and λ_2 are the length of de Broglie waves in regions 1 and 2, respectively.

Using Equations (1.18) and (1.19) Equations (1.16) and (1.17) take the form

$$\frac{d^2\Psi_1}{dx^2} + k_1^2\Psi_1 = 0, \tag{1.20}$$

$$\frac{d^2\Psi_2}{dx^2} + k_2^2\Psi_2 = 0. \tag{1.21}$$

General solutions of these equations can be written as

$$\Psi(x) = \begin{pmatrix} A_1 e^{ik_1 x} + B_1 e^{-ik_1 x}, \ldots\ldots\ldots \; x \le 0 \\ A_2 e^{ik_2 x} + B_2 e^{-ik_2 x}, \ldots\ldots\ldots \; x \ge 0 \end{pmatrix} \tag{1.22}$$

The wave function of the particle can be considered as two plane waves that move in opposite directions.

Let's consider the features of the electron passing from region 1 to region 2 in two situations: when the total electron energy E is higher than its potential energy U_2 in region 2 (Figure 1.5a) and when $E < U_2$ (Figure 1.5b).

1.2.1.1 Case 1: $E > U_2$

Since the motion of an electron is a plane de Broglie wavelength, then at the regions border 1–2 the wave should be partly reflected and partly penetrated in region 2, or, in other words, moving from one region to another, the electron has a chance to reflect and a chance to go to another region (Figure 1.5a). Determination of these probabilities is the answer to the question about the peculiarities of the electron passing through a potential barrier. Remember that a particular solution to Equation (1.20) *exp(*ik_1x*)* characterizes the wave traveling toward the positive axis of X, that is, the incident wave, and the particular solution *exp(*$-ik_1x$*)* corresponds to the reflected wave. Similar assertions hold for partial solutions $exp(\pm ik_2x)$ Equation (1.22) for the second region ($x>0$). When $x<0$ both the incident and reflected waves extend, so we need to consider the general solution of Equation (1.22) where $|A_1|^2$ is the intensity of the incident wave, and $|B_1|^2$ is the intensity of the reflected waves.

The physical constraints on the allowable solutions are essential for solving this problem. First, $B_2=0$, since no particles are incident from the right (barrier). Second, at $x=0$ the required continuity of $\Psi(x)$ implies $A_1+B_1=A_2$. Third, at $x=0$ the derivatives, $d\Psi/dx=A_1ik_1exp(ik_1x)-B_1ik_1exp(-ik_1x)$ on the left, and $d\Psi/dx=A_2ik_2exp(ik_2x)$, on the right, must be equal. Thus

$$A_1+B_1=A_2, \tag{1.23}$$

and

$$k_2A_2=k_1(A_1-B_1). \tag{1.24}$$

Equations (1.23) and (1.24) are equivalent to

$$B_1=\frac{k_1-k_2}{k_1+k_2}A_1 \text{ and } A_2=\frac{2k_1}{k_1+k_2}A_1. \tag{1.25}$$

The reflection and transmission probabilities, R and T, respectively, for the particle flux are then Equation (1.15)

$$R=\frac{k_1|B_1|^2}{k_1|A_1|^2}=\left(\frac{k_1-k_2}{k_1+k_2}\right)^2 \tag{1.26}$$

and

$$T=\frac{k_2|A_2|^2}{k_1|A_1|^2}=\frac{4k_1k_2}{(k_1+k_2)^2}. \tag{1.27}$$

The same results can be obtained with using transfer matrix technique. In this case

$$\begin{pmatrix}A_1\\B_1\end{pmatrix}=M_1^{12}\begin{pmatrix}A_2\\B_2\end{pmatrix}. \tag{1.28}$$

Taking into account the continuity of the wave function and its first derivative at the interface we obtain

$$A_1+B_1=A_2+B_2, \tag{1.29}$$

$$k_1(A_1-B_1)=k_2(A_2-B_2). \tag{1.30}$$

We can determine the connection between coefficients that determine the amplitude of wave processes in region 1 (before barrier) and in region 2 (in the barrier).

$$A_1=\frac{1}{2}\left(1+\frac{k_2}{k_1}\right)A_2+\frac{1}{2}\left(1-\frac{k_2}{k_1}\right)B_2 \tag{1.31}$$

$$B_1=\frac{1}{2}\left(1-\frac{k_2}{k_1}\right)A_2+\frac{1}{2}\left(1+\frac{k_2}{k_1}\right)B_2 \tag{1.32}$$

and

$$M_1 = \begin{pmatrix} M_{11}, M_{12} \\ M_{21}, M_{22} \end{pmatrix} = \frac{1}{2} \begin{pmatrix} 1 + \dfrac{k_2}{k_1}, 1 - \dfrac{k_2}{k_1} \\ 1 - \dfrac{k_2}{k_1}, 1 + \dfrac{k_2}{k_1} \end{pmatrix}. \tag{1.33}$$

Taking into account that the particle moves from the left to the right and assume that amplitude of falling wave is equal to 1 ($A_I = 1$) we obtain for refraction coefficient of wave amplitude

$$r = \frac{B_1}{A_1} = B_1 = \frac{M_{21}}{M_{11}} = \frac{k_1 - k_2}{k_1 + k_2} \tag{1.34}$$

and for transmission coefficient of wave amplitude

$$t = \frac{A_2}{A_1} = A_2 = \frac{1}{M_{11}} = \frac{2k_1}{k_1 + k_2}. \tag{1.35}$$

At this the transmission coefficient for the particles is the ratio of particles that go through the barrier to the particles that fall on the barrier.

$$T = \frac{(\hbar k_2/m^*)}{(\hbar k_1/m^*)}|t|^2 = \frac{4k_1 k_2}{(k_1 + k_2)^2}. \tag{1.36}$$

So far as we assumed $m_1^* = m_2^* = m^*$.

The refractive coefficient for the particles is the ratio of particles that reflect from the barrier to the particles that fall on the barrier.

$$R = \frac{(\hbar k_1/m^*)}{(\hbar k_1/m^*)}|r|^2 = \frac{(k_1 - k_2)^2}{(k_1 + k_2)^2}. \tag{1.37}$$

It is easy to see that

$$T + R = 1. \tag{1.38}$$

Substituting in Equations (1.37) and (1.36) the wave vectors of de Broglie wave from Equations (1.18) and (1.19), we determine the reflection R and transmission T coefficients (Figure 1.6) depending on the ratio between the total energy E and potential U_2:

$$R = \left(\frac{1 - \sqrt{1 - \frac{U_2}{E}}}{1 + \sqrt{1 - \frac{U_2}{E}}} \right)^2 \tag{1.39}$$

and

$$T = 4\frac{\sqrt{1 - \frac{U_2}{E}}}{\left(1 + \sqrt{1 - \frac{U_2}{E}}\right)^2}. \tag{1.40}$$

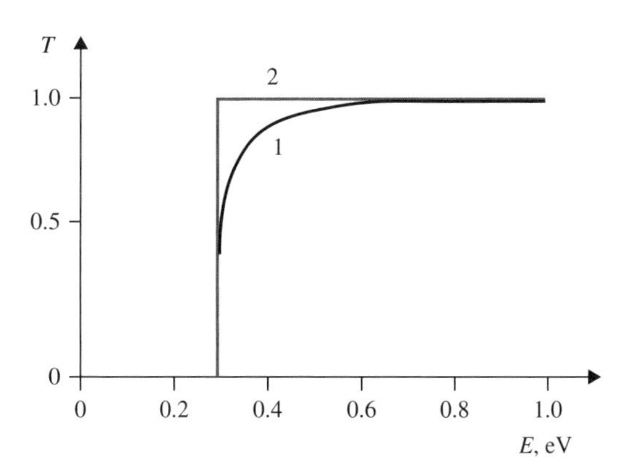

Figure 1.6 Energy dependence of the transmission coefficient of quantum particle (1) at moving over the potential step of 0.3 eV height. Curve 2 is the transmission coefficient in classical case

As can be seen from Equations (1.39) and (1.40), at $E = U_2$ $T = 0$, that is, the particle does not penetrate the barrier. At the electron energy E, twice the barrier, the reflection coefficient has reached quite appreciable value about 3%. These results are very different from the classical ones. In classical mechanics, a particle with energy $E \gg U_2$ always penetrate into the region 2 (at $E = U_2$ kinetic energy E_k is zero). But according to quantum mechanics the particle with $E > U_2$ has finite probability of electron reflection from the barrier.

1.2.1.2 Case 2: $E < U_2$

The only change is that now $E - U_2$ is negative, making k_2 an imaginary number (Figure 1.5b). For this reason k_2 is now written as $k_2 = i\alpha_2$, where

$$k_2 = i\alpha_2 = i\sqrt{2m_2^*(U_2 - E)/\hbar^2}; \tag{1.41}$$

α_2 is a real decay constant. Now the solution for the positive x becomes

$$\Psi(x) = A_2 \exp(-\alpha_2 x) + B_2 \exp(\alpha_2 x), \tag{1.42}$$

where

$$\alpha_2 = \sqrt{2m_2^*(U_2 - E)/\hbar^2}. \tag{1.43}$$

In this case, $T = 0$, to prevent the particle from unphysical collecting at large positive x. Equations (1.36) and (1.37) and Equation (1.25) remain valid setting $k_2 = i\alpha_2$.

$$T = \frac{(\hbar k_2/m^*)}{(\hbar k_1/m^*)}|t|^2 = \frac{4k_1 i\alpha_2}{(k_1 + i\alpha_2)^2}. \tag{1.44}$$

$$R = \frac{(\hbar k_1/m^*)}{(\hbar k_1/m^*)}|r|^2 = \frac{(k_1 - i\alpha_2)^2}{(k_1 + i\alpha_2)^2}. \tag{1.45}$$

It is seen that $R = 1$, because the numerator and denominator in Equation (1.45) are complex conjugates of each other, and thus have the same absolute value.

Thus, when $E < U_2$ reflection coefficient is 1, that is, the reflection is complete, however, despite the fact that the transmission coefficient $T = 0$, there is a nonzero probability of finding an electron in region 2. The solution for positive x is now an exponentially decaying function, and is not automatically zero in the region of negative energy. In other words, reflection does not occur at the boundary of two regions, while the electrons go at reflection at a certain depth in region 2, then return to the region 1. Indeed, at the imaginary value of k_2 the solution of Schrödinger Equation (1.22) for region 2 becomes

$$\Psi(x) = A_2 \exp(ik_2 x) = A_2 \exp(-\alpha_2 x), \tag{1.46}$$

and the probability of finding an electron per unit length in region 2 will be

$$\Psi_2 \Psi_2^* = |\Psi_2|^2 = A_2^2 \exp(-2\alpha_2 x). \tag{1.47}$$

Taking into account Equation (1.43) we obtain

$$|\Psi_2|^2 = A_2^2 \exp\left(-\frac{2}{\hbar}\sqrt{2m\left(U_2 - E\right)} \times x\right), \tag{1.48}$$

that is, there is a definite probability of finding the particle in region 2 at a depth of x from the boundary of two regions. However, this probability decreases exponentially with distance from the interface. Thus, when $x = 0.1$ nm and $U_2 - E = 1$ eV the probability of finding an electron is equal to about 0.3, while at $x = 1$ nm the probability is already an order of 10^{-8}. Electron passes into the barrier and turns back, so that the total flux of particles in region 2 is zero. From the wave point of view, this effect is similar to the case of total internal reflection of light, when even at angles greater than critical in the less dense medium is the wave field with exponentially decreasing amplitude, but the flow of energy through the interface over a sufficiently long period of time is equal to zero.

We can determine $|A_2|^2$ from Equation (1.35) assuming $A_l = 1$, setting $k_2 = i\alpha_2$, and forming $|A_2|^2 = A_2 A_2^*$. It is the probability to find the particle at interface ($x = 0$).

$$|A_2|^2 = \frac{4k_1^2}{(k_1^2 + \alpha_2^2)} = \frac{4E}{U_2}, \tag{1.49}$$

where $E = (\hbar^2 k_l^2/2\,m) < U_2$. Note that $|A_2|^2 = 0$ for an infinite potential. Also, this expression agrees in the limit $E = U_2$ with Equation (1.35).

Thus, the probability of finding the particle in the forbidden region of positive x is

$$P(x > 0) = \frac{4E}{U_2}\int_0^{\infty} \exp(-2\alpha_2 x)dx = \frac{4E}{U_2}\int_0^{\infty} \exp\left(-\frac{2}{\hbar}\sqrt{2m\left(U_2 - E\right)} \times x\right)dx = \frac{2E}{\alpha_2 U_2}, \tag{1.50}$$

where $E < U_2$.

1.2.2 The Particle Moves above the Potential Barrier

In this case the structure is more complicated because the potential barrier has finite width (Figure 1.7). In contrast to the infinitely wide barrier (potential step), the reflection of electrons will take place both on the border of regions 1 and 2, and on the boundary of regions 2 and 3. Solutions of Schrödinger equations for these regions can be written as

$$\Psi(x) = \begin{cases} A_1 e^{ik_1 x} + B_1 e^{-ik_1 x}, \ldots\ldots\ldots x < x_1 \\ A_2 e^{ik_2 x} + B_2 e^{-ik_2 x}, \ldots\ldots\ldots x_1 < x < x_2 \\ A_3 e^{ik_3 x} + B_3 e^{-ik_3 x}, \ldots\ldots\ldots x > x_2 \end{cases}. \tag{1.51}$$

The particle has energy $E > U_2$. Then wave vectors of particle in region 1, region 2, and region 3 are correspondingly

$$k_1 = \sqrt{2m_1^* E/\hbar^2} \tag{1.52}$$

$$k_2 = \sqrt{2m_2^*(E - U_2)/\hbar^2} \tag{1.53}$$

$$k_3 = \sqrt{2m_1^* E/\hbar^2}. \tag{1.54}$$

To determine the reflection R and transmission T coefficients, we must first find the waves amplitudes A_j and B_j. For this we use the boundary conditions: continuity of Ψ function and its derivative at the boundaries of regions 1–2 and 2–3, that is, at $x = x_1 = 0$ and $x = x_2 = L$. These conditions can be written as

$$(\Psi_1)_{x=0} = (\Psi_2)_{x=0}, \quad \left(\frac{d\Psi_1}{dx}\right)_{x=0} = \left(\frac{d\Psi_2}{dx}\right)_{x=0}, \tag{1.55}$$

$$(\Psi_2)_{x=L} = (\Psi_3)_{x=L}, \quad \left(\frac{d\Psi_2}{dx}\right)_{x=L} = \left(\frac{d\Psi_3}{dx}\right)_{x=L}. \tag{1.56}$$

Solving the system Equations (1.55) and (1.56), we can find an expression for the A_3 because it determines the transmittance T (at $A_1 = 1$):

$$A_3 = \frac{4k_1 k_2 e^{ik_1 d}}{(k_2 + k_1)^2 e^{-ik_2 d} - (k_2 - k_1)^2 e^{ik_2 d}}. \tag{1.57}$$

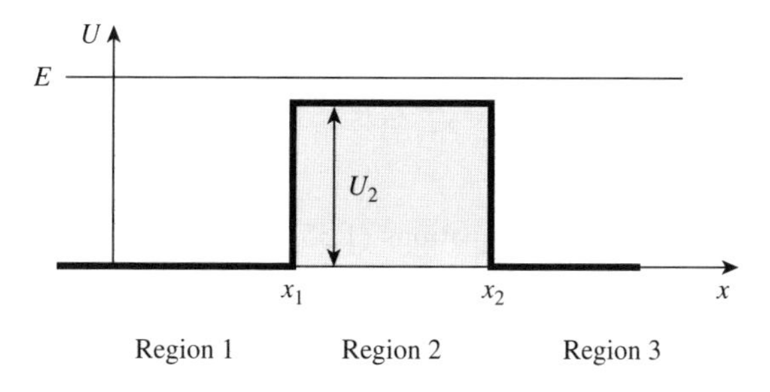

Figure 1.7 The particle moves above the potential barrier

Transmission coefficient is equal to

$$T = \frac{k_3|A_3|^2}{k_1|A_1|^2} = |A_3|^2 = A_3A_3^*, \tag{1.58}$$

where $k_1 = k_3$.

Transfer matrix method simplifies the procedure. In this case

$$\begin{pmatrix} A_1 \\ B_1 \end{pmatrix} = M_1M_2M_3\begin{pmatrix} A_3 \\ B_3 \end{pmatrix} \tag{1.59}$$

where matrix M_1 describes the transition of the border 1–2 from region 2 to region 1 (point x_1)

$$M_1 = \begin{pmatrix} M_{11}, M_{12} \\ M_{21}, M_{22} \end{pmatrix} = \frac{1}{2}\begin{pmatrix} 1+\dfrac{k_2}{k_1}, 1-\dfrac{k_2}{k_1} \\ 1-\dfrac{k_2}{k_1}, 1+\dfrac{k_2}{k_1} \end{pmatrix}. \tag{1.60}$$

Diagonal matrix M_2 describes the phase changing of Ψ_2 during the transition of region 2 (barrier).

$$M_2 = \begin{pmatrix} e^{-ik_2d}, 0 \\ 0, e^{ik_2d} \end{pmatrix} \tag{1.61}$$

and M_3 describes the transition of the border 2–3 from region 3 to region 2 (point x_2)

$$M_3 = \frac{1}{2}\begin{pmatrix} 1+\dfrac{k_3}{k_2}, 1-\dfrac{k_3}{k_2} \\ 1-\dfrac{k_3}{k_2}, 1+\dfrac{k_3}{k_2} \end{pmatrix}. \tag{1.62}$$

Multiplication of the matrices gives such expression for the final matrix

$$M = \begin{pmatrix} M_{11}^{(13)}, M_{12}^{(13)} \\ M_{21}^{(13)}, M_{22}^{(13)} \end{pmatrix} = \frac{1}{4k_1k_2}\begin{pmatrix} (k_2+k_1)^2e^{-ik_2d} - (k_2-k_1)^2e^{ik_2d}, (k_2^2-k_1^2)e^{-ik_2d} - (k_2^2-k_1^2)e^{ik_2d} \\ -(k_2^2-k_1^2)e^{-ik_2d} + (k_2^2-k_1^2)e^{ik_2d}, -(k_2-k_1)^2e^{-ik_2d} + (k_2+k_1)^2e^{ik_2d} \end{pmatrix} \tag{1.63}$$

We assume $k_1 = k_3$. According to Equations (1.35) and (1.36) the transmission coefficient can be represented as

$$T = \frac{1}{|M_{11}|^2} = \left|\frac{4k_1k_2}{(k_2+k_1)^2e^{-ik_2d} - (k_2-k_1)^2e^{ik_2d}}\right|^2 \tag{1.64}$$

It is possible to write the final result as

$$T = \left(1 + \frac{\left(k_2^2 - k_1^2\right)^2}{4k_1^2 k_2^2} \sin^2 k_2 d\right)^{-1} \tag{1.65}$$

Note that for integer values of k_2d/π the transmission coefficient, as can be seen from Equation (1.65), equals to 1, that is, the above barrier reflection of the particle is absent.

In this case, twice the length of the potential barrier fits the de Broglie wavelength of the particle $\lambda = 2\pi/k_2$ an integer number of times. These waves cancel each other. At given particle energy the transmission coefficient T as the function of barrier thickness d changes periodically from $T_{min} = 4k_1{}^2k_2{}^2/(k_1{}^2 + k_2{}^2)^2$ to $T_{max} = 1$ with a period of $\lambda/2$.

In this case the refractive coefficient R is equal to

$$R = \left(\frac{M_{21}}{M_{11}}\right)^2 = \left|\frac{-\left(k_2^2 - k_1^2\right) e^{-ik_2 d} + (k_2^2 - k_1^2)e^{ik_2 d}}{(k_2 + k_1)^2 e^{-ik_2 d} - (k_2 - k_1)^2 e^{ik_2 d}}\right|^2 \tag{1.66}$$

and

$$T + R = 1 \tag{1.67}$$

We rewrite the Equations (1.65) and (1.66) the using the Equations (1.52)–(1.54) in energy view.

The transmission coefficient is equal to

$$T = \left(1 + \frac{U_2^2}{4E\left(E - U_2\right)} \sin^2 k_2 d\right)^{-1}, \tag{1.68}$$

and the reflection coefficient

$$R = 1 - T = \left(1 + \frac{4E\left(E - U_2\right)}{U_2^2 \sin^2 k_2 d}\right)^{-1}. \tag{1.69}$$

Equations (1.68) and (1.69) show that at $T = T_{min}$ the reflection coefficient is $R = R_{max}$.

The most interesting consequence of Equations (1.68) and (1.69) is the appearance of oscillations of transmission and reflection coefficients in dependence on the electron energy E. The oscillation period corresponds to the condition

$$\sin^2(k_2 d) = 0 \text{ or } k_2 d = n\pi, \tag{1.70}$$

where $n = 1, 2, 3$, and so on.

At this condition the transmission coefficient of an electron with the wave vector k_2 is $T = 1$, and the reflection coefficient $R = 0$. In this case the integer of half de Broglie wave is placed on the barrier width d for electrons with the wave vector k_2, or with a given energy $E_n = E - U_2$. Indeed, substituting $k_2 = 2\pi/\lambda_2$ in Equation (1.70) we have

$$\frac{2\pi}{\lambda_2} d = n\pi, \text{ or } d = n\frac{\lambda_2}{2}. \tag{1.71}$$

Semiclassically, this can be interpreted as the result of interference of waves reflected from the boundaries of the barrier, and the incident waves. The last expression can be used to determine the electron energy above the potential barrier

$$E_n = E - U_2 = \frac{mv^2}{2} = \frac{h^2}{2m\lambda_2^2}, \tag{1.72}$$

where $\lambda_2 = h/mv$. Substituting the λ from Equation (1.71), we have

$$E_n = \frac{n^2h^2}{8md^2}. \tag{1.73}$$

The energy E_n, over the barrier coincides with the energy n-th level of an electron localized inside the potential well of width d with infinitely high walls [1].

During the change of electron energy the transmission coefficient oscillates and the maximum value of T_{max} (resonant values) occurs at the condition (1.70). The minimum values of transmittance T_{min} and the corresponding values of energy $E_n' = E' - U_2$, called antiresonant, can be estimated from the condition

$$\sin^2(k_2d) = 1. \tag{1.74}$$

Hence

$$T_{\min} = \left(1 + \frac{U_2^2}{4E'\left(E' - U_2\right)}\right)^{-1}, \tag{1.75}$$

and

$$E_n' = \frac{h^2}{8md^2}\left(n + \frac{1}{2}\right)^2, \tag{1.76}$$

here $n = 1, 2, 3$, and so on.

With increasing the resonance number n and decreasing the barrier width d the minimum transmission coefficient T_{min} increases rapidly, so that the oscillations are smoothed out. Increasing the barrier height U_2, in contrast, reduces the transmission coefficient, increasing the amplitude of the oscillation [5]. The transmission coefficient of electrons above the potential barrier on their energy dependences at different values of n is shown in Figure 1.8.

It is quite difficult to observe the quantum oscillations of the above barrier electron transmission probability in semiconductor structures experimentally because the oscillation amplitude decreases rapidly with the increasing of the energy, while at low energies the oscillations become blurred due to thermal fluctuations.

1.2.3 The Particle Moves above the Well

In this case the particle also has energy $E > U_2$ (Figure 1.9). Then wave vectors of particle in region 1, region 2, and region 3 are correspondingly

$$k_1 = \sqrt{2m_1^*E/\hbar^2} \tag{1.77}$$

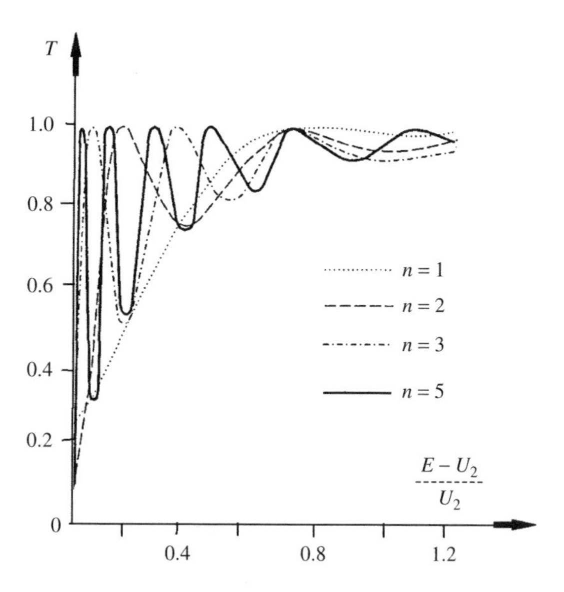

Figure 1.8 Transmission coefficient on energy dependences at moving of the particle above the barrier at $n = 1, 2, 3, 5$

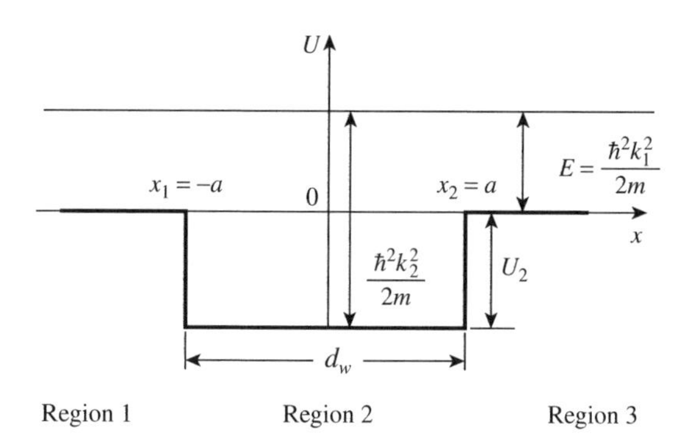

Figure 1.9 The particle moves above the well

$$k_2 = \sqrt{2m_2^*(E + U_2)/\hbar^2} \tag{1.78}$$

$$k_3 = \sqrt{2m_1^* E/\hbar^2}, \tag{1.79}$$

where E is the particle energy, U_2 is the depth of potential well (with the thickness $d_w = 2a$).

Using the procedure described in Section 1.2.2 we obtain

$$T = \left(1 + \frac{(k_2 - k_1)^2}{4k_1^2 k_2^2} \sin^2 k_2 d_w\right)^{-1} \tag{1.80}$$

At integer values of $k_2 d_w/\pi$ the transmission coefficient becomes equal to 1.

The refractive coefficient is

$$R = 1 - T.$$

The transmission coefficient in this case Equation (1.80) is described by the same formulas as in case movement over the barrier Equations (1.65) and (1.68) by replacing U_2 on $-U_2$. As in the case of the potential barrier, as well as in the case of the potential well the oscillations of T have the same nature, namely, semiclassical oscillations can be interpreted as the result of interference of electron waves reflected from the potential jumps at the boundaries of the barrier or well. However, there is a noticeable difference. For equal values of thickness, d, for the barrier and width, d_w, for the well and the same potential energy $|U_2|$ the scale of oscillations of T in the case of passage of the electrons above the barrier are significantly higher than during the passage above the well.

It is possible to find wave functions for such structure in all regions Figure 1.10 [5].

As can be seen the wave function amplitude in region 2 (well) is significantly smaller. It means that at small particle energy $E = \hbar^2 k^2/2m \ll U_2$ the density of probability to find the particle in the well region is significantly lower than outside.

It is more accessible to observe the oscillations of the transmission coefficient at an electron moving above the potential well than at moving above the barrier on experiment, since in this case it is possible to use electrons with relatively small energy.

1.2.4 *The Particle Moves through the Potential Barrier*

In this case (Figure 1.11) at $E < U_2$ the wave function are

$$\Psi(x) = \begin{cases} A_1 e^{ik_1 x} + B_1 e^{-ik_1 x}, \ldots\ldots\ldots \ x \le x_1 \\ A_2 e^{-a_2 x} + B_2 e^{a_2 x}, \ldots\ldots\ldots \ x_1 \le x \le x_2 \\ A_3 e^{ik_2 x} + B_3 e^{-ik_2 x}, \ldots\ldots\ldots \ x \ge x_2 \end{cases} \tag{1.81}$$

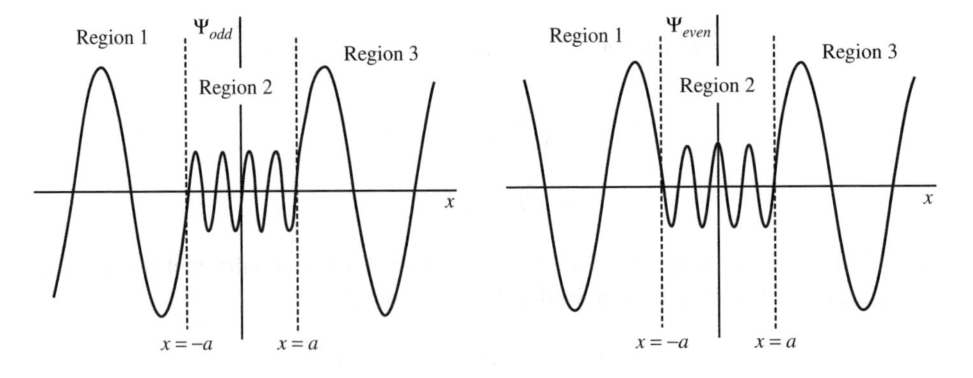

Figure 1.10 The waves of the particle moving above the well

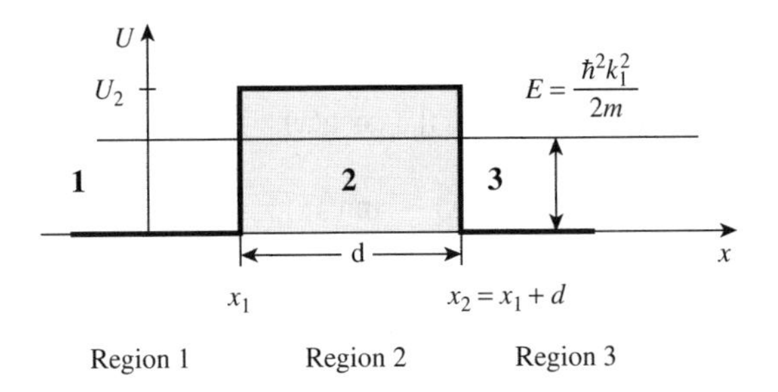

Figure 1.11 The particle moves through the potential barrier

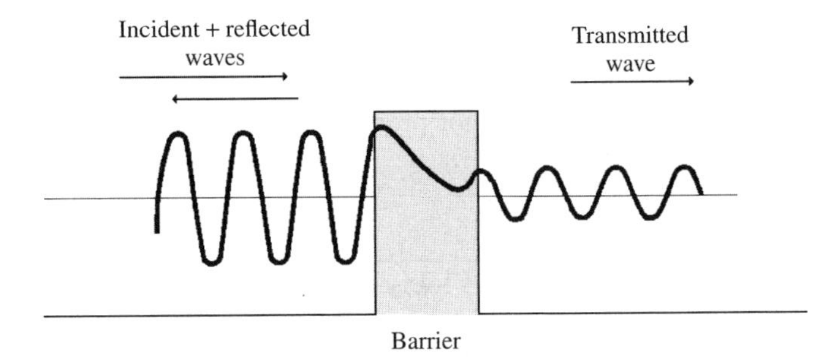

Figure 1.12 Transition of electron waves through the barrier

Then wave vectors of particle movement in region 1, region 2, and region 3 are, respectively:

$$k_1 = \sqrt{2m_1^* E/\hbar^2} \tag{1.82}$$

$$k_2 = i\alpha_2 = i\sqrt{2m_2^*(U_2 - E)/\hbar^2} \tag{1.83}$$

$$k_3 = \sqrt{2m_1^* E/\hbar^2}. \tag{1.84}$$

The schematic image of electron waves at transition through the potential barrier is shown in Figure 1.12.

The procedure for obtaining the transmission coefficient is as in Section 1.2.2 according to Equations (1.59)–(1.63), but in this case we used $k_2 = i\alpha_2$.

As a result the transmission probability is

$$T = \frac{1}{|M_{11}|^2} = \frac{16\alpha_2^2 k_1^2}{[(\alpha_2 + ik_1)^2 e^{-\alpha_2 d} - (\alpha_2 - ik_1)^2 e^{\alpha_2 d}]^2} \tag{1.85}$$

After additional transformation where we take into account that

$$\sinh(x) = \frac{1}{2}[\exp(x) - \exp(-x)], \;\; \cosh(x) = \frac{1}{2}[\exp(x) + \exp(-x)], \tag{1.86}$$

$$\cosh^2(x) - \sinh^2(x) = 1, \tag{1.87}$$

we obtain

$$T = \left(1 + \frac{\left(\alpha_2^2 + k_1^2\right)^2}{4k_1^2\alpha_2^2}\sinh^2\alpha_2 d\right)^{-1} \tag{1.88}$$

The penetration of the particle with energy E through the potential barrier U at condition $E < U$ is the well-known tunnel effect. Electron transport through the potential barrier is not associated with the loss of electron energy: the electron leaves the barrier with the same energy with which entry into a barrier. As can be seen from Equation (1.88) in the case of significantly thick and high barrier $\alpha_2 d \gg 1$ the transmission probability T is small enough and exponentially decreases with growth of $\alpha_2 d$ parameter:

$$T = \frac{16k_1^2\alpha_2^2}{(\alpha_2^2 + k_1^2)^2}e^{-2\alpha_2 d}. \tag{1.89}$$

In this case refractive coefficient R is equal to

$$R = \left(\frac{M_{21}}{M_{11}}\right)^2 = \left|\frac{-\left(-a_2^2 - k_1^2\right)e^{\alpha_2 d} + (-\alpha_2^2 - k_1^2)e^{-\alpha_2 d}}{[(\alpha_2 + ik_1)^2 e^{-\alpha_2 d} - (\alpha_2 - ik_1)^2 e^{\alpha_2 d}]^2}\right|^2$$

$$= \left|\frac{\left(a_2^2 + k_1^2\right) \times (e^{\alpha_2 d} - e^{-\alpha_2 d})}{[(\alpha_2 + ik_1)^2 e^{-\alpha_2 d} - (\alpha_2 - ik_1)^2 e^{\alpha_2 d}]^2}\right|^2. \tag{1.90}$$

Formula (1.89) for the transmission coefficient for rectangular barrier can be generalized to the barrier of arbitrary shape (Figure 1.13)

$$T = T_0 \exp\left[-\frac{2\sqrt{2m_2^*}}{\hbar}\int_{x_1}^{x_2}\sqrt{\left(U_2(x) - E\right)}dx\right], \tag{1.91}$$

where T_0 is the constant, order of the unity.

The generalized dependence of transmission probability (through the barrier and above the barrier) on particle energy is shown in Figure 1.14 [6].

In the general case the reverse transfer matrix can be presented as

$$M_p^{(p,p+1)} = \begin{pmatrix} M_{11}^{(p,p+1)}, M_{12}^{(p,p+1)} \\ M_{21}^{(p,p+1)}, M_{22}^{(p,p+1)} \end{pmatrix} = \frac{1}{2}\begin{pmatrix} 1 + \frac{k_{p+1}}{k_p}, 1 - \frac{k_{p+1}}{k_p} \\ 1 - \frac{k_{p+1}}{k_p}, 1 + \frac{k_{p+1}}{k_p} \end{pmatrix}. \tag{1.92}$$

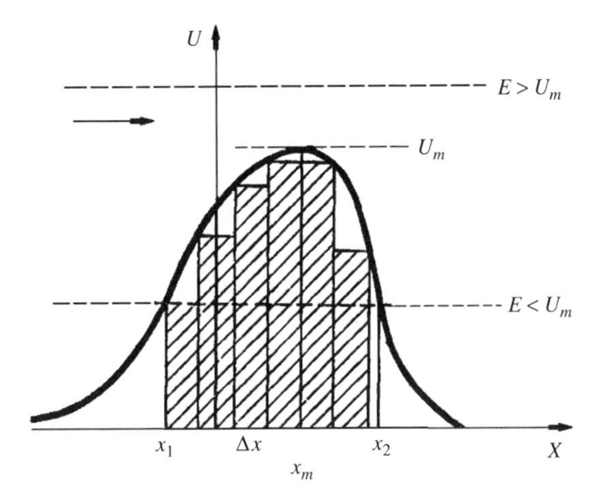

Figure 1.13 Potential barrier of arbitrary shape

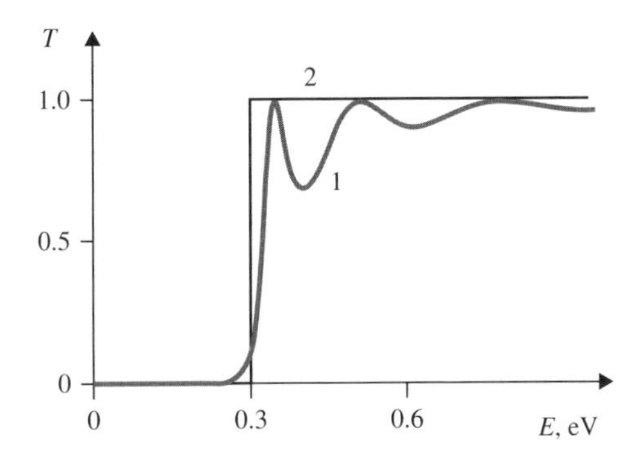

Figure 1.14 Energy dependence of the transmission coefficient of quantum particle (1) at moving through the barrier (AlGaAs) of 0.3 eV height and 10 nm width in GaAs-AlGaAs-GaAs structure. Curve 2 is the transmission coefficient in the classical case

As was summarized in Ref. [7], to describe the transition of the particle through multilayer structure containing barriers and wells based on the transfer matrix technique it is necessary to know four different types of matrices, namely, those respective joint points: within classically allowed regions (M_A), below the barrier (M_B), across discontinuity in the direction from a classically allowed region into the barrier (M_{in}), and across a discontinuity in the direction from the barrier into a classically allowed region (M_{out}).

$$M_A = \begin{pmatrix} e^{-ikw}, 0 \\ 0, e^{ikw} \end{pmatrix}, \tag{1.93}$$

$$M_B = \begin{pmatrix} e^{-\alpha d}, 0 \\ 0, e^{\alpha d} \end{pmatrix}, \tag{1.94}$$

$$M_{in} = \frac{1}{2}\begin{pmatrix} 1 - i\frac{k}{\alpha}, 1 + i\frac{k}{\alpha} \\ 1 + i\frac{k}{\alpha}, 1 - i\frac{k}{\alpha} \end{pmatrix} \tag{1.95}$$

$$M_{out} = \frac{1}{2}\begin{pmatrix} 1 + i\frac{\alpha}{k}, 1 - i\frac{\alpha}{k} \\ 1 - i\frac{\alpha}{k}, 1 + i\frac{\alpha}{k} \end{pmatrix}. \tag{1.96}$$

As can be seen the reverse matrix to M_{in} is M_{out} and vice versa. It was pointed out that the above matrices were particular cases of more general forms which could be derived by exploiting the wave function properties with respect to conjugation and conservation of probability current [3].

In reference [8] the authors approximated the arbitrary potential well by multistep function and then used a matrix method to determine the transmission coefficient. The position dependence of electron effective mass, $m_n{}^*$, and permittivity were also approximated by multistep functions. Despite the fact that the matrix method is straightforward, some authors have applied other approaches during the calculation of transmission coefficient. In reference [9] the arbitrary potential well was approximated by piecewise linear functions and then there was used a numerical method to calculate the transmission coefficient. Another method was applied in Ref. [10]. To determine the transmission probability and other parameters required to investigate the system they used the method of logarithmic derivate.

1.3 Tunneling through Triangular Barrier at Electron Field Emission

If we apply to a metal or semiconductor large electric field ($\sim 10^7$ V/cm) so that it is the cathode, then such a field pulls the electrons: it generates an electric current. This phenomenon is called electron field emission or "cold emission." Let us consider, for simplicity, the emission from the metal. We turn first to the picture of the motion of electrons in metal without an external electric field. To remove an electron from the metal, we need to do some work. Consequently, the potential energy of an electron in the metal is less than outside the metal. The simplest way this can be expressed is if we assume that the potential energy $U(x)$ inside the metal is equal to zero, while outside the metal it is equal to $U > 0$, so that the potential energy has the form shown in Figure 1.15. Simplifying in such manner the view of the potential energy, we actually operate with the average field in the metal. In fact, the potential inside the metal varies from point to point with a period equal to the lattice constant. Our approximation corresponds to the hypothesis of free electrons, since, as $U(x) = 0$ inside the metal there are no forces acting on an electron.

At such energy distribution of the electron gas the vast majority of electrons have the energy $E < U$ (at absolute zero temperature the electrons fill all the energy levels of $E = 0$ to $E = E_F < U$), where E_F is Fermi level. Let us denote the flow of electrons of the metal, falling from inside the metal on its surface, by J_0. Since the electrons have an energy $E < U$, then the flow is totally reflected by the jump in potential U, which takes place at the metal-vacuum interface (see Section 1.2.1).

The applied electric field F is directed toward the metal surface. Then the potential energy of an electron in the constant field of F, equal to qFx (electron charge equal to q) was added to

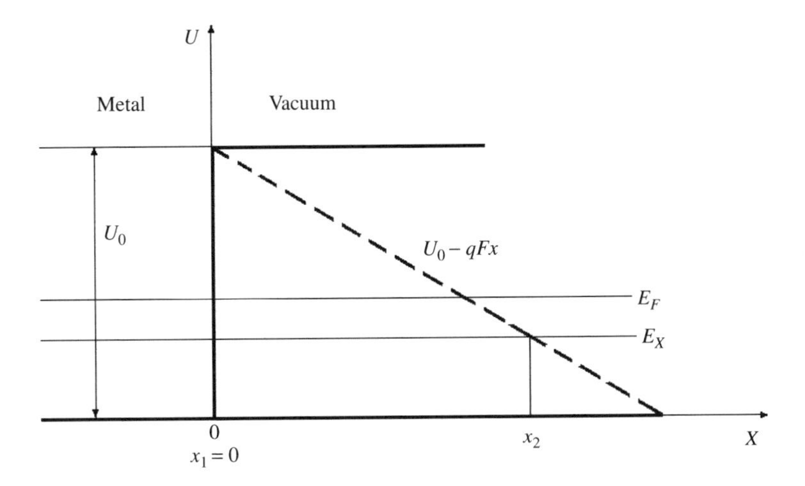

Figure 1.15 Band diagram of metal-vacuum interface without (solid line) and with (dashed line) applied electric field

the potential energy *U(x)* (Figure 1.15). Now the full potential energy will be

$$U^*(x) = U(x) - qFx = U_0 - qFx, (x > 0),$$
$$U^*(x) = 0, (x < 0). \tag{1.97}$$

Potential energy curve now has another view. It is shown in Figure 1.15 with dashed line. Note that large field cannot be created inside the metal, so the change of the *U(x)* takes place only outside the metal.

As it can be seen the triangular potential barrier is created. According to classical mechanics, an electron could pass through the barrier only if its energy is $E > U$. Such electrons are very little (they cause small thermionic emission). Therefore, according to classical mechanics the electron current is absent when the field is applied. However, if F is sufficiently large, the barrier is narrow, we have to deal with abrupt change of potential energy and classical mechanics is inapplicable: the electrons pass through the potential barrier.

Let us calculate the transmittance of the barrier for electrons with energy E_x moving along the *OX* axis. According to Equation (1.91) we have to calculate the integral

$$S = \int_{x_1}^{x_2} \sqrt{2m[U^*(x) - E_x]}dx, \tag{1.98}$$

where x_1 and x_2 are the coordinates of the turning points. The first turning point is (see Figure 1.15) obviously $x_1 = 0$, since for every energy $E_x < U$ the horizontal line E_x, representing the motion energy along *OX*, intersects the potential energy curve at $x = 0$. The second turning point is obtained, as can be seen from the figure, at

$$E_x = U_0 - qFx, \tag{1.99}$$

hence

$$x_2 = \frac{U_0 - E_x}{qF}, \tag{1.100}$$

consequently,

$$S = \int_0^{\frac{U_0 - E_x}{qF}} \sqrt{2m[U_0 - qFx - E_x]}dx. \tag{1.101}$$

Let us introduce the variable of integration

$$\xi = \frac{qF}{U_0 - E_x}x. \tag{1.102}$$

Then we get

$$S = \sqrt{2m}\frac{(U_0 - E_x)^{3/2}}{qF}\int_0^1 \sqrt{1-\xi}d\xi = \frac{2}{3}\sqrt{2m}\frac{(U_0 - E_x)^{3/2}}{qF} = \frac{2}{3}\sqrt{2m}\frac{(\Phi_0 - E_x)^{3/2}}{qF}. \tag{1.103}$$

Thus the transmission coefficient T for electrons with the energy of motion along the OX axis, equal to E_x, is

$$T(E_x) = T_0 e^{-\frac{4}{3}\frac{\sqrt{2m}}{\hbar}\frac{(\Phi_0 - E_x)^{3/2}}{qF}}. \tag{1.104}$$

This is the well-known Fowler–Nordheim equation [11].

The transmission coefficient is somewhat different for the different E_x, but as $E_x < U$, the average (in electrons energy) coefficient can be presented in the form

$$\overline{T} = \overline{T_0} e^{-\frac{F_0}{F}}, \tag{1.105}$$

where T_0 and F_0 are the constants depending on the type of the metal.

1.4 Effect of Trapped Charge in the Barrier

The influence of trapped charge on electron tunneling through the barrier has been intensively investigated in connection with the degradation of metal- oxide-semiconductor (MOS) structures with an ultra-thin oxide layer due to the carrier injection. The created charge in the oxide causes instability of MOS devices and oxide breakdown [12–15]. In the case where charges are trapped in the oxide with areal density Q_{ox} and centroid position X_b as referred to the cathode interface, the effective oxide electric field (F_{ox}) is no more equal to the cathode electric field (F)

$$F_{ox} = F + \frac{Q_{ox}}{\varepsilon_{ox}}\left(1 - \frac{X_b}{d_{ox}}\right). \tag{1.106}$$

where ε_{ox} is the oxide permittivity and d_{ox} is the oxide thickness.

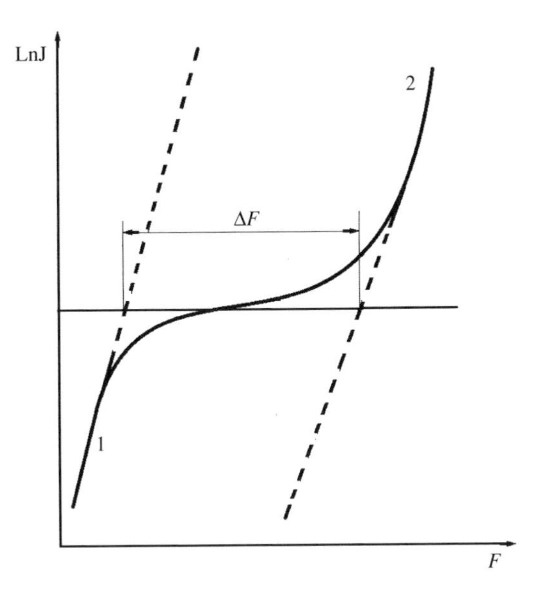

Figure 1.16 Schematic illustration of I-V curves shift due to charge trapping: (1) before charge trapping and (2) after charge trapping. Reproduced with permission from Ref. [16]. Copyright (1977), AIP Publishing LLC

Trapping of the charge will cause the shift of I-V characteristics (Figure 1.16).

Let's assume that negative charge (electrons) has been trapped. Trapping of the charge modifies the barrier shape significantly and as a result modifies the tunnel transparency (Figure 1.17) [13, 17–19]. To analyze the changing transmission probability and tunneling current changing due to charge we take into account that the trapped charge is localized at $x = X_b$ in the barrier (oxide) when $V_g < 0$ (Figure 1.17).

The transmission probability $T(E_x)$ for an electron at energy E_x is given by the following relationship [20]:

$$T(E_x) = T_0 \ \exp(-2 \int_{X=0}^{X_t} \sqrt{\frac{2m_{ox}}{\hbar^2}(U(x) - E_x)dx}), \tag{1.107}$$

where E_x is the perpendicular to the barrier electron energy (E) component, X_t is the tunnel distance in the oxide for the electron with energy (E_x), m_{ox} is the effective mass of the electron in oxide, $\hbar$ is the reduced Plank constant, q is the electron charge, and $U(x)$ is the potential barrier in oxide.

The Fowler–Nordheim (F–N) tunnel current density J_{FN}, which crosses the structure for given voltage V_g, is obtained by summing the contribution to the current of electrons at all energies E_x. The current density is given by the following expression [20]:

$$J_{FN} = \frac{4\pi q m_0}{\hbar^3} \int_{E_x} T(E_x)dE_x \int_{E_x}^{\infty} f(E, T)dE, \tag{1.108}$$

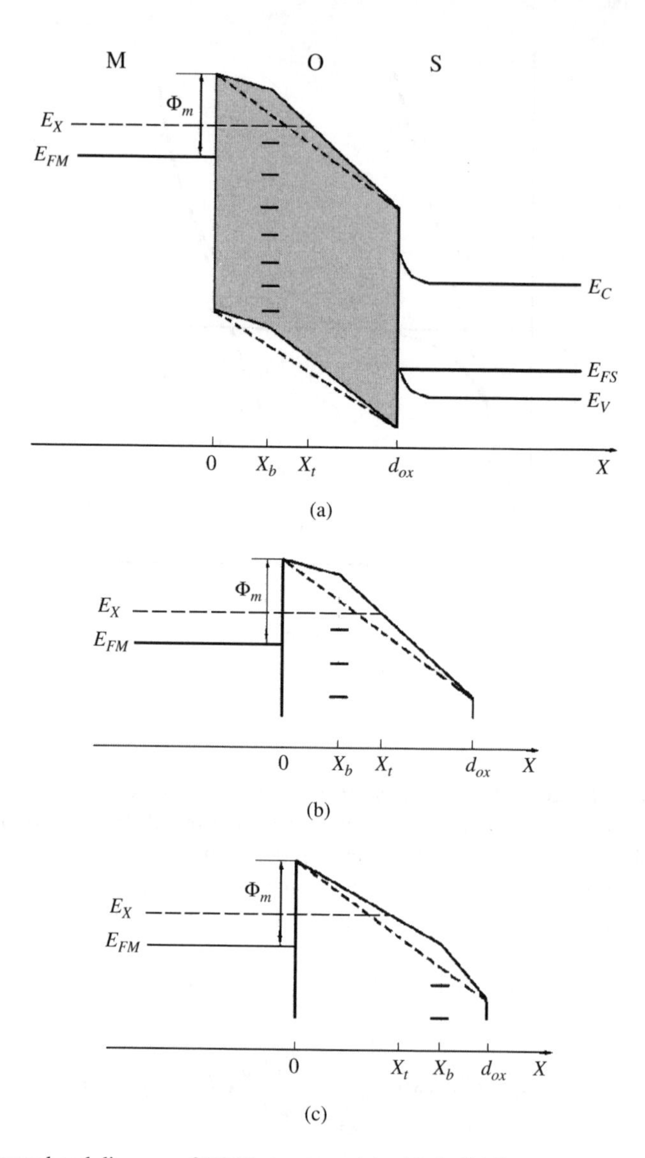

Figure 1.17 Energy band diagram of MOS structures (a) with (solid lines) and without (dashed lines) of the captured negative charge. Different location of trapped charge: (b) $X_b < X_t$ and (c) $X_b > X_t$

where m_0 is the mass of free electron and $f(E,T)$ is the Fermi–Dirac distribution of electrons that depends on the temperature T [21, 22].

With respect to the trapped charges with density N_1 at $X = X_b$, there exist two fields for given voltage V_g in the oxide: one F_1 between the metal and X_b, the other F_2 between X_b and silicon [22]. Using the Gauss equation, one can determine the field E_1 as the function of the charge density N_1 and field F_2:

$$F_1 = \frac{qN_1}{\varepsilon_{ox}} + F_2. \tag{1.109}$$

The potential barrier $U(x)$ distribution can be obtained for given voltage V_g, by solving the Poisson equation:

$$U(x) = -\frac{q}{\varepsilon_0 \varepsilon_{ox}}(N_1 x) - qF_2 x + U(0). \text{ if } 0 \le x \le X_b, \tag{1.110}$$

$$U(x) = -\frac{q}{\varepsilon_0 \varepsilon_{ox}}(N_1 X_b) - qF_2 x + U(0). \text{ if } X_b \le x \le d_{ox}, \tag{1.111}$$

where ε_0 is the permittivity of vacuum, ε_{ox} is the relative permittivity of the oxide, $U(0)$ is the metal/oxide interface barrier (input barrier Φ_m), q is the absolute value of electron charge.

The expressions for tunneling probability and tunnel current depend on trapped charge location (X_b) in relation to length of tunneling path (X_t) (Figure 1.17). Taking into account Equations (1.109)–(1.111), the transmission probability $T(E_x)$ for an electron with energy E_x can be expressed, as a function of the electric fields F_1, by the expressions [17]:

1. if the charge centroid is localized in the tunnel distance X_t $(X_t > X_b)$:

$$T(E_x) = T_0 \exp\left[-\frac{4}{3}\frac{(2m_{ox})^{1/2}}{\hbar q F_1} \times \left\{(\Phi_m - E_x)^{3/2} + \left(\frac{qN_1}{\varepsilon_{ox}F_1 - qN_1}\right)(\Phi_m - qF_1 X_b - E_x)^{3/2}\right\}\right]; \tag{1.112}$$

2. if the charge centroid is localized outside the tunnel distance X_t $(X_t < X_b)$:

$$T(E_x) = T_0 \exp\left[-\frac{4}{3}\frac{(2m_{ox})^{1/2}}{\hbar q F_1}(\Phi_m - E_x)^{3/2}\right]. \tag{1.113}$$

For the given voltage V_g, Equations (1.107)–(1.113) yield the potential barrier distribution in oxide, the transmission probability $T(E_x)$ and the current density J_{FN}.

In this case if the trapped charge is distributed on oxide thickness the shape of potential barrier is complicated significantly and calculation of transmission probability and tunnel current are more difficult.

The transient component of the current connected with charge trapping/detrapping processes can be observed [23]. It was shown that positive oxide charge assisted tunneling current also exhibits transient effect [24]. The transient behavior arises from the positive oxide charges, which help electron to tunnel through oxide, and they can escape to the Si substrate. As a result, the transient current should consist of three components in general, I_e, I_h, and I_t (Figure 1.18), if both positive and negative oxide charges are created [25].

I_e represents the negative oxide charge detrapping induced current, I_h is the positive oxide charge detrapping current, and I_t denotes the positive oxide charge assisted electron tunneling current. I_e and I_h have t^{-1} time dependence on the tunneling front model [23] while I_t has t^{-n} time dependence in the certain range of the measurement field. The power factor n is dependent on effective electron and hole tunneling barrier heights and tunneling carrier masses [24].

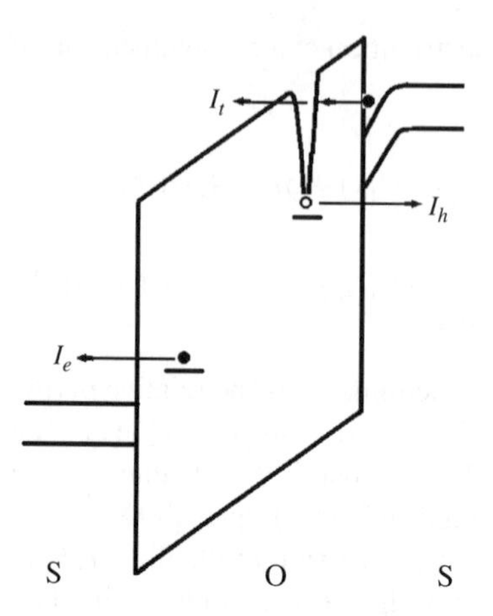

Figure 1.18 The schematic illustration of three transient components, I_e, I_h, and I_t. Reproduced with permission from Ref. [25]. Copyright (1999), AIP Publishing LLC

1.5 Transmission Probability in Resonant Tunneling Structures: Coherent Tunneling

At resonance tunneling the tunneling electron goes through double barriers via quantized states in the well [26]. Resonant tunneling of electrons represents sharp increase in the transmission coefficient of the quantum structure for the electrons which energy E coincides with the energy of one of the resonance levels E_n in the potential well. Despite the fact that at $E = E_n \ll U_1$, U_2 the transmission coefficients of the barriers T_1, $T_2 \ll 1$, the electron near the resonance "ignores" the barriers, passing through the entire structure without reflections. Resonant tunneling appears due to the interference of electron waves reflected from the barriers. As a result of interference at the resonance condition there are only the incident and transmitted boundary electron waves, and the reverse wave is completely extinguished. During this process the amplitude of the wave function inside the potential well is much greater than in the barriers. The mechanism of resonant tunneling from corpuscular positions can be represented as a delay of an electron inside the potential well on the time of its life τ_n (in the absence of scattering), during which the electron $\nu = \tau_n/L$ times encounter with barrier. Therefore, the probability of electron tunneling from the well increases in the ν times.

In resonant tunneling, the Schrodinger equation has to be solved simultaneously in three regions – emitter, well, and collector. Because of the quantized states within the well, the tunneling probability exhibits peak when the energy of the incoming particle coincides with one of the quantized levels. In the structure with barriers of finite thickness d_1 and d_2, the electron wave function is not located entirely within the well, but is smeared over the entire space. Nevertheless, there are selected values of energy, similar in magnitude to discrete resonant levels in completely isolated potential well, in which the amplitude of the wave function inside the well due to the interference of reflected from the barriers electron waves is much higher than the amplitude of the wave function outside the well. In this coherent-tunneling consideration, if the incoming energy does not coincide with any of the quantized levels, the global tunneling probably T_G is a product of the individual probability between the well and the emitter T_1, and that between the well and the collector T_2,

$$T_{Gnr} = T_1 T_2. \tag{1.114}$$

However, when the incoming energy matches one of the quantized level, the wave function builds up within the well similar to a Fabry-Perot resonator, and the transmission probability becomes [7, 27].

$$T_{Gres} = \frac{4T_1 T_2}{(T_1 + T_2)^2}. \tag{1.115}$$

Applying the transfer matrix method, it is possible to calculate the transmission probability in resonant tunneling structures (RTSs) or double barrier resonant tunneling structures (DBRTSs). The potential energy diagram of such a structure in the general case with rectangular barriers is shown in Figure 1.19. The exact analytical solution is possible in this case.

According to our consideration in this case

$$\begin{pmatrix} A_1 \\ B_1 \end{pmatrix} = M_1 M_2 M_3 M_4 M_5 M_6 M_7 \begin{pmatrix} A_5 \\ B_5 \end{pmatrix}, \tag{1.116}$$

where M_1, M_5 are the input matrices Equation (1.95), M_2, M_6 are the barrier matrices Equation (1.94), M_3, M_7 are the output matrices Equation (1.96), and M_4 is well matrix Equation (1.93).

The final result can be presented as

$$T_G = \frac{T_1 T_2}{1 - 2\sqrt{R_1 R_2}\cos\theta + R_1 R_2}. \tag{1.117}$$

where T_1, T_2 are the transmission coefficient for first and second barriers; correspondingly, R_1, R_2 are the refractive coefficients for first and second barriers, respectively, and θ is the round-trip phase shift in the quantum well.

$$\theta = 2kw \tag{1.118}$$

where k is the wave vector in the well and w is the well width.

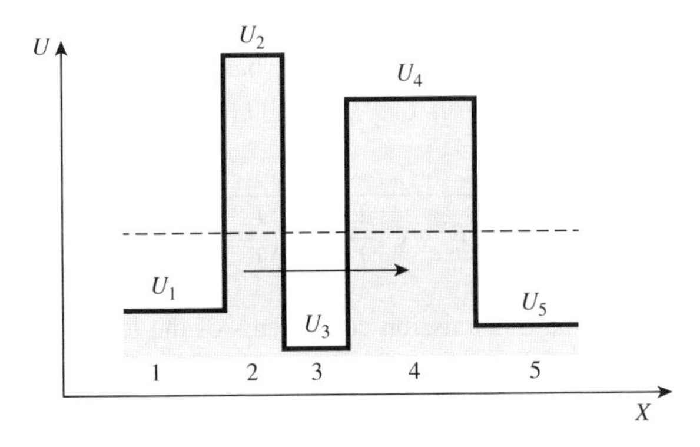

Figure 1.19 Potential-energy diagram of the double rectangular barrier case

As can be seen, the global transmission coefficient T_G through all structure is exactly derived.

We can rewrite Equation (1.117)

$$T_G = \frac{T_1 T_2}{(1 - \sqrt{R_1 R_2})^2 + 4\sqrt{R_1 R_2}\sin^2(\theta/2)}. \tag{1.119}$$

Here we use the trigonometric ratio

$$\cos\theta = 1 - 2\sin^2(\theta/2). \tag{1.120}$$

The condition of resonance is $\theta/2 = n\pi$ or $n\lambda = 2w$, where λ is the length of wave and n is the integer number.

At resonance $sin^2(\theta/2) = 0$ and we obtain the next equation for transmission coefficient

$$T_{Gres} = \frac{T_1 T_2}{(1 - \sqrt{R_1 R_2})^2}. \tag{1.121}$$

In real two barrier structures the transmission coefficients are small ($T \ll 1$). Then

$$\begin{aligned}(1 - \sqrt{R_1 R_2})^2 &= (1 - \sqrt{(1 - T_1)(1 - T_2)})^2 = (1 - \sqrt{(1 - T_1 - T_2 + T_1 T_2})^2 \\ &= (1 - \sqrt{(1 - T_1 - T_2})^2 = (1 - 1 + (T_1 - T_2)/2)^2 = \frac{(T_1 + T_2)^2}{4}.\end{aligned} \tag{1.122}$$

Here we used $T_1 T_2 = 0$ and Tailor series expansion

$$\sqrt{(1 - (T_1 + T_2))} = 1 - (T_1 + T_2)/2. \tag{1.123}$$

As a result in the case of resonant tunneling we have [7]

$$T_{Gres} = \frac{4T_1 T_2}{(T_1 + T_2)^2}. \tag{1.124}$$

The global transmission coefficient for double barrier resonance tunneling structure (Figure 1.19) can be put in the following general form [7, 28]:

$$T_G = \frac{C_0}{C_1 T_1 T_2 + C_2 \frac{T_1}{T_2} + C_3 \frac{T_2}{T_1} + C_4 \frac{1}{T_1 T_2}} \tag{1.125}$$

where T_1 and T_2 represent the transmission coefficients of the left and right barrier respectively which are exponentially dependent on energy. The C coefficients in Equation (1.125) are phase factors exhibiting much weaker energy dependence and, at first approximation, can be used as constants if T coefficients are small ($T \ll 1$) ("strong localization" case). Under these conditions Equation (1.125) can be simplified, in the dominator the last term is dominated.

$$T_{Gnr} = \frac{C_0}{C_4} T_1 T_2 = T_1 T_2 \quad (1.126)$$

In this case then, the presence of potential energy well between two barriers has, in practice, little or no effect. The view of the transmission probability will be the same if the well is absent. The effect produced by the well is the reduction of the phase changing path of the total barrier.

For some special energy C_4 and C_1 go to zero. The main term is, consequently, canceled out and the resonance occurs. In this case, as easily seen from Equation (1.125), the global transmission coefficient T_{Gres} becomes

$$T_{Gres} = \frac{C_0}{C_2 \frac{T_1}{T_2} + C_3 \frac{T_2}{T_1}} = \frac{C_0 T_1 T_2}{C_2 T_1^2 + C_3 T_2^2} = \frac{T_1 T_2}{T_1^2 + T_2^2}. \quad (1.127)$$

Here we have assumed $C_0 = C_2 = C3 = 1$.

The more precise formula may be seen in Equation (1.115). For a symmetric structure, $T_1 = T_2$, and $T_{Gres} = 1$. Away from the resonance, the value T_G quickly drops by many orders of magnitude.

If $T_1 \gg T_2$ or $T_2 \gg T_1$ the relations (1.124) and (1.127) transform into

$$T_{Gres} = C \frac{T_{\min}}{T_{\max}} = \frac{T_{\min}}{T_{\max}}. \quad (1.128)$$

where T_{min} and T_{max} represent the smaller and larger among T_1 and T_2, respectively, while C is either C_0/C_2 or C_0/C_3 depending on whether or not $T_{max} = T_1$. In case of using as an initial formula (1.125) $C = 4$.

The comparison of Equations (1.126) and (1.128) show that resonance always implies an increased transmission coefficient since it is

$$\frac{T_{Gres}}{T_{Gnr}} = \frac{1}{T_{\max}^2}, \quad (1.129)$$

where T_{Gnr} represents the nonresonance (without resonance) value of T_G, that is, if no well has been presented between the two barriers.

Such an increase is, therefore, larger for smaller T_{max} and has vanished in the limiting case of $T_{max} \to 1$ (which, on the other hand, is incompatible with the assumption of strong localization). Equation (1.128) shows that regardless of how small T_1 and T_2 are, T_{Gres} can be order of unity under the only condition $T_1 = T_2$ while Equation (1.129) clearly indicates that the transmission coefficient can easily increase at the several orders of magnitude for arbitrary small changes in energy producing resonance.

Another important aspect of resonance concerns the wave function as schematically presented in Figure 1.20.

Without resonance the wave function $\Psi(x)$ monotonically and exponentially decreases within the classically forbidden regions thus reflecting the multiplication of the single barrier

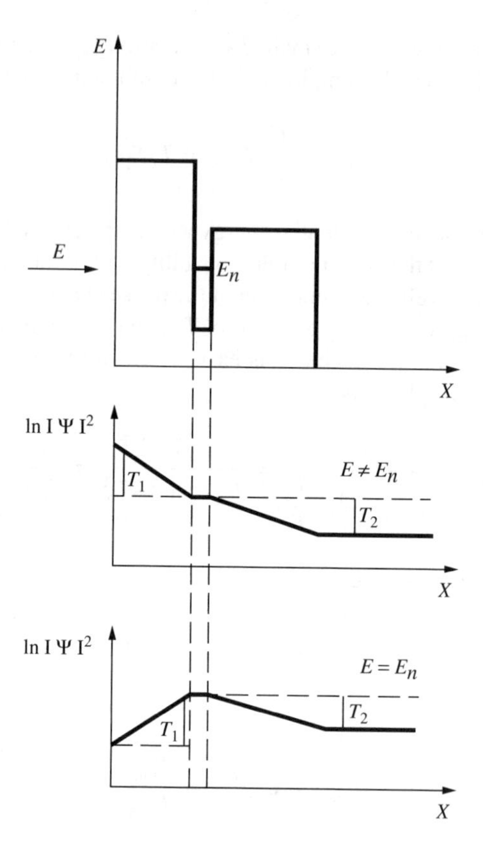

Figure 1.20 Schematic images of the wave function in the cases of resonance and without resonance. Reprinted with permission from Ref. [7]. Copyright (1984) by the American Physical Society

transmission coefficient (Equation (1.126)). At resonance, instead, the tunneling particle finds its eigenstate in the well where, consequently, the wave function has to be peaked with an exponential decrease on both sides (Figure 1.20). Since we have assumed both T_1 and $T_2 \ll 1$, this implies that the state is strongly localized. Because T_1 and T_2 are not zero, the localized states are, strictly speaking, quasi eigenstates with finite lifetime and energy width. The increase in transmission coefficient at resonance is a consequence of the wave function being peaked within the well. The typical dependence of transmission probability at tunneling through double-barrier resonant-tunneling structure is shown in Figure 1.21. The sharp resonance peaks are observed at $E = E_n$.

1.6 Lorentzian Approximation

Analysis shows that transmission coefficients through double barrier resonance tunneling structure are described by Equations (1.126) and (1.124) for nonresonant and resonant conditions, respectively. The sharp peak is observed at specific value of E_n. It allows performing above described approximations. For more precise description we have to use full

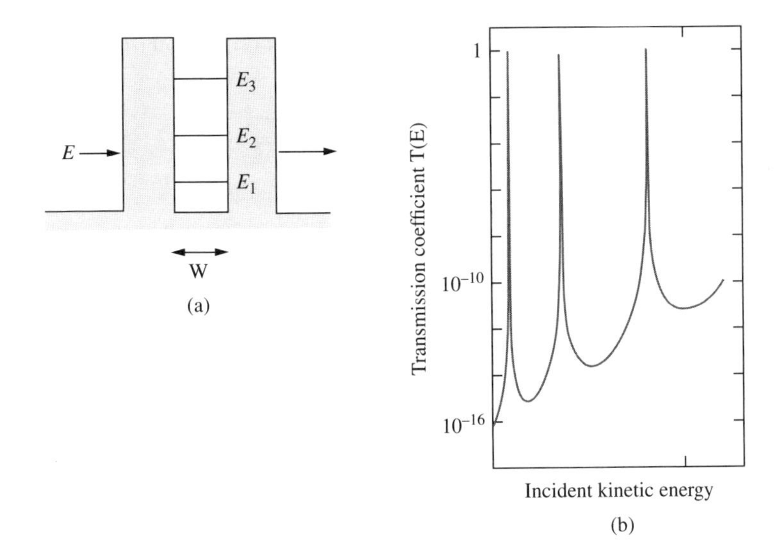

Figure 1.21 (a) Double barrier resonant tunneling structure. (b) Transmission coefficient of electron with energy E through a double barrier via coherent resonant tunneling. Transmission peaks occur when E aligns with E_n

Equation (1.117) and additionally analyze the influence of the wave function phase θ in well. We rewrite Equation (1.117), assuming R_1, $R_2 = 1$ [2]:

$$T_G = \frac{T_1 T_2}{1 - 2\sqrt{R_1 R_2}\cos\theta + R_1 R_2} = \frac{T_1 T_2}{(1 - \sqrt{R_1 R_2})^2 + 2\sqrt{R_1 R_2}(1 - \cos\theta)}$$
$$= \frac{T_1 T_2}{\left(\frac{T_1 + T_2}{2}\right)^2 + 2(1 - \cos\theta)}. \quad (1.130)$$

The sharpness of the resonance coefficient arises from the fact that R_1, $R_2 = 1$, T_1 and T_2 are very small, so that the denominator in Equation (1.130) is very small every time round-trip phase shift θ is close to the multiple of 2π.

At resonance $cos\theta = 1$ (see Equation (1.120)) we obtain Equation (1.124).

Close to resonance value we can expand the cosine function in Equation (1.130) in the Taylor series

$$(1 - \cos\theta) = \frac{1}{2}\theta^2 = \frac{1}{2}\left(\frac{d\theta}{dE}\right)^2 (E - E_n)^2 \quad (1.131)$$

and rewrite the transmission coefficient (Equation (1.130)) as

$$T_G = \frac{T_1 T_2}{\left[\frac{T_1 + T_2}{2}\right]^2 + \left(\frac{d\theta}{dE}\right)^2 (E - E_n)^2}. \quad (1.132)$$

Multiplying the denominator and numerator in Equation (1.132) on $(dE/d\theta)^2$ we obtain

$$T_G = \frac{\Gamma_1\Gamma_2}{\left[\frac{\Gamma_1+\Gamma_2}{2}\right]^2 + (E-E_n)^2}, \tag{1.133}$$

where

$$\Gamma_1 \equiv \frac{dE}{d\theta}T_1 \text{ and } \Gamma_2 \equiv \frac{dE}{d\theta}T_2. \tag{1.134}$$

This approximate result is often used (neglecting by energy dependence of Γ_1 and Γ_2) in place of the exact result (Equation (1.117)) for analytical calculations.

Multiply the denominator and numerator in Equation (1.133) on $(\Gamma_1+\Gamma_2)$ we have

$$T_G = \frac{\Gamma_1\Gamma_2(\Gamma_1+\Gamma_2)}{(\Gamma_1+\Gamma_2)\times\left[(E-E_n)^2+\left(\frac{\Gamma_1+\Gamma_2}{2}\right)^2\right]} = \frac{\Gamma_1\Gamma_2}{\Gamma_1+\Gamma_2}\times\frac{(\Gamma_1+\Gamma_2)}{(E-E_n)^2+\left(\frac{\Gamma_1+\Gamma_2}{2}\right)^2}$$

$$= \frac{\Gamma_1\Gamma_2}{\Gamma_1+\Gamma_2}\times\frac{\Gamma}{(E-E_n)^2+\left(\frac{\Gamma}{2}\right)^2} = \frac{\Gamma_1\Gamma_2}{\Gamma_1+\Gamma_2}\times A(E-E_n), \tag{1.135}$$

where $A(E-E_n)$ is a Lorentzian function:

$$A(E-E_n) = \frac{\Gamma}{(E-E_n)^2+\left(\frac{\Gamma}{2}\right)^2}, \tag{1.136}$$

where $\Gamma \equiv \Gamma_1 + \Gamma_2$.

Total transmission function through n resonant energy levels can be presented as

$$T_T = \sum_n T_{Gn} = \frac{\Gamma_1\Gamma_2}{\Gamma_1+\Gamma_2}\sum_n A(E-E_n). \tag{1.137}$$

where n is the number of resonance level (Figure 1.21).

The magnitude of transmission is determined by the parallel combination of Γ_1 and Γ_2 while the width of the peak depends on the sum of Γ_2 and Γ_2.

The Lorentzian approximation for the transition function is often used for analytical calculations. It is reasonably accurate close to the resonance, but should not be used far from the resonance.

1.7 Time Parameters of Resonant Tunneling

In general, there are several time scales of importance in resonance tunneling structures: (1) the traversal time, the time needed to tunnel through a barrier; (2) the resonant state lifetime; and (3) the escape time [1]. All these factors influence the overall temporal response of the device.

A crucial aspect usually overlooked in experiments [29–35] is that, depending on initial conditions, nonnegligible time might be required before the high conductivity resonant state is fully established [7]. As a rule the analysis of resonant tunneling is based on the time-independent Schrodinger equation, which hence describes the stationary situation. This requires the carrier wave function at resonance to be strongly localized within the well. In this case it is possible to obtain Equation (1.117) for description of the resonance.

In resonant tunneling the main contribution to the characteristic time is from the well region of the device. In resonant tunneling, the electrons become trapped in a quasibound state and persist for some time before they "leak" out of the well through the second barrier. Resonant levels are metastable, that is, the average electron lifetime τ_{life} on them is finite. As a result, the resonant state lifetime can be appreciably larger than the barrier traversal time and the escape time. Therefore, we estimate the characteristic time by calculating the resonant state lifetime of the RTD (resonant tunneling diode).

The resonant state lifetime or, equivalently, the lifetime of the quasibound state can be estimated as follows. For simplicity it is assumed that the quantization direction is along the z axis. The velocity of the electron in this direction can be estimated as

$$v_z = \sqrt{\frac{2E_n}{m}}, \tag{1.138}$$

where E_n is the energy level of the quantized state. An attempt frequency can be defined as

$$f_{att} = \frac{v_z}{2L}, \tag{1.139}$$

where L is the effective one-way distance the electron travels in the well. Notice that the attempt frequency simply represents how often the electron encounters a boundary while reflecting back and forth within the well. The effective length L is given as

$$L = w + \frac{1}{k_{b1}} + \frac{1}{k_{b2}}, \tag{1.140}$$

where w is the width of the well and k_{b1} and k_{b2} are the imaginary wave vectors within the barriers. They represent the electron travel while partially penetrating the barriers. The probability per unit time of the electron escaping depends on the product of the attempt frequency (how often the electron encounters a boundary) and the transmissivity of each boundary, denoted as T_1 and T_2 (how likely it is for the electron to tunnel through the boundary). The lifetime is proportional to the inverse of the probability per unit time of the electron escaping from the quasibound level. The lifetime τ_{life} is then given as

$$\tau_{life} = \frac{1}{f_{att}(T_1 + T_2)}. \tag{1.141}$$

If it is further assumed that the electron can escape only from the second barrier, which is usually the case when the RTD is under high bias, then the lifetime becomes

$$\tau_{life} = \frac{1}{f_{att}T_2}. \tag{1.142}$$

The lifetime can also be estimated from the *uncertainty principle,* which states that

$$\Delta E \Delta t \geq \frac{\hbar}{2}. \quad (1.143)$$

Since the state is assumed to be quasibound, it has a finite lifetime. That lifetime is simply Δt. Therefore, the resonant lifetime is given as

$$\Delta t = \tau_{life} = \frac{\hbar}{2\Delta E}, \quad (1.144)$$

where ΔE is the half-maximum width of the transmission peak, $\Gamma_r/2$. Equating Equations (1.144) and (1.141) yields

$$\frac{\hbar}{2\Delta E} = \frac{1}{f_{att}(T_1 + T_2)}. \quad (1.145)$$

Using Equations (1.138)–(1.140), f_{att} can be written as

$$f_{att} = \frac{v_z}{2(w + 1/k_{b1} + 1/k_{b2})} = \frac{\sqrt{2E_n/m}}{2(w + 1/k_{b1} + 1/k_{b2})}. \quad (1.146)$$

Substituting Equation (1.146) into Equation (1.145) yields

$$\Gamma_r = 2\Delta E = \frac{\hbar\sqrt{2E_n/m}(T_1 + T_2)}{2(w + 1/k_{b1} + 1/k_{b2})}. \quad (1.147)$$

Therefore, the resonant lifetime is simply

$$\tau_{life} = \frac{2(w + 1/k_{b1} + 1/k_{b2})}{\sqrt{2E_n/m}(T_1 + T_2)}. \quad (1.148)$$

It is interesting to note that the resonant state lifetime describes both fully sequential and fully resonant conditions of good approximation. The resonant lifetime can be determined in somewhat different manner using a wavelike picture of the electron [1, 2].

Let's apply lifetime consideration to analysis of Γ_l and Γ_2 parameters in Lorentzian approximation. One advantage of this approximation is that the entire physics is now characterized by just two parameters Γ_l and Γ_2, which are defined in Equation (1.134). Physically Γ_l and Γ_2 (divided by $\hbar$) represent the rate at which an electron placed between the barriers would leak out through the barriers into emitter, $\Gamma_l/\hbar$, and collector, $\Gamma_2/\hbar$, respectively.

It is possible to write the round-trip phase shift as $\theta = 2kL$ where L is effective width of the well (see Equation (1.140)) which includes also phase shifts associated with the reflections at the barriers.Then

$$\frac{dE}{d\theta} = \frac{1}{2L}\frac{dE}{dk} = \hbar f_{att}, \quad (1.149)$$

where $v \equiv dE/\hbar dk$ is the velocity with which an electron moves back and forth between the barriers. The quantity f_{att} means the number of times per second that the electron impinges on one of the barriers (that is, attempt to escape). It is equal to the inverse of the time that the electron takes to travel from one barrier to another and back.

The physical significance of Γ_1 and Γ_2 is easy to see. From Equations (1.134) and (1.149) we can write

$$\Gamma_1 \equiv \frac{dE}{d\theta}T_1 = \hbar f_{att}T_1 = \frac{\hbar\sqrt{2E_n/m}T_1}{2(w+1/k_{b1}+1/k_{b2})}, \quad (1.150)$$

$$\Gamma_2 \equiv \frac{dE}{d\theta}T_2 = \hbar f_{att}T_2 = \frac{\hbar\sqrt{2E_n/m}T_2}{2(w+1/k_{b1}+1/k_{b2})}, \quad (1.151)$$

$$\Gamma = \Gamma_1 + \Gamma_2 = 2\Delta E = \hbar f_{att}(T_1+T_2) = \frac{\hbar\sqrt{2E_n/m}(T_1+T_2)}{2(w+1/k_{b1}+1/k_{b2})}. \quad (1.152)$$

Fraction T_1 of the attempts on barrier 1 is successful while fraction T_2 of the attempts on barrier 2 are successful. Hence $\Gamma_1/\hbar$ and $\Gamma_2/\hbar$ tell us the number of times per second that an electron succeeds in escaping through barrier 1 and 2 respectively.

Finite lifetime of an electron at the resonance level causes broadening of the level (natural broadening), equal to

$$\Gamma_n = \frac{\hbar}{\tau_n}. \quad (1.153)$$

Electron lifetime and broadening are strongly dependent on the height and thickness of barriers and width of potential well, which determines the energy of the resonant level E_n, relative to the bottom of the well (Figure 1.22).

Various electron scattering processes that violate the coherence of electron waves within the layers and on the borders cause broadening of the resonance levels. This so-called collisional broadening of Γ_φ is associated with relaxation time τ_φ by relation similar to Equation (1.153):

$$\Gamma_\varphi \equiv \frac{\hbar}{\tau_\varphi}. \quad (1.154)$$

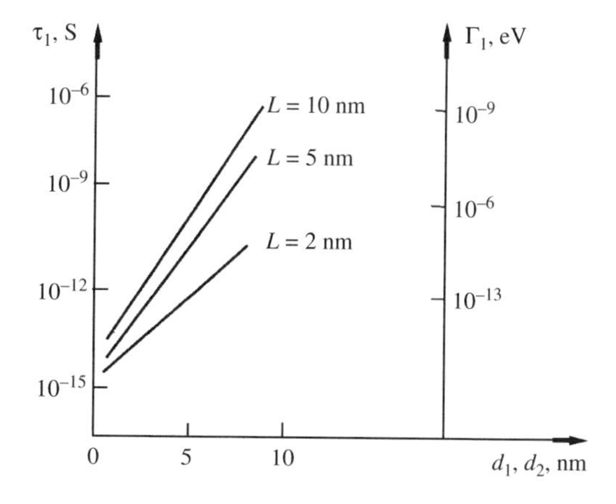

Figure 1.22 The lifetime τ_1 and natural width Γ_1 of low (the first) resonant level in quantum well on width L and barrier thicknesses d_1, d_2 dependences

Relaxation time τ_φ (see Equation (1.154)) decreases with increasing of impurities concentration, structural defects, and with increasing temperature. So, in pure gallium arsenide at room temperature $\tau_\varphi \sim 3\times 10^{-13}$ s, and $\Gamma_\varphi \sim 2$ meV; at liquid nitrogen temperature (77 K) $\tau_\varphi \geq 10^{-11}$ s, and $\Gamma_\varphi \leq 0.10$ meV.

1.8 Transmission Probability at Electric Fields

From an experimental point of view it is very important to consider the double barrier resonance tunneling structure subjected to applied external electric fields (Figure 1.23). Under electric fields the potential barriers change their shape. And even the initial structure has the same potential barriers as the applied voltage destroys the symmetry of the barriers whose transmission coefficients are no longer equal. As can be seen, the shape of potential barriers has changed from rectangular to trapezoidal.

We use the transmission coefficient for one rectangular barrier to obtain the coefficient for trapezoidal barriers under an electric field. According to Equation (1.85) in the case of significantly thick and high barrier ($\alpha d \gg 1$):

$$T = \frac{1}{|M_{11}|^2} = \frac{16\alpha^2k^2}{[-(\alpha - ik)^2 e^{+\alpha d}]^2} = \frac{16\alpha^2k^2}{(\alpha - ik)^4}e^{-2\alpha d} = \frac{16\alpha^2k^2}{(\alpha^2 + k^2)^2}e^{-2\alpha d} = T_0 e^{-2\alpha d} \tag{1.155}$$

The barrier changes its shape under an electric field

$$\alpha = \frac{1}{\hbar}[2m^*(U(x) - E)]^{1/2}. \tag{1.156}$$

Denote

$$U(x) - E = \Phi(x). \tag{1.157}$$

Under an electric field

$$\Phi = \Phi(x) = \Phi_0 - qFx. \tag{1.158}$$

We approximate the barrier by multistep function as some rectangular barriers with width dx and height $U(x)$. The transition coefficient for each element of barrier is

$$T^* = T_0^{\,*} e^{-2\alpha dx} \tag{1.159}$$

Then for all barrier

$$T = T_0 e^{\int_{x1}^{x2} -2\alpha dx} = T_0 e^{-\frac{2}{\hbar}\int_{x1}^{x2} [2m*(U(x)-E)]^{1/2} dx} = T_0 e^{-\frac{2}{\hbar}\int_{x1}^{x2} [2m*(\Phi_0 - qFx)]^{1/2} dx}, \tag{1.160}$$

where m^* is the constant (in barrier) and F is the electric field in the barrier.

In our case $x_1 = 0$; $x_2 = d$.

$$T = T_0 e^{-\frac{2}{\hbar}\int_{x1}^{x2} [2m*(\Phi_0 - qFx)]^{1/2} dx} = T_0 e^{-\frac{2}{\hbar}\sqrt{2m^*}\int_0^d (\Phi_0 - qFx)^{1/2} dx}. \tag{1.161}$$

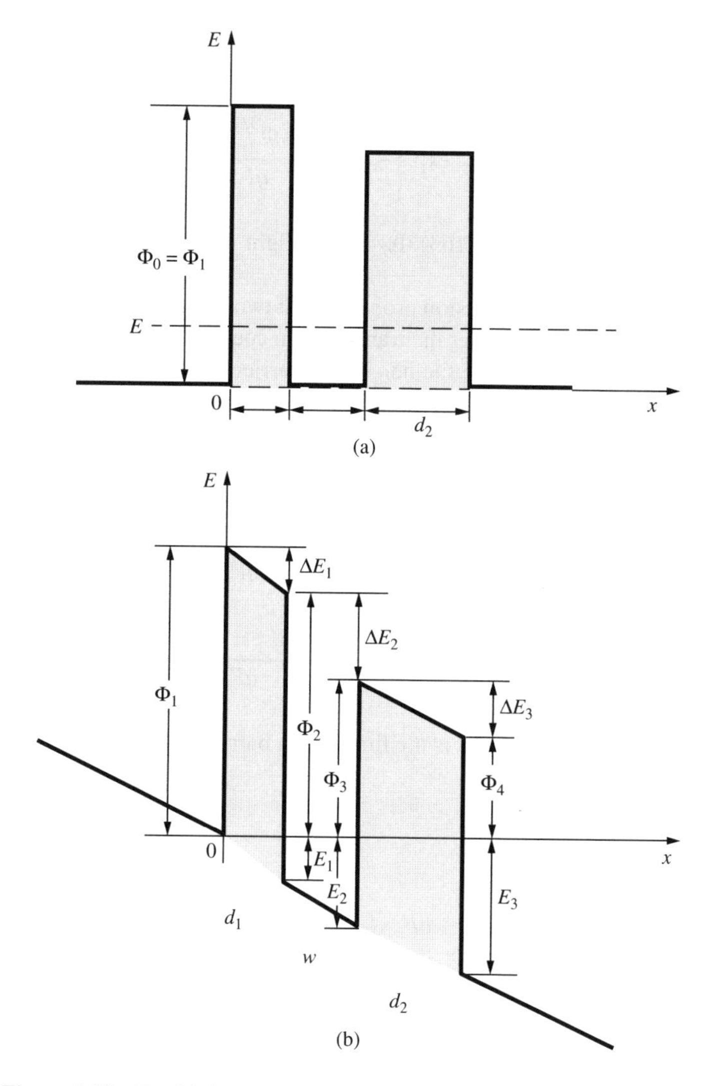

Figure 1.23 Double barrier structure without (a) and with (b) applied field

Take the integral in Equation (1.161)

$$\int_{0}^{d} [(\Phi_0 - qFx)]^{1/2} dx = \frac{\Phi_0^{3/2} - (\Phi_0 - qFd)^{3/2}}{3/2qF} \tag{1.162}$$

and we obtain the following equation for transmission probability of trapezoidal barrier.

$$T = T_0 \exp\left[-\frac{4}{3} \frac{\sqrt{2m^*}}{\hbar} \frac{\Phi_0^{3/2} - \left(\Phi_0 - qFd\right)^{3/2}}{qF} \right] \tag{1.163}$$

In the case of triangular barrier the second member in exponent of the function is equal to zero and we obtain a well-known Fowler–Nordheim equation [11].

$$T = T_0 \exp\left[-\frac{4}{3}\frac{\sqrt{2m^*}}{\hbar}\frac{\Phi_0^{3/2}}{qF}\right]. \tag{1.164}$$

As can be seen from Equation (1.163) Φ_0 is the height of barrier input and $(\Phi_0 - qFd)$ is the height of barrier output.

During the analysis of the transmission probability of two barrier resonance tunneling structure (Figure 1.23) we have to consider the transmission coefficients through the first (T_1) and the second (T_2) barriers. In this case the heights of barriers for coming electrons are significantly changed under electric field. It is necessary to take into account voltage dropping on each barrier and well. Based on Equation (1.163) we rewrite the transmission coefficients for the first and second barrier correspondingly

$$T_1 = T_{01} \exp\left[-\frac{4}{3}\frac{\sqrt{2m_1{}^*}}{\hbar}\frac{\Phi_1^{3/2} - \Phi_2{}^{3/2}}{qF_{b1}}\right], \tag{1.165}$$

$$T_2 = T_{02} \exp\left[-\frac{4}{3}\frac{\sqrt{2m_2{}^*}}{\hbar}\frac{\Phi_3^{3/2} - \Phi_4{}^{3/2}}{qF_{b2}}\right], \tag{1.166}$$

where Φ_1 is the first input barrier, Φ_2 is the first output barrier, Φ_3 is the second input barrier, and Φ_4 is the second output barrier.

They are expressed as:

$$\Phi_1 = \Phi_0 \tag{1.167}$$

$$\Phi_2 = \Phi_0 - qF_{b1}d_1 \tag{1.168}$$

$$\Phi_3 = \Phi_0 - qF_{b1}d_1 - qF_w w \tag{1.169}$$

$$\Phi_4 = \Phi_0 - qF_{b1}d_1 - qF_{b2}d_2 - qF_w w, \tag{1.170}$$

where F_{b1}, F_{b2}, and F_w represent the electric field in the barriers and well region, respectively, and they are

$$F_{b1} = V_a / \left(d_1 + \frac{\varepsilon_{b1}}{\varepsilon_{b2}}d_2 + \frac{\varepsilon_{b1}}{\varepsilon_w}w\right) \tag{1.171}$$

$$F_{b2} = V_a / \left(d_2 + \frac{\varepsilon_{b2}}{\varepsilon_{b1}}d_1 + \frac{\varepsilon_{b2}}{\varepsilon_w}w\right) \tag{1.172}$$

$$F_w = V_a / (w + \frac{\varepsilon_w}{\varepsilon_{b1}}d_1 + \frac{\varepsilon_w}{\varepsilon_{b2}}d_2. \tag{1.173}$$

Here ε_{b1}, ε_{b2}, and ε_w denote the dielectric constants of the barriers and well material while d_1, d_2, and w denote the barriers and well width, respectively, and V_a is the applied electric voltage.

Using Equations (1.165) and (1.166) it is possible to calculate the general transmission probability through two barrier resonance tunneling structure according to Equations (1.114), (1.115), or (1.135).

The dependence of the transmission coefficient, T_G, on applied electric field for resonance tunneling structure AlN-GaN-AlN calculated according to Equation (1.135) is shown in Figure 1.24. During the calculation of such parameters the double barrier resonance tunneling structure has been used: $d_1 = 2.5$ nm, $d_2 = 2.5$ nm, $w = 10$ nm, $\Phi_0 = \Phi_1 = \Phi_{GaN-AlN} = 2.0$ eV, $\varepsilon_{GaN} = 10.4\varepsilon_0$, $\varepsilon_{AlN} = 8.5\varepsilon_0$, $\varepsilon_0 = 8.85 \times 10^{-14}$ F/cm [34].

The analysis shows that even in the case of symmetrical structure with the same barriers the transmission coefficients of the first and second barriers are different. Using the nonsymmetrical structure with the left (first) barrier thinner, then the right (second) one, the condition for $T_1 = T_2$ might be recreated thus enhancing the resonance effects (see Equation (1.128)) looked for in experiments. But the possibility to realize the optimized condition (i.e., $T_G = 1$) is not guaranteed and even if this can be achieved, it would be true only for a particular resonance peak.

At high electric fields or in the case of some special structures, for example, at emission into vacuum (the second barrier is the vacuum) the shape of the second barrier is triangular (Figure 1.25).

In this case Equation (1.164) is applied for calculation of transmission probability through the second barrier.

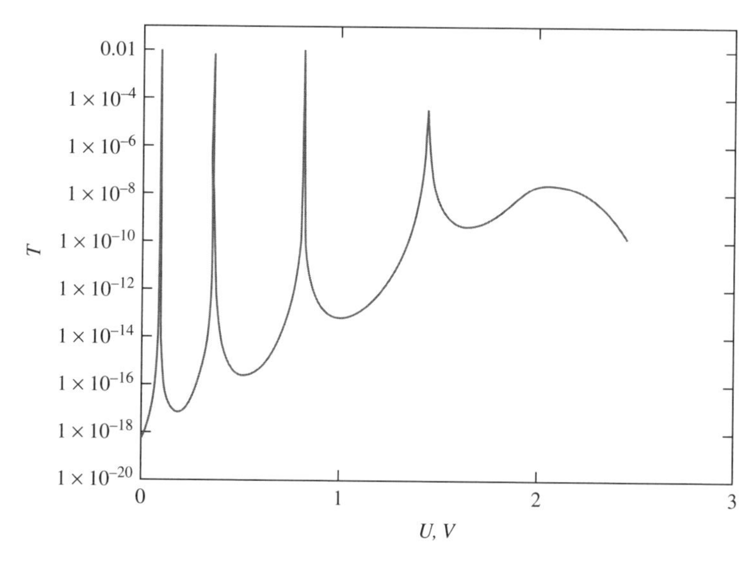

Figure 1.24 Calculated transmission coefficient for the AlN-GaN-AlN double barrier resonant tunneling structure

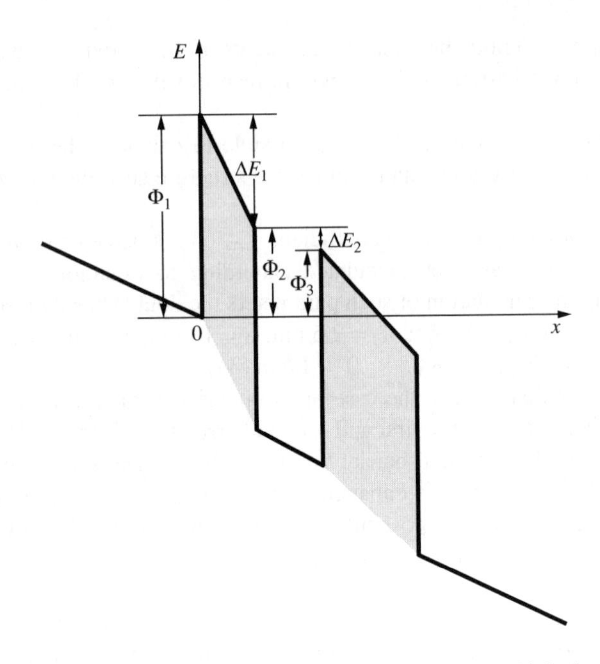

Figure 1.25 Two-barrier resonant tunneling structure under high electric field

1.9 Temperature Effects

1.9.1 One Barrier

Temperature dependence on transmission probability (Equation (1.117)) and tunnel current is caused by temperature induced carrier energy spreading. The Boltzmann distribution function and Fermi–Dirac distribution function, respectively, include the temperature

$$f(E) = \sqrt{\frac{E}{\pi k_B T}} \exp\left(-\frac{E}{k_B T}\right). \tag{1.174}$$

$$f(E) = 1/\{1 + \exp[(E - E_F)/k_B T]\}. \tag{1.175}$$

The F–N current density through the barrier (for example, SiO_2) can be calculated, as a function of temperature, under the assumption that the electrons in the emitting electrode (for example, Si) can be described by three-dimensional Fermi gas, according to classical approach while neglecting the Schottky effect as [11, 36]

$$J_{F-N}(T) = \frac{q m_{Si}^* k_B T}{2\pi^2 \hbar^3} \int_0^{\Phi} \ln\left[1 + \exp\left(\frac{E_F(T) - E}{k_B T}\right)\right] \times \exp\left(-\frac{4\sqrt{2m_{ox}^*}(\Phi - E)^{3/2}}{3\hbar q F}\right) dE, \tag{1.176}$$

where q is the absolute electron charge, m^*_{Si} and m^*_{ox} are the effective electron mass into the Si and SiO_2, respectively, $k_B T$ is the thermal energy, $\hbar$ is the reduced Planck constant, F is the electric field across the insulator, and E_F is the Fermi level.

In the low-temperature approximation the conventional temperature independent F–N analytical expression can be derived from relation (1.176) in the form [36]

$$J^0_{F-N} = AF^2 \exp\left(-\frac{B}{F}\right), \tag{1.177}$$

where A and B are the so-called pre-exponential and exponential F–N coefficients defined as

$$A = \frac{q^3 m^*_{Si}}{16\pi^2 \hbar m^*_{ox} \Phi}, \tag{1.178}$$

$$B = \frac{4\sqrt{2m^*_{ox}}\Phi^{3/2}}{3\hbar q}. \tag{1.179}$$

Although relation (1.177) has been derived under low-temperature approximation, it can be used to empirically describe the temperature dependence of the F–N current in MOS structure after extraction of the effective coefficients $A(T)$ and $B(T)$ [37].

As a rule the next analytical but approximate formula that has been proposed to account for the variation with temperature of the F–N emission current are used [20, 38]

$$J_{F-N}(T) = \frac{\pi c k_B T}{\sin(\pi c k_B T)} J^0_{F-N}, \tag{1.180}$$

where $c = 2\sqrt{2m_{ox}\Phi}/q\hbar F$.

This analytical approximation is not applicable at high temperatures and/or low electric field such that $ck_BT \geq 1$.

In order to overcome this limitation in Ref. [37] the new series expansion (Sommerfeld expansion) of the F–N current versus temperature derived in Equation (1.176) has been obtained

$$\frac{J_{F-N}(T)}{J_{F-N}(0)} = 1 + \frac{\pi^2}{6} T(E_F)(k_B T)^2 + \frac{7\pi^2}{360} T^{//}(E_F)(k_B T)^4 + \ldots + C_{2n} T^{(2n-2)}(E_F)(k_B T)^{2n} + \ldots. \tag{1.181}$$

The advantage of the given analytical formula is that (1) it does not diverge for any critical condition as it is in the case for relation (1.180), and (2) the desired accuracy of the analytical expression can be controlled by the polynomial order up to which the expansion is done. Figure 1.26 illustrates the capability of the proposed in Ref. [37] analytical expression (1.181) expanded to the sixth order and that of relation (1.180) to approximate the exact F–N current Equation (1.174). In this example, the maximum error given by relation (1.181) does not exceed 15% for 400 °C, whereas it is larger than a factor 2 for relation (1.180).

The experimental results on F–N tunneling [37] confirm (1) the strong impact of temperature on the F–N current amplitude, especially for temperatures above 250 °C; and (2) the good linearity of the F–N plots whatever the temperature is. The last point demonstrate that the low temperature analytical model of relation (1.180) can still be applied up to 400 °C, but with the temperature-dependent effective pre-exponential and exponential F–N coefficients $A(T)$ and $B(T)$.

At electron injection from degenerated silicon into oxide the good agreement of experimental and calculated results has been achieved after the assumption on linear variation of Φ with

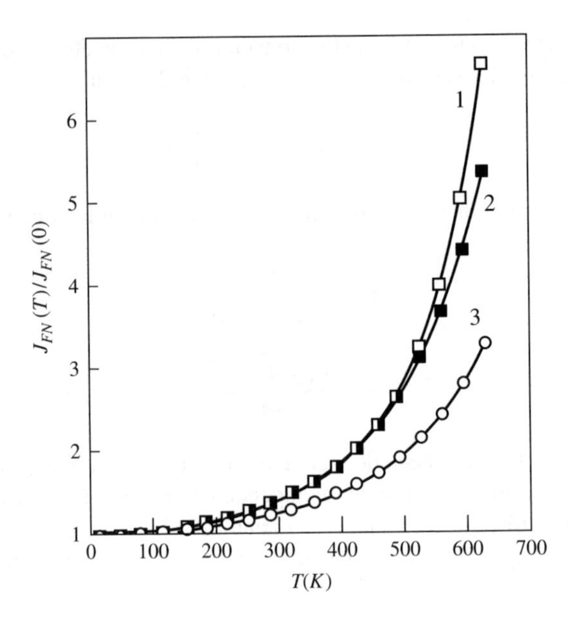

Figure 1.26 Theoretical relative variations of the F–N currents as obtained using the exact formula and analytical approximations: (1) exact dependence, (2) analytical approximation according to formula (1.181), and (3) analytical approximation according to formula (1.180). Reproduced with permission from Ref. [37]. Copyright (1995), AIP Publishing LLC

temperature with $d\Phi/dT = -2.4 \times 10^{-4}$ eV/°C, but at emission from metal electrode $d\Phi/dT = -2.67 \times 10^{-4}$ eV/°C [20, 39, 40]. It is well established that temperature induces decrease in the barrier height at the Si/SiO_2 interface [41]. As the tunneling probability exponentially depends on Φ this effect is large enough. For a given temperature, the decrease of 0.1 eV of Φ leads to nearly one decade of increase in J_{FN} for $F = 8$ MV/cm. The theoretical F–N coefficients A and B have been extracted for different temperatures. They are in good agreement with the experimental ones. This clearly demonstrates that the temperature behavior of the F–N tunnel current can be reasonably well interpreted by the classical model of relation (1.176), which accounts satisfactorily for the temperature-induced carrier energy statistical spreading. It is worth noting that the effective barrier height is always found to be smaller than the "actual" barrier height deduced from the relation, which takes into account the Fermi level variation with temperature over the whole temperature range [37]:

$$E_F(T) = E_{F0}\left[1 - \frac{\pi^2}{12}\left(\frac{k_B T}{E_{F0}}\right)^2\right], \tag{1.182}$$

where $E_{F0} = \hbar^2(3\pi^2 N_s)^{2/3}/(2m_{Si}^*)$.

This result clearly indicates that all the approaches based on only low-temperature F–N approximation Equation (1.177) cannot explain the behavior with temperature of the F–N emission by including only the temperature dependence through Fermi level variation.

It is now clear that the temperature-induced carrier energy distribution accounted for 3D model is the dominant mechanism for the F–N current changing with temperature. The strong effect of temperature experimentally observed [37, 42, 43] is rather due to the temperature effect on the barrier height caused by carrier energy distribution.

1.9.2 Double-Barrier Resonance Tunneling Structure

The temperature can effect the resonance time. The key point is that the thermal motion of the atoms in any sample contributes in making the potential energy time dependent. As far as this effect is concerned, from a qualitative point of view different cases can be distinguished. If the variations of potential energy E_p are very small or/and very slow (compare to τ_0), then E_p can be considered not to depend on time to all practical purposes. If, instead, E_p significantly varies in values on a time scale comparable or smaller than τ_0, then a more complicated analysis is required.

Overall it is expected that the temperature will give rise to broadening of the resonance peaks and decrease in their effects on the current measured in experiments.

It is very important to consider the effect due to the electron thermal population [7]. The important conclusion is reached that at resonance a variety of current-versus-T relationship can result depending on the relative position of the resonance state and the Fermi energy (E_F). In particular, currents increasing as well as decreasing with T and complicated nonmonotonic temperature behavior are possible. Each state gives rise to its own (individual) J vs. T dependence according to its energy position. In real samples where many such states are presented, different (individual) current behavior is to be expected at each resonant peak. This is in agreement with experiments showing that the conductance at peak is proportional to $exp[(T_0/T)^{1/2}]$ where T_0 is individual for the considered peak [44].

As the temperature varies the cathode carrier concentration at the resonant energy also varies and so does the current J measured in experiments. At the same time the carrier thermal velocity also increases with T and, with semiconductor or metal cathode, this implies an increase in the electron flux hitting the barrier, hence J. This latter is, however, only a minor effect (because the thermal velocity depends on $T^{1/2}$) with respect to that mentioned earlier whose temperature dependence comes from the exponential factor in the Fermi distribution function.

Because we essentially deal with the carrier concentration within the definite narrow energy window (the width of the resonant eigenstate centered on E_n) the effects to be expected depend on its position relative to E_n. If E_n and E_F are close (compared with k_BT), an increase of T spreading out the distribution function can only lead to decrease of particle concentration at the resonance energy, hence to decrease of tunneling current. In this case J exhibits a metallic type of behavior.

In any case, for large increase of temperature a subsequent increase in current may occur since, as the distribution function spreads out, the carrier concentration can become nonnegligible at other, higher eigenstates whose contribution will rapidly become important. If, on the other hand, the distance between E_n and E_F is large, an increase of current is first expected to occur as a consequence of the increase in carrier concentration available for resonant tunneling. Here too, however, a subsequent metallic type of behavior can arise for the same reasons as given above [7].

References

1. K.F. Brennan and A.S. Brown. *Theory of Modern Electronic Semiconductor Devices* (New York, John Wiley & Sons, Inc., 2002).
2. S. Datta. *Electronic Transport in Mesoscopic Syst.ems*, Ed. H. Ahmed, M. Pepper, A. Broers (Cambridge, Cambridge University Press, 1995), pp. 247–75.
3. M. Ya. Azbel, Eigenstates and properties of random systems in one dimension at zero temperature, *Physical Review B* **28**, 4106 (1983).
4. A.F. Kravchenko and V.N. Ovsyuk. *Electron Processes in Solid-State Low-Dimensional System* (Novosibirsk University Press, 2000), 448p (in Russian).
5. S.N. Lykov, V.G. Gasumyantz, and S.A. Rykov. *Quantum Mechanics, Part 2*, (St Petersburg, St Petersburg State Technical University, 2004) (in Russian).
6. O.V. Tretyak and V.Z. Lozovskyy. *Foundation of Semiconductor Physics* Vol. 2 (Kyiv University Press, 2009) (in Ukrainian).
7. B. Ricco and M.Ya. Azbel, Physics of resonant tunneling. The one-dimensional double-barrier case, *Physical Review B* **29**, 1970 (1984).
8. Y. Ando and T. Itoh, Calculation of transmission tunneling current across arbitrary potential barriers *Journal of Applied Physics* **61**, 1497 (1987).
9. W.W. Liu and M. Fukuma, Exact solution of the Schrodinger equation across an arbitrary one-dimensional piecewise-linear potential barrier, *Journal of Applied Physics* **60**, 1555 (1986).
10. A.F.M. Anwar and M.M. Jahan, Resonant tunneling devices. *Encyclopedia of Nanoscience and Nanotechnology*. Ed. H.S. Nalwa (American Scientific Publisher, Vol. 9, pp. 357–70, 2004).
11. R.H. Fowler and L. Nordheim, Electron emission in intense electric fields, *Proceedings of the Royal Society of London Series A* **119**, 173 (1928).
12. P. Bellutti and N. Zorzi, High electric field induced positive charges in thin gate oxide, *Solid-State Electronics* **45**, 1333 (2001).
13. I.C. Chen, S.E. Holland and C. Hu, Electrical breakdown in thin gate and tunneling oxides, *IEEE Transactions on Electron Devices* **32**, 413 (1985).
14. Y. Hokari, Stress voltage polarity dependence of thermally grown thin gate oxide wearout, *IEEE Transactions on Electron Devices* **35**, 1299 (1988).
15. P. Samanta and C.K. Sarkar, Analysis of positive charge trapping in silicon dioxide of MOS capacitors during Fowler-Nordheim stress, *Solid-State Electronics* **32**, 507 (1989).
16. P. Solomon, High-field electron trapping in SiO_2, *Journal of Applied Physics* **48**, 3843 (1977).
17. S.J. Oh and Y.T. Yeow, Voltage shifts of Fowler-Nordheim tunneling J-V plots in thin gate oxide MOS structures due to trapped charges, *Solid-State Electronics* **32**, 507 (1989).
18. J. Lopez-Villanueva, J. Jimenez-Tejada, P. Cartujo, J. *et al.* Analysis of the effects of constant-current Fowler–Nordheim-tunneling injection with charge trapping inside the potential barrier, *Journal of Applied Physics* **70**, 3712 (1991).
19. P.S. Ku and D.K. Schroder, Charges trapped throughout the oxide and their impact on the Fowler-Nordheim current in MOS device, *IEEE Transactions on Electron Devices* **41**, 1669 (1994).
20. M. Lenzlinger and E.H. Snow, Fowler-Nordheim tunneling into thermally grown SiO_2, *Journal of Applied Physics* **40**, 278 (1969).
21. S.M. Sze. *Physics of Semiconductor Devices* (John Wiley & Sons, Inc., Hoboken, NJ 1981).
22. A. Aziz, K. Kassmi, Ka Kassmi, and F. Olivie, Modelling of the influence of charges trapped in the oxide on the $I(V_g)$ characteristics of metal-ultra-thin oxide-semiconductor structures, *Semiconductor Science and Technology* **19**, 877 (2004).
23. D.J. Dumin and J.R. Maddux, Correlation of stress-induced leakage current in thin oxides with trap generation inside the oxides, *IEEE Transactions on Electron Devices* **40**, 986 (1993).
24. T. Wang, N.K. Zous, J.L. Lai, and C. Huang, Hot hole stress induced leakage current (SILC) transient in tunnel oxides, *IEEE Electron Device Letters* **19**, 411 (1998).
25. N.K. Zous, T. Wang, C.-C. Yeh, and C.W. Tsai, Transient effects of positive oxide charge on stress-induced leakage current in tunnel oxides, *Applied Physics Letters* **75**, 734 (1999).
26. S. Luryi and A. Zaslavsky, Quantum-effect and hot-electron devices, in S.M. Sze, Ed, *Modern Semiconductor Device Physics* (John Wiley & Sons, Inc, New York, 1998), pp. 253–341.
27. S.M. Sze and K.K. Ng. *Physics of Semiconductor Devices*, 3rd edn (John Wiley & Sons, Inc., New York 2007).

28. O. Kane. *Tunneling Phenomena in Solids*, ed. E. Burnstein and D. Lundquist (Plenum, New York, 1969), pp. 79–92.
29. L.L. Chang, L. Esaki, and R. Tsu, Resonant tunneling in semiconductor double barriers, *Applied Physics Letters* **24**, 593 (1974).
30. M. Hirose, M. Morita, and Y. Osaka, Resonant tunneling through Si/SiO_2 double barriers, *Japanese Journal of Applied Physics* **16**, 561 (1977).
31. T.C.L.G. Sollner, W.D. Goodhue, P.E. Tannenwald, C.D. Parker, and D.D. Peck, Resonant tunneling through quantum wells at frequencies up to 2.5 THz, *Applied Physics Letters* **43**, 588 (1983).
32. M. Buttiker and R. Landauer, Traversal time for tunneling, *Physical Review Letters* **49**, 1739 (1982).
33. M. Buttiker, Larmor precession and the traversal time for tunneling, *Physical Review B* **27**, 6178 (1983).
34. Yilmazoglu, O., Considine, L., Joshi, R. *et al.* (2011) Resonant electron-emission from a flat surface AlN/GaN system with carbon nanotube gate electrode. In *Technical Digest of International Vacuum Nanoelectronics Conference*, Wuppertal, Germany, July 18–22, 2011, pp. 216–217.
35. B. Ricco, M.Ya. Azbel, and M.N. Brodsky, Novel mechanism for tunneling and breakdown of thin SiO_2 films, *Physical Review Letters* **51**, 1795 (1983).
36. J.J. O'Dwyer, *The Theory of Electrical Conduction and Breakdown in Solids Dielectrics*, (Clarendon, Oxford, 1973).
37. G. Pananakakis, G. Ghibaudo, and R. Kies, Temperature dependence of the Fowler–Nordheim current in metal-oxide-degenerate semiconductor structures, *Journal of Applied Physics* **78**, 2635 (1995).
38. R.H. Good and W. Muller, *Field emission, Hanbuch der Physik*, Vol. 21 (Springer, Berlin, 1956).
39. M.O. Aboelfotoh, Schottky-barrier behavior of a Ti-W alloy on Si(100), *Journal of Applied Physics* **61**, 2558 (1987).
40. K.S. Kim and M.E. Lines, Temperature dependence of chromatic dispersion in dispersion-shifted fibers: Experiment and analysis, *Journal of Applied Physics* **73**, 2069 (1993).
41. A. Hadjiadji, G. Salace, and C. Petit, Fowler–Nordheim conduction in polysilicon (n^+)-oxide–silicon (p) structures: Limit of the classical treatment in the barrier height determination, *Journal of Applied Physics* **89**, 7994 (2001).
42. G. Salace, A. Hadjiadji, C. Petit, and M. Jourdain, Temperature dependence of the electron affinity difference between Si and SiO_2 in polysilicon (n^+)–oxide–silicon (p) structures: Effect of the oxide thickness, *Journal of Applied Physics* **85**, 7768 (1999).
43. G. Salace, A. Hadjiadji, C. Petit, and Dj. Ziane, The image force effect on the barrier height in MOS structures: correlation of the corrected barrier height with temperature and the oxide thickness, *Microelectronics Reliability* **40**, 763 (2000).
44. A.B. Fowler, A. Hartstein, and R.A. Webb, Conductance in restricted-dimensionality accumulation layers, *Physical Review Letters* **48**, 196 (1982).

2

Supply Function

The supply function as an important part of electron field emission description is analyzed in this Chapter for three-, two-, one-, and zero dimension cases. The density of electron states and Fermi distribution function are considered for nanostructures. As the examples of such analyses the two-dimensional electron gas in GaN/AlGaN heretojunction and quantum sized semiconductors films are presented.

2.1 Effective Mass Approximation

In principle, the density of states could be determined from band theory calculations for a given material [1]. Such calculations, however, would be rather involved and impractical. Fortunately, an excellent approximation for the density of states near the band edges, the region of the bands normally populated by carriers, can be obtained through a much simpler approach. This is visualized in Figure 2.1a. The electrons in the conduction band are essentially free to roam throughout the crystal. For electrons near the bottom of the band, the band itself forms a pseudo-potential well, as shown in Figure 2.1b. The well bottom lies at E_c and the termination of the band at the crystal surfaces forms the walls of the well. Since the energy of the electrons relative to E_c is typically small compared with the surface barriers, one effectively has a particle in a three-dimensional box. The density of states near the band edges can therefore be equated to the density of states available to a particle of mass m^* in a box with the dimensions of the crystal.

Electronic conduction in semiconductors can take place either through electrons in the conduction band or through holes in the valence band. However, most experiments on mesoscopic conductors and especially the electron field emission involve the flow of electrons in the conduction band. The dynamics of electrons in the conduction band can be described by an equation of the form [2]

$$\left[E_c + \frac{\left(i\hbar\nabla - q\bar{A}\right)^2}{2m^*} + U(\bar{r})\right]\Psi(\bar{r}) = E\Psi(\bar{r}), \tag{2.1}$$

Vacuum Nanoelectronic Devices: Novel Electron Sources and Applications, First Edition.
Anatoliy Evtukh, Hans Hartnagel, Oktay Yilmazoglu, Hidenori Mimura and Dimitris Pavlidis.

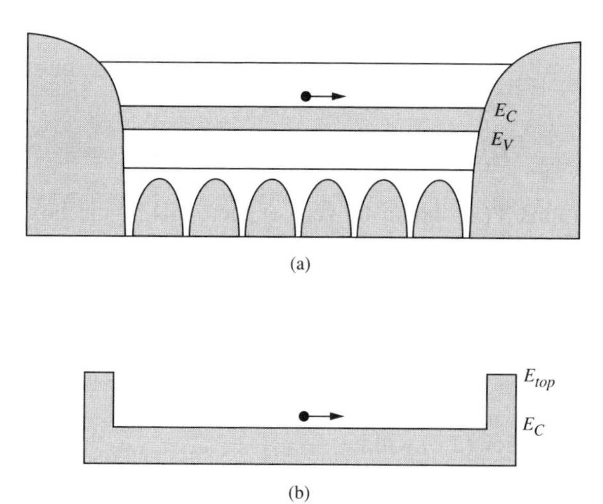

Figure 2.1 (a) Visualization of a conduction band and electron moving in a crystal; (b) idealized pseudo-potential well formed by the crystal surfaces and the band edges

where $U(\bar{r})$ is the potential energy due to space-charge, and so on, $\bar{A}$ is the vector potential, and m^* is the effective mass. Although Equation (2.1) looks just like the Schrodinger equation it is really what is called a single-band effective mass equation. The lattice potential, which is periodic on an atomic scale, does not appear explicitly in Equation (2.1); its effect is incorporated through the effective mass m^* which we assume to be spatially constant. Any band discontinuity ΔE_c at heterojunctions is incorporated by letting E_c be position-dependent.

It should be noted that the wave functions that we calculate from Equation (2.1) are not the true wave functions but are smoothed out versions that do not show any rapid variations on the atomic scale. This can easily be seen by considering a homogeneous semiconductor with $U(\bar{r}) = 0$ and $E_c = \text{constant}$. The wave functions satisfying Equation (2.1) have the form of plane waves

$$\Psi(\bar{r}) = \exp(i\bar{k}\bar{r}) \tag{2.2}$$

and not that of Bloch waves

$$\Psi(\bar{r}) = U_{\bar{k}}(\bar{r}) \exp(i\bar{k}\bar{r}), \tag{2.3}$$

since the lattice potential is not included. The simplified description based on the single-band effective mass equation, Equation (2.1), is usually adequate for conduction band electrons at low fields.

2.2 Electron in Potential Box

Let us consider an electron in a box, on the borders of which at $x=0$ and $x=L$ the potential energy increases sharply to infinity. Inside the box the potential energy is constant and equal to zero [3]. Obviously, the electron does not go beyond the box. Schrodinger equation for one-dimensional electron motion along the X axis has the form

$$\frac{d^2\Psi}{dx^2} + \frac{8\pi^2 m}{h^2}(E - E_0)\Psi = 0. \tag{2.4}$$

In our case

$$E_0 = 0, \ldots\ldots 0 < x < L, \tag{2.5a}$$

$$E_0 \to \infty, \ldots\ldots x = 0, \ \ldots \text{ and } \ldots \ x = L. \tag{2.5b}$$

Under these conditions $\Psi(x)$ should be zero at the walls of the box. Thus, the problem reduces to the integration of

$$\frac{d^2\Psi}{dx^2} + \frac{8\pi^2 m}{h^2} E\Psi = 0, \tag{2.6}$$

with boundary conditions

$$(\Psi)_{x=0} = 0 \text{ and } (\Psi)_{x=L} = 0. \tag{2.7}$$

Eigenfunctions of Equation (2.6) will be

$$\Psi_n = \sin\left(n\frac{\pi x}{L}\right) \tag{2.8}$$

on condition that coefficient at Ψ in Equation (2.6) is equal to

$$\frac{8\pi^2 m}{h^2} E = 4\pi^2 \left(\frac{n}{2L}\right)^2, \tag{2.9}$$

where $n = 1, 2, 3 \ldots$

From Equation (2.9) the energy eigenvalues are

$$E = n^2 \frac{h^2}{8mL^2}, \tag{2.10}$$

where $n = 1, 2, 3 \ldots$

Thus, the boundary conditions of the problem will be satisfied only for a discrete number of energy values (Equation (2.10)), that is, an electron, placed in a potential box, has the quantized values of energy. The quantization becomes noticeable, of course, only for potential boxes of atomic dimensions.

Let us analyze in more detail the properties of wave functions of the electron, described by Equation (2.8).

First, the eigenfunctions are orthogonal, that is,

$$\int_{-\infty}^{+\infty} \Psi_m(x)\Psi_n(x)dx = 0 \text{ at } m \neq n. \tag{2.11}$$

Indeed, due to boundary conditions the function $\Psi(x)$ vanishes outside the potential box and the integration between infinite limits reduces to the integration from 0 to L:

$$\int_{-\infty}^{+\infty} \Psi_m(x)\Psi_n(x)dx = \int_0^L \sin\left(m\frac{\pi x}{L}\right) \sin\left(n\frac{\pi x}{L}\right) dx$$

$$= \frac{1}{2}\int_0^L \left[\cos\left((m-n)\frac{\pi x}{L}\right) - \cos\left((m+n)\frac{\pi x}{L}\right)\right] dx \equiv 0. \tag{2.12}$$

Secondly, since the above integral does not vanish when $m=n$, then the function Ψ_n can be normalized to unity, that is, one can find such a factor N_n, that the next condition will be performed.

$$N_n^2 \int_{-\infty}^{+\infty} \Psi_n^2(x)dx = 1. \tag{2.13}$$

Let us find the value of this factor N_n

$$N_n^2 = \frac{1}{\int_{-\infty}^{+\infty} \Psi_n^2(x)dx} = \frac{1}{\int_0^L \sin^2\left(n\frac{\pi x}{L}\right)dx}, \tag{2.14}$$

but

$$\int_0^L \sin^2\left(n\frac{\pi x}{L}\right)dx = \frac{L}{2}, \tag{2.15}$$

then

$$N_n = \sqrt{\frac{2}{L}}, \tag{2.16}$$

that is, the normalizing factor is the same for all functions Ψ_n and the normalized eigenfunctions have the form

$$\Psi_n = \sqrt{\frac{2}{L}} \sin\left(n\frac{\pi x}{L}\right). \tag{2.17}$$

As is known from quantum mechanics, in the case of a conservative field, when the potential energy E_0 does not depend explicitly on the time, the time dependence of the eigenfunction Ψ is expressed by the factor of $\exp\left(-i\frac{2\pi}{h}Et\right)$.

As a result, the function that describes the electron state with energy E in any time has the following form:

$$\Psi_n(x,t) = \Psi_n(x)\exp\left(-i\frac{2\pi}{h}Et\right) = \sin\left(n\frac{\pi x}{L}\right)\exp\left(-i\frac{2\pi}{h}Et\right). \tag{2.18}$$

The physical meaning only has the square of modulus of the complex Ψ function that satisfies the Schrodinger equation,

$$|\Psi_n(x,t)|^2 = \Psi_n^*(x,t)\Psi_n(x,t) = \Psi_n^2(x) = \sin^2\left(n\frac{\pi x}{L}\right), \tag{2.19}$$

which in the general case (three-dimensional space) determines the density of probability to find an electron in a volume dV around the point with x, y, z coordinates. The probability is proportional to $\Psi\Psi^* dV$.

The curves of the eigenfunctions (Equation (2.8)) and the squares of their moduli (Equation (2.19)) are shown in Figure 2.2. The meaning of squares of the Ψ-function moduli in a potential well is that they characterize the distribution of probability to find an electron per unit length in a particular location within the box at various electron energies. The figure shows

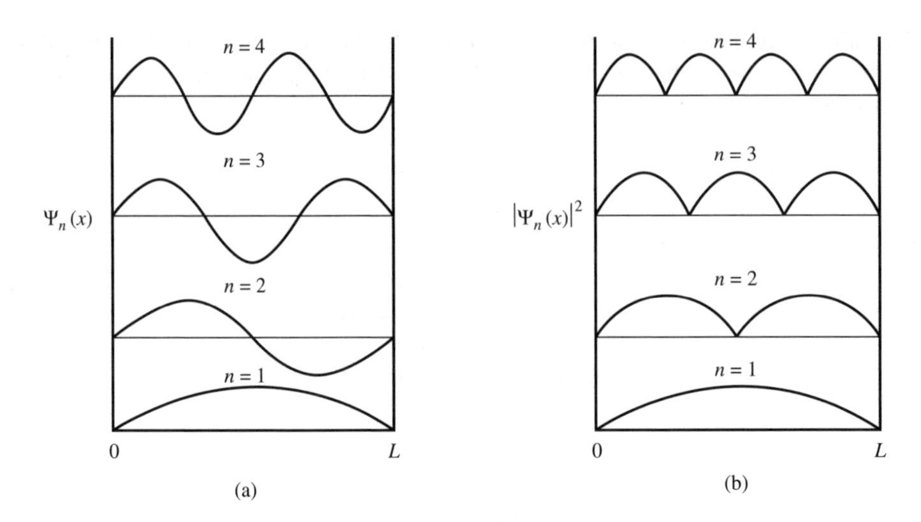

Figure 2.2 Electron in the potential box: (a) the eigenfunctions; (b) squares of their modules

that at the lowest energy state ($n = 1$) the electron can be found in the middle of the box with the greatest probability, and only at higher energies the uniform distribution of the electron along the length of the potential box arises that corresponds to the macroscopic particles.

2.3 Density of States

The density of states is required as the first step in determining the carrier concentrations and energy distributions of carriers within a semiconductor. Integrating the density of states function, *g(E),* between two energies E_1 and E_2 tells one the number of allowed states available to electrons in the given energy range per unit volume of the crystal.

2.3.1 Three-Dimension (3D) Case

Let us consider the determination the density of electronic states in the conduction band of bulk semiconductors or metals [3]. Let us suppose that a crystal has the form of a box with dimensions of edges L_x, L_y, L_z along the selected axes *X, Y, Z.* All dimensions L_x, L_y, L_z are significantly higher than the de Broglie wavelength

$$\lambda = \frac{h}{m_n^* v} = \frac{h}{\sqrt{2m_n^* E}}. \tag{2.20}$$

The state of free electrons is described in geometric space by vector with radius $\overline{r}(x, y, z)$, and in pulse space by the vector of pulse $\overline{p}(p_x, p_y, p_z)$, or the wave vector $\overline{k} = \overline{p}/\hbar$. The value of wave vector is $k = 2\pi/\lambda$.

Let us form the elementary volume of the imaginary six-dimensional phase space

$$d\Omega = dx \times dy \times dz \times dp_x \times dp_y \times dp_z, \tag{2.21}$$

$$d\Omega = d^3x \times d^3p. \tag{2.22}$$

In accordance with the conditions of the von Karman–Born the phase volume $d\Omega$ contains $d\Omega/h^3$ phase cells, each of which contains one electron state, but in a cell volume of h^3 two electrons with opposite spins can be placed. Thus, the total number of electronic states in the elementary phase volume $d\Omega$ is equal to

$$dN = \frac{d\Omega}{h^3} = \frac{d^3x \times d^3p}{h^3}. \tag{2.23}$$

The number of electrons dn that are located in an elementary volume $d\Omega$ of phase space is the product of the number of states on the probability of finding the electrons in these states:

$$dn = 2\frac{d\Omega}{h^3} f(\overline{r}, \overline{p}, t). \tag{2.24}$$

Factor 2 shows that in every phase cell there can be two electrons with opposite spins.

Now we transit from the phase space of pulses to the space of wave vectors, k-space, using the relations

$$p_x = \hbar k_x,\ p_y = \hbar k_y,\ p_z = \hbar k_z, \tag{2.25}$$

then (Equation (2.23)) takes the form

$$dN = \frac{d^3x \times d^3k}{8\pi^3}. \tag{2.26}$$

Geometric coordinates and the coordinates of k_i are independent, and therefore the number of states throughout the crystal volume V, in which the vectors k are placed in the phase volume d^3k, are equal to

$$\int_V dN = \frac{d^3k}{8\pi^3} \int_V d^3x = V\frac{d^3k}{8\pi^3}. \tag{2.27}$$

In unity of the geometric volume of the number of states with wave vectors k, whose ends are placed in the volume d^3k, will be $d^3k/8\pi^3$.

The number of electronic states per unit volume, for example, in 1 cm^3, with the energy from 0 to the given value E, or, respectively, with wave number between 0 and k, is given by

$$N = \frac{1}{8\pi^3} \int_{V_E} d^3k, \tag{2.28}$$

where V_E is the volume of the figure in the k-space bounded by a surface of constant energy equal to E.

Let us consider two particular examples.

2.3.1.1 Spherical Isoenergetic Surfaces with E_{min} at the Center of the Brillouin Band

In this case

$$\int_{V_E} d^3k = \frac{4}{3}\pi k^3, \tag{2.29}$$

and the number of electronic states per unit volume with energies from 0 to E

$$N(E) = \frac{1}{8\pi^3}\frac{4}{3}\pi k^3 = \frac{k^3}{6\pi^2} \tag{2.30}$$

then the density of states $g(E)$, equal to the number of states in the unit interval of energy for unit volume of the crystal has the form

$$g(E) = \frac{dN}{dE} = \frac{k^2}{2\pi^2}\frac{dk}{dE}. \tag{2.31}$$

Let us take

$$E(k) = E_C + \frac{\hbar^2 k^2}{2m_n^*}. \tag{2.32}$$

We express k in terms of E:

$$k^2 = \frac{2m_n^*}{\hbar^2}[E(k) - E_C], \tag{2.33}$$

from this

$$2kdk = \frac{2m_n^*}{\hbar^2}dE \tag{2.34}$$

and

$$\frac{dk}{dE} = \frac{m_n^*}{\hbar^2 k}. \tag{2.35}$$

Substituting Equations (2.33) and (2.35) in Equation (2.31), we obtain

$$g(E) = 2\pi\left(\frac{2m_n^*}{h^2}\right)^{3/2}\sqrt{E - E_C}. \tag{2.36}$$

Thus, the density of states is proportional to $\sqrt{E}$ (Figure 2.3).

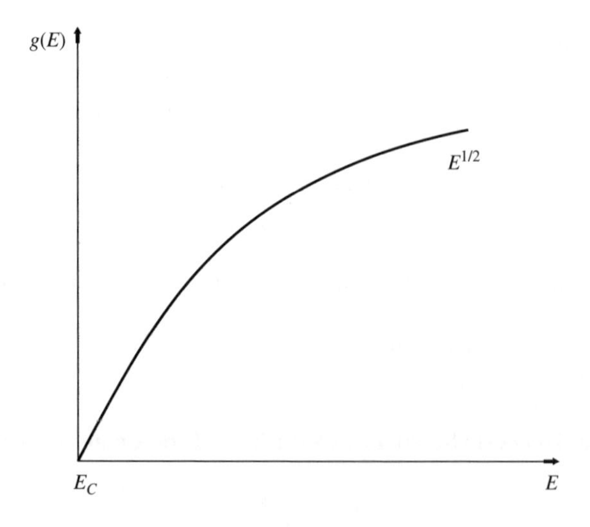

Figure 2.3 Density of electron states on energy dependence

2.3.1.2 Ellipsoidal Energy Surfaces

Let's consider a more general case, where energy surfaces have the form of an ellipsoid:

$$E(\bar{k}) = E_C + \frac{\hbar^2}{2}\left(\frac{k_x^2}{m_x^*} + \frac{k_y^2}{m_y^*} + \frac{k_z^2}{m_z^*}\right), \tag{2.37}$$

or, in canonical form,

$$\frac{k_x^2}{A_x^*} + \frac{k_y^2}{A_y^*} + \frac{k_z^2}{A_z^*} = 1, \tag{2.38}$$

where the axis of the ellipsoid are

$$A_i = \sqrt{\frac{2m_i^*(E - E_C)}{\hbar^2}}, \ldots .i = x, y, z. \tag{2.39}$$

The volume of one ellipsoid with axes A_x, A_y, A_x is

$$\int_{V_E} d^3k = \frac{4\pi}{3}A_xA_yA_z = \frac{4\pi}{3}\frac{\sqrt{8m_x^*m_y^*m_z^*}}{\hbar^3}(E - E_C)^{3/2}. \tag{2.40}$$

The number of states per unit volume with energies from 0 to E is

$$N = \frac{1}{8\pi^3}\int_{V_E} d^3k = \frac{\sqrt{8m_x^*m_y^*m_z^*}}{6\pi^2\hbar^3}(E - E_C)^{3/2}, \tag{2.41}$$

then the density of states

$$g(E) = M\frac{dN}{dE} = M\frac{\sqrt{2m_x^*m_y^*m_z^*}}{2\pi^2\hbar^3}(E - E_C)^{1/2}, \tag{2.42}$$

where M is the number of equivalent minima.

We remind that the number of equivalent energy minima in the conduction band of germanium is $M=4$, and in silicon $M=6$. Thus, the density of states is proportional to $\sqrt{(E - E_C)}$ and $\sqrt{m_x^*\, m_y^*\, m_z^*}$, where m_x^*, m_y^*, m_z^* are the components of the effective mass (in the principal axes of the ellipsoid of constant energy).

If we put

$$M(m_x^*m_y^*m_z^*)^{1/2} = (m_{ng}^*)^{3/2}, \tag{2.43}$$

where m_{ng}^* is called the effective mass for density of states. Then, in the general case, the density of states can be represented as follows

$$g(E) = 2\pi\left(\frac{2m_{ng}^*}{h^2}\right)^{3/2}(E - E_C)^{1/2}. \tag{2.44}$$

This coincides with the first case (Equation (2.36)), when spherical energy surfaces with isotropic mass and one minimum of energy have been considered.

It must be emphasized that the general expression (2.44) and a special case of Equation (2.36) are valid only when the energy E is a quadratic function of wave vector k, that is, for states near the energy minimum.

It is easy to show that the density of states for holes in the valence band is expressed by a similar formula

$$g(E) = 2\pi\left(\frac{2m^*_{pg}}{h^2}\right)^{3/2}(E_V - E)^{1/2}, \tag{2.45}$$

where the E_V is the energy of the valence band top and m^*_{pg} is an effective density of states for holes.

2.3.1.3 The Restriction of the Crystal Size

Considering in Section 2.2 the electron placed into a potential box with size L, we have found that its normalized wave function is

$$\Psi_n = \sqrt{\frac{2}{L}}\sin\left(n\frac{\pi x}{L}\right), \tag{2.46}$$

and the energy

$$E_n = \frac{n^2h^2}{8mL^2}, ...n = 1, 2, 3 \ldots \tag{2.47}$$

that is, there is the quantization of electron energy.

In our three-dimensional model, in three-dimensional potential box with edges of L_x, L_y, L_z along the selected axes X, Y, Z, the wave function can be represented as the product of three functions, each of which depends only on one coordinate:

$$\Psi(x, y, z) = \Psi(x)\Psi(y)\Psi(z), \tag{2.48}$$

or

$$\Psi(x, y, z) = \sqrt{\frac{8}{L_xL_yL_z}}\sin\left(n_1\frac{\pi x}{L_x}\right)\sin\left(n_2\frac{\pi y}{L_y}\right)\sin\left(n_3\frac{\pi z}{L_z}\right), \tag{2.49}$$

where $n_1, n_2, n_3 = 1, 2, 3 \ldots$, and the energy can only take the following discrete set of values:

$$E(n_1, n_2, n_3) = E_{n1x} + E_{n2y} + E_{n3z} = \frac{h^2}{8}\left(\frac{n_1^2}{m_x^*L_x^2} + \frac{n_2^2}{m_y^*L_y^2} + \frac{n_3^2}{m_z^*L_z^2}\right). \tag{2.50}$$

Eigenfunctions (Equation (2.49)) are standing waves with $n_1 + 1$ nodal planes parallel to coordinate plane YZ, $n_2 + 1$ nodal planes parallel to thc plane of the XZ, and $n_3 + 1$ nodal planes parallel to the XY plane.

According to Equation (2.50) the energy values create discrete sequence, with each triplet of integers n_1, n_2, n_3 that corresponds to certain level of energy. It is necessary to emphasize a very important feature of the model. Suppose, for simplicity, that we have a cubic box, so that

$L_x = L_y = L_z = L$. In this case,

$$\Psi(x, y, z) = \sqrt{\frac{8}{L^3}} \sin\left(n_1 \frac{\pi x}{L}\right) \sin\left(n_2 \frac{\pi y}{L}\right) \sin\left(n_3 \frac{\pi z}{L}\right), \tag{2.51}$$

$$E = \frac{h^2}{8mL^2}(n_1^2 + n_2^2 + n_3^2). \tag{2.52}$$

It is clear that each set of integers n_1, n_2, n_3 will fit the particular state, its wave function $\Psi(x, y, z)$, but for all states for which the sum of ${n_1}^2 + {n_2}^2 + {n_3}^2$ is the same, the value of energy will match. These states, which have the same energy, are called degenerated. For example, when $n_1 = 2$, $n_2 = 1$, $n_3 = 3$ energy is equal to

$$E = \frac{h^2}{8mL^2}(2^2 + 1^2 + 3^2) = \frac{h^2}{8mL^2}14, \tag{2.53}$$

and six different combinations of n_1, n_2, n_3, or six different states (degree of degeneracy is 6) correspond to this value of energy.

It is interesting to estimate the value of discrete energy spectrum of electrons from the crystal size. The lowest energy level will be at $n_1 = n_2 = n_3 = 1$. We estimate its magnitude in the direction of the X:

$$E_x = \frac{h^2}{8m_x^* L_x^2}, \tag{2.54}$$

we set $m_x^* = 0.1\ m_0$, $L_x = 1$ cm, then

$$E_{x1} = 3.76 \times 10^{-14}\text{eV} \ \text{ and } \ E_{x2} = 1.5 \times 10^{-13}\text{eV}. \tag{2.55}$$

The energy difference between levels is

$$\Delta E_{x1,x2} = 1.28 \times 10^{-13}\text{eV}. \tag{2.56}$$

With decreasing of the size of semiconductor quantum confinement increases markedly, and at $L_x = 10^{-6}$ cm

$$\Delta E_{x1,x2} = 128\ \text{meV}. \tag{2.57}$$

To observe the phenomenon of quantum confinement experimentally, it is necessary to fulfill the following conditions:

$$\Delta E_{2,1} \gg k_B T, \tag{2.58}$$

$$\Delta E_{2,1} \gg \frac{\hbar}{\tau}. \tag{2.59}$$

At liquid nitrogen temperature $k_B T \approx 6.6$ meV, $\hbar/\tau = q\hbar/\mu m^*_n \approx 1.17$ meV, if the electron mobility $\mu \approx 10^4$ cm^2/V s and $m_n^* = 0.1\ m_0$. Therefore, only when $\Delta E_{2,1} \geq 10$ meV, which corresponds to the size of the order of the tens of nanometers, is it is possible to observe quantization at 77 K.

2.3.2 Two-Dimension (2D) Case

We define the density of electronic states in the crystal, the electron motion which is limited in one direction, for example, along the *Z* axis. Such a situation can easily create in inversion or accumulation layers of the space charge regions of the semiconductor and on the interfaces of heterojunctions.

Electron motion will be determined by periodic potential of the crystal lattice and the additional potential *U(z)*, limiting the electron motion along the *Z* axis [3]:

$$U(z) = \begin{cases} 0, \ldots\ldots\ldots\ldots 0 < z < L_z \\ \infty, \ldots\ldots\ldots\ldots z \leq 0, z \geq L_z \end{cases}. \tag{2.60}$$

Considering the energy states of the conduction band with isotropic dispersion law, you can use the effective mass approximation. In this case the motion of conduction electrons along the *X* and *Y* axis is free, but along the *Z* axis the pulse is limited by potential *U(z)*, that is, we will have two-dimensional electron gas, or *2D* electrons. In such a *2D* system, the normalized electron wave functions can be represented as

$$\Psi(x, y, z) = \sqrt{\frac{8}{L_x L_y L_z}} \sin\left(n\frac{\pi z}{L_z}\right) \exp(ik_x x) \exp(ik_y y), \tag{2.61}$$

and the energy of the electrons will be

$$E(k_x, k_y, n) = \frac{n^2 h^2}{8m_n^* L_z^2} + \frac{\hbar^2}{2m_n^*}(k_x^2 + k_y^2). \tag{2.62}$$

Remember that the L_x, L_y are the size of the crystal along the *X* and *Y* axis and $L_x, L_y \gg L_z$.

As can be seen from Equation (2.62), the energy spectra of electrons in *2D* system is broken in separate overlapping two-dimensional subbands $E_n(k_x, k_y)$, corresponding to specific value of an integer *n*. Constant energy surfaces are transformed into a circle (Figure 2.4). The count of energy begins from the bottom of conduction band of the bulk crystal.

Discrete quantum number *n* suites to the wave vector $\bar{k}$ in the direction of *Z:*

$$k_z = n\frac{\pi}{L_z}. \tag{2.63}$$

In this case the volume of *k*-space bounded by spherical constant energy surface $E = const$, will be divided into several sections of the sphere, corresponding to specific values of $n = 1, 2, 3 \ldots$ (Figure 2.5).

Knowing the distribution of electronic states in *k*-space, we define the density of states on energy dependence in the *2D* system. For this aim, we find the area of the zone *S* between two isoenergetic circles corresponding to the energies *E* and $E + dE$ (Figure 2.6):

$$S = 2\pi k dk, \tag{2.64}$$

where $k = \sqrt{k_x^2 + k_y^2}$ and *dk* is the width of the zone.

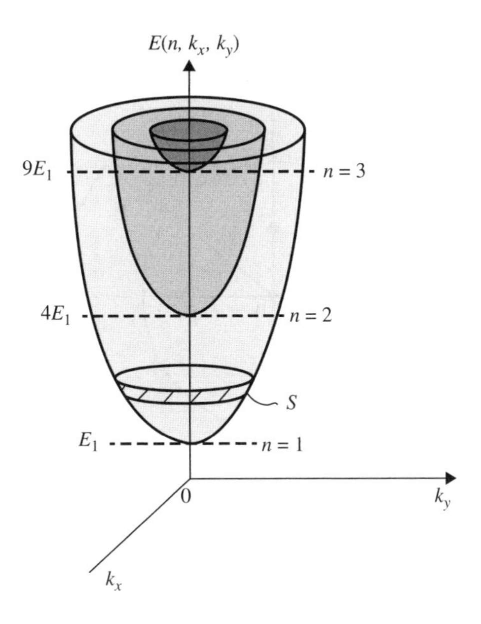

Figure 2.4 Energy spectrum of $2D$ electron gas

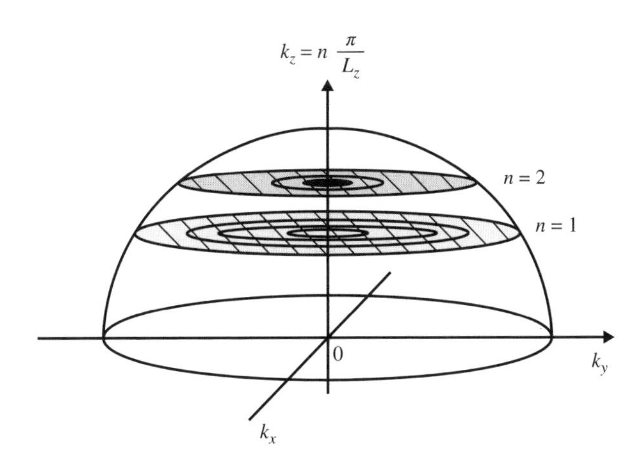

Figure 2.5 Distribution of electron states in k-space of $2D$ system

The area at the zone in k space that corresponds to one state is

$$dS = \frac{(2\pi)^2}{L_x L_y}. \tag{2.65}$$

The number of electronic states in the zone, calculated per unit volume of the crystal film is

$$dN = \frac{S}{VdS} = \frac{kdk}{2\pi L_z}, \tag{2.66}$$

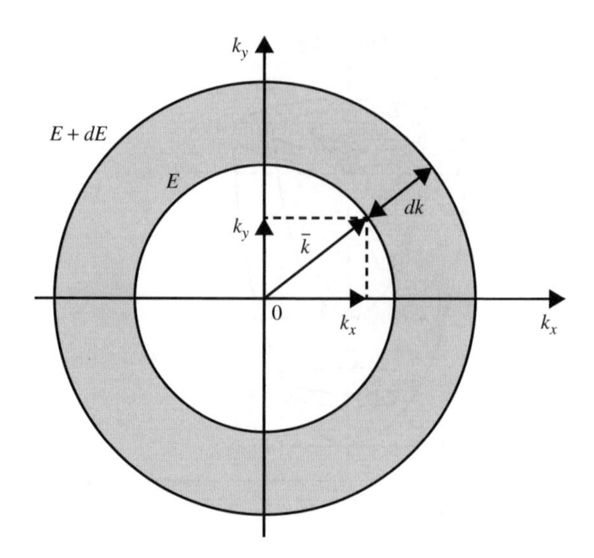

Figure 2.6 Isoenergetic circuits in k-space

or

$$dN = \frac{dk^2}{4\pi L_z}. \tag{2.67}$$

The expression (2.67) is written without the spin. When we are going to determine the concentration of conduction electrons, we introduce a factor of 2, indicating the possibility of finding in one state of two electrons with opposite spins.

The expression for the total energy (Equation (2.62)) can be written as

$$E = E_n + \frac{\hbar^2}{2m_n^*}(k_x^2 + k_y^2), \tag{2.68}$$

or

$$E = E_n + \frac{\hbar^2}{2m_n^*}k^2, \tag{2.69}$$

where the energy corresponding to the bottom of the nth subband is

$$E_n = \frac{h^2}{8m_n^* L_z^2}n^2 = \frac{\hbar^2\pi^2}{2m_n^* L_z^2}n^2, \tag{2.70}$$

then from Equation (2.69)

$$k^2 = \frac{2m_n^*}{\hbar^2}(E - E_n). \tag{2.71}$$

Taking into account Equations (2.67) and (2.71) it is possible to write the formula for the density of states in the 2D case (film)

$$g(E) = \frac{\sum_n dN}{dE} = \frac{m_n^*}{2\pi L_z \hbar^2}\sum_n \Theta(E - E_n), \tag{2.72}$$

where $\Theta(E-E_n)$ is the function of the inclusion, or the Heaviside function, characterized by the following property:

$$\Theta(E-E_n) = 1, ...E \geq E_n \tag{2.73a}$$

$$\Theta(E-E_n) = 0, ...E < E_n. \tag{2.73b}$$

The summation in Equation (2.72) is performed using a number of subbands, the bottom of which lies below the given energy *E*.

Equation (2.72) shows that the density of states in each subband is constant, regardless of energy, and each subband gives the same contribution to the total density of states *g(E)*. Whenever the energy *E* coincides with the bottom of the next subband, that is, when $E = E_n = E_1 n^2$, a jump of the density of states arrives, as depicted in Figure 2.7.

For ease of comparison with three-dimensional electron gas the expression (2.72) can be written as

$$g(E) = \frac{m_n^*}{2\pi L_z \hbar^2}\left[\sqrt{\frac{E}{E_1}}\right], \tag{2.74}$$

where $\left[\sqrt{\frac{E}{E_1}}\right]$ is the integer part of $\sqrt{\frac{E}{E_1}}$, in other words the number of subbands, the bottom of which is below the given energy *E*.

Since, according to Equation (2.70)

$$E_1 = \frac{\pi^2\hbar^2}{2m_n^* L_z^2}, \tag{2.75}$$

then at the given energy *E* with increasing film thickness L_z and, consequently, a decreasing of E_1 the density of states (Eq. (2.74)) will fall as long as the bottom of the next higher subband does not coincide with the given energy *E*. Density of states of the film on its thickness is shown in Figure 2.8. The value L_{z1} represents the film thickness at which the bottom of the lowest

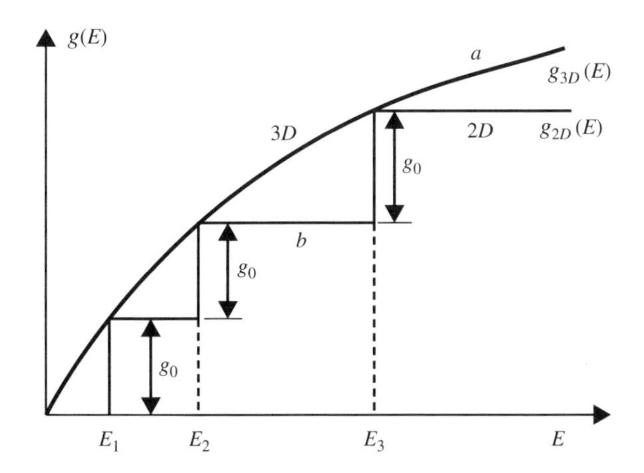

Figure 2.7 Density of states on energy dependences: (a) for 3*D* system; (b) for 2*D* system

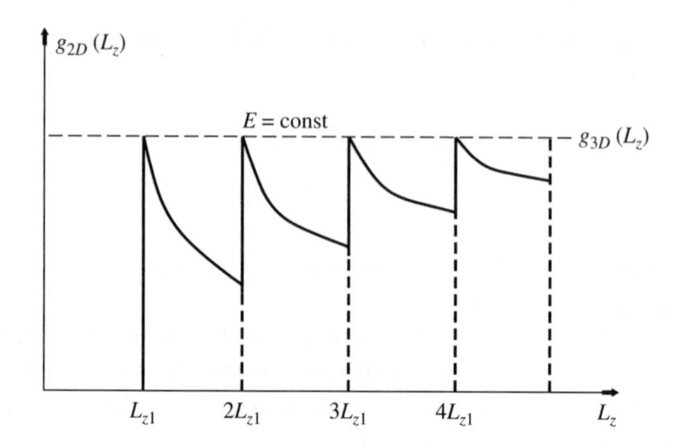

Figure 2.8 Density of electron states in the film on its thickness dependence

subband coincides with the given energy E. Thus, the density of states in two-dimensional electron gas is a periodic function of thickness, while the oscillation period is equal to

$$L_{z1} = \frac{\pi\hbar}{\sqrt{2m_n^* E_1}} = \frac{h}{2m_n^* v_1} = \frac{\lambda}{2}, \tag{2.76}$$

where λ is de Broglie wavelength of an electron with energy E_1.

The density of states at the lower quantum level, where $E = E_1$ according to Equation (2.74) per unit volume (area) is

$$g(E_1) = \frac{m_n^*}{2\pi L_z \hbar^2}. \tag{2.77}$$

This is the value of g_0 step in the density of states in Figure 2.7.

$$g_0 = \frac{m_n^*}{2\pi\hbar^2 L_z}. \tag{2.78}$$

2.3.3 One-Dimension (1D) Case

Let's consider one-dimensional potential box in the form of a long and narrow sample, in which the electron motion is restricted in two directions, namely Y and Z and is free along the X axis. Such structures are called one-dimensional systems ($1D$ system) or quantum wires.

The motion of electrons in one-dimensional model is determined by the periodic potential of crystal lattice and the additional potential of the form

$$V(z) = \begin{cases} 0, \ldots\ldots\ldots\ldots 0 < z < L_z \\ \infty, \ldots\ldots\ldots\ldots z \leq 0, z \geq L_z \end{cases}. \tag{2.79a}$$

$$V(y) = \begin{cases} 0, \ldots\ldots\ldots\ldots 0 < y < L_y \\ \infty, \ldots\ldots\ldots\ldots y \leq 0, y \geq L_y \end{cases}. \tag{2.79b}$$

that restricts the movement of an electron along the Y and Z axes.

Near the edge of a nondegenerated band with isotropic dispersion the normalized wave functions and energy spectrum of electrons can be written as

$$\Psi_{n,m}(x, y, z) = \sqrt{\frac{8}{L_x L_y L_z}} \sin\left(n\frac{\pi z}{L_z}\right) \sin\left(m\frac{\pi y}{L_y}\right) \exp(ik_x x), \tag{2.80}$$

$$E(k_x, m, n) = \frac{\hbar^2}{2m_n^*}(k_x^2 + k_y^2 + k_z^2). \tag{2.81}$$

The wave vectors k_y and k_z are quantized:

$$k_y = m\frac{\pi}{L_y}, k_z = n\frac{\pi}{L_z}, \tag{2.82}$$

where n, $m = 1, 2, 3 \ldots$ are the positive number characterizing the quantum subbands; wave vector k_x is the quasi-continuous quantity, then Equation (2.81) will be

$$E(k_x, m, n) = \frac{\hbar^2}{2m_n^*}k_x^2 + \frac{\hbar^2}{2m_n^*}\left[m^2\frac{\pi^2}{L_y^2} + n^2\frac{\pi^2}{L_z^2}\right]. \tag{2.83}$$

In accordance with Equation (2.83) the energy spectrum of quantum wire ($1D$ system) is divided into separate overlapping subbands $E(k_x, n, m)$, each of which is determined by the specific values of n and m (Figure 2.9).

The distance between adjacent quantum subbands are

$$E_{m+1} - E_m = \frac{h^2}{8m_n^* L_y^2}(2m + 1), \tag{2.84a}$$

$$E_{n+1} - E_n = \frac{h^2}{8m_n^* L_z^2}(2n + 1). \tag{2.84b}$$

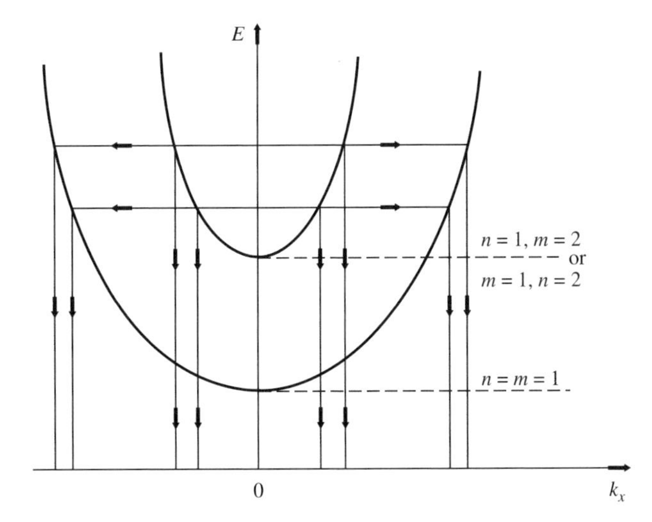

Figure 2.9 Energy spectrum of electrons in one-dimensional system

Let's find the density of states on energy dependence.

At the interval dk_x, calculated per unit volume, we have dN quantum states

$$dN = \frac{dk_x}{V\left(\frac{2\pi}{L_x}\right)} = \frac{dk_x}{2\pi L_y L_z}, \tag{2.85}$$

but from Equation (2.83)

$$k_x = \sqrt{\frac{2m_n^*}{\hbar^2}(E - E_{n,m})}, \tag{2.86}$$

where

$$E_{n,m} = \frac{\hbar^2}{2m_n^*}\left[m^2\frac{\pi^2}{L_y^2} + n^2\frac{\pi^2}{L_z^2}\right] \tag{2.87}$$

is the energy corresponding to the bottom of the subband with quantum numbers n and m.

After differentiating Eq. (2.86) we obtain

$$dk_x = \frac{\sqrt{2m_n^*}dE}{2\hbar\sqrt{E - E_{n,m}}}. \tag{2.88}$$

In the one-dimensional system the density of states on the energy dependence calculated per unit volume, with taking into account Equations (2.85) and (2.86) can be written as

$$g(E) = \frac{\sqrt{2m_n^*}}{4\pi\hbar L_y L_z}\sum_n\sum_m \frac{\Theta(E - E_{n,m})}{\sqrt{E - E_{n,m}}}. \tag{2.89}$$

The expression takes into account the spin degeneracy of the state and the fact that two intervals of wave vector $\pm dk_x$ for each subband correspond to the same interval of energy, for which $E \gg E_{n,m}$ (Figure 2.9).

It is interesting to note that although the density of states $g(E) \sim \frac{1}{\sqrt{E}}$, the total number of states per unit length with energy less than E is proportional to $\sqrt{E}$.

The energy in Equation (2.89), as usual, is measured from the bottom of the conduction band of three-dimensional model. The dependence of $g(E)$ is shown in Figure 2.10, where the numbers on the curves are the quantum numbers n and m, respectively.

Thus, the density of states within a subband decreases with increasing energy as $\frac{1}{\sqrt{E-E_{n,m}}}$. When $E = E_{n,m}$ the density $g(E) \to \infty$. If $L_z = L_y$ the subbands with quantum numbers $n \neq m$ are doubly degenerate. The total density of states is a sum of decreasing functions of individual subbands, displaced in energy.

2.3.4 *Zero Dimension (0D) Case*

Let's consider the situation where the electron motion is restricted in all three directions *X, Y, Z*. This model is called a quantum dot, quantum island, or 0*D* system.

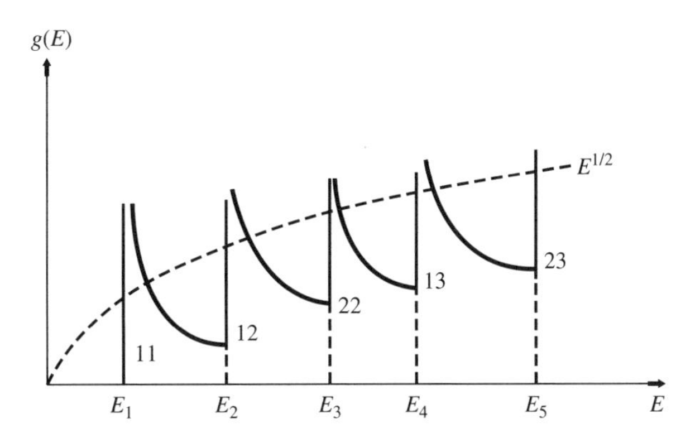

Figure 2.10 Density of electron states on energy dependence at $L_z = L_y$ for quantum wire (1*D* system)

In this case, along with the periodic lattice potential the additional potential *V (x, y, z)* acts on the electron. Its value is determined by the following conditions:

$$V(z) = \begin{cases} 0, \ldots\ldots\ldots\ldots 0 < z < L_z \\ \infty, \ldots\ldots\ldots\ldots z \leq 0, z \geq L_z \end{cases}. \tag{2.90a}$$

$$V(y) = \begin{cases} 0, \ldots\ldots\ldots\ldots 0 < y < L_y \\ \infty, \ldots\ldots\ldots\ldots y \leq 0, y \geq L_y \end{cases}. \tag{2.90b}$$

$$V(x) = \begin{cases} 0, \ldots\ldots\ldots\ldots 0 < x < L_x \\ \infty, \ldots\ldots\ldots\ldots x \leq 0, x \geq L_x \end{cases}. \tag{2.90c}$$

In the bottom of the isotropic nondegenerate conduction band the normalized wave functions and energy spectrum of electrons can be represented as

$$\Psi_{l,m,n}(x, y, z) = \sqrt{\frac{8}{L_x L_y L_z}} \sin\left(n\frac{\pi z}{L_z}\right) \sin\left(m\frac{\pi y}{L_y}\right) \sin\left(l\frac{\pi x}{L_x}\right), \tag{2.91}$$

$$E(l, m, n) = \frac{\pi^2 \hbar^2}{2m_n^*}\left(\frac{l^2}{L_x^2} + \frac{m^2}{L_y^2} + \frac{n^2}{L_z^2}\right), \tag{2.92}$$

where L_x, L_y, L_z, as before, are the size of the quantum box, cell, point along the respective axes *X, Y, Z; l, m, n* = 1, 2 … are the number of subbands.

Equation (2.92) shows that the energy spectrum of electrons in a quantum dot is a set of discrete values of energy determined by the *l, m, n* integers. The number of electronic states corresponding to one set of *l, m, n* numbers (including spin degeneracy), calculated per unit volume is $\frac{2}{L_x L_y L_z}$, and the total number of states *N* per unit volume with the same energy, is dependent on crystal symmetry and in general can be represented by the expression

$$N = \frac{2}{L_x L_y L_z} G, \tag{2.93}$$

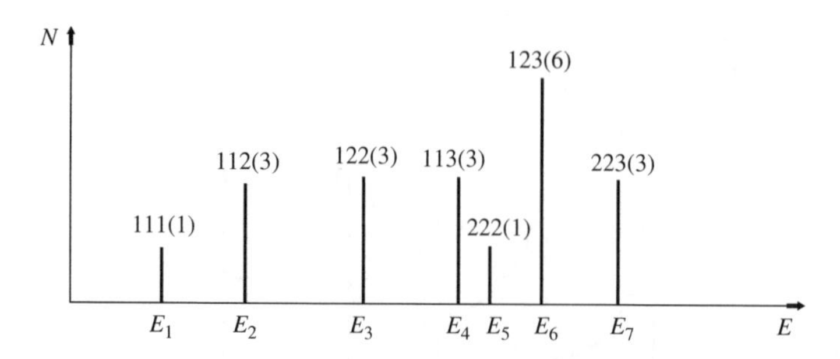

Figure 2.11 Distribution of number of allowable states N in conduction band for quantum dot at $L_x = L_y = L_z$

where G is the degeneracy of the energy level. For example, if $n = m = l$, then the degeneracy factor due to symmetry, $G = 1$, if $l = m \neq n$, then $G = 3$, and finally, if all the numbers l, m, n are distinct, then the degeneracy factor $G = 6$. Distribution of the number of allowed states of N in the conduction band in a quantum dot are presented schematically in Figure 2.11. The numbers above the straight lines indicate the quantum numbers l, n, m, and the figure in parentheses is the degeneracy factor of G.

2.4 Fermi Distribution Function and Electron Concentration

The Fermi-Dirac function, $f_{FD}(E)$, is a probability distribution function that tells one the ratio of filled to total allowed states at the given energy E. As we will see, statistical arguments are employed to establish the general form of the function [1]. Basically, the electrons are viewed as indistinguishable "balls" that are being placed in allowed-state "boxes." Each box is assumed to accommodate a single ball. The boxes themselves are grouped into rows, the number of boxes per row corresponding to the allowed electronic states at the given energy. The numerical occurrence of all possible arrangements of balls per row yielding the same overall system energy is determined statistically, and the most likely arrangement is identified. Finally, the Fermi function is equated to the most likely arrangement of balls (electrons) per row (energy). This arrangement, it turns out, occurs more often than all other arrangements combined. Moreover, the distribution of arrangements is highly peaked about the most probable arrangement. Thus, it is reasonable to use the Fermi function – the most probable arrangement – to describe the filling of allowed states in actual electronic systems.

The final form of the Fermi-Dirac function has the view

$$f_{FD}(E, T) = \frac{1}{1 + \exp\left(\dfrac{E - E_F}{k_B T}\right)}, \tag{2.94}$$

where E_F is the electrochemical potential or Fermi energy of the electrons in the solid, $k_B = 8.617 \times 10^{-5}$ eV/K is Boltzmann's constant and T is the system temperature.

A sample plot of the Fermi function versus $E - E_F$ for a selected number of temperatures is displayed in Figure 2.12.

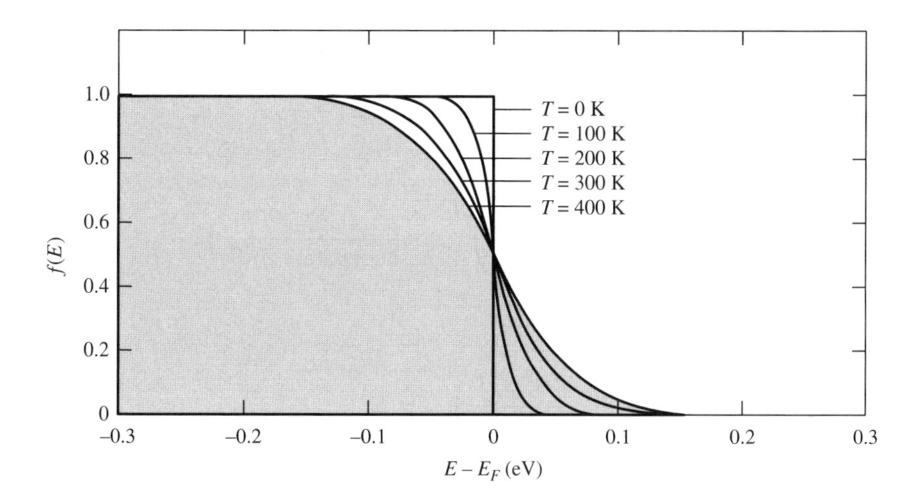

Figure 2.12 Value of Fermi function versus energy with the system temperature as a parameter

Having established the energy distribution of available band states $g(E)$ and the ratio of filled to total states under equilibrium conditions $f_{FD}(E)$, one can now easily deduce the distribution of carriers in the conduction and valence bands. The desired distribution is obtained by simply multiplying the appropriate density of states by the appropriate occupancy factor: $g_c(E)f_{FD}(E)$ yields the distribution of electrons in the conduction band and $g_v(E)[1-f_{FD}(E)]$ yields the distribution of holes (unfilled states) in the valence band.

2.4.1 *Electron Concentration for 3D Structures*

Let's determine the electron concentration in conduction band for 3D structure [4–7]. In the general case the equation for free electron concentration is given as:

$$n = 2\int_{E_C}^{E_{\max}} g(E)f_{FD}(E,T)dE, \tag{2.95}$$

where $f_{FD}(E,T)$ is distribution function of Fermi–Dirac for the equilibrium states, E_F is a Fermi level.

Since the function of the Fermi–Dirac decreases very rapidly with increasing energy, the contribution to the total electron density of the upper energy states will be small and the expression (2.44) can be used for all energies.

Taking into account this circumstance, the expression (2.95) can be written as

$$n = 2\int_{E_C}^{\infty} \frac{2\pi\left(\dfrac{2m_{ng}^*}{h^2}\right)^{3/2}\sqrt{E-E_C}}{1+\exp\left(\dfrac{E-E_F}{k_BT}\right)}dE. \tag{2.96}$$

Let's introduce for simplicity the dimensionless coefficients

$$\frac{E - E_C}{k_B T} = \varepsilon, \ldots\ldots dE = \frac{1}{k_B T} d\varepsilon, \ldots\ldots \frac{E_F - E_C}{k_B T} = \eta, \tag{2.97}$$

in this case,

$$n = 4\pi \left(\frac{2m_{ng}^* k_B T}{h^2} \right)^{3/2} \int_0^\infty \frac{\varepsilon^{1/2} d\varepsilon}{1 + \exp(\varepsilon - \eta)}. \tag{2.98}$$

Now we introduce the following notation:

$$2\left(\frac{2\pi m_{ng}^* k_B T}{h^2} \right)^{3/2} = N_C, \tag{2.99}$$

$$\int_0^\infty \frac{\varepsilon^{1/2} d\varepsilon}{1 + \exp(\varepsilon - \eta)} = F_{1/2}(\eta). \tag{2.100}$$

The value of N_C is called the effective density of states in the conduction band, $F_{1/2}(\eta)$ is Fermi integral of order ½ [4].

With these notations

$$n = \frac{2}{\sqrt{\pi}} N_C F_{1/2}(\eta). \tag{2.101}$$

The concentration of free electrons is a function of temperature and Fermi level. Fermi integral in the general form cannot be presented in elementary functions; however, there are various approximate expressions that are valid for different values of the argument.

2.4.1.1 Case 1: $-\infty < \eta < -1$

For this case we have

$$F_{1/2}(\eta) = \frac{\sqrt{\pi}}{2} \exp(\eta). \tag{2.102}$$

This approximation corresponds to the classical Boltzmann statistics (Figure 2.13).

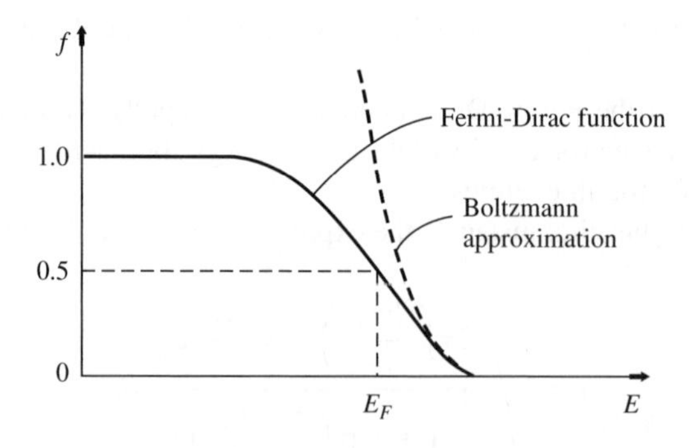

Figure 2.13 Fermi-Dirac function and Boltzmann approximation

From the condition (2.102) it implies

$$E_F < E_C - k_B T, \tag{2.103}$$

that is, the semiconductor is not degenerate and (Figure 2.14)

$$n = \frac{2N_C}{\sqrt{\pi}} F_{1/2}(\eta) \approx N_C \exp(\eta), \tag{2.104}$$

or

$$n = 2\left(\frac{2\pi m_{ng}^* k_B T}{h^2}\right)^{3/2} \exp\left(-\frac{E_C - E_F}{k_B T}\right). \tag{2.105}$$

2.4.1.2 Case 2: $5 < \eta < \infty$

We can represented $F_{1/2}(\eta)$ as

$$F_{1/2}(\eta) = \frac{2}{3}\eta^{3/2}, \tag{2.106}$$

or

$$F_{1/2}(\eta) = \frac{2}{3}\left(\frac{E_F - E_C}{k_B T}\right)^{3/2}, \tag{2.107}$$

$$E_F > E_C + 5k_B T. \tag{2.108}$$

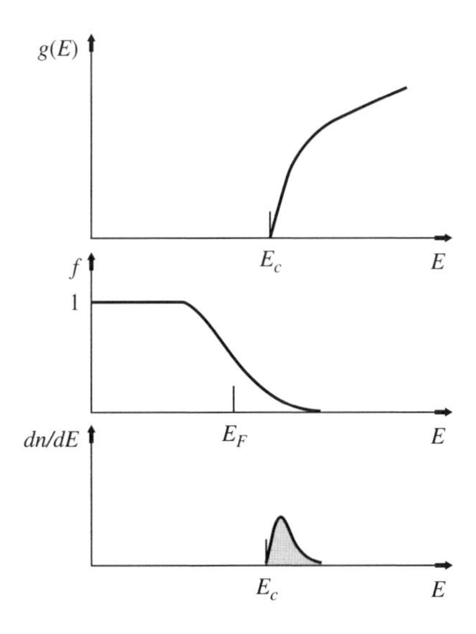

Figure 2.14 Schematic images of $g(E)$, $f_{FD}(E,T)$, and dn/dE functions in n-type nondegenerate semiconductor

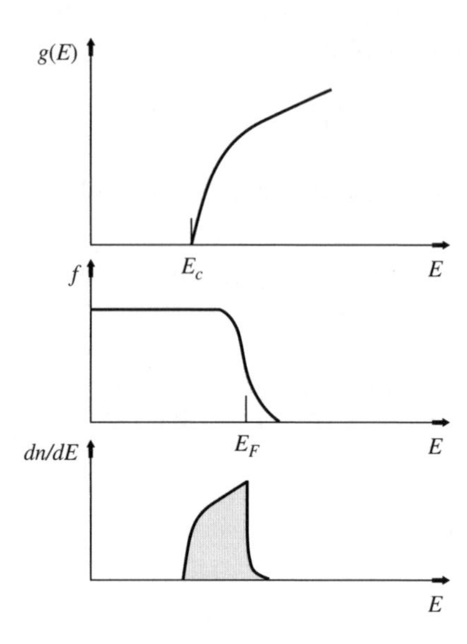

Figure 2.15 Schematic images of *g(E)*, $f_{FD}(E,T)$, and *dn/dE* functions in *n*-type degenerate semiconductor

We are dealing with completely degenerate semiconductors. The Fermi level lies above the conduction band at least $5k_BT$. The concentration of electrons in a semiconductor will be

$$n = \frac{2N_C}{\sqrt{\pi}} F_{1/2}(\eta) = \frac{8\pi}{3}\left(\frac{2m^*_{ng}}{h^2}\right)^{3/2}(E_F - E_C)^{3/2}, \tag{2.109}$$

that is, the concentration of free electrons does not depend on temperature (Figure 2.15).

2.4.1.3 Case 3: $-1 < \eta < 5$

In the transition region from nondegenerate to completely degenerate semiconductor we have

$$F_{1/2}(\eta) = \frac{2\sqrt{\pi}}{1 + 4\exp(-\eta)}, \tag{2.110}$$

and

$$n = 4N_C \frac{1}{1 + 4\exp\left(\dfrac{E_C - E_F}{k_BT}\right)}. \tag{2.111}$$

Expressions obtained for the electrons can be easily converted to holes.

The important observations should be made with reference to Figures 2.14 and 2.15. The first relates to the general form of the carrier distributions. All of the carrier distributions are zero at the band edges, reach a peak close to E_c (or E_v) and then decay very rapidly toward zero as one moves upward into the conduction band or downward into the valence band. In other words, most of the carriers are grouped energetically in the near vicinity (within a few k_BT) of the band edges.

2.4.2 *Electron Concentration for 2D Structures*

Using the value of $g(E)$ for $2D$ structures (Equation (2.77)), it is easy to determine the concentration of free $2D$ electrons. In the general case

$$n = 2\int_{E_j}^{E_m} g(E) f_{FD}(E,T) dE, \tag{2.112}$$

where the factor 2 is introduced to account the location of two electrons with opposite spins in one state,

$$f_{FD}(E,T) = \frac{1}{1+\exp\left(\dfrac{E-E_F}{k_BT}\right)}. \tag{2.113}$$

For the first quantum subband and near the center of the Brillouin band

$$n = 2\int_{E_1}^{E_m} g_0 \frac{1}{1+\exp\left(\dfrac{E-E_F}{k_BT}\right)} dE. \tag{2.114}$$

If the Fermi level E_F is located below the bottom of the lowest quantum subband E_1 then

$$n \cong g_0 k_B T \exp\left(\frac{E_F - E_1}{k_BT}\right), \tag{2.115}$$

we have nondegenerated $2D$ electron gas.

If the level E_F is above E_1, then

$$n \cong g_0(E_F - E_1), \tag{2.116}$$

the $2D$ electron gas is degenerated.

2.5 Supply Function at Electron Field Emission

During the calculation of emission current it is necessary to know the distribution function of electron in the emitter $f(k)$. Integrand for current calculation is proportional to the energy distribution in E_x of the transmitted electrons. The electrons' energy is characterized by a Fermi distribution function $f_{FD}(E): f_{FD}(E<\mu)=1$ *at* $T=0$, and is zero otherwise. Here μ is the chemical potential. At finite temperatures, μ is no longer the energy of the most energetic electron, but rather the chemical potential defined such that integral over the Fermi distribution function

$f_{FD}(E)$ over all momentum space at any temperature reproduces the electron density [8]. μ is obtained from the electron (number) density n through Fermi-Dirac distribution according to [9] (see Section 2.4)

$$n = 4\frac{M_c}{\sqrt{\pi}}\left(\frac{mk_BT}{2\pi\hbar^2}\right)^{3/2}\int_0^\infty \frac{\sqrt{\varepsilon}d\varepsilon}{1+\exp\left(\varepsilon - \frac{\mu}{k_BT}\right)} = N_c\frac{2}{\sqrt{\pi}}F_{1/2}\left(\frac{\mu}{k_BT}\right), \tag{2.117}$$

where M_c is a number of equivalent minima in the conduction band (e.g., one for metals, six for silicon, etc.), $F_{1/2}$ is Fermi-Dirac integral, and m is the effective mass of the electron (equal to the rest mass m_0 for metals). Given the high electron density for metals, band bending under the influence of high applied electric fields is negligible (but not for semiconductor), in which case μ is, to a good approximation, equal to its bulk value. Equation (2.117) holds for any temperature: at 0 K, $\mu = \mu_0$ is the Fermi energy E_F [10,11]. If $M_c = 1$ and $T = 0$ K, then $n = k_F{}^3/(3\pi)^2$ [9]. For large value of μ_0/k_BT, asymptotic expansions which allow iterative solution may be employed [9, 12, 13]

$$\frac{2}{3}\left(\frac{\mu_0}{k_BT}\right)^{3/2} = \int_0^\infty dx\frac{\sqrt{x}}{1+\exp\left(y - \frac{\mu}{k_BT}\right)} \approx \frac{2}{3}\left(\frac{\mu_0}{k_BT}\right)^{3/2} \times\left(1+\frac{1}{8}\left(\frac{\pi k_BT}{\mu}\right)^2 + \frac{7}{640}\left(\frac{\pi k_BT}{\mu}\right)^4\right), \tag{2.118}$$

which, upon inversion, yields

$$\mu(T) = \mu_0\left[1 - \frac{1}{12}\left(\frac{\pi k_BT}{\mu_0}\right)^2 - \frac{1}{80}\left(\frac{\pi k_BT}{\mu_0}\right)^4 + \ldots\right] \tag{2.119}$$

The supply function $f(k_x) \equiv f(k)$ is obtained from the Fermi distribution by (in cylindrical coordinates $dk_ydk_z = 2\pi k' dk'$) [14]

$$f(k) = \frac{s}{(2\pi)^2}\left(\int_0^\infty \frac{2\pi k' dk'}{\exp\left(\frac{q(E-\mu)}{k_BT}\right)+1}\right) = \frac{mk_BT}{\pi\hbar^2}\ln(1+\exp\left(\frac{(\mu - E(k))}{k_BT}\right), \tag{2.120}$$

k' is identified with the parallel momentum to the surface (in cylindrical coordinates) and $s = 2$ account for electron spin. At $T = 0$ K, Equation (2.120) reduces to $f(k) = (k^2{}_F - k^2)/(2\pi)$ for $k \lesssim k_F$. The current integrand is sharply peaked at $k \approx k_F$.

The dependences of supply function on energy for T = 500, 1000, and 2000 K at $\mu = \Phi = 2$ eV are shown in Figure 2.16. The "knee" of the supply function marks the approximate location of the chemical potential.

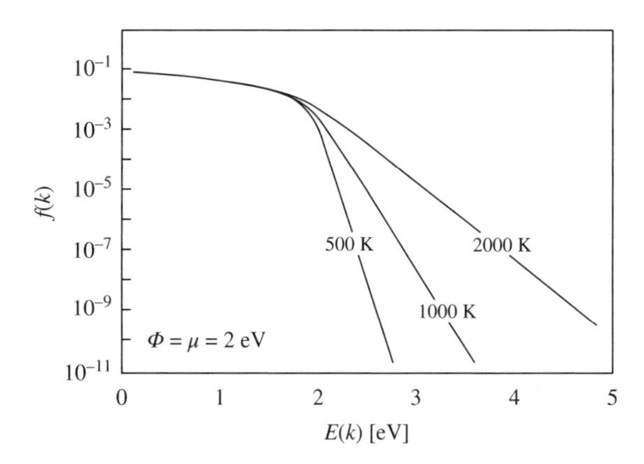

Figure 2.16 Supply function for temperatures (500, 1000, and 2000 K) [14]. Copyright (2001) by John Wiley & Sons, Inc. This material is reproduced with permission of John Wiley & Sons, Inc

2.6 Electron in Potential Well

We have examined the behavior of electrons in a rectangular potential box with infinitely high walls (see Section 2.2). Using the boundary conditions, we immediately obtain a system of discrete energy levels

$$E = \frac{n^2 h^2}{8mL^2}. \tag{2.121}$$

Now suppose that the electrons are in the potential field shown in Figure 2.17. This field can also be called a potential well or a potential box, but with walls of finite height E_0. Consider the case where the total electron energy E is lower than the E_0 [3].

We divide the whole field into three areas:

$$\begin{aligned} &\text{in region 1} \quad x < 0, E = E_0 = \text{ const}, \\ &\text{in region 2} \quad 0 < x < L, E_0 = 0, \\ &\text{in region 3} \quad x > L, E = E_0 = \text{ const}. \end{aligned}$$

We write the Schrodinger equation for these areas

$$\frac{d^2\Psi_{1,3}(x)}{dx^2} - k^2\Psi_{1,3}(x) = 0, \tag{2.122}$$

$$\frac{d^2\Psi_2(x)}{dx^2} - k_2^2\Psi_2(x) = 0, \tag{2.123}$$

where are used the following notations

$$k = \frac{1}{\hbar}\sqrt{2m(E_0 - E)}, k_2 = \frac{1}{\hbar}\sqrt{2mE}. \tag{2.124}$$

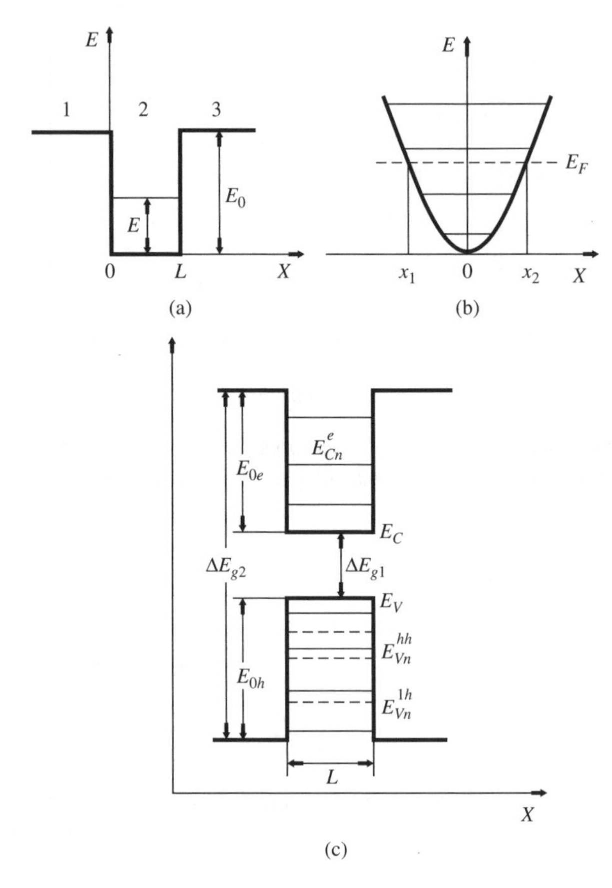

Figure 2.17 (a) One dimensional potential well; (b) potential energy of linear harmonic oscillator; (c) energy model of structure with quantum well

The solution of the Schrodinger equation, respectively, for these three areas can be written as

$$\Psi_1 = A_1 \exp(kx) \text{ in region 1, } x < 0, \tag{2.125a}$$

$$\Psi_2 = A_2 \exp(ik_2x) + B_2 \exp(-ik_2x) \text{ in region 2, } 0 < x < L, \tag{2.125b}$$

$$\Psi_3 = B_3 \exp(-kx) \text{ in region } 3, x > L. \tag{2.125c}$$

Taking into account the continuity of the wave functions and their derivatives at the boundaries of the quantum well (QW):

$$(\Psi_1)_{x=0} = (\Psi_2)_{x=0}, \left(\frac{d\Psi_1}{dx}\right)_{x=0} = \left(\frac{d\Psi_2}{dx}\right)_{x=0} \tag{2.126a}$$

$$(\Psi_2)_{x=L} = (\Psi_3)_{x=L}, \left(\frac{d\Psi_2}{dx}\right)_{x=L} = \left(\frac{d\Psi_3}{dx}\right)_{x=L} \tag{2.126b}$$

you can define the relations between the coefficients A_2 and B_2.

From the boundary conditions at $x = 0$

$$B_2 = -\frac{k - ik_2}{k + ik_2} A_2, \tag{2.127}$$

and from the boundary conditions at $x = L$

$$B_2 = -A_2 \frac{k + ik_2}{k - ik_2} \exp(i2k_2 L). \tag{2.128}$$

As the last two expressions are compatible, it is obviously necessary that

$$\frac{k - ik_2}{k + ik_2} = \frac{k + ik_2}{k - ik_2} \exp(i2k_2 L). \tag{2.129}$$

From this

$$\frac{k - ik_2}{k + ik_2} \exp(-ik_2 L) = \frac{k + ik_2}{k - ik_2} \exp(ik_2 L). \tag{2.130}$$

After the transformation we have

$$(k^2 - k_2^2) \sin(k_2 L) = -2kk_2 \cos(k_2 L), \tag{2.131}$$

or

$$tg(k_2 L) = -\frac{2kk_2}{k^2 - k_2^2}. \tag{2.132}$$

Returning to the energy parameters E and E_0, we obtain the transcendental equation:

$$tg\left(\frac{2\pi}{h}\sqrt{2mEL}\right) = -2\frac{\sqrt{E(E_0 - E)}}{E_0 - 2E}. \tag{2.133}$$

We can determine from Equation (2.133) the values of energy E, in which the wave function $\Psi(x)$ satisfies the boundary conditions of continuity.

This equation can be solved graphically by constructing curves of left and right sides of the equation on energy. Points of intersections of the curves give a discrete set of values of the electron energy in a quantum well.

For comparison with the quantization of the energy spectrum of electrons in a box with infinitely high walls, we consider the distribution of the lowest energy levels in the well at sufficiently high walls, that is, let $E_0 \gg E$, then

$$tg\left(\frac{2\pi}{h}\sqrt{2mEL}\right) = -2\sqrt{\frac{E}{E_0}} \approx 0. \tag{2.134}$$

Consequently

$$\frac{2\pi}{h}\sqrt{2mEL} \approx n\pi, \tag{2.135}$$

from this

$$E = \frac{n^2h^2}{8mL^2}, \tag{2.136}$$

that is, the distribution of the lowest energy levels of electrons in a potential well practically coincides with the distribution in a box with infinitely high walls.

2.6.1 *Quantum Well with Parabolic Shape of the Potential*

Modern technology of molecular beam epitaxy (MBE) allows realizing different potential profile of quantum wells. In particular, the parabolic shape of the quantum well potential can implement a system of equidistant energy levels. Let's consider the one-dimensional case of this potential (Figure 2.17b) [3]. The problem of finding the stationary electron energy states is reduced almost to the problem of linear quantum oscillator. The potential energy of a harmonic oscillator, as it is known, is given by

$$E_0 = \frac{1}{2}m\omega^2x^2, \tag{2.137}$$

where x is the distance measured from the center of the well; ω is angular frequency of particle oscillation.

In contrast to the potential well with vertical walls here it is a feature due to which the mathematical problem becomes much more complicated. Within the well potential energy has no constant value throughout, but varies according to the parabolic law, so the de Broglie wavelength

$$\lambda = \frac{h}{\sqrt{2m(E - E_0)}} \tag{2.138}$$

is not constant in the well, and increases toward the edges and decreases toward the center.

Stationary Schrodinger equation in the case of linear quantum oscillator has the form

$$\frac{d^2\Psi(x)}{dx^2} + \frac{8\pi^2m}{h^2}(E - E_0)\Psi(x) = 0, \tag{2.139}$$

where the wave function $\Psi(x)$ must satisfy the requirement: $x \to \pm\infty$ $\Psi(x) = 0$. Oscillators have no boundary conditions analogous to the infinite rectangular well, and the electron motion is not limited to any impermeable wall.

The solution of the Schrodinger equation for a linear oscillator, taking into account the mentioned conditions, leads to the energy eigenvalues and electron wave functions:

$$E = \hbar\omega\left(n + \frac{1}{2}\right), \quad \text{n} = 0,\ 1,\ 2,\ 3\ \ldots \tag{2.140}$$

$$\Psi_n(x) = \frac{H_n\left(\frac{x}{a}\right)}{\sqrt{2^n n!\sqrt[4]{\pi a^2}}}\exp\left(-\frac{x^2}{2a^2}\right), \tag{2.141}$$

where $a = \sqrt{\frac{\hbar}{m\omega}}$, $H_n\left(\frac{x}{a}\right)$ are the Hermit polynomials.

In particular, $H_0\left(\frac{x}{a}\right) = 1$, $H_1\left(\frac{x}{a}\right) = 2\frac{x}{a}$, $H_2\left(\frac{x}{a}\right) = 4\left(\frac{x}{a}\right)^2 - 2$, and so on.

Thus, the only functions $\Psi(x)$ that correspond to the number of discrete energy values simultaneously satisfy both the Schrodinger equation for the oscillator and the boundary conditions.

In contrast to the potential box (well), where the quantization of electron energy was the result of boundary conditions, in the case of the linear oscillator the quantization was the result of natural conditions for the finiteness of the wave function throughout the space.

An important feature of the spectrum of the oscillator is that the energy levels are located at equal distances

$$\Delta E = E_{n+1} - E_n = \hbar\omega. \tag{2.142}$$

The selection rules allow, in this case, only transitions between neighboring levels, and the radiation of the oscillator is only on one frequency, which coincides with the frequency of a classical oscillator.

Let's pay attention to the fact that in the lowest quantum state (for $n = 0$) the energy of the quantum oscillator is not zero, but equal to $(1/2)\hbar\omega$. This energy value is called zero-point energy, since it does not disappear at absolute zero temperature. Zero-point energy is of the minimum energy, which, at least, should have a quantum oscillator in the ground state to be satisfied with the uncertainty relation. The presence of the smallest nonzero value of the energy of the quantum oscillator is a significant difference from the classical oscillator.

Energy diagram schematically representing a potential well for electrons (Figure 2.17a), can display a certain approximation of three-layer thin-film heterostructures, the middle layer of which is a narrow-gap semiconductor, while the outer layers are the wide-gap (Figure 2.17). Sharpness of the heterojunction at the boundaries of the potential well (the band gap changes abruptly within a single monolayer) allows us to assume a rectangular well. If the thickness of the narrow-gap semiconductor is commensurate with de Broglie wavelength ($L \approx h/mv$), the conditions of the quantum size effect are realized in the potential well. The motion of electrons in it is restricted in one direction (axis X) by potential barriers arising at the boundaries of the layers because of different band gaps. In the potential quantum well the motion of both electrons and holes is quantized.

Since the value of longitudinal pulse $p_y = \hbar k_y$, $p_z = \hbar k_z$ and electron energy E_y, E_z in this well are arbitrary, then the electron gas is two-dimensional. In the effective mass approximation the problem of quantization of the carrier motion in the QW essentially boils down to the quantum-mechanical problem on the behavior of an electron in a potential box (see Section 2.2). If the motion of electrons in a quantum well is restricted in one direction, for example, along the X axis, the electron energy spectrum consists of several subbands:

$$E = E_C + E_{Cn} + \frac{\hbar^2(k_y^2 + k_z^2)}{2m_n^*}, \tag{2.143}$$

where E_C is the energy of the conduction band; k_y, k_z are the projection of the electron wave vector on the Y and Z axis, respectively; E_{Cn} are the initial levels of the subbands of electrons with quantum number n ($n = 1, 2, 3 \ldots$), measured from the bottom of conduction band.

The value $\frac{\hbar^2(k_y^2+k_z^2)}{2m_n^*}$ represents the kinetic energy of an electron moving along the potential barriers in the quantum well. Quantization levels E_{Cn} determined from the Schrodinger equation for a quantum well with barrier height E_{0e} and width L, whose solution, taking into account the continuity of the electron wave function and its derivative at the boundaries of the quantum well, is reduced to a transcendental equation

$$E_{Cn} = \frac{h^2}{8m_n^* L^2}\left(n - arctg\sqrt{\frac{E_{0e}}{E_{Cn}} - 1}\right)^2. \tag{2.144}$$

In general, the allowed values of the electron energy in a quantum well can be found by solving Equation (2.144) numerically or graphically.

In the limiting case of infinitely high potential barriers ($E_{0e} \to \infty$) the latter expression reduces to the value (Eq. (2.10)) for the potential box. The normalized wave functions of electrons in a quantum well have the form

$$\Psi_{Cn} = \sqrt{\frac{2}{L}} \sin\left(n\frac{\pi x}{L}\right) \Psi_0 \exp\{i(k_y y + k_z z)\}, \tag{2.145}$$

where Ψ_0 is periodic in the crystal lattice part of the Bloch function. Electron wave vector k has components only in the YZ plane.

Similarly, the behavior of holes in the quantum well is considered. But it is necessary to take into account the degeneracy of the valence band: the presence of heavy and light holes.

Now let's shortly summarize the peculiarities of electron behavior in quantum box and quantum well.

In case of quantum well with significantly thick barriers, namely

$$d_1 \gg \frac{2\pi}{k_1} \text{ and } d_3 \gg \frac{2\pi}{k_3}, \tag{2.146}$$

where the wave vectors

$$|k_1| = \frac{1}{\hbar}|2m_n^*(U_1 - E)|^{1/2} \text{ and } |k_3| = \frac{1}{\hbar}|2m_n^*(U_3 - E)|^{1/2}, \tag{2.147}$$

and the electron with energy $E < U_1, U_3$ is localized in a quantum well. The energy of its pulse along the X axis

$$E = E_x = \frac{\hbar^2 k_x^2}{2m_n^*} \tag{2.148}$$

is quantized, taking discrete values of E_n. In the simplest case, the potential box (infinity high barriers) (Figure 2.18a), where $U_1, U_2 \to \infty$,

$$E_n = \frac{n^2 h^2}{8m_n^* L^2}, \tag{2.149}$$

where $n = 1, 2, 3 \ldots$

These values correspond to the resonant energy levels

$$\frac{2L}{\lambda_n} = \frac{k_n L}{\pi} = n. \tag{2.150}$$

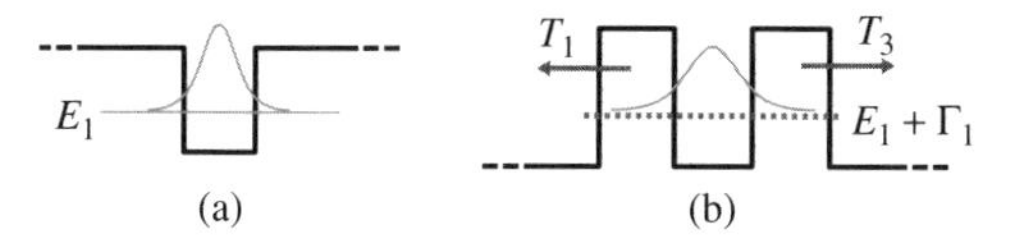

Figure 2.18 (a) Single quantum well with infinite barriers; (b) spreading of quantum energy level into subband in double barrier structure with finite barrier heights and widths

In these energy levels the integer numbers of half wavelength $\lambda_n/2$ of de Broglie waves are inside the box.

State of the electron in the nth energy level is described by the wave function

$$\Psi = A\sin(k_n x), \tag{2.151}$$

which can be considered as a result of interference of propagating traveling electron waves with wave vectors $k_n = \pm n\pi/L$ and reflected from the walls of the potential box.

The quantum well of finite depth $U_1 = U_3 = U$ hosts inside a limited number N of resonant energy levels E_n, that are shifted relatively to Equation (2.149) to lower energies.

Values of E_n are solutions to the equation [15]

$$k_n L = n\pi - 2\arcsin\left(\frac{E_n}{U}\right)^{1/2}, \tag{2.152}$$

and the wave functions Ψ_n, corresponding to these energy values are localized within the well and exponentially decaying in the barriers.

If

$$L\sqrt{2m_n^* U} < \pi\hbar, \tag{2.153}$$

then inside the well can be only one resonant energy level with energy

$$E_1 \approx U\left(1 - \frac{m_n^* L^2}{2\hbar} U\right). \tag{2.154}$$

There are also virtual energy levels $E_n > U$, which, although they do not correspond to localized electron states in the well, they however determine the resonant transmission of electrons over the well.

It must be stressed that the strictly discrete resonant energy levels in the quantum well are only in the case of infinitely thick barriers and lack of electrons scattering inside the well and on its borders (Figure 2.18b). Only under these conditions, the coherence of the wave function is preserved indefinitely.

2.7 Two-Dimensional Electron Gas in Heterojunction GaN-AlGaN

Heterojunctions have attracted additional attention of researchers after demonstration of the possibility of increasing the mobility of two-dimensional electron gas at the borders of the heterojunction [1]. The idea was to use selective doping for separation of the donor impurity from free electrons and thus reduce the Coulomb scattering of electrons.

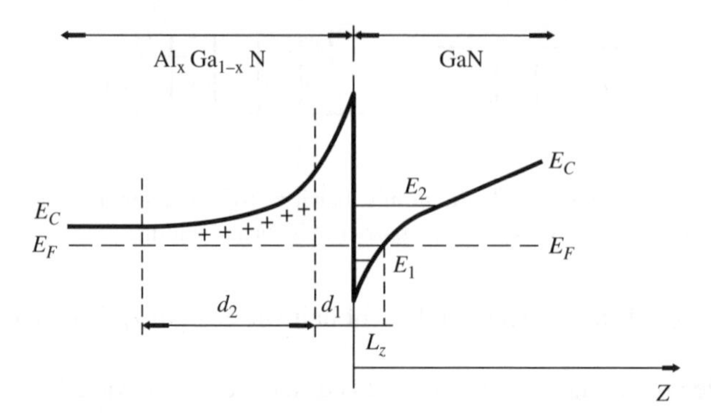

Figure 2.19 Edge of conduction band of n^+-$Al_xGa_{1-x}N$-i-GaN structure with layer of 2D electrons at the heterojunction interface. d_1 is the undoped neutral layer, d_2 is the doped depletion layer

If on the surface of undoped GaN the layer of doped AlGaN is grown, then on their interface a narrow potential well with the 2*D*-electron gas (2DEG) can be formed due to the difference values of the electron affinity of the materials (Figure 2.19). In the structure shown in the figure, the mobile electrons in GaN near the heterojunction are separated from the created by the positively charged donors in AlGaN. As a result, the impurity scattering of electrons decreases and their mobility increases. Electron motion in 2*D* gas is quantized in the direction perpendicular to the heterojunction plane, since the de Broglie wavelength is the order of the potential well width. Remind that for GaN at $T = 300$ K, the length of the electron wave is 0.018 μm and increases with decreasing the temperature.

Electron motion in the 2*D* gas is described by a wave function that is a solution of the Schrodinger equation [3],

$$\frac{\hbar^2}{2m_n^*}\frac{d^2\Psi_j}{dz^2} + [E_j(k) - E_n(z)]\Psi_j = 0, \tag{2.155}$$

where m_n^* is the effective mass of electrons in the conduction band of bulk GaN; $E_j(k)$ is the quantized electron energy in the *j*th dimensional subband; E_n is the potential energy corresponding to band bending at the interface; and z is the distance from the surface to the GaN layer depth.

Potential energy of electron is determined by solving Poisson's equation

$$\frac{d^2E_n}{dz^2} = -\frac{\rho(z)}{\varepsilon_0\varepsilon_s}. \tag{2.156}$$

The density of surface charge $\rho(z)$ is determined by the electrons concentration at the heterojunction interface and doping level of GaN:

$$\rho(z) = q(N_d^+ - N_a^-) - q\sum_{j=0}^{\infty} n_j|\Psi_j(z)|^2, \tag{2.157}$$

$$n_j = \frac{m_n^* k_0 T}{\pi\hbar^2}\ln\left[1 + \exp\left(\frac{E_F - E_j}{k_B T}\right)\right]. \tag{2.158}$$

Here, N_d^+ and N_a^- are the concentration of ionized donors and acceptors, respectively, in the GaN layer.

To solve Equations (2.155)–(2.158), we assume that for $z<0$ the barrier is infinitely high ($U \to \infty$); and for $z>0$, the potential varies linearly on z near the interface of the heterojunction:

$$U/q = V = F_s z, \tag{2.159}$$

where F_s is the electric field on the interface.

In other words, we use the approximation of a triangular potential well. In this case, we find that the energy of the bottom of the subbands according to equation is

$$E_j \approx \left(\frac{\hbar^2}{2m_n^*}\right)^{1/3} \left(\frac{3\pi q F_s \left(j + \frac{3}{4}\right)}{2}\right)^{2/3}, \tag{2.160}$$

and the total surface concentration of free electrons is

$$n_s = \sum_{j=0}^{\infty} n_j = \left(\frac{m_n^* k_0 T}{\pi \hbar^2}\right) \sum_{j=0}^{\infty} \ln\left[1 + \exp\left(\frac{E_F - E_j}{k_B T}\right)\right]. \tag{2.161}$$

If the Fermi level E_F of the structure passes through a potential well, it defines the use of the Fermi–Dirac distribution function.

Blurring the boundaries (i.e., a finite thickness of transition layer) leads to expansion of the potential well, in which 2DEG is localized. It in turn reduces the energy confinement subbands, and therefore increases the electron density at the same position of Fermi level. At the same time, blurring of the boundaries of the heterojunction reduces the magnitude of the potential barrier, which leads to a decrease in the concentration of $2D$ electrons, so there is some optimum width of the transition layer, which provides high parameters of devices based on heterojunctions.

The mobility of $2D$ electrons in heterostructures at low temperatures is mainly determined by impurity scattering on remote ionized donors. Impurity potential is screened by $2D$ electron gas and has the effect of electron of neutral layer of AlGaN. The latter can be interpreted as image forces caused by a charge induced by ionized donors in the neutral layer of AlGaN. The induced charge is removed from the border between the neutral and the depletion region at some distance, whose value is the order of screening length in the AlGaN. Thus, the effect of this charge on the 2DEG decreases.

At low temperatures ($T = 10$ K) the influence of phonon scattering can be neglected. At temperatures $T > 100$ K, the main mechanism of scattering of 2D electrons is the polar optical scattering.

At temperatures around 77 K acoustic (deformation-potential) and the piezoelectric scattering mechanisms are dominated. They determine the mobility that is roughly proportional to $1/T$. Finally, at low temperature ($T < 10$–20 K) the mobility is determined by impurity scattering that is weakly temperature-dependent in this temperature range.

In general, the mobility of the 2DEG can be represented as

$$\frac{1}{\mu} = \frac{1}{\mu_i} + \frac{1}{\mu_o} + \frac{1}{\mu_a}, \tag{2.162}$$

where μ_i, μ_o, μ_a are the mobilities, determined by impurity scattering, optical phonon scattering, acoustic (deformation-potential), and the piezoelectric scattering respectively.

2.8 Electron Properties of Quantum-Size Semiconductor Films

There are significant changes in the physical properties of the films when their thickness leads to the quantum motion of the charge carriers.

In metals, the electron wave length is the order of the lattice spacing and therefore in the real metal films containing many atomic layers, quantum size effects appear relatively weak. A more favorable situation occurs in the semiconductor and semimetallic films, as de Broglie wavelength of the electrons there can be several orders of magnitude greater than the interatomic distance. Under certain conditions, size quantization in the films becomes significant already for thicknesses $L_z \sim 10^{-5}$ cm. Quantum size effects were observed in many materials, including InSb, PbTe, Bi, and so on.

Because of the limited film in one direction, for example, along the axis *z*, the electron energy in the film is determined by the longitudinal projections of the quasipulse (k_x, k_y), and the discrete quantum number *n*, substituting k_z:

$$E = E(k_x, k_y, n). \tag{2.163}$$

For a fixed value of *n* the energy varies quasi-continuously in a certain range, called subband. Subbands usually overlap, since the energy spacing at fixed k_x, k_y is usually less than the width of the subband. As a result, the electron energy spectrum in the film is quasi-discrete and consists of overlapping subbands.

Due to the scattering of electrons the quasi-discrete spectrum of electrons in the film is partially blurred. Obviously, for the preservation of quasi-discrete spectrum it is necessary that its blurring $\hbar/\tau$ to be less than the distance between adjacent subbands:

$$\frac{\hbar}{\tau} \ll E_{n+1} - E_n, \tag{2.164}$$

where τ is the time of relaxation, taking into account all carriers scattering mechanisms in the films.

If the film is approximated by an infinitely high box, then

$$E_n = \frac{n^2 h^2}{8 m_n^* L^2}, \tag{2.165}$$

where $n = 1, 2, 3, \ldots$, and m_n^* is the effective mass in the direction perpendicular to the film.

To observe the effects of quantization, except for restrictions (Equation (2.164)) on the magnitude of relaxation time (or electron mobility), there are restrictions on the temperature and concentration of free carriers.

The temperature conditions is quite obvious: it is necessary that the thermal spread to be small compared to the distance between the subbands

$$k_B T < E_{n+1} - E_n = \frac{h^2(2n+1)}{8 m_n^* L^2}. \tag{2.166}$$

With regard to the free carrier concentration, it is desirable that the number of populated subbands to be small, the quantum limit is one populated subband.

In the real semiconductor films the condition on the concentration and temperature can easily run to thickness of about 0.1 μm. At the effective electron mass 0.01 m_0 the condition (2.166) is satisfied even at room temperature. For a population of only one of the lowest subbands with the same thickness of 0.1 μm the concentration should not exceed 10^{16} cm^{-3}, which is also easily achievable. The most severe condition for observing the discrete energy spectrum in the films is condition (2.164) imposed on the carrier mobility. This condition is fulfilled only in the sufficiently perfect and pure films. Thus, when the film thickness of the order of 0.05 μm the mobility should be more than 10^3 cm^2/(V s).

The quantitative calculation of the energy spectrum and wave functions in the films is very complicated because of the uncertainty of self-consistent potential in the films and due to complicated boundary conditions [3].

The wave function has the Bloch form:

$$\Psi_k(\bar{r}) = U_k(\bar{r}) \exp(i\bar{k}\bar{r}), \tag{2.167}$$

where $U_k(\bar{r})$ has two-dimensional translational symmetry and $\bar{k}$ is the quasipulse in the two-dimensional Brillouin zone.

At quasipulse changing in the two-dimensional Brillouin band the energy runs a continuous range of values that form the subband. Flat Brillouin band is determined by two-dimensional translational symmetry of the film structure. For example, in Ge and Si during the growth of the film in the (111) direction, their structure has the translational symmetry of a planar hexagonal lattice, in connection with which the Brillouin band will have the form of a regular hexagon. Reducing the symmetry in the transition from three-dimensional crystal to the film removes the degeneracy.

In particular, the Ge and Si films degeneracy of the valence band at $k=0$, observed in bulk crystals, is completely eliminated. In the conduction band of germanium the four-valley minimum energy is split into three-valley and single-valley minima, and the single-valley minimum lies at the center of a two-dimensional Brillouin band; the points that mark the position of three-valley minimum are placed in the middle of the sides of a hexagon, forming a two-dimensional Brillouin band (Figure 2.20).

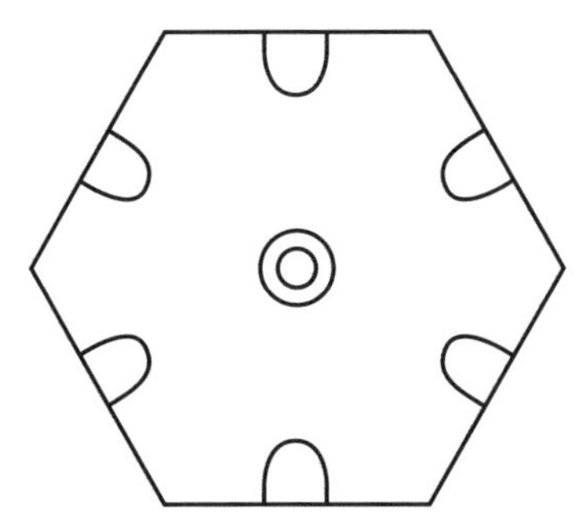

Figure 2.20 Structure of Brillouin band in Ge film

The dispersion relation in the effective mass approximation can be written as

$$E(k_x,k_y,n) = \frac{\hbar^2}{2m_x^*}k_x^2 + \frac{\hbar^2}{2m_y^*}k_y^2 + E_n. \tag{2.168}$$

Using Equation (2.168), it is easy to estimate the density of states for one subband

$$\rho_1^0(E) = \frac{M\sqrt{m_x^* m_y^*}}{\pi\hbar^2}S, \tag{2.169}$$

where M is the number of two-dimensional valleys and S is the area of the film.

The density of states per unit volume of film

$$\rho_1(E) = \frac{\rho_1^0(E)}{V} = \frac{M\sqrt{m_x^* m_y^*}}{\pi\hbar^2 L}, \tag{2.170}$$

where L is the thickness of the film.

It is important to note that the density of states (Equation (2.170)) in the subband does not depend on energy and number of subbands and is inversely proportional to film thickness.

The total density of states $\rho(E)$ in the film at a given energy E is equal to the sum of $\rho_1(E)$ over all the subbands for which $E_n < E$:

$$\rho(E) = n_{\max}(E,L)\rho_1(E), \tag{2.171}$$

where n_{max} is the maximum value of n (number of subbands). To determine $n_{max}(E,L)$ it is necessary to specify the potential in the film. The simplest model is to approximate the film rectangular potential box with infinitely high walls. In this model

$$E_n = \frac{n^2h^2}{8m_n^*L^2}. \tag{2.172}$$

Expression (2.171) determines the total density of states in the film as a function of thickness and energy. With increasing thickness and growth of the energy the number of occupied subbands increases abruptly.

Density of states on energy dependence $\rho(E)$ at constant film thickness is shown in Figure 2.21. It should be noted that the density of states is nonzero only for energies $E > E_1$, at condition that the energy is measured from the conduction band in a bulk semiconductor. The appearance of the minimum energy $E_1 \neq E_C$, where E_C is the conduction band bottom, is caused by the uncertainty principle.

Taking into account (Equation (2.172)), we find that the density of states in the film at $E = E_n$ is equal to the density of states in bulk semiconductor at the same energy.

Let us consider the density of states on the film thickness dependence, since this dependence determines the oscillatory nature of transport coefficients in the films.

With increasing the film thickness L the density of states ρ in Fermi level E_F is reduced as $1/L$, while the number of populated subbands remains constant. At a certain thickness L_n the new subband begins to fill, and the density of states increases abruptly on the amount of $\rho_1(E_n)$ resulting in the oscillatory dependence of the density of states on the thickness (Figure 2.22).

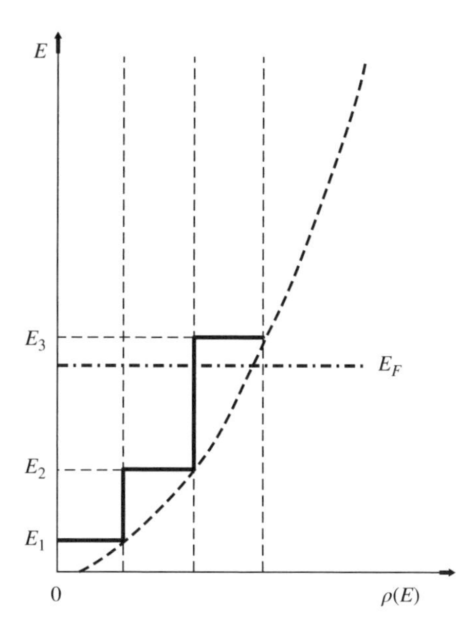

Figure 2.21 Density of electron state on energy dependence at constant film thickness

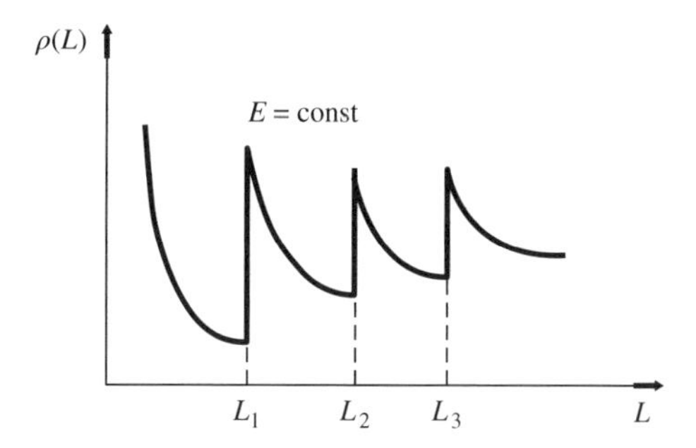

Figure 2.22 Density of electron state on film thickness dependence

Averaged over the thickness of the film the density of electrons in the conduction band is given by

$$n(T) = \sum_n \int_{E_n}^{\infty} \frac{\rho_1(E)dE}{1 + \exp\left(\frac{E - E_F}{k_0 T}\right)}, \tag{2.173}$$

where $\rho_1(E)$ is the density of states in one subband (Equation (2.169)).

The integration is over the nth subband, and the summation over all subbands.

If only one subband is filled, and the donors N_d are weakly ionized, then

$$n(T) = \left(\frac{m_n^*}{\pi\hbar^2}\right)^{1/2} (k_0 T)^{1/2} \exp\left(-\frac{\Delta E_d}{2k_0 T}\right), \tag{2.174}$$

where ΔE_d is the ionization energy of donors, measured from the bottom of the conduction band in the film.

The density of states and the type of electron wave functions change during the transition to the thin film effect primarily on the kinetic characteristics. Thus, the carrier mobility μ in the case of degeneracy is determined by the density of states at the Fermi surface, which oscillates with the film thickness. This leads to an oscillatory dependence of mobility on thickness and, accordingly, to oscillations of resistivity.

Resistance of the film oscillates not only with the change in thickness, but with the change in the longitudinal electric field. Resistance oscillations occur for values of the electric field when the energy obtained by an electron at free path motion becomes equal to the distance between the quantum energy levels.

References

1. R.F. Pierret. *Advanced Semiconductor Fundamentals*. Modular Series on Solid State, ed. R.F. Pierret and G.W. Neudeck (Pearson Education Inc, Upper Saddle River, NJ, vol. 6, 2004) pp. 50–86.
2. S. Datta. *Electronic Transport in Mesoscopic Systems,* ed. H. Ahmed, M. Pepper, and A. Broers (Cambridge University Press, Cambridge, 1995), pp. 247–75.
3. A.F. Kravchenko and V.N. Ovsyuk. *Electron Processes in Solid-State Low-Dimensional System* (Novosibirsk University Press, 2000 (in Russian).
4. S.M. Sze and K.K. Ng. *Physics of Semiconductor Devices*, 3rd edn (John Wiley & Sons, Inc., New York 2007).
5. D.N. Neamen. *Semiconductor Physics and Devices. Basic Principles*. 3rd edn (McGrawHill Education 2003).
6. V.L. Bonch-Bruyevich and S.G. Kalashnikov. *Physics of Semiconductors* (Moscow University Press, 1977).
7. J.-P. Colinge and C.A. Colinge. *Physics of Semiconductor Devices* (Kluwer Academic Publisher, 2000).
8. R. Kubo, H. Ichimura, T. Usui, and N. Hashitsume, *Statistical Mechanics* (North-Holland: New York, 1965).
9. S.M. Sze, *Physics of Semiconductor Devices*, 2nd edn (John Wiley & Sons, Inc.: New York, 1981).
10. R.P. Feynman, *Statistical Mechanics: A Set of Lectures* (W. A. Benjamin Inc.: Reading, MA, 1972).
11. C. Kittel, *Introduction to Solid State Physics*, 7th edn (John Wiley & Sons, Inc.: New York, 1996).
12. K.L. Jensen, Exchange-correlation, dipole, and image charge potentials for electron sources: Temperature and field variation of the barrier height, *Journal of Applied Physics* **85**, 2667 (1999).
13. K.L. Jensen and A.K. Ganguly, Numerical simulation of field emission and tunneling: A comparison of the Wigner function and transmission coefficient approaches, *Journal of Applied Physics* **73**, 4409 (1993).
14. K.L. Jensen, Theory of field emission, in *Vacuum Microelectronics*. Ed. W. Zhu (John Wiley & Sons, Inc., Hoboken, NJ, 2001), pp. 33–104.
15. O.V. Tretyak and V.Z. Lozovskyy. *Foundation of Semiconductor Physics*, Vol. 2 (Kyiv University Press, 2009) (in Ukrainian).

3

Band Bending and Work Function

Important parameters during the analysis of electron field emission from semiconductors are the band bending and work function. They are considered in this chapter. Surface space-charge region, quantization of electron energy spectra at semiconductor surface, and image charge potential are analyzed in detail. The work function as the most important parameter at electron field emission and its changing at cathode coating with film, under electric field, temperature, and surface adatoms are also under consideration here.

3.1 Surface Space-Charge Region

Let's describe the formation of space charge region (SCR) on the surface of a semiconductor under the influence of an applied electric field and derive the relations between the surface potential, space charge, and electric field [1]. These relations are used, in particularly, in the analysis of electron field emission from semiconductors [2–4].

A detailed band diagram at the surface of a p-type semiconductor is shown in Figure 3.1. The potential $\psi_p(x)$ is defined as the potential $E_i(x)/q$ with respect to the bulk of the semiconductor:

$$\psi_p(x) = -\frac{[E_i(x) - E_i(\infty)]}{q}. \tag{3.1}$$

At the semiconductor surface, $\psi_p(x) \equiv \psi_s$, and ψ_s is called the surface potential. The electron and hole concentrations as a function of ψ_p are given by the following relations:

$$n_p(x) = n_{p0} \exp\left(\frac{q\psi_p}{k_B T}\right) = n_{p0} \exp(\beta\psi_p), \tag{3.2}$$

$$p_p(x) = p_{p0} \exp\left(\frac{-q\psi_p}{k_B T}\right) = p_{p0} \exp(-\beta\psi_p), \tag{3.3}$$

where ψ_p is positive when the band is bent downward (as shown in Figure 3.1), n_{po} and p_{po} are the equilibrium densities of electrons and holes, respectively, in the bulk of the semiconductor,

Vacuum Nanoelectronic Devices: Novel Electron Sources and Applications, First Edition.
Anatoliy Evtukh, Hans Hartnagel, Oktay Yilmazoglu, Hidenori Mimura and Dimitris Pavlidis.

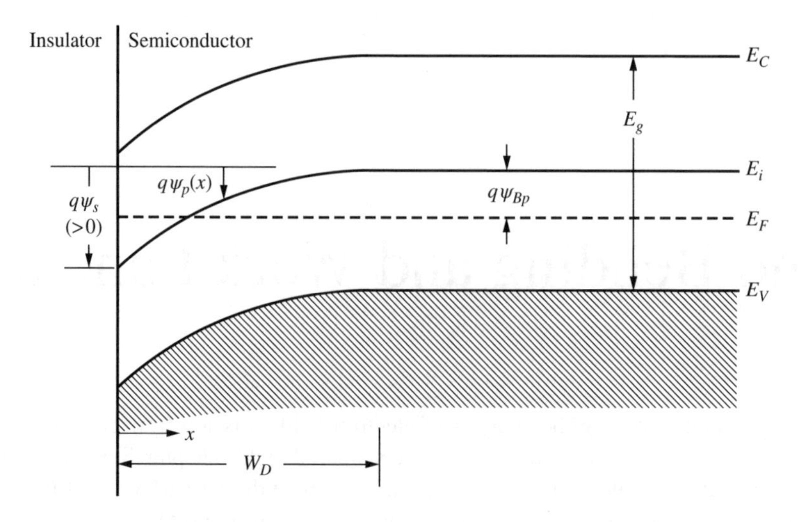

Figure 3.1 Energy-band diagram at the surface of a p-type semiconductor. The potential energy $q\psi_p$, is measured with respect to the intrinsic Fermi level E_i in the bulk. The surface potential ψ_s is positive as shown. Accumulation occurs when $\psi_s < 0$. Depletion occurs when $\psi_{Bp} > \psi_s > 0$. Inversion occurs when $\psi_s > \psi_{Bp}$ [1]. Copyright (1981) by John Wiley & Sons, Inc. This material is reproduced with permission of John Wiley & Sons, Inc.

and $\beta \equiv q/k_BT$. At the surface the densities are

$$n_p(0) = n_{p0} \ \exp(\beta\psi_s), \tag{3.4}$$

$$p_p(0) = p_{p0} \ \exp(-\beta\psi_s). \tag{3.5}$$

The following regions of surface potential can be distinguished [1]:

$\psi_s < 0$: Accumulation of holes (bands bending upward).
$\psi_s = 0$: Flat-band condition.
$\psi_{Bp} > \psi_s > 0$: Depletion of holes (bands bending downward).
$\psi_s = \psi_{Bp}$: Fermi-level at midgap, $E_F = E_i(0)$, $n_p(0) = p_p(0) = n_i$.
$2\psi_{Bp} > \psi_s > \psi_{Bp}$: Weak inversion (electron enhancement, $n_p(0) > p_p(0)$).
$\psi_s > 2\psi_{Bp}$: Strong inversion ($n_p(0) > p_{p0}$ or N_A).

The potential $\psi_p(x)$ as a function of distance can be obtained by using the one-dimensional Poisson equation

$$\frac{d^2\psi_p}{dx^2} = -\frac{\rho(x)}{\varepsilon_s}, \tag{3.6}$$

where $\rho(x)$ is the total space-charge density given by

$$\rho(x) = q(N_D^+ - N_A^- + p_p - n_p), \tag{3.7}$$

$N_D{}^+$ and $N_A{}^-$ are the densities of ionized donors and acceptors, respectively. Now, in the bulk of semiconductor, far from the surface, charge neutrality must exist. Therefore at $\psi_p(\infty)=0$, we have $\rho(x)=0$ and

$$N_D^+ - N_A^- = n_{p0} - p_{p0}. \tag{3.8}$$

The resultant Poisson equation to be solved within the depletion region is therefore

$$\frac{d^2\psi_p}{dx^2} = -\frac{q}{\varepsilon_s}(n_{p0} - p_{p0} + p_p - n_p) = -\frac{q}{\varepsilon_s}\{p_{p0}[\exp(-\beta\psi_p) - 1] - n_{p0}[\exp(\beta\psi_p) - 1]\}. \tag{3.9}$$

Integrating Equation (3.9) from the surface toward the bulk [5]

$$\int_0^{d\psi_p/dx} \left(\frac{d\psi_p}{dx}\right) d\left(\frac{d\psi_p}{dx}\right) = -\frac{q}{\varepsilon_s}\int_0^{\psi_p} \{p_{p0}[\exp(-\beta\psi_p) - 1] - n_{p0}[\exp(\beta\psi_p) - 1]\}d\psi_p \tag{3.10}$$

gives the relation between the electric field $(F \equiv -d\psi_p/dx)$ and the potential ψ_p:

$$F^2 = \left(\frac{2k_BT}{q}\right)^2 \left(\frac{qp_{p0}\beta}{2\varepsilon_s}\right) \left\{ [\exp(-\beta\psi_p) + \beta\psi_p - 1] + \frac{n_{p0}}{p_{p0}}[\exp(\beta\psi_p) - \beta\psi_p - 1] \right\}. \tag{3.11}$$

We shall use the following abbreviations:

$$L_D \equiv \sqrt{\frac{k_BT\varepsilon_s}{q^2p_{p0}}} \equiv \sqrt{\frac{\varepsilon_s}{qp_{p0}\beta}}, \tag{3.12}$$

$$f\left(\beta\psi_p, \frac{n_{p0}}{p_{p0}}\right) \equiv \left\{ [\exp(-\beta\psi_p) + \beta\psi_p - 1] + \frac{n_{p0}}{p_{p0}}[\exp(\beta\psi_p) - \beta\psi_p - 1] \right\}^{1/2} \geq 0, \tag{3.13}$$

where L_D is the extrinsic Debye length for holes. (Note that $n_{p0}/p_{p0} = exp(-2\beta\psi_{Bp})$.) Thus the electric field is given by

$$F(x) = \pm\frac{\sqrt{2}k_BT}{qL_D} f\left(\beta\psi_p, \frac{n_{p0}}{p_{p0}}\right), \tag{3.14}$$

with positive sign for $\psi_p > 0$ and negative sign for $\psi_p < 0$. To determine the electric field at the surface F_s, we let $\psi_p = \psi_s$:

$$F_s = \pm\frac{\sqrt{2}k_BT}{qL_D} f\left(\beta\psi_s, \frac{n_{p0}}{p_{p0}}\right). \tag{3.15}$$

From this surface field we can deduce the total space charge per unit area by applying Gauss' law:

$$Q_s = -\varepsilon_s F_s = \mp\frac{\sqrt{2}\varepsilon_s k_BT}{qL_D} f\left(\beta\psi_s, \frac{n_{p0}}{p_{p0}}\right). \tag{3.16}$$

To determine the change in the hole density, Δp, and electron density, Δn, per unit area when ψ_p at the surface is shifted from zero to the final value ψ_s, it is necessary to evaluate the following expressions [6]:

$$\Delta p = p_{p0}\int_0^\infty [\exp(-\beta\psi_p) - 1]dx = \frac{qp_{p0}L_D}{\sqrt{2}k_BT}\int_{\psi_s}^0 \frac{[\exp(-\beta\psi_p) - 1]}{f\left(\beta\psi_p, \frac{n_{p0}}{p_{p0}}\right)}d\psi_p, \tag{3.17}$$

$$\Delta n = n_{p0}\int_0^\infty [\exp(\beta\psi_p) - 1]dx = \frac{qn_{p0}L_D}{\sqrt{2}k_BT}\int_{\psi_s}^0 \frac{[\exp(\beta\psi_p) - 1]}{f\left(\beta\psi_p, \frac{n_{p0}}{p_{p0}}\right)}d\psi_p. \tag{3.18}$$

Typical variation of the space-charge density Q_s as a function of the surface potential ψ_s, is shown in Figure 3.2, for a p-type silicon with $N_A = 4\times10^{17}$ cm^{-3} at room temperature. Note that for negative ψ_s, Q_s is positive and it corresponds to the accumulation region. The function f is dominated by the first term in Equation (3.13), that is, $Q_s \sim \exp(q|\psi_s|/2k_BT)$. For $\psi_s = 0$,

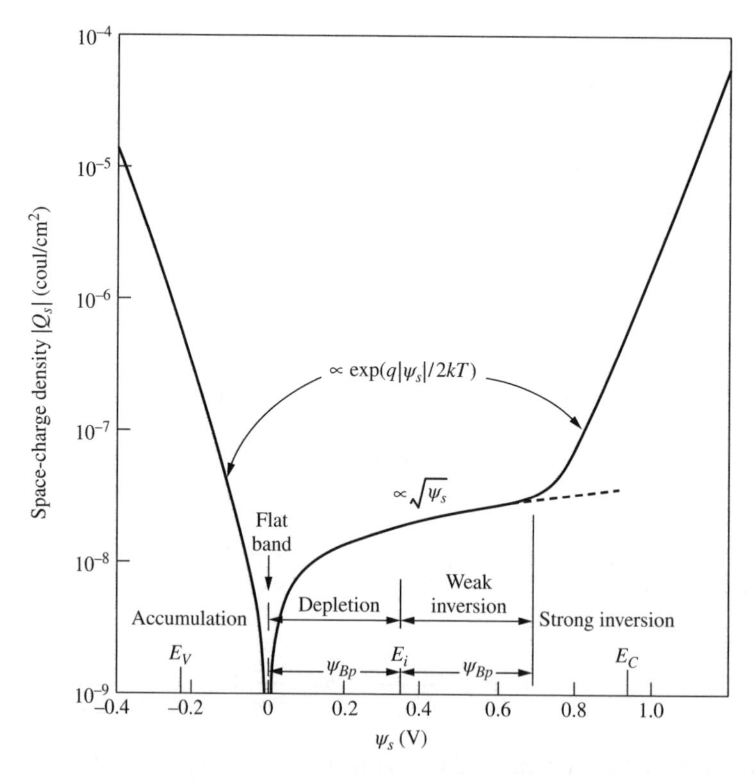

Figure 3.2 Variation of space-charge density in the semiconductor as a function of the surface potential ψ_s, for p-type silicon with $N_A = 4\times10^{15}$ cm^{-3} at room temperature [1]. Copyright (1981) by John Wiley & Sons, Inc. This material is reproduced with permission of John Wiley & Sons, Inc.

we have the flat-band condition and $Q_s = 0$. For $2\psi_B > \psi_s > 0$, Q_s is negative and we have the depletion and weak-inversion cases. The function *f* is now dominated by the second term, that is, $Q_s \sim \sqrt{\psi_s}$. For $\psi_s > 2\psi_B$, we have the strong inversion case with the function *f* dominated by the fourth term, that is, $Q_s \sim \exp(q\psi_s/2k_BT)$. Also note that this strong inversion begins at a surface potential,

$$\psi_s = 2\psi_{Bp} = \frac{2k_BT}{q}\ln\left(\frac{N_A}{n_i}\right). \tag{3.19}$$

It is important to know the distance of electric field distribution in semiconductor (depletion region, W_D) (Figure 3.1). For *p*-type semiconductors, under the abrupt approximation that $\rho = qN_A$, for $x < W_D$ and $\rho = 0$ and $F = 0$ for $x > W_D$, where W_D is the depletion width and N_A is the impurity concentration in semiconductor, after integration of the Poisson's equation we obtain

$$W_D = \sqrt{\frac{2\varepsilon_s}{qN_A}(\psi_s)}. \tag{3.20}$$

The schematic images of the energy band diagrams of the semiconductor at different conditions on the surface are shown in Figure 3.3. The appearance of SCR can be caused as the charge on the surface and applied voltage.

Applying the electric voltage some part of it drops on vacuum (insulator) region (V_{vac}) another on semiconductor ($V_s \equiv \psi_s$)

$$V = V_{vac} + V_s. \tag{3.21}$$

The diagram of potential spreading is shown in Figure 3.4. According to Gauss' law at absence of charge at the semiconductor surface electric field in vacuum (F_{vac}) and at the surface of semiconductor (F_s) are connected such simple relations

$$\varepsilon_0 F_{vac} = \varepsilon_s F_s,\ \ F_{vac} = \frac{\varepsilon_s F_s}{\varepsilon_0},\ \ V_{vac} = F_{vac}L = \frac{\varepsilon_s F_s L}{\varepsilon_0} = \frac{|Q_s|L}{\varepsilon_0}, \tag{3.22}$$

where ε_0 and ε_s are the vacuum and semiconductor permittivity respectively and L is the anode-semiconductor (emitter) distance.

3.2 Quantization of the Energy Spectrum of Electrons in Surface Semiconductor Layer

In the case of sufficiently large band bending at the surface the potential well in the SCR either for electrons (positive bending down), or for holes (negative bending up) is quite narrow, resulting in confinement quantization of energy spectrum of the free carriers [7].

In the quantization in Z axis that is directed perpendicularly to the semiconductor surface, the movement of carriers along the surface remains semiclassical and the corresponding to this motion *E(k)* dependence is determined by a quasi-continuous set of wave numbers k_x and k_y.

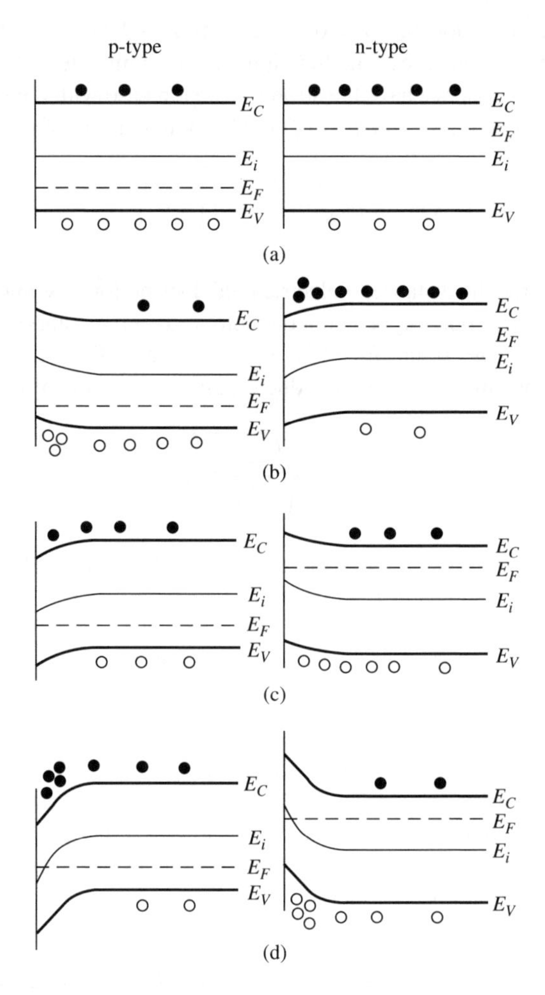

Figure 3.3 Energy-band diagrams of semiconductor surfaces under bias for conditions of: (a) flat band, (b) accumulation, (c) depletion, and (d) inversion

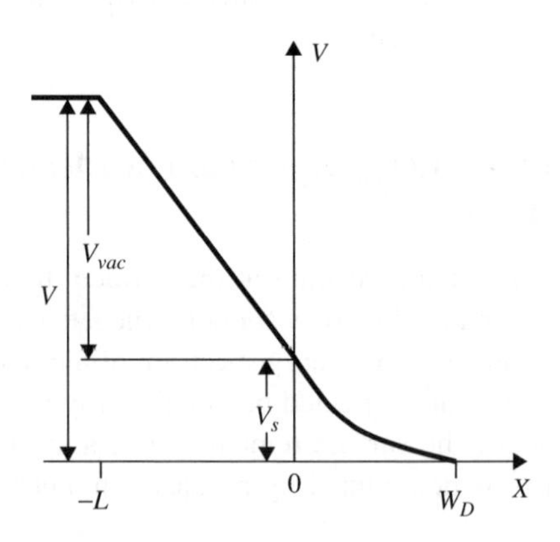

Figure 3.4 Potential distribution (relative to the semiconductor bulk)

The full dependence of $E(k)$ and wave functions of carriers $\Psi_j(x, y, z)$ can be represented as

$$E(k) = \frac{\hbar^2}{2}\left(\frac{k_x^2}{m_x^*} + \frac{k_y^2}{m_y^*}\right) + E_j, \tag{3.23}$$

$$\Psi_j(x, y, z) = \Omega_j(z)\ \exp[i(k_x x + k_y y)], \tag{3.24}$$

where the values of E_j and components $\Omega_j(z)$ are the energy eigenvalues and eigenfunctions of the Schrodinger equation for a given potential $V(z)$.

In the one-electron approximation and the effective mass approximation:

$$\frac{d^2\Omega_j(z)}{dz^2} + \frac{2m_z^*}{\hbar^2}[E_j - qV(z)]\Omega_j(z) = 0. \tag{3.25}$$

Exact solution of the problem requires joint self-consistent solution of Schrödinger and Poisson equations, because the potential in SCR $V(z)$ is determined by the charge distribution along the coordinate z.

For positive bending in p-semiconductor the Poisson equation can be written as

$$\frac{d^2V(z)}{dz^2} = -\frac{q}{\varepsilon_0\varepsilon_s}[N_A + \rho_n(z)], \tag{3.26}$$

where the density of electrons in the inversion layer $\rho_n(z)$ is expressed as a sum

$$\rho_n(z) = \sum_j \Gamma_{nj}, \tag{3.27}$$

where Γ_{nj} is the population of the jth quantum subband and it is equal to

$$\Gamma_{nj} = \beta_j \frac{m_j^* k_B T}{\pi\hbar^2} \ln\left[1 + \exp\left(\frac{E_F - E_j}{k_B T}\right)\right], \tag{3.28}$$

where β_j is the multiplicity of valley degeneracy; $m_j^* = \sqrt{m_{xj}^* m_{yj}^*}$ is the effective mass of the density of states in each subband; m_{xj}^*, m_{yj}^* are the effective mass in the X and Y directions parallel to the surface; and E_F is the Fermi level.

If $E_F < E_j$ we have

$$\Gamma_{nj} \sim \exp\left(\frac{E_F - E_j}{k_B T}\right), \tag{3.29}$$

which corresponds to the Boltzmann distribution of energy, and when $E_F > E_j$

$$\Gamma_{nj} \sim E_F - E_j, \tag{3.30}$$

and the electron gas is degenerated.

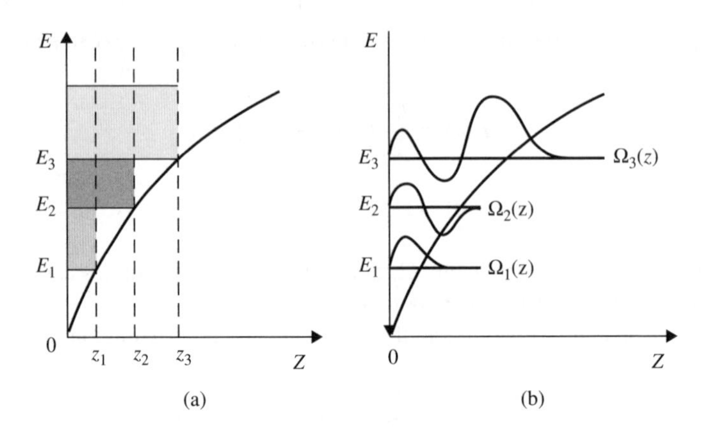

Figure 3.5 Energy band diagram of SCR with (a) quantum subbands and (b) form of wave functions $\Omega_j(z)$ for three first subbands

The qualitative view of the energy diagram for the SCR of the n-type inversion channel and forms of the wave functions for the first three subbands are shown in Figure 3.5.

On the surface, at $z = 0$, the wall of the potential well is infinitely high and $\Omega_j = 0$. On the second boundary of the potential well the wave functions $\Omega_j(z)$ decay exponentially during the transition in the classically inaccessible region.

In the range of $E_2 - E_1$ for the first quantum subband the electrons energies have quasi-continuous energy spectrum that can describe their motion along the surface in a semiclassical way. Energy $E = E_1$ corresponds to the energy of "resting" electron and at the energies $E > E_1$ the electron has kinetic energy $E - E_1$ associated with some of its group velocity v along the surface of the semiconductor. Electrons of the first subband with any kinetic energy have the same wave function $\Omega_1(z)$, that is, they have the same region of localization in Z axis.

At energy $E > E_2$ near the surface of the semiconductor it is possible to find electrons as the first and the second subbands of energies.

For the second subband the situation is similar, namely, the electron with kinetic energy $E - E_2$ corresponds to some energy $E > E_2$ and the wave function $\Omega_2(z)$ as it is shown in Figure 3.5b for the level of E_2.

The solution of Equations (3.25) and (3.26) is carried out numerically. It's the most time-consuming when you want to record the populations of several quantum confinement subbands, and is significantly simplified in the approximation of the so-called quantum limit.

The quantum limit corresponds to the degenerate case and the populations of only one, the first quantum subband. This approximation is well satisfied, for example, for germanium and silicon at liquid helium temperature. It is this case that is now the subject of more systematic experimental and theoretical studies.

In the quantum limit approach a variation method for the solution of Equations (3.25) and (3.26) is used in which the wave function of $\Omega_1(z)$ is approximated by a trial function

$$\Omega_1(z) = \sqrt{\frac{b^3}{2z}} \exp\left(-\frac{bz}{2}\right), \tag{3.31}$$

and the parameter b is found from the condition of electrons energy minimization in the channel.

Another approximation is the approximation of the "triangular well" in which the potential well near the surface of the semiconductor is limited by the walls of $V \to \infty$ at $z=0$ and $V=F_s z$ at $z>0$.

The magnitude of surface electric field F_s is determined by the total quantity of charge in SCR:

$$F_s = \frac{q\Gamma_n}{\varepsilon_0 \varepsilon_s}. \tag{3.32}$$

Schrodinger equation (3.25) for this case takes the form

$$-\frac{\hbar^2}{2m_z^*}\frac{d^2\Omega_j(z)}{dz^2} + qF_s z\Omega_j(z) = E_j\Omega_j(z), \tag{3.33}$$

or

$$\frac{d^2\Omega_j(z)}{dz^2} - \left[\frac{2qm_z^*}{\hbar^2}F_s\left(z - \frac{E_j}{qF_s}\right)\right]\Omega_j(z) = 0. \tag{3.34}$$

By changing the variables $z \to \eta$ [7], where

$$\eta = \frac{2qm_z^* F_s}{\hbar^2}\left(z - \frac{E_j}{qF_s}\right), \tag{3.35}$$

the Schrodinger equation (3.33) transforms into the Airy equation.

$$\frac{d^2\Omega(\eta)}{d\eta^2} - \eta\Omega(\eta) = 0. \tag{3.36}$$

Solutions of this equation are the Airy functions

$$\Phi(\eta) = \frac{1}{\pi}\int_0^\infty \cos\left(\eta t + \frac{t^3}{3}\right) dt, \tag{3.37}$$

and, as a result, we have

$$\Omega(\eta) = A\Phi(\eta), \tag{3.38}$$

where A is the normalization factor.

From the boundary condition $\Omega(z)=0$ at $z=0$ the transcendental equation follows

$$\Phi = \left\{E_j \frac{\left(2m_z^*\right)^{1/3}}{(q\hbar E_s)^{2/3}}\right\} = 0, \tag{3.39}$$

which implies that the argument of the Φ function in Equation (3.39) is equal to the roots of the Airy function a_j, taken in increasing order. Airy functions are well studied and tabulated, and the first four roots are 2.34, 4.09, 5.52, and 6.79 respectively.

We will obtain from Equation (3.39) the position of quantum subbands E_j

$$E_j = a_j\left(\frac{q^2\hbar^2F_s^2}{2m_z^*}\right)^{1/3}. \tag{3.40}$$

With the increasing of j number the levels density grows, and when $j \gg 1$ the next approximation is good.

$$E_j = \frac{1}{2}\left(9\pi^2q^2\hbar^2\frac{1}{m_z^*}F_s^2\right)^{1/3}\left(j+\frac{3}{4}\right)^{3/2}, \tag{3.41}$$

It already leads to the semiclassical energy distribution of states density.

Equation (3.40), after substituting the numerical coefficients for the first level gives

$$E_1 = 1.7\times 10^{-5}\left(\frac{m_0}{m_z^*}\right)^{1/3}F_s^{2/3}, \tag{3.42}$$

where E_1 is expressed in eV and F_s is in V/cm.

The position of the second level is defined by

$$E_2 = \frac{a_2}{a_1}E_1. \tag{3.43}$$

The region of electrons localization in the subbands z_j is obtained from Equation (3.35) in the form

$$z_j = a_j\left(\frac{\hbar^2}{2m_z^*qF_s}\right)^{1/3}. \tag{3.44}$$

For example, for germanium at field $F_s = 10^5$ V/cm, the total excess charge in the SCR is 8.85×10^{11} cm^{-2}, while at $F_s = 10^6$ V/cm the charge is 8.85×10^{12} cm^{-2}. In the first case, at $m_z^* = 0.2\,m_0$, we have $E_1 = 0.062$ eV and $E_2 = 0.11$ eV, while in the second case we have $E_1 = 0.29$ eV and $E_2 = 0.49$ eV [7]. As it can be seen, in this range of the electric field the detachment of the first subband from the conduction band bottom is a few units k_BT already at room temperature.

3.3 Image Charge Potential

A charge particle outside the surface of metal induces charge distribution on the surface to screen the charged particle's electric field inside the bulk. The surface charge acts as though a charged particle of equal and opposite sign is located equidistant from the surface as the original charge. The image-force induces the lowering of the potential barrier for charge carrier emission when an electric field is applied [1, 8]. Let's consider the metal-vacuum system. The minimum energy necessary for the electron to escape into vacuum from the initial energy at the Fermi level is defined as the work function Φ_m (Figure 3.6). For metal, Φ_m is of the order of a few electron volts and varies from 2 to 6 eV [9]. The values of Φ_m are generally sensitive to surface contamination.

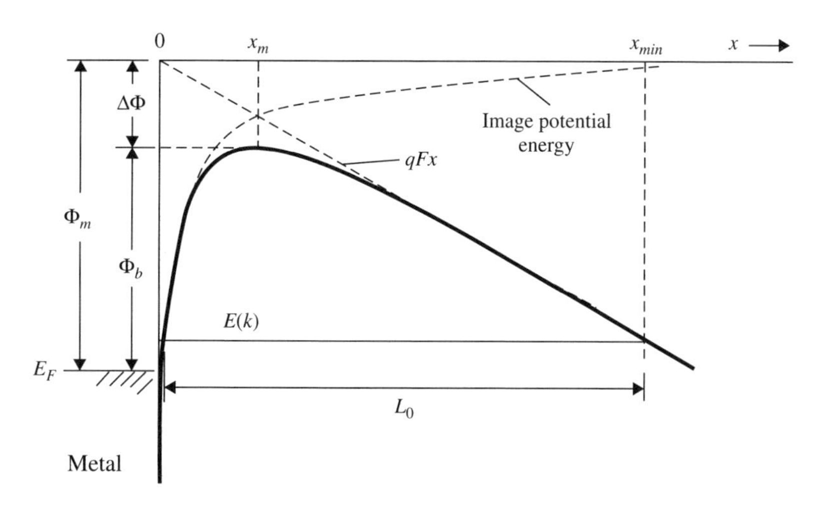

Figure 3.6 Energy band diagram between the metal surface and vacuum

When the electron is at the distance x from the metal, a positive charge will be induced on the metal surface (Figure 3.7). The force of attraction between the electron and the induced positive charge is equivalent to the force that would exist between the electron and equal positive charge located at $-x$. This positive charge is referred to as the image charge. The attractive force (F_f), called the image force, is given by

$$F_f = \frac{-q^2}{4\pi\varepsilon_0(2x)^2} = \frac{-q^2}{16\pi\varepsilon_0 x^2}, \tag{3.45}$$

where ε_0 is the permittivity of free space. The work done by an electron in the case of its transfer from infinity to the point x is given by

$$E(x) = \int_{\infty}^{x} F_f dx = \frac{q^2}{16\pi\varepsilon_0 x}. \tag{3.46}$$

The energy in Equation (3.46) corresponds to the potential energy of an electron at a distance x from the metal surface (Figure 3.6) and is measured downwards from x axis. The results of this classical description are retained in a form in which quantum mechanical considerations are taken into account [10]: the image charge contribution affects the barrier near the surface and the form of Equation (3.46) holds to within a few angstroms of the metal/vacuum interface.

When an external field F is applied, the total potential energy U as a function of distance (measured downward from the x axis) is given by the sum

$$U = \frac{q^2}{16\pi\varepsilon_0 x} + qFx. \tag{3.47}$$

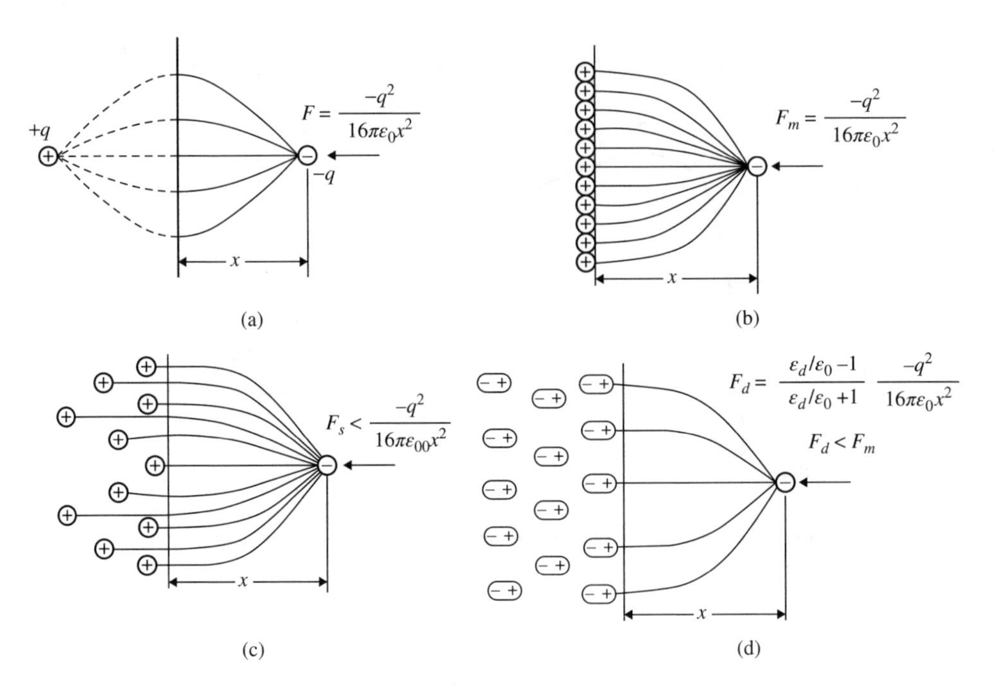

Figure 3.7 Image forces (a) and influence of electric fields on metal (b), semiconductor (c), and dielectric (d) surfaces

The Schottky barrier lowering (also referred to as image force lowering) $\Delta\Phi$ and the location of the maximal lowering x_m (Figure 3.6) are given by the condition $dU(x)/dx = 0$, or

$$x_m = \sqrt{\frac{q}{16\pi\varepsilon_0 F}}, \tag{3.48}$$

$$\Delta\Phi = \sqrt{\frac{q^3 F}{4\pi\varepsilon_0}} = 2qFx_m. \tag{3.49}$$

It is possible to determine $\Delta\Phi$ and x_m for different applied fields from Equations (3.48) and (3.49): $\Delta\Phi = 0.12$ eV and $x_m = 6$ nm for $F = 10^5$ V/cm; and $\Delta\Phi = 1.2$ eV and $x_m = 1$ nm for $F = 10^7$ V/cm [1]. Thus, at high fields the Schottky barriers are considerably lowered, and the effective metal work function Φ_b is reduced.

These results can also be applied to metal-semiconductor and metal-dielectric systems (Figure 3.7). However, the field should be replaced by the maximum field at the interface, and free-space permittivity ε_0 should be replaced by an appropriate permittivity ε_s characterizing the semiconductor (dielectric) medium, that is

$$\Delta\Phi = \sqrt{\frac{q^3 F}{4\pi\varepsilon_s}}. \tag{3.50}$$

The value ε_s may be different from the semiconductor static permittivity. If during the emission process the electron transit time from the metal-semiconductor interface to the barrier maximum x_m is shorter than the dielectric relaxation time, the semiconductor medium does not have enough time to be polarized, and smaller permittivity than the static value is expected.

In the case of field emission asymptotically, for $F=0$, the difference between the barrier height (U_0) and the chemical potential (μ) is the work function, so $\phi=\Phi$. The image charge potential for field and thermionic emission then takes the form [11] (Figure 3.6).

$$U(x) = \begin{cases} \mu + \Phi - qFx - \frac{Q}{x} \dots\dots\dots\dots x \le x_{\min} \\ 0 \dots\dots\dots\dots\dots\dots\dots\dots\dots\dots\dots\dots x > x_{\min} \end{cases}, \tag{3.51}$$

where $U(x_{min})=0$ and $Q=q^2/(16\pi\varepsilon_0)$. The zeros of $U(x)-E(k)$ occur at locations

$$x_{\pm} = \frac{h \pm \sqrt{h^2 - 4qQF}}{2qF}, \tag{3.52}$$

where ($\pm$) subscript is associated with the larger and smaller root, respectively and h is given by $h(E)=\mu+\phi-E(k)$. The triangular barrier is covered in the limit $Q=0$. The difference $x_+ - x_-$ shall be designated L_0 (Figure 3.6).

The image charge approximation given in Equation (3.51) is classical in origin, but it is nevertheless a good approximation for $x > 3$ Å [12–14] (depending on the metal surface). Close to the surface, the quantum mechanical nature of the electron degrades the approximation.

3.4 Work Function

It is possible to give the following definition of electron bond energy in a solid, particularly in a metal, which doesn't depend on the model of this solid. The fact of the stationary existence of electrons in it points out that the system containing N_p ions and $N_e=N_p$ electrons is in equilibrium at $T=0$, and has lower energy than the same N_p ions with $N_e^{/}=N_e-n$ electrons at the same temperature in equilibrium. Defining the first system energy as $E(N_p, N_e)$, and second as $E(N_p, N_e-n)$ it is possible to write the energy changing when removing one electron, that is, work function at $T=0$ will be as follows:

$$\Phi = \frac{1}{n}[E(N_p, N_e - n) - E(N_p, N_e)]_{N_p=N_e} = \frac{dE}{dN_e}. \tag{3.53}$$

This definition of work function is analogous to the ionization work of neutral unexcited atom.

At T > 0 the definition Equation (3.53) becomes undetermined. Really, the difference in the energy of equilibrium state of the system (N_p, N_e-n) and the equilibrium state of the system (N_p, N_e) depends on the conditions of transition, namely whether the transition is adiabatic (without exchange of the heat with environment, $dQ=0$) or isothermal ($T=$ const). In thermodynamic the real work function is the chemical potential with opposite sign (if the energy is accounted from the vacuum level)

$$\Phi = -\mu = \left(\frac{\partial E}{\partial N_e}\right)_{S=const, V=const}, \tag{3.54}$$

where S is the system entropy and V is the system volume.

The thermodynamic definition of work function given above in Equation (3.53) is beyond the band theory of the solids, because it includes the change of full energy of the solid that includes the change of energy as electron system and lattice energy in removing the electron.

In the frame of band theory of solids each electron has a definite value of energy E_i and the full energy of electron gas is $E=\Sigma E_i$ where the sum includes all electrons in the solid. Therefore, the change of the energy E at removing of ith electron according to band theory is $-E_i$. And it may be supposed that each electron has its own work function $\Phi_i=-E_i$. But removing the ith electron with work function Φ_i leaves the electron gas in an unequilibrium state. There is only one energy level when the removing of the electron from it doesn't change the equilibrium of rest electron gas. At $T=0$ it is $E_{max}=-\Phi_0$ level. The removing of the electron from the level with energy $E_i<E_{max}$ will require to give this electron energy $-E_i$, but to realize the equilibrium state of rest gas at $T=0$ it is necessary to take energy from gas $E_{max}-E_i=\Phi_i-\Phi_0$, that is, the changing of electron gas energy is also equal to Φ_0.

At T > 0 if we remove the electron from level

$$-\Phi^{/} = \mu - T\left(\frac{\partial \mu}{\partial T}\right)_{N_e,V}, \tag{3.55}$$

the heat equilibrium realized in the rest electron gas will correspond to the same temperature T. To realize the equilibrium it is necessary to give some (positive or negative) heat to the solid. At removing the electron from any another level the equilibrium temperature of the solids changes to the value $(\Phi^{/}-\Phi_i)/C_v$, where Φ_i is the work for electron removing from the given level and C_v is the metal heat capacity. In this case the removing of electron from chemical potential level $(\Phi_i=-\mu)$ increases the temperature on $\frac{T}{C_v}\left(\frac{\partial \mu}{\partial T}\right)_{N_e,V}$. But the formation of equilibrium state doesn't require external heat.

The work function of the metal can be written as

$$\Phi = E_{vac} - \mu = E_{vac} - E_F \tag{3.56}$$

where E_{vac} is the vacuum level.

One of the first explanations of the physical nature of the work function considers it as work against interaction forces between the electrons and induced by it charge on the surface of metal, that is, against polarization forces [8]. It is a so-called polarization part of the work function. After removing one electron from the metal the rest of the electrons turn out in newer conditions than before removing. It causes a change of state in the rest of the electrons and an increase in the energy on the value of polarization part of the work function.

Another part of the work function is connected with relaxation phenomena [8]. In removing the electron from the ideal metal the exceeded positive charge that has appeared in the metal at each moment in time is distributed on the metal surface according to image forces theory. On removing the electron from the ideal dielectric crystal the charge is localized in that part of the surface from which each electron has gone out. If the emitter is the body with small electrical conductivity, then after removing the electron the localized charge will be neutralized by some currents in the body, and the final condition is the condition of the surface charged body. The difference in energy states will not be determined only by the surface structure as in ideal metal because the currents generate some Joule heat in the metal. This heat has an influence on the oscillations of the lattice. The full change of body energy is the sum of changing the electrons system energy and lattice energy.

It is obvious that the change in lattice energy caused by the relaxation phenomena has the same sign as in removing the electron from the body and at the introduction of the electron from outside, that is, it is an irreversible part of the work function. At transition to the metal with finite electrical conductivity the relaxation effects don't disappear, though they move significantly faster. Therefore, even for such metal the relaxation irreversible part of the work function is unequal to zero though its role is significantly lower than in bodies with low electrical conductivity.

It is simple to understand the nature of this irreversible part of work function from the point of view of the image forces. Charge q located before the conductive plane causes according to electrostatic laws the distributed surface charge $\sigma_0(x,y)$ in such value that the sum of the fields of outside charge q and surface charges is equal to zero inside the metal. If the charge is moving before the conductive plane those surface charges have to change in time, as a result of which there have to be currents inside the metal that cause the σ change, that is, there has to exist an electric field $F(x,y,z,t)$ that creates these currents. Therefore the instantaneous distribution of surface charges $\sigma(x,y,t)$ has to be different from electrostatic one $\sigma_0(x,y,t)$ corresponding to instantaneous location of charge q. The force of interaction of the charge q with charges σ on the surface will not be determined by relation $-q^2/(4\pi\varepsilon_0 x^2)$ that corresponds to interaction with charges of $\sigma_0(x,y,t)$. It is possible to show that at a small distinction of σ and σ_0, that is, the last is equivalent to interaction of charge q equal to the charge of opposite sign $-q$, located not in the place of image, corresponding to instantaneous location of charge q, but in the place corresponding to the earlier position of charge q (Figure 3.8). The moving of electrical image delays on real charge moving. The value of this delay depends both on charge q moving velocity and on electrical conductivity of the metal that determines the velocity of relaxation processes in it. The work $q\varphi$ against interaction forces with a delay in the electrical image differs in calculation based on static image forces.

To understand the separate parts of the potential barrier and work function it is necessary to take into account the electron wave nature and density. The environment in which electrons propagate as plane waves has so far been assumed to be uniform background potential and a step function of height $U_0 = \mu + \phi$ at the metal/vacuum interface. The origin of the step function is the change of electron density from metallic value (= 10^{22} electrons/cm^{-3}) to a

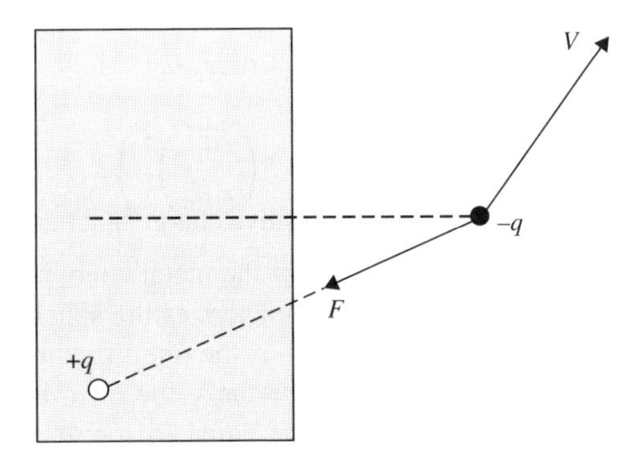

Figure 3.8 Electrical image of moving charge

negligible fraction of that in the vacuum, plus a dipole introduced by electron wave function penetration into the barrier, plus the effects of the ionic core of the metal atom lattice constituting the crystal in which the electrons propagate. Mathematically, these three terms are captured respectively in the equation for the barrier height [11, 15]

$$U_0 = -\frac{\partial}{\partial \rho}[\rho \varepsilon_{xc}(\rho)] + \Delta\phi + \varepsilon_{ion} = U_{xc} + \Delta\phi + \varepsilon_{ion}, \tag{3.57}$$

where ρ is the electron density, ε_{xc} is the exchange-correlation energy per particle [16] according to many-body effects: because of the interaction of the electrons, there is a reduction in probability of electrons being near each other in addition to the Pauli exclusion principle, and both account for the many-body interaction [17]. According to Fermi-Dirac distribution electron (number) density ρ is [1]

$$\rho = 4\frac{M_c}{\sqrt{\pi}}\left(\frac{mk_BT}{2\pi\hbar^2}\right)^{3/2}\int_0^\infty \frac{\sqrt{y}dy}{1+\exp\left(y-\frac{\mu}{k_BT}\right)} = N_c\frac{2}{\sqrt{\pi}}F_{1/2}\left(\frac{\mu}{k_BT}\right), \tag{3.58}$$

where M_c is a number of equivalent minima in the conduction band (e.g., 1 for metals, 6 for silicon, etc.), $F_{1/2}$ is Fermi-Dirac integral, and m is the effective mass of the electron (equal to the rest mass m_0 for metals).

3.4.1 Energy of Ionic Cores (ε_{ion})

The replacement of the ionic lattice constituting the crystal which forms the metal by the uniform positive background ("jellium" model) is predicated on the fact that the electron concentration does not build up in the vicinity of the ionic core: the conducting electrons are not bound to the cores by the same barrier which confines the core electrons, and so conducting electrons spend only a small portion of their time there [14]. Nevertheless, the stability of simple metals in comparison to free atoms is because of the decrease in the ground state energy due to ionic cores, thereby increasing the binding energy of the electrons [18]. The satisfactory treatment involves the use of pseudo-potentials [19] to model the contribution of the ion lattice [13]. The estimation of the ionic core term, based on Wigner-Seitz model, proceeds by demanding that the derivative of the electron wave function vanishes at the boundary for spherically symmetric potential which vanishes for $r \geq r_s a_0$ [15], for which [11]

$$\varepsilon_{ion} = \frac{6}{5}\frac{\alpha_{fs}\hbar c}{r_s a_0}\left(1-5\left(\frac{a_i}{r_s a_0}\right)^2\right), \tag{3.59}$$

where a_i is of the same order as the ionic radius of the metal atom, but is nevertheless different, r_s is the dimensionless parameter which characterizes the sphere per one electron $r_s a_0$, a_0 is the Bohr radius 0.529 Å, and $\alpha_{fs} = 1/137.036$ is the fine structure constant. Ignorance of ε_{ion} has been shifted to ignorance of a_i, but once the value of a_i has been found by other means (e.g., forcing U_0 in Equation (3.57) to be equal to $\mu + \Phi_{exp}$, where "exp" denotes "experimental"), investigations may proceed for different temperatures and field configurations. Even so, the conceptual motivation for introducing Equation (3.59), namely, the conduction band characteristics are dictated by *s–p* electron behavior and not complicated by *d* electrons behavior, is adequate only for simple bulk metal and not true in general. Side-stepping of these complications are warranted only because the present interest is to develop generalizations equation, which nominally account for many-body effects.

3.4.2 Exchange-Correlation Potential (U_{xc})

The exchange-correlation potential requires great attention. The total energy of electron gas E divided by the number of electrons N is denoted as ε and is composed of three parts: a kinetic term ε_{ke}, an exchange term ε_{ex} (which accounts for the same spin electrons' tendency not to occupy the same space), and a correlation term ε_{corr} (which is simply the difference between the first two terms and the "true" energy). For a uniform electron gas, the kinetic energy, and density terms are [11]

$$\rho = \frac{g}{(2\pi)^3}\int f_{FD}(E)d^3k = \frac{g}{(2\pi)^3}\int_0^{k_F} 4\pi k^2 dk$$

$$\varepsilon_{ke} = \frac{g}{(2\pi)^3}\int E(k)f_{FD}(E)d^3k = \frac{g}{(2\pi)^3}\int_0^{k_F} \frac{\hbar^2k^2}{2m}4\pi k^2 dk, \tag{3.60}$$

where $g=2$ is the number of spins and $d^3k=4\pi k^2dk$. In Equation (3.60), $f_{FD}(E)$ is the Fermi-Dirac distribution in the zero-temperature limit, $f_{FD}(E)|_{T=0K}=\theta(\mu - E)=\theta(k_F - k)$, where θ is Heaviside step function. Thus

$$\varepsilon_{ke} = \frac{3}{5}\mu\rho = \frac{3}{5}\left(\frac{9\pi}{4}\right)^{2/3}\frac{1}{r_s^2}Ry, \tag{3.61}$$

where $Ry \equiv \alpha\hbar c/(2a_0) = 13.6045$ eV (Rydberg energy) and $a_0 = 0.529$ (Bohr radius).

Evaluation of the potential energy requires greater detail. The following expressions for ε_{ex} and ε_{corr} have been obtained [11, 15]:

$$\varepsilon_{ex} = \frac{3}{2\pi}\left(\frac{9\pi}{2g}\right)^{1/3}\frac{1}{r_s}Ry \tag{3.62}$$

$$\varepsilon_{corr} = -\left(\frac{130.78}{r_s^6 + 4388.4} + \frac{0.87553}{r_s + 3.0016\sqrt{r_s} + 5.6518}\right)Ry, \tag{3.63}$$

where the first term in parentheses is designed to achieve the minimum value of $\varepsilon_{corr}(r_s)$ as found in Ref. [20].

3.4.3 Dipole Term ($\Delta\phi$)

The other term involved in the creation of barrier height is dipole. The existence of a double electric layer at metal/vacuum interface is explained by the modern theory of solid states. The density of electron wave inside the crystal is a periodical function of coordinates with period equal to the lattice constant. At this for metal crystal the heavy center of electron cloud in each cell coincides with the atom nuclear and therefore the dipole moment of these cells is equal to zero. The situation is different for the cells on the metal interface because the electron cloud is located asymmetrically in relation to atom nuclei. It is wider on the outside to the value of the order of the difference of atom radii in gas and in crystal lattice. Therefore each surface cell of the crystal has the dipole momentum p, and their sum creates the double electrical layer with momentum $pS = \Delta\phi$, where S is the surface density of atoms. As dipole momentum of separate cell p and their surface density depend on surface structure of crystal, the dipole momentum of the different faces of the same crystal can be different.

The dipole term $\Delta\phi$ is obtained from Poisson's equation once the electron density as a function of position is known. We have considered in Chapter 1 the penetration of an electron wave into a barrier. In terms of normalized wave function $\psi_k(x)$, the quantum mechanical density $\rho(x)$ is given by

$$\rho(x) = \frac{1}{2\pi}\int_0^\infty f(k)|\psi_k(x)|^2 dk, \tag{3.64}$$

where $f(k)$ is the supply function (see Chapter 2).

For $T=0$ and high and/or wide barrier, an analytical expression has been obtained in Ref. [18] which is identical to the triangular potential case for high barriers and/or low fields:

$$\lim_{\beta,L,U_0\to\infty} \rho(x) = \frac{gk_F^3}{6\pi^2}\left[1 + 3\frac{\cos(\xi)}{\xi^2} - 3\frac{\sin(\xi)}{\xi^3}\right], \tag{3.65}$$

where $\beta \equiv 1/k_B T$, L is the barrier width, the coefficient is the bulk density, $\xi \equiv 2k_F(x - x_0)$, and $x_0 = 1/k_0$. $\xi(x)$, therefore, depends on both work function (due to k_F) and barrier height (due to x_0). The trigonometric functions in Equation (3.65) are responsible for the "Friedel oscillations" in the electron density near the interface, visible in Figure 3.9 [11] for varying chemical potential and work function. Changes in the chemical potential have greater impact on the profile due to the wavelength associated with the most energetic electrons. Equation (3.65) holds as the limiting case for both triangular and rectangular potentials. This is significant because of the following: (i) for metals, the thermal Fermi-Dirac distribution is close to the $T=0$ K limit, and so $\rho(T,x) = \rho(0,x)$; (ii) to leading order, the density profile is dependent on barrier height, but not on shape, as in Figure 3.9b; and (iii) to leading order, changes in the potential barrier height result in a translation of density profile by $x_0 = \hbar/(2mU_0)^{1/2}$,where $U_0 = \mu + \Phi$; that is the curves in Figure 3.9b would overlap if plotted versus $(x + x_0)$ instead of x.

Spilling out of electrons into a vacuum leads to a dipole-induced potential difference $\Delta\phi$ at the interface. A crude approximation of $\Delta\phi$, suggested by the variation approach [21], is obtained using a hyperbolic tangent approximation to the density [11]

$$\rho_{aprx}(x) = \frac{1}{2}\left(\frac{gk_F^3}{6\pi^2}\right)\{1 - \tanh[\lambda k_F(x - x_i)]\}. \tag{3.66}$$

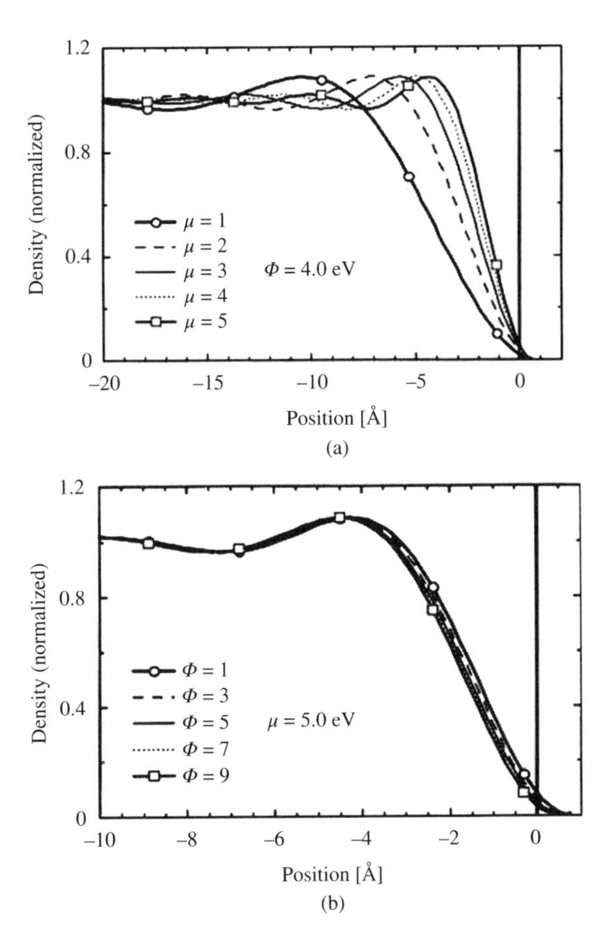

Figure 3.9 (a) Electron density as the function of position for various chemical potentials for a work function of 4.0 eV and a rectangular barrier profile. Undulations in the density are associated with Friedel oscillations. (b) Same as part (a) but for various work functions with chemical potential of 5.0 eV [11]. Copyright (2001) by John Wiley & Sons, Inc. This material is reproduced with permission of John Wiley & Sons, Inc.

x_i is chosen such that $\xi_0 \equiv \xi(x = x_i)$ and $\rho(x_i)$ is half of its bulk value (artificial) symmetry of Equation (3.66) about the ion origin. λ is obtained by setting $\partial_x \rho = \partial_x \rho_{aprx}$ at $x = x_i$.

The crude hyperbolic approximation of Equation (3.66) allows for the analytical integration of Poisson's equation to give

$$\Delta\phi = \frac{\alpha\pi\hbar c g k_F}{36\lambda^2}; \quad x_i = -\frac{5}{4k_F} + x_0. \tag{3.67}$$

If Equation (3.65) for the density is used, a better approximation is obtained [15]: the insistence that an explicit integration over the ion and electron densities give zero net charge (plus a judicious neglect of smaller order terms) gives the following approximations

$$\Delta\phi = \frac{a_0}{\pi x_0}\left(\left(4 - \pi k_F x_0\right) k_F x_0 - (1 - k_F^2 x_0^2)\ln\left[\frac{1 + k_F x_0}{1 - k_F x_0}\right]\right) Ry;$$

$$x_i = \frac{84 x_0}{168 - 36 k_F^2 x_0^2 - 5 k_F^4 x_0^4}\ln\left(\frac{17}{25}k_F^2 x_0^2\right). \tag{3.68}$$

The evaluation of the barrier height U_0 must be iteratively obtained due to the dependence of x_0 on U_0 (and x_i on x_0), and must proceed in conjunction with determining the value a_i in Equation (3.59), as ε_{ion} is sensitive to a_i's value.

The electron penetration into the barrier will be enhanced if the barrier height decreases or is thinner (i.e., the applied field is increased), or if a greater portion of electrons can be induced to higher energy (i.e., the temperature is raised), both affecting the apparent barrier height. Therefore, the "work function" in the F–N equation will be temperature and field dependent.

For convenience the work function is often expressed as a sum of two contributions. The first contribution is related to the bulk properties of the crystal (terms 1 and 3 in Equation (3.57)) and is determined by the electrostatic potential in the interior volume (in this case the Fermi level is taken as zero). The second term (term 2 in Equation (3.57)) is sensitive to surface properties, which is expressed by the difference of electrostatic potential energy:

$$\Delta U = U(+\infty) - U(-\infty) = 4\pi\varepsilon_0 \int_{-\infty}^{+\infty} q|n(z) - n_+(z)|dz, \tag{3.69}$$

where $n(z)$ is the density distribution of the electron charge and $n_+(z)$ is the density of the positive charge background.

This term is the electrostatic dipole barrier, since it corresponds to the work that has to be done to move an electron through the surface dipole layer.

The magnitude of the dipole moment is a characteristic of the surface and it varies from one surface to another. For the same metal tightly packed (atomically smooth) faces usually have a larger dipole moment compared to the "loose" (atomically rough) edges.

3.4.4 Work Function of Semiconductor

All the above presented analysis is also suitable for semiconductor. But in the case of semiconductor we have to take into account the band bending and general view of work function which it is possible represent as

$$\Phi = E_{vac} - \mu = E_{vac} - E_F = \chi + qV_s + (E_C - E_F), \tag{3.70}$$

where χ is the electron affinity (difference between the vacuum level and the bottom of the conduction band), qV_s is the semiconductor band bending, and $(E_C - E_F)$ is the energy difference between the bottom of the conduction band in the volume and Fermi level. The dipoles on the surface influence on qV_s, but the third member is determined by the type of semiconductor and the level of its doping with impurities (Figure 3.10).

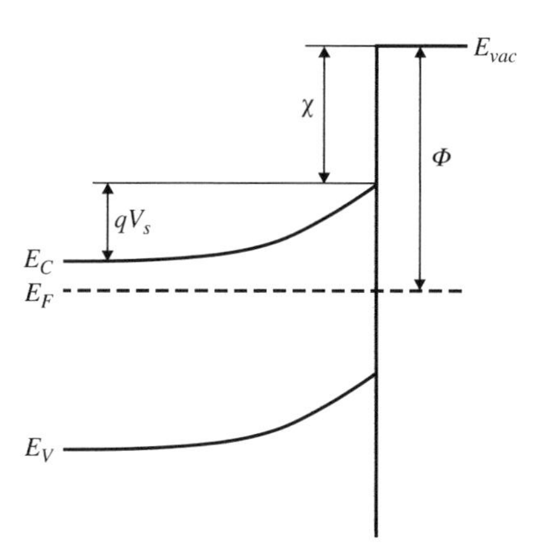

Figure 3.10 Schematic energy band diagram of semiconductor surface: χ is the electron affinity, qV_s is the band bending, E_C is the bottom of conduction band, E_V is the top of valence band, E_F is the Fermi level, and E_{vac} is the vacuum level

3.4.5 Work Function of Cathode with Coating

It is important to know the depth of the surface layer of the emitter, which determines the value of the work function of the surface [8]. If one piece of material with work function $\Phi_A = q\varphi_A$ (Figure 3.11a,c) covers a sufficiently thick shell of another substance with the work function $\Phi_B = q\varphi_B$ (Figure 3.11b), the electric double layer of the contact potential difference appears on the interface between two substances. This double layer is localized in the border areas with the depth of the order of the Debye radius L_{DA} and L_{DB} for the A and B substances. The potential drop is equal to $\Delta V_c = \varphi_A - \varphi_B$. As a result of the potential jump the average potential energy of electrons in the body A in relation to the energy of the electron that rests outside the body rises (if $\varphi_A > \varphi_B$) on the amount $(\Phi_A - \Phi_B)$. As a consequence, all full electrons energies in the body A will raise by the same amount in the relation to the energy of the electron rested outside the system $(A + B)$. In particular, the energy of the conduction band bottom in the body A increases. The level of the electrochemical potential of μ_A increases to align it with the level of μ_B (Figure 3.11d). Electron work function of the system $(A + B)$ will be equal to $\Phi_B = -\mu_B$.

In decreasing the shell thickness h the situation will continue until such is the case for the small h, at which the potential drop at the interface of A and B is equal to $\Delta V_c = \varphi_A - \varphi_B$, that is, to thicknesses greater than the localized electric double layer in the material B. It is the order of L_{DB}. Experience has shown that the coating of metal with the material thickness in the two atomic layers leads to $\Phi = \Phi_B$ [22]. This is in agreement with the value of Debye radius for metals, where $n_0 \sim 10^{22}\ \text{cm}^{-3}$ (Equation (3.12)).

The work function of the substance depends on the atomic structure of the surface. Naturally, the work function of the body, coated with two or three layers of metal, will be equal to the work function of the surface of metal with the same atomic structure as that of the layer. In addition, it is obvious that all these provisions are valid for these metals and on such substrates when the atomic solid coating can be formed. In the case of semiconductor coatings, obviously, the layer thickness is greater than for metals [23]. The layer of material at $h \ll L_D$, of course, can change the work function of the substrate, but does not make it equal to the work function of the continuous coating film.

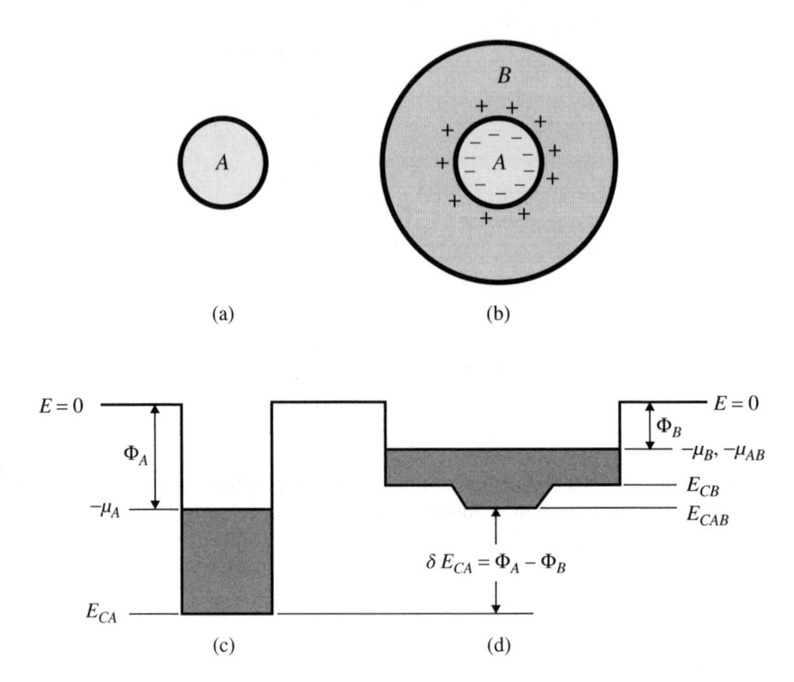

Figure 3.11 The work function changing at coating of the emitter: (a) and (b) uncoated and coated with film emitters; (c) and (d) energy band diagrams of uncoated and film coated emitters respectively

It is necessary to note here that it is not clear whether it is always possible to speak about semiconductor properties of thin, about the monoatomic layer of matter, solid samples which are semiconductors, in the case when this layer is located on the metal substrate. All electrons of emitter form a single system. The wave functions of the metal do not break off on its borders, but penetrate into the coating layer, the amplitude of the waves with energies that correspond to the forbidden band E_g of the semiconductor coating material decreases from the interface to the outer surface of the layer. Therefore, for the sufficiently thick layer of semiconductor, this amplitude becomes almost zero at some depth less than the thickness of the layer. There will be only the electrons with energies outside the range of E_g in the outer part of the coating layer. For the electrons with energies within E_g interval there will be the forbidden energy band. In the case of very thin layers the metal wave functions are not vanishing small in the whole thickness of the coating. The band gap energy will be absent in full thickness of the coating and the coating will not be semiconductor, at least in the usual sense of the term.

Thus, the work function of the emitter is not determined by its composition and structural properties but, first of all, by the nature of its surface layer. Note that the system state of the electrons within the emitter (the wave functions describing the electrons states, the density of the electron gas, the density of states, Fermi level position relative to the bottom of the conduction band, etc.) will be determined by the bulk properties of the emitter, and they will be different in different substances in the presence at the surface of the same coating and even $h > L_D$. The surface layer defines different position in different bodies of the electrons energy levels with respect to the energy level $E = 0$ of an electron at rest outside the emitter and shifts the level system so that μ_A is equal to $-\Phi_B$.

3.5 Field and Temperature Dependences of Barrier Height

The image charge potential requires modification to accommodate many-body effects [11]. The primary effect of barrier height is to shift the origin of the electron density ($\xi(x)$) in Equation (3.65) by an amount x_0. The image charge potential will depend on the origin of the electron density, so the term Q/x in Equation (3.51) should be replaced by $Q/(x + x_0)$, a conclusion which can be justified through more rigorous theory [24–26]. Here, $x = 0$ is identified as the electron coordinate system origin. The potential will resemble $U(x) = \mu + \Phi(T, F) - qFx - Q/(x + x_0)$.

Because x_0 depends, to the first order, on barrier height, but not on shape (as for the sufficiently abrupt potentials of thermal and field emission), approximating the image charge potential by the triangular barrier potential of the same height is adequate to obtain the electron density. From that density, Poisson's equation may be numerically solved to obtain the dipole and exchange-correlation potential for arbitrary field (once a_i is determined from the $F = 0$ case), and the process iterated until the predicted barrier height is convergent with the assumed barrier height [15]. Due to the relation of the dipole term to the penetration of electrons into the barrier, the apparent height of the barrier will increase with temperature and field.

There is no a priori reason to expect the ratio of the increase in barrier height $\Phi(F) - \Phi_0$ (where $\Phi_0 = \Phi(F = 0)$) and the product Fx_0 to be linear: in fact, $\Phi(F) - \Phi_0 = (qFx_0)^{1-\delta}$. In practice, δ is numerically found to be small so that pursuing its evaluation is unnecessarily fastidious (e.g., for molybdenum and cesium parameters with $Fx_0 \leq 0.11$, $\delta_{Mo} = 0.15$, and $\delta_{Cs} = 0.12$, respectively). The simple approximation $\delta = 0$ is, therefore, adopted, and the effective potential is

$$U(x) = \mu + \Phi_0(T) - qF(x - x_0) - Q/(x + x_0), \tag{3.71}$$

where $\mu + \Phi_0$ is the asymptotic barrier height in the absence of the applied field. In the same way, the simplest approximation for the temperature dependence of $\Phi(T)$ is to mimic empirical fits and let $\Phi_0(T) = \Phi_0(0) + \alpha T$, where the constant α is not the fine structure constant [27]. In the range of temperatures typical of thermionic cathodes (T = 1000–1400 K), α is approximately constant and of the order of $k_B = 8.617 \times 10^{-5}$ eV/K.

Finally, consider the effect of shifting the origin of the background positive charge by x_i. From classical electrostatics, a sheet of charge generates a field of magnitude $\sigma/2\varepsilon_0$, where σ is the surface charge density [28]. In the present case, there is global charge neutrality, to the extent that this field acts as finite. In an ad hoc manner, the field from a sheet of charge $\sigma = \rho x_i$ acting over the length scale x_i gives rise to a potential difference $\rho x^2{}_i/2\varepsilon_0 = 8Qk^3{}_F x^2{}_i/(3\pi)$. Incorporating this effect into the potential of Equation (3.71) gives rise to the analytical image charge potential for approximating the effects of dipole and exchange-correlation potentials in the classical image charge model:

$$U_{analytic}(x > 0) = \mu(T) + \Phi_0(T) + \frac{8}{3\pi} Q k_F^3 x_i^2 - qF(x - x_0) - \frac{Q}{(x + x_0)}. \tag{3.72}$$

For $x \leq 0$, a reasonable approximation is to assume that $U(x)$ for $x < 0$ decays in a manner compatible with Equation (3.66) (recall that electron density and $U(x)$ are related via Poisson's equation), but that obscures the ripples in $U(x)$ (much reduced compared to the rectangular barrier) due to the Friedel oscillations in the density. Equation (3.72) allows for a transparent assessment of various effects on the F–N equation. By shifting to a coordinate system $y = x + x_0$, the classical image charge potential is recovered as long as an "effective" work function is defined [29]

$$\Phi_{eff}(T) = \Phi_0(T) + \frac{8}{3\pi} Q k_F^3 x_i^2 + 2qFx_0. \tag{3.73}$$

Clearly, Φ_{eff} is to be identified with the experimental work function, and Φ_0 is found such that $\Phi = \Phi_{exp}$ at $F = 0$ eV/Å. Use of Equations (3.72) and (3.73) mimics a self-consistently evaluated exchange-correlation potential (simultaneous iterative solution of Poisson's and Schrodinger's equations) and the dipole term is rather well for both high (Mo) and low (Cs) values of the work function.

In contrast to Equation (3.73), the application of the F–N equation in the literature assumes that Φ is a constant value. Let's consider the effects that arise by neglecting the other factors in Equation (3.73). If the barrier height is not allowed to vary, the changes due to F and T will be attributed to the field enhancement factor $\beta_g = F/V$, where V is the potential of the anode or extraction grid/gate. The slope of the F–N equation is then proportional to $\Phi^{3/2}/\beta_g$. An error $\delta\Phi$ in work function estimates will generate an error of $\delta\beta_g = (3\beta_g\delta\Phi)/(2\Phi)$ in field enhancement factor estimates (and vice versa). Consider, as an example, Mo values such that when the modifications indicated in Equation (3.73) are neglected, then $\delta\Phi/\Phi = 0.15$ for typical fields, giving $\delta\beta_g/\beta_g = 0.23$ [11].

A fair portion of $\delta\Phi$ is traceable to the image charge term. The sanctity of the classical image charge approximation is so widely embraced that suggesting Q/x needs modification requires support. In effect, the electron density profile at high fields is such that the barrier is not being lowered by the $\sqrt{4qQF}$ factor predicted by the classical image charge potential, but by a lesser factor. That finding is, in fact, compatible with the analytical image charge potential, which likewise predicts that the barrier will not lower to the extent predicted by classical image charge theory, when comparison of the numerical solution with the analytical and classical image charge models has been performed [28].

In terms of current density, the slope and intercept of $J(F)$ on a F–N plot (i.e., $ln(J(F)/F^2)$ vs. $1/F$) is given in Figure 3.12 for fields between 0.15 and 1.0 eV/Å. At high fields, the analytical image charge model deviates toward the triangular barrier (no image) approximation. At lower fields, $J(F)$ approaches the classical image charge case, indicating that the barrier more closely matches that encountered in thermionic emission, where an applied field does reduce the apparent barrier height by the amount $\sqrt{4qQF}$.

In Ref. [30] it has been argued that the tunneling barrier is dynamically related to the behavior of the electron density near the interface, due to the nature of the exchange-correlation and dipole potentials, assuming a static equilibrium-based potential, such as the classical image charge approximation, is provisional.

3.6 Influence of Surface Adatoms on Work Function

Adsorption of foreign atoms on the surface of the body, particularly the metal alters the electron clouds from the surface and causes changes in the work function. At adsorption of atoms or molecules the dipole moment is changed and hence the work function of the sample is changed. The additional dipole part D has to be taken into account. The adsorbed atom (adatom) can be found on the surface not only in the neutral state, but in the state of partial ionization [31].

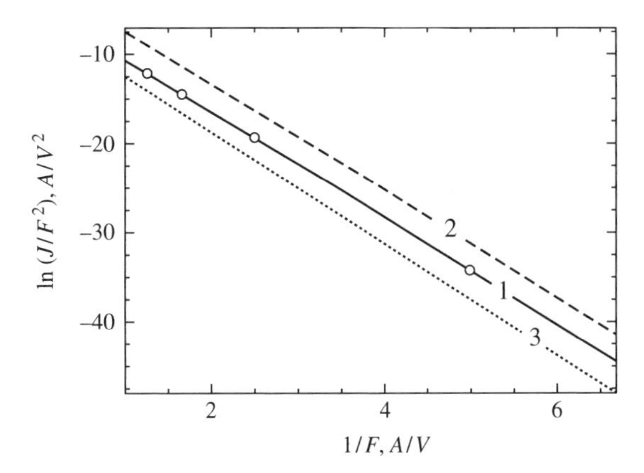

Figure 3.12 Comparison of the current density as the function of applied field in F–N coordinates (*ln(J(F)/F*^{2}*)* vs. *1/F*) for the triangular barrier (3), classical image charge (2), and analytical image charge (1) potential barriers (Molybdenum: $\Phi_{exp}=4.41$ eV, $\mu(300\ \text{K})=5.87$ eV) [11]. Copyright (2001) by John Wiley & Sons, Inc. This material is reproduced with permission of John Wiley & Sons, Inc.

Let us consider the charge of the atom adsorbed on the metal surface [8, 32–34]. At the analysis of redistribution of the electron density between the atom adsorbed on the ideal homogeneous surface and electron levels of the substrate an approximation of the local density of states (LDOSs), as well as one-dimensional band model have been used. For definiteness, we consider the adsorption of atoms or molecules on the surface of simple metal approximated by the "jelly" model. In the approximation of LDOS only the ion-electron interactions and the average energy of the electrostatic interaction between the electrons are taken into account. Exchange-correlation contribution to these interactions is determined from the data for the volume of the crystal. Before the interaction the atom (or molecule) has been characterized by the ionization energy I and electron affinity, χ, and metal–jelly with the thermodynamic work function of the Φ are independent quantum-mechanical systems (Figure 3.13a). On approaching the metal surface the ionization and affinity levels of the atom are widened (Figure 3.13b). After their interaction and formation of the adsorption complex of the metal and the adsorbed particles they are of a single system, in which the adsorbed atom (molecule) corresponds to the resonant surface electron state (Figure 3.13c). Due to the electron tunneling the initial narrow discrete levels of valence electron of free particle are broadened. If the $I<\Phi$ due to transition of a valence electron in the metal the positive ion is formed, if $\chi>\Phi$ the negative ion is formed.

The population of resonance levels depends on their position relative to the Fermi level of the metal. If they lie above (below) E_F, then the complete electron transition from adatom (metal) to metal (molecule) is realized and vice versa, so the ionic bond is formed. If the resonance level is close to E_F, then the connection will be covalent, electrons of the molecule are collectivized. The level of the valence electron rises on the height of ΔE, and level of an additional captured (excited) electron drops to the value of $\Delta\chi$ (Figure 3.13c). In the first case, the maximum density of states above E_F and the adsorbed particle have excess charge $+\delta$, in the second the excess charge is $-\delta$. Such estimates can allow only knowing the direction of charge transfer and qualitatively estimating the fraction of transferring charge. For example, [35], for the alkali metals Li, Na, K, Cs, and so on. ($I<\Phi$) on the metal–jelly surface the positive ions

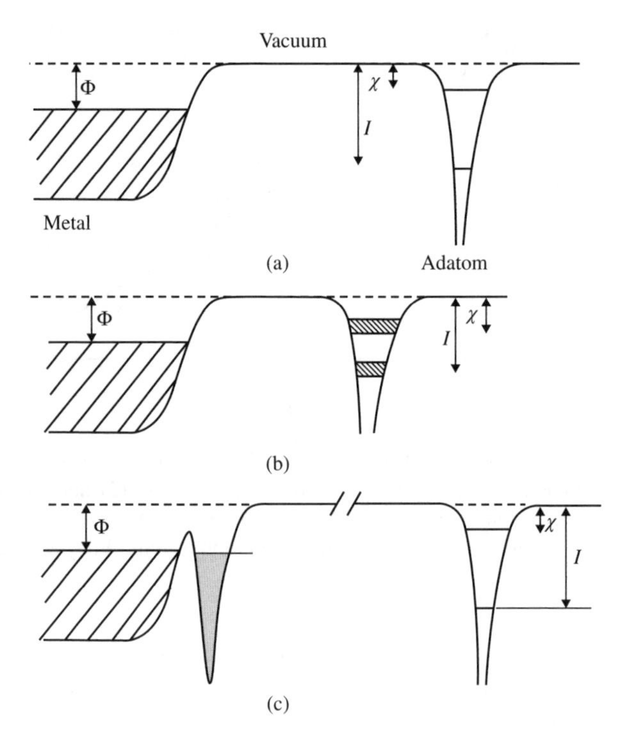

Figure 3.13 Energy diagram of the complex metal-adsorbed atom (molecule): (a) before interaction, (b) weak interaction, and (c) strong interaction. Φ is the work function, I is the ionization potential, χ is the electron affinity

are formed (Figure 3.14) [35]. The case of negative ions formation is characteristic for the halogens (Figure 3.14). The resonance levels $3sLi$ and $3pCl$ lie above and below E_F in the metal. At this the typical ionic bonds are formed. In the case of adsorption of Si the covalent bond is formed as a result of interaction of the electron of *s* and *p* orbitals with the jelly.

Adatom and metal form a single system, and the electrons of the adatom and the metal do not belong to the metal or adatom, and to the whole system. Their electron clouds are distributed both in the volume of the metal and in the volume of the adatom. In this state, the electrons corresponding to the deep energy levels (*K*-, *L*-shells as the adatom and the metal atoms) do not differ from states in the isolated atoms, that is, corresponding electrons are almost localized in either the adatom or the metal atoms. But the states corresponding to high values of energy, for example, almost continuous spectrum of energies E_n of the conduction band, have the electron clouds whose density $\rho(x, y, z)$ differs from zero as inside the metal and in the region of the adatom. The integral $v_n = \int \rho_n(x, y, z)d\tau$, taken over the volume of the adsorbed atom is proportional to the probability of finding of the electron with energy E_n in the range of the adatom, and qv_n gives the fraction of the electron charge that neutralizes the positive charge of the atomic core of the adatom (Figure 3.15). This share for the E_n depends on the type of ρ_n for metal-adatom system, that is, on the nature of the metal and the adatom. For each such system v_n has the greatest value for some electron state corresponding to the energy E_m and decreases for larger and smaller E_n. In other words, the curve of probability of finding the electron within

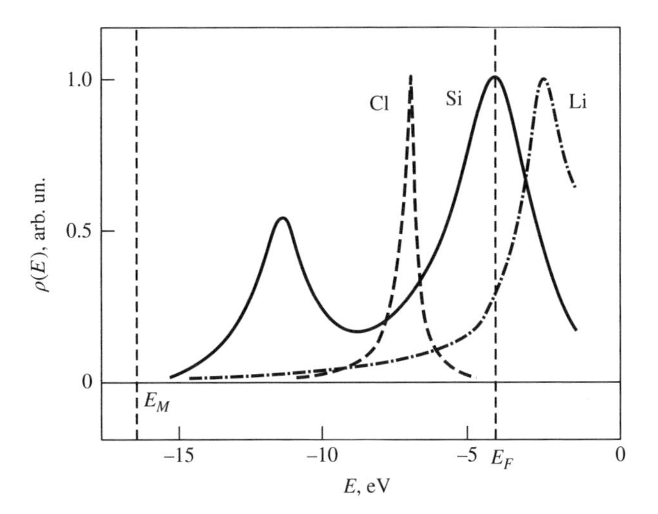

Figure 3.14 Changing of the density of states $\rho(E)$ as a result of chemisorption of the atoms, Cl, Si, and Li on the metal. E is the energy with respect to vacuum, E_M is the band edge in the metal, E_F is the Fermi energy

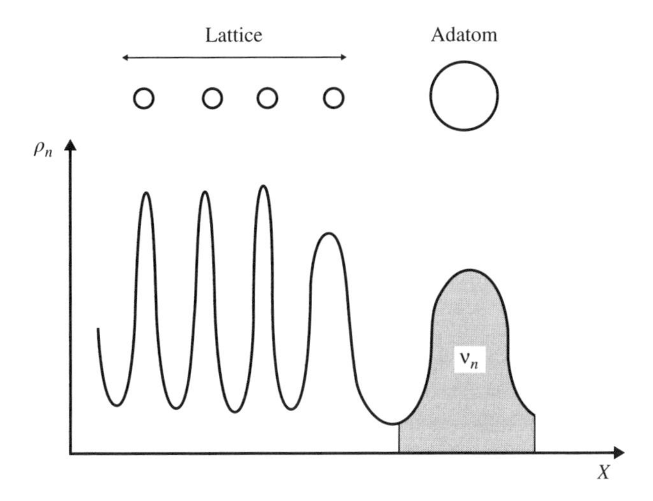

Figure 3.15 Charge exchange during the interaction of adatom with crystal lattice

the adatom on the energy E_n has a maximum at $E_n = E_m$, whose position is determined by the nature of the adatom and the metal (Figure 3.16).

The full negative charge that neutralizes the charge of atomic core is equal to the sum of the above integrals, that is,

$$q\sum_n \int \rho_n d\tau, \tag{3.74}$$

where the summation is over all n, corresponding to states in which there are electrons in this system, that is, over all states occupied by electrons at given conditions.

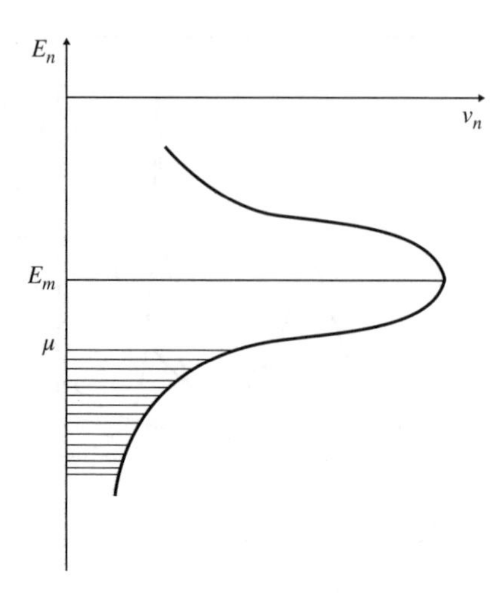

Figure 3.16 The probability of the electron location in region of adatom in dependence on energy

Such a sum of integrals can be both smaller and higher than electron charge $-q$. In the first case, the adatom is positively charged, but in the second it is negatively charged and its charge can be any portion of the electron charge.

The level of E_m coincides with the work of ionization of the adatom and the filling of electron levels to the E_m corresponds to value of $q\sum_n \int \rho_n d\tau$ equal to q. Therefore, if the level of the electrochemical potential of the metal μ lies above the E_m (i.e., in first approximation, the atomic ionization potential V_i is higher than metal work function, Φ), the adatom is negatively charged. In the case of $V_i < \Phi$ the charge of the atom is positive (Figure 3.16). In the first case negative ions together with their electrical images create the double charge layer and increase the metal work function. In the second case the metal work function is decreased.

If the energy level of electron in the adatom, qV_i, lies below the bottom of the conduction band of the solid body, then in the metal–the adatom system it will be, on the one hand, the electron states corresponding to the conduction band of metal, for which $\rho_n(x,y,z)$ is different from zero only in a solid (i.e., giving $v = 0$), and, on the other hand, a single state, corresponding to $E_a = -qV_i$. The density of electron cloud ρ_a of this state is different from zero only in region of adatom, that is, the corresponding integral $v_a = 1$, and the adatom will be neutral. However, if the adatoms have electron affinity equal to χ, so that the level $-\chi$ is lying against the levels of the conduction band, the values of v_s corresponding to the states of electrons in this region again are different from zero. In this case, the sum of $q\Sigma v_s$, taken over the states of the conduction band occupied by electrons, gives the excess of negative charge in the excess of the value $qv_a = q$, required to neutralize the positive charge of the atomic core, and the adatom will be negatively charged.

Indeed, the atoms such as oxygen and halogens have appreciable electron affinity, with adsorption on metals increasing their work function.

As an example, Figure 3.17 shows the changes in the work function due to adsorption of chlorine and cesium on the surface of Cu(111). Since electronegative chlorine accepts electrons of the metal, the total dipole moment of the surface increases, leading to an increase in the work

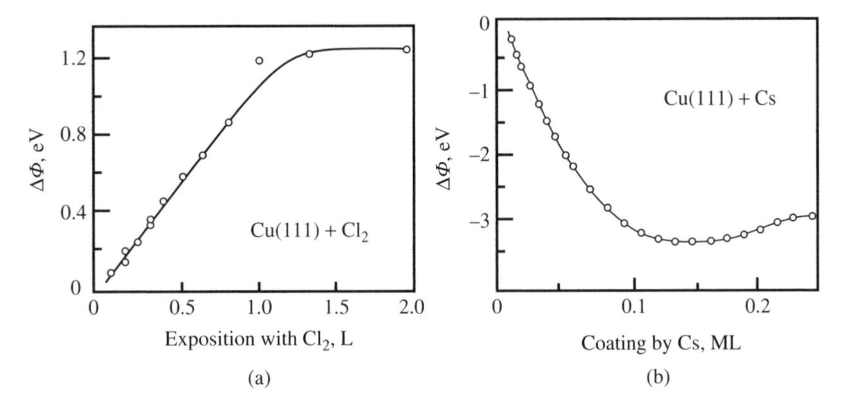

Figure 3.17 Changing in work function due to adsorption: (a) chlorine and (b) Cs on the surface of Cu (111). The work function of clean surface of Cu (111) is 4.88 eV. Reprinted from Ref. [36]. Copyright (1977), with permission from Elsevier. Reprinted with permission from Ref. [37]. Copyright (1980) by the American Physical Society

function (Figure 3.17). In contrast, electropositive Cs gives the electrons in the metal, where they remain in close proximity to the surface and screen the adsorbed ions. The dipole moment induced by the adsorbate is directed in the opposite direction relative to the dipole moment of the clean surface of Cu(111), resulting in the decrease in work function (Figure 3.17b). The form of dependence for Cs is typical for the adsorption of alkali metals on the metal surfaces. After its fast (almost linear) decrease of the initial work function it should be minimum and then the slight increase at the saturation of the monolayer. This behavior can be explained simply as follows. Initially, each adsorbed ion gives its personal contribution to the work function. Therefore, from the slope of the dependence of $\Delta\Phi$ on the coverage we can estimate the dipole moment of the single atom. As the dipoles begin to cover the surface they begin the "feel" of the field of neighboring dipoles. The result of the interaction becomes depolarization of density-packed dipoles.

The jelly model does not apply to the transition metals, in which the strictly oriented *d(f)* orbitals have to be considered. At adsorption there is a strong shift of the wave functions of adsorbed molecules and metal and the characteristic peaks of the density of states $\rho(E)$ become blurred. In these cases the cluster numerical methods for calculating the electronic structure of surface complexes are used.

At the study of photoemission spectra in some wide bandgap materials there was observed even negative electronic affinity. Negative electron affinity χ first observed in the crystal plane of (111) in diamond [38]. The surface was covered with the hydrogen (H-terminated). Later negative χ was watched when covered with hydrogen (100) faces of diamond [39–42], as well as other wide-gap materials [43–45]. The adsorption of hydrogen on the surface of the material contributes to the negative electron affinity. The adsorbed on the surface atoms form electric dipoles [46]. Change in potential due to dipole layer changes as the work function and electron affinity on the same amount. Taking into account the impact of dual-charged dipole layer (D) the work function can be written as

$$\Phi = \Phi_0 + D. \tag{3.75}$$

Namely, the influence of dipole layer explains the dependence of the work function on the orientation of the surface. During the adsorption of more electronegative atoms (such as oxygen for diamond and diamond-like carbon (DLC) films) the increase in work function and electron affinity is observed (*D* is positive). If there is the adsorption of electropositive atoms (hydrogen, barium, cesium) the work function and electron affinity are reduced (*D* is negative).

As a rule, the negative χ is observed for wide bandgap materials. The adsorption of electropositive atoms can lead to negative value of the effective χ. In this case the initial value of χ (energy difference between the bottom of the conduction band and vacuum level) of the material is small and the potential dipole layer influence is enough to bend a band of the semiconductor surface and to realize the negative effective χ. In the case of a semiconductor with average value of E_g the value of χ is larger and band bending due to the dipole layer potential is not enough for the negative effective electronic affinity. Similar behavior of the band diagram can be obtained by charging the surface states of semiconductor with the positive charge. By changing the nature of adsorbed atoms it is possible to change χ and work function. For example, in Ref. [46] based on measurements of photoemission it was shown that the change of adsorbed atoms on the surface of diamond changed the χ in the range from -1.27 to 0.38 eV.

To reduce the electronic affinity and work function in Ref [47] two effects, namely the influence of the dipole layer and the influence of charge on the surface states have been used. The DLC films (a-C: H) has been deposited in plasma with high content of hydrogen. Therefore, it is naturally to assume the presence of C-H bonds on their surface. Hydrogen is electropositive relative to carbon and therefore the double charge (dipole) C^{-}-H^{+} surface layer is formed. The dipoles are directed from the surface and therefore in the expression (3.75) the value of *D* is negative, that is, the effective value of χ and Φ are reduced. Let's estimate the influence of the formed negative dipole layer on the surface on the value of Φ and χ. For this particular approach we use semi-empirical formula for elementary metals [48]

$$D = 0.445\ (2.21\ /\ r_e^2 - 0.458\ /\ r_e) \tag{3.76}$$

where, r_e is the distance between charges in angstroms. In our case, r_e is the length of C-H bonds, $r_e = 1.07$ Å [49]. The calculated value of the contribution of surface dipoles according to Equation (3.76) is $D = 0.67$ eV. Furthermore, as the results of photoemission research show the illumination of DLC films with electrons and vacuum UV leads to emission of electrons and positive charging of the surface. The positive charge on the surface states of semiconductor bends the energy bands down in the same direction as negative (directed from the surface) dipoles. The joint effect of the dipole layer and the positive charge on the surface states leads to significant decrease in work function and electron affinity of DLC films.

For emission of photoexcited electron from valence band into vacuum it has to obtain minimum energy $\Phi_0 = \chi_0 + E_g$, called the optical work function or photoemission threshold. Optical work function is independent of the surface potential and near-surface band bending. The minimum photon energy for photoemission is determined by $h\nu_{min} = \Phi_0$. Unlike the optical work function the thermodynamic work function, which is defined as the energy required to release an electron from the Fermi level in the vacuum depends on the band bending at the surface [38, 50]. The sign and the degree of band bending depend on the characteristics of surface states and the external electric field. If you bend bands down (positive potential on the surface) the work function decreases, and vice versa, at bending the bands upward it increases.

References

1. S.M. Sze, *Physics of Semiconductor Devices* (John Wiley & Sons, Inc. New York 1981).
2. R. Stratton, Energy distributions of field emitted electrons, *Physical Review* **135**, A794 (1964).
3. G.N. Fursey, Early field emission studies of semiconductors, *Applied Surface Science* **94**, 44 (1996).
4. V. Litovchenko, A. Evtukh, Yu. Kryuchenko, *et al.*, Quantum-size resonance tunneling in the field emission phenomenon, *Journal of Applied Physics* **96**, 867 (2004).
5. C.G.B. Garrett and W.H. Brattain, Physical theory of semiconductor surfaces, *Physical Review* **99**, 376 (1955).
6. R.H. Kingston and S.F. Neustadter, Calculation of the space charge, electric field, and free carrier concentration at the surface of a semiconductor, *Journal of Applied Physics* **26**, 718 (1955).
7. A.F. Kravchenko and V.N. Ovsyuk. *Electron Processes in Solid-State Low-Dimensional System* (Novosibirsk, Siberian University Press, 2000) (in Russian).
8. L.N. Dobretsov and M.V. Gomoyunova. *Emission Electronics* (Nayka, Moskow, 1966) (in Russian).
9. H.B. Michaelson. Relation between an atomic electronegativity scale and the work function, *IBM Journal of Research and Development* **22**, 72 (1978).
10. M.K. Harbola and V. Sahni, Quantum-mechanical origin of the asymptotic effective potential at metal surfaces, *Physical Review B* **39**, 10437 (1989).
11. K.L. Jensen, Theory of field emission, in *Vacuum Microelectronics*, ed. W. Zhu (New York, John Wiley & Sons, Inc., 2001), pp. 33–104.
12. J. Bardeen, The image and Van der Waals forces at a metallic surface, *Physical Review* **58**, 727 (1940).
13. (a) N.D. Lang and W. Kohn, Theory of metal surfaces: charge density and surface energy, *Physical Review B* **1**, 4555 (1970);Lang, N.D. and Kohn, W. Theory of metal surfaces: Induced surface charge and image potential, *Physical Review B* **7**, 3541 (1973).
14. N.D. Lang, The density-functional formalism and the electronic structure of metal surfaces, in *Solid State Physics*, Vol. 28, H. Ehrenreich, F. Seitz, and D. Turnbull (eds) (Academic Press: New York, p. 225, 1973).
15. K.L. Jensen, Exchange-correlation, dipole, and image charge potentials for electron sources: Temperature and field variation of the barrier height, *Journal of Applied Physics* **85**, 2667 (1999).
16. R.P. Feynman, *Statistical Mechanics: A Set of Lectures* (W. A. Benjamin Inc.: Reading, MA, 1972).
17. J. Bardeen, Theory of the work function. II. The surface double layer, *Physical Review* **49**, 653 (1936).
18. C. Kittel, *Introduction to Solid State Physics*, 7th edn (John Wiley & Sons, Inc.: New York, 1996).
19. A. Kiejna, Surface properties of simple metals in a structureless pseudopotential model, *Physical Review B* **47**, 7361 (1993).
20. A. Ishihara and E.W. Montroll, A note on the ground state energy of an assembly of interacting electrons, *Proceedings of the National Academy of Sciences of the United States of America* **68**, 3111 (1971).
21. W. Jones and N.H. March, *Theoretical Solid State Physics*, Vol. II: Non-equilibrium and Disorder (Dover: New York, Section 10.11.1, 1985).
22. N.D Morgulis, Theory of thermoionic monometer, *Journal of Technical Physics* **1**, 51 (1931).
23. K.B. Tolpygo, About thermionic emission from thin semiconductor films, *Journal of Technical Physics* **19**, 1301 (1949).
24. J.A. Appelbaum and D.R. Hamann, Variational calculation of the image potential near a metal surface, *Physical Review B* **6**, 1122 (1972).
25. A.G. Eguiluz and W. Hanke, Evaluation of the exchange-correlation potential at a metal surface from many-body perturbation theory, *Physical Review B* **14**, 10433 (1989).
26. L.G. Il'chenko and Y.V. Kryuchenko, External field penetration effect on current-field characteristics of metal emitters, *Journal of Vacuum Science and Technology B* **13**, 566 (1995).
27. G.A. Haas and R.E. Thomas, Thermionic emission and work function, in *Techniques of Metals Research*, Part 1, Vol. VI, R.F. Bunshah (ed.) (John Wiley & Sons, Inc.: New York, 1972), pp. 91–262.
28. L. Eyges, *The Classical Electromagnetic Field* (Dover: New York, 1972).
29. K.L. Jensen, Exchange-correlation, dipole, and image charge potentials for electron sources: Temperature and field variation of the barrier height, *Journal of Applied Physics* **85**, 4455 (2000).
30. M.G. Ancona, Density-gradient analysis of field emission from metals, *Physical Review B* **46**, 4874 (1992).
31. R.W. Gurney, Theory of electrical double layers in adsorbed films, *Physical Review* **47**, 479 (1935).
32. V.K. Kisilyov, S.N. Kozlov, A.V. Zoteyev *Fundamentals of Solid State Surface Physics*, Nauka (Moskow, 2000) (in Russian).

33. M. Prutton. *Introduction in Surface Physics* (Clarendon Press, Oxford, 1994).
34. K. Oura, V.G. Lifshits, A.A. Saranin, *et al.*, *Surface Science*, Springer, (2006).
35. E. Zenguil, *Surface Physics* (Mir, Moskow, 1990) (in Russian).
36. P.J. Goddard and R.M. Lambert, Adsorption-desorption properties and surface structural chemistry of chlorine on Cu(111) and Ag(111), *Surface Science* **67**, 180 (1977).
37. S.A. Lindgren and L. Wallden, Electronic structure of clean and oxygen-exposed Na and Cs monolayers on Cu (111), *Physical Review B* **22**, 5967 (1980).
38. F.J. Himpsel, J.A. Knapp, van Vechten, J.A., and D.E. Eastman, Quantum photoyield of diamond(111) – A stable negative-affinity emitter, *Physical Review B* **20**, 624, (1979).
39. N.S. Xu, Y. Tzeng, and R.V. Latham, Similarities in the "cold" electron emission characteristics of diamond coated molybdenum electrodes and polished bulk graphite surfaces, *Journal of Physics D* **26**, 1776 (1993).
40. J. Weide, Z. Zhang, P.K. Baumann, *et al.*, Negative-electron-affinity effects on the diamond (100) surface, *Physical Review B* **50**, 5803 (1994).
41. P.K. Baumann and R.J. Nemanich, Negative electron affinity effects on H plasma exposed diamond (100) surfaces, *Diamond and Related Materials* **4**, 802 (1995).
42. R.E. Thomas, T.P. Humphreys, C. Pettenkofer, *et al.*, Influence of surface terminating species on electron emission from diamond surfaces, *Proceeding of Materials Research Society* **416**, 263 (1996).
43. R.J. Nemanich, P.K. Baumann, M.C. Benjamin, *et al.*, Negative electron affinity surfaces of aluminum nitride and diamond, *Diamond and Related Materials* **5**, 790 (1996).
44. M.J. Powers, M.C. Benjamin, L.M. Poster, *et al.*, Observation of a negative electron affinity for boron nitride, *Applied Physics Letters* **67**, 3912 (1995).
45. M.C. Benjamin, C. Wang, R.F. Davis, and R.J. Nemanich, Observation of a negative electron affinity for heteroepitaxial AlN on α(6H)-SiC(0001), *Applied Physics Letters* **64**, 3288 (1994).
46. J.B. Cui, J. Ristein, and L. Ley, Low-threshold electron emission from diamond, *Physical Review B* **60**, 16135 (2000).
47. V.G. Litovchenko, A.A. Evtukh, Yu.M. Litvin, and M.I. Fedorchenko, The model of photo- and field electron emission from thin DLC films, *Materials Science and Engineering A* **353**, 47 (2003).
48. M. Cardona and L. Ley, *Photoemission in Solids I. General Principles*, Springer-Verlag (Berlin, Heidelberg, New York, 1978).
49. J. Robertson, Diamond-like amorphous carbon, *Materials Science and Engineering R* **37**, 129 (2002).
50. R.G. Forbes, Low-macroscopic-field electron emission from carbon films and other electrically nanostructured heterogeneous materials: hypotheses about emission mechanism, *Solid State Electronics* **45**, 779 (2001).

4

Current through the Barrier Structures

This chapter is devoted to the consideration of current transport through the barrier nanostructures at electron field emission (EFE). It combines three previous chapters in obtaining one of the most important EFE parameters, namely current. The currents through one barrier, at EFE from semiconductors are considered. Special attention is paid to current transport through double barrier resonant tunneling structure. Coherent, sequential, and single electron tunneling is analyzed thoroughly. The theoretical consideration of novel current transport mechanisms at EFE, namely resonant tunneling, two-step electron tunneling in nanoparticle and single-EFE through nanoparticles are presented.

4.1 Current through One Barrier Structure

Consider the simple potential barrier with metal or heavy doped semiconductor emitter and collector. Positive bias *V*, which has been applied to the right contact of the barrier, reduces the amount of energy $-qV$ (Figure 4.1). In the case of equilibrium the distribution of electrons is given by the Fermi function and is determined by the Fermi level (i.e., chemical potential). If the bias is applied to the system then we have nonequilibrium distribution. But if the deviation from the equilibrium distribution is not large, we can apply the concept of Fermi quasi-level. Each unit of the contact is now characterized by its Fermi quasi-level (hereafter we will call them simply the Fermi level). The potential bias determines their difference. That is, $E_{FL} - E_{FR} \equiv \mu_L - \mu_R = qV$.

Consider now the currents in the case of one-dimensional motion of electrons (carriers). We will be computing currents as streams of charged particles that leave the left contact (emitter) and come to the right contact (collector) of the system and in the opposite direction. Thus expressions will differ only by different values of Fermi level.

In the more complicated case of quantization of the energy spectrum of electrons in emitter (for example, in surface semiconductor layer (see Chapter 3)) to calculate the current we note

Vacuum Nanoelectronic Devices: Novel Electron Sources and Applications, First Edition.
Anatoliy Evtukh, Hans Hartnagel, Oktay Yilmazoglu, Hidenori Mimura and Dimitris Pavlidis.

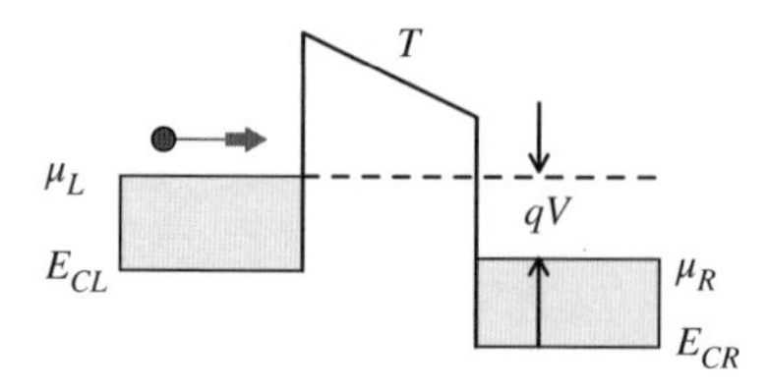

Figure 4.1 Electron tunneling through one barrier

that the states belong to different transverse modes or subbands [1]. Each mode has a dispersion relation *E(N,k)* with a cut-off energy below which it cannot propagate:

$$E_n = E(N, k). \tag{4.1}$$

The number of transverse modes at energy *E* is obtained by counting the number of modes having cut-off energies smaller than *E:*

$$M(E) = \sum_N \vartheta(E - E_n), \tag{4.2}$$

where $\vartheta(E-E_n)$ is the step function ($\vartheta(E-E_n)=1$, if $E > E_n$ and $\vartheta(E-E_n)=0$, if $E < E_n$).

We can evaluate the current carried by each transverse mode separately and add them up.

Consider a single transverse mode whose $+k$ states are occupied according to some function $f_1(E)$. We use the well-known expression for current density

$$J = qvn, \tag{4.3}$$

where n is the density of uniform electron gas per unit length and v is the velocity of carries.

Since the electron density associated with single k state in a emitter layer of length L is *(1/L)* we can write the current I_L carried by the $+k$ states as

$$I_L = \frac{q}{L}\sum_k vf_1(E(k))T(k), \tag{4.4}$$

where *T(k)* is the transmission coefficient describing the probability of electron going through the barrier.

Assuming periodic boundary conditions and converting the sum over k into the integral according to the usual prescription [1]

$$\sum_k \to \int 2\frac{L}{2\pi}dk \tag{4.5}$$

(where 2 is used for taking into account the spin) we obtain

$$I_L = 2q\int_0^{\infty} \frac{1}{2\pi} f_1(E(k))v(k)T(k)dk. \tag{4.6}$$

Let's transit from integration over the momentum to the integration over energy.

$$dk = \frac{dk}{dE}dE = \left(\frac{1}{\frac{dE}{dk}}\right)dE = \left(\frac{1}{\frac{d}{dk}\left(\frac{\hbar^2 k^2}{2m}\right)}\right)dE = \frac{1}{\hbar v}dE \tag{4.7}$$

Here we used $k = \sqrt{2mE/\hbar^2} = \frac{mv}{\hbar}$.

Substituting this expression (4.7) in formula (4.6) we obtain

$$J_L = \frac{2q}{h}\int_{E_{CL}}^{\infty} f_1(E, \mu_L)T(E)dE. \tag{4.8}$$

Similarly we can write the current that creates the flow of electrons emitted from the right side and moving to the left side of the system. This should only take into account the fact that movement occurs in the opposite direction and the right side is characterized by its value of Fermi quasi-level E_{FR} and bottom of the conduction band E_{CR}

$$J_R = -\frac{2q}{h}\int_{E_{CR}}^{\infty} f_2(E, \mu_R)T(E)dE. \tag{4.9}$$

Note that the transmission coefficient from the left side to the right is equal to the coefficient of transition from the right side to the left. This fact was used in writing the formula (4.7). Now we can write the complete current as the sum of two currents Equations (4.8) and (4.9) while we can define the lower limit of integration value E_{CL}. As can be seen from Figure 4.1 the electrons with energies that lie in the range from E_{CR} to E_{CL} do not give a contribution to the current because these energy states are occupied and lie deep in the band and they cannot participate in transport processes. So

$$J = J_R + J_L = \frac{2q}{h}\int_{E_{CL}}^{\infty} [f_1(E, \mu_L) - f_2(E, \mu_R)]T(E)dE. \tag{4.10}$$

Quantitative calculations of the *I-V* characteristics can be performed using the general expression for the current [1]:

$$I = \frac{2q}{h}\int \overline{T}(E)[f_1(E) - f_2(E)]dE, \tag{4.11}$$

where $f_1(E)$ and $f_2(E)$ are the Fermi functions in the two contacts and $\overline{T}(E)$ is the transmission function obtained by summing the transmission probability $T_{nm}(E)$ over all input and output modes:

$$\overline{T}(E) = \sum_m \sum_n T_{nm}(E). \tag{4.12}$$

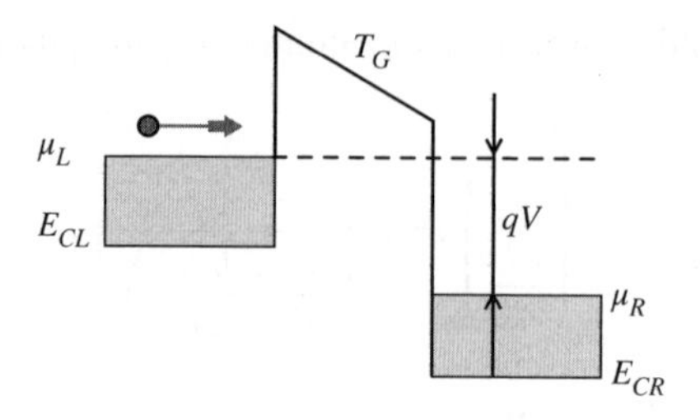

Figure 4.2 Electron tunneling through barrier at high bias

The transmission function is usually calculated from the Schrodinger equation, neglecting all scattering processes. The first part of the current-voltage curve is described fairly well by this approach.

Now let's consider some limited cases that allow simplifying the analysis [2].

4.1.1 *Case 1: High Bias*

In this case the energy band diagram has the view as in Figure 4.2.

That is, all available states of the right side of the system lie below the states of left side and, thus, do not give contribution to transport. In this case the distribution function $f_2(E,\mu_R)$ can be removed in expression (4.10). The current will be completely described by the formula (4.8).

4.1.2 *Case 2: High Bias and Low Temperature*

In the case of low temperatures when electrons are highly degenerated, the distribution function can be approximated by step function. That is, only electrons with energy interval $\mu_L - \mu_R$ give contribution to the current.

$$J = \frac{2q}{h}\int_{\mu_R}^{\mu_L} T(E)dE. \tag{4.13}$$

In the case of large values of bias, the electrons from the right side of the system will not give contribution to the current, and the lower limit of integration should change on E_{CL}. So instead of Equation (4.13) we obtain

$$J = \frac{2q}{h}\int_{E_{CL}}^{\mu_L} T(E)dE. \tag{4.14}$$

4.1.3 *Case 3: Small Bias: Linear Response*

If the bias is very small, we can decompose the distribution function on the small parameter qV. That is, you can put that and $\mu_L = E_F + qV/2$ and $\mu_R = E_F - qV/2$, where E_F is the Fermi level in equilibrium state. Then

$$f(E,\mu_L) - f(E,\mu_R) = qV\frac{\partial f(E,E_F)}{\partial E_F}. \tag{4.15}$$

Taking into account that the energy and chemical potential is included in the distribution function only as a difference $E - E_F$, the right part of the formula (4.15) can be rewritten in another form

$$f(E, \mu_L) - f(E, \mu_R) = qV\frac{\partial f(E, E_F)}{\partial E}. \tag{4.16}$$

Then we obtain in this case that current is proportional to bias, that is, corresponds to Ohm law.

$$J = \frac{2q^2V}{h}\int_{\mu_L}^{\infty}\left(-\frac{\partial f}{\partial E}\right)T(E)dE. \tag{4.17}$$

Then we obtain the (nonzero temperature) linear response formula for conductivity.

$$G = \frac{2q^2}{h}\int_{\mu_L}^{\infty}T(E)\left(-\frac{\partial f}{\partial E}\right)dE. \tag{4.18}$$

4.1.4 *Case 4: Small Bias and Low Temperature*

At low temperatures the Fermi distribution function is almost step-function. And its derivatives with high accuracy can be considered as a delta function

$$-\frac{\partial f}{\partial E} = \delta(E - E_F). \tag{4.19}$$

Then the integral Equation (4.17) is taken easy and we have a simple result for the current

$$J = \frac{2q^2V}{h}T(E_F) \tag{4.20}$$

and

$$G = \frac{2q^2}{h}T(E_F). \tag{4.21}$$

This expresses the fact that the conductivity is provided by the states that lie near the Fermi level in the narrow band of $\sim k_BT$.

4.2 Field Emission Current

The theory of EFE was described in many works [3–18] and it was thoroughly considered and analyzed in detail in Ref. [19]. Here we consider its main features.

The Fowler–Nordheim (F–N) emission theory [3] is the background of description of the EFE. It was developed for the metal, one-dimensional emission and zero temperature.

The Fowler–Nordheim considered the triangular potential barrier, namely $U(x) = U_0 - qFx$ for $x \geq 0$ and 0 otherwise. The obtained relation is

$$J(F) = \frac{q}{4\pi^2\hbar}\frac{\sqrt{\mu}}{[\mu + \Phi]\sqrt{\Phi}}F^2\exp\left(-\frac{4\sqrt{2m}}{3\hbar}\frac{\Phi^{3/2}}{qF}\right), \tag{4.22}$$

where J is the current density, F is the electric field, μ is the chemical potential, Φ is the barrier height, m is the electron mass, q is the elementary charge, and $\hbar$ is the reduced Planck constant. The parameter Φ represents the height of the barrier above μ, that is, $\Phi = U_0 - \mu$.

The relation (4.22) was obtained basing on some simplifying assumptions, namely: (1) it was developed for metal; metal has band structure with free electrons; (2) electrons are in thermodynamical equilibrium and correspond to Fermi-Dirac statistics; (3) zero temperature; (4) plane flat emitting surface; (5) local work function is homogeneous at all emitting surface and does not depend on external electric field; (6) homogeneous electric field above emitting surface; (7) interaction between emitting electron and surface represented by classical image forces; and (8) transmission coefficient is estimated with using Wentzel-Kramers-Brillouin (WKB) approximation.

Equation (4.22) was later modified by Murphy-Good [4].

$$J(0) = \frac{q^3 F^2}{8\pi h \Phi t^2(y)} \times \exp\left[-\frac{4\sqrt{2m}\Phi^{3/2}}{3\hbar q F} v(y)\right], \tag{4.23}$$

where $v(y)$ and $t(y)$ are the elliptic integrals of $y = \Delta\Phi/(\Phi - E)$.

In some cases it is possible in good enough approximation to represent $v(y)$ and $t(y)$ as

$$t^2(y) = 1.1 \text{ and } v(y) = 0.95 - y^2. \tag{4.24}$$

Most of the tunneling theories are predicated on the WKB approximation using a classical image charge potential.

Excluding one or more assumptions from F–N theory it is possible to obtain an equation for description of EFE of various views and with different degrees of precision [17, 18, 20, 21]. All of them can be represented in the form of the generalized F–N equation

$$I = \lambda A a \Phi^{-1} F^2 \times \exp(-sb\Phi^{3/2}/F), \tag{4.25}$$

where λ and s are the generalizing correction factors.

There were many attempts for further development of F–N theory. They took into account the influence of (1) features of the energy band structure and surface electron waves [10]; (2) effects caused by curvature of the surfaces and atomic sharp surfaces [15, 22]; and (3) alternative methods for estimation of barrier penetration (instead of WKB approximation) [13], and so on. In work [23] the dependence of work function on electric field has been supposed

$$\Phi(F) = \Phi_0(1 + \eta F), \tag{4.26}$$

where Φ_0 is the work function at zero electric field, η is the constant with dimension of cm/V.

In the general case the total current density J through the barrier is obtained by summing up the currents of all the individual electron wave functions with momentum $0 \le k \le k_F$. At finite temperatures, μ is no longer the energy of the most energetic electron, but rather the chemical potential defined such that the integral over the Fermi distribution function $f_{FD}(E)$ over all momentum space at any temperature reproduces the electron density [19, 24]. The electrons' energy is characterized by a Fermi distribution function $f_{FD}(E) : f_{FD}(E < \mu) = 1$ at $T = 0$ K, and is zero otherwise. The maximum energy and momentum at $T = 0$ are μ and k_F such that $\mu = E(k_F)$, where μ is the Fermi energy (at $T = 0$) and k_F is the Fermi momentum.

The three-dimensional (3-D) nature of $\bar{k}$-space has so far been ignored. Considering it $E(\bar{k}) = \sum_i (\hbar k_i)^2/2m$, where the sum is over the momentum components along the x, y, and z axes. The current density through the barrier is given by the integral of the product of the electron charge, velocity, distribution function $f(k_x)$, and tunneling probability $T(k_x)$, or

$$J(T,d) = \frac{q}{2\pi}\int_0^{\infty} \frac{\hbar k_x}{m} f(k_x)T(k_x)dk_x, \tag{4.27}$$

where the electron velocity into the barrier is given by $\hbar k_x/m$, d is the barrier thickness. If we let $E_x \equiv (\hbar k_x)^2/2m$, then it is clear that the integrand of Equation (4.27) is proportional to the energy distribution in E_x of the transmitted electrons. The integral of Equation (4.27) is often represented as $(q/2\pi\hbar)\int f(E_x)T(E_x)dE_x$, a form particularly convenient for deriving the F–N equation.

The distribution function $f(k_x) \equiv f(k)$ is obtained from the Fermi distribution by

$$f(k) = \frac{mk_BT}{\pi\hbar^2}\ln\left[1+\exp\left(\frac{\mu - E(k)}{k_BT}\right)\right]. \tag{4.28}$$

At $T = 0$ K, Equation (4.28) reduces to $f(k) = (k^2{}_F - k^2)/(2\pi)$ for $k \le k_F$.

Beyond the famous F–N equation (rigorously only valid at zero temperature) several approximate formulas have been proposed to take into account the F–N current changes with temperature [25–28]. Under the assumption that the electrons in the emitting electrode can be described by three-dimensional Fermi gas, and if the image force effect is neglected, the F–N current density J_{FN} is expressed as [29]:

$$J_{FN}(T) = \frac{qm_EkT}{2\pi^2\hbar^3}\int_0^{\Phi}\ln\left[1+\exp\left(\frac{E_F(T)-E}{k_BT}\right)\right]\times\exp\left(-\frac{4\sqrt{2m_0}(\Phi-E)^{3/2}}{3\hbar qF}\right)dE, \tag{4.29}$$

where q is the absolute electron charge, m_E and m_0 are the effective electron mass into the emitter and into the vacuum, respectively, k_BT is the thermal energy, $\hbar$ is the reduced Planck constant, F is the electric field across the vacuum, and E_F is the Fermi level.

The image force will modify the triangular potential barrier. Consequently, the barrier height will be lowered by $\Delta\Phi$ [4]:

$$\Delta\Phi = \left(\frac{q^3F}{4\pi\varepsilon_0}\right)^{1/2}, \tag{4.30}$$

where ε_0 is the vacuum permeability. The integral Equation (4.29) becomes [29]

$$J_{F-N}(T) = \frac{qm_EkT}{2\pi^2\hbar^3}\int_0^{\Phi-\Delta\Phi}\ln\left[1+\exp\left(\frac{E_F(T)-E}{kT}\right)\right]\times\exp\left(-\frac{4\sqrt{2m_0}(\Phi-E)^{3/2}}{3\hbar qF}\right)v(y)dE. \tag{4.31}$$

The last term $v(y)$ in Equation (4.31) can be evaluated in terms of complete elliptic integrals of $y = \Delta\Phi/(\Phi - E)$ [4].

Beginning from Equation (4.27) the general expression for thermal-field emission from the metal has been obtained in Ref. [10].

$$J(T,F) = \frac{m}{2\pi^2\beta\hbar^3}\exp\left(-\frac{b_{FN}}{F}\right)\int_0^{\infty}\ln\left(1+e^{\beta(\mu-E)}\right)\exp[-c_{FN}(\mu-E)]d,$$

$$= \frac{m}{2\pi^2\hbar^3 c_{EN}^2}\exp\left(-\frac{b_{FN}}{F}\right)\left[\frac{\pi c_{FN}}{\beta\sin\left(\frac{\pi c_{FN}}{\beta}\right)}\right], \quad (4.32)$$

where $\beta \equiv 1/k_B T$,

$$b_{FN}(F) = \frac{4}{3\hbar}\sqrt{2m\Phi^3},$$

$$c_{FN}(F) = \frac{2}{\hbar F}\sqrt{2m\Phi}.$$

Use has been made of the exact integral (applicable when $exp(-\beta\mu) = 0$, that is, the lower limit of the integral in Equation (4.32) can be neglected):

$$\int_0^{\infty} x^{\mu-1}\ln(1+x)dx = \frac{\pi}{\mu\sin(\mu\pi)}, \quad (4.33)$$

where $0 < Real(\mu) < 1$ [30]. The term in square brackets in Equation (4.32) is close to unity for field emission at room temperature, and so it is often neglected. In this case we have

$$J(F) = a_{FN}F^2\exp\left(-\frac{b_{FN}}{F}\right). \quad (4.34)$$

But if fields become low or temperature becomes high, a positive convexity is introduced on a F–N plot. Equation (4.32) is related to the elliptical integral functions *v(y)* and *t(y)* commonly used through the identification

$$a_{FN}(F) = \frac{1}{16\pi^2\hbar\Phi t(y)^2},$$

$$b_{FN}(F) = \frac{4}{3\hbar}\sqrt{2m\Phi^3}v(y),$$

$$c_{FN}(F) = \frac{2}{\hbar F}\sqrt{2m\Phi}t(y), \quad (4.35)$$

where $y^2 = 4QqF/\Phi^2$, and $Q = q^2/16\pi\varepsilon_0$. The *v(y)* and *t(y)* have been tabulated in [17, 18]. In terms of the *R(s)* function, they are given by

$$v(y) - 3(1-y^2)^{3/4}R\left(\frac{1}{2}\frac{1-\sqrt{1-y^2}}{\sqrt{1-y^2}}\right),$$

$$t(y) = \left(1-\frac{2}{3}y\frac{\partial}{\partial y}\right)v(y). \quad (4.36)$$

All the coefficients in Equation (4.35) are either explicitly or (through the definition of y) implicitly dependent on applied field: linearity of Equation (4.32) on a F–N plot is not a priori guaranteed, though it may be so by linearization $v(y)$ in F and approximating $t(y)$ by a constant [19]. For $T=0$ K, a widely used approximation due to Ref. [31] is $t(y)=\sqrt{1.1}$ and $v(y)=v_0-v_1y^2$, where they chose $v_0=0.95$, and $v_1=1$. A slightly better approximation based on the demand that $v_1\equiv 1$ gives $t(y)=\sqrt{1.1164}$ and $v_0=0.93685$ [32]. Quadratic $v(y)$ approximations are reasonable for intermediate values of y.

The approximation in Equation (4.32) is predicated on $\mu\beta\gg 1$, which is manifestly true for the metal. For semiconductors, the lower limit cannot be casually neglected. As shown in Refs. [7, 32], including the lower limit results in the replacement

$$\frac{\pi c_{FN}/\beta}{\sin\left(\frac{\pi c_{FN}}{\beta}\right)} \rightarrow \frac{\pi c_{FN}/\beta}{\sin\left(\frac{\pi c_{FN}}{\beta}\right)} - (1+c_{FN}\mu)e^{-c_{FN}\mu}. \tag{4.37}$$

4.3 Electron Field Emission from Semiconductors

In comparison to metals the electronic picture of the phenomenon of field emission in semiconductors is much more complicated [33]. Qualitative features of the processes occurring in semiconductors at EFE are caused mainly by the presence of internal electric fields. These fields can arise due to the penetration of an external electric field in the surface layer of the semiconductor, due to surface states, as well as due to the flow of the emission current through the emitter. The electric field penetration into semiconductor can be estimated using the relation $F=\sigma/\varepsilon_0$, where $\sigma=\rho W$ is the surface density of the charge and ρ is the volume density of the charge, W is the distance of electric field penetration. The comparison of the typical metal with the electron concentration of 5.0×10^{22} cm^{-3} and highly doped semiconductor with electron density of 10^{18} cm^{-3} shows that field penetration effects will extend far into the semiconductor in comparison to metals.

On the one hand the field penetration into the semiconductor leads to the uncertainty of the forces acting on the electrons on the surface of the body and, on the other hand, to shift the energy levels of electrons in the surface layer down (at required in the case electronic field emission polarity), that is, so-called band bending (Figure 4.3a). The field penetration depth is

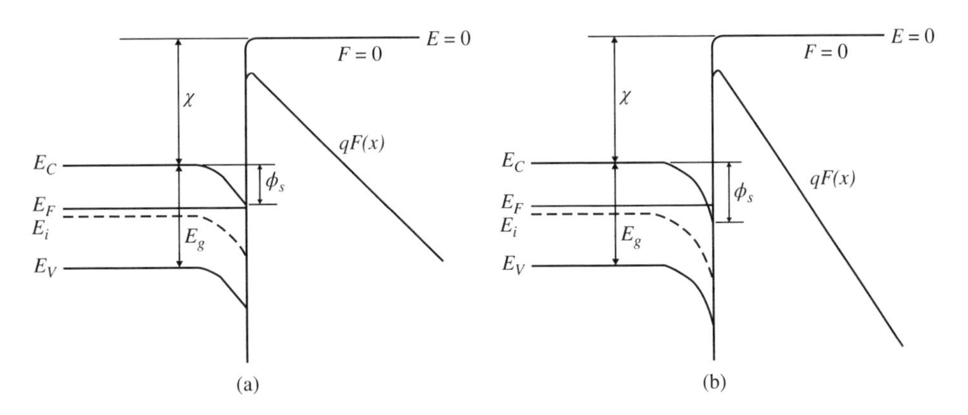

Figure 4.3 Energy band diagrams at electron field emission from semiconductors: (a) nondegenerated electron gas and (b) degenerated electron gas

determined by the Debye radius L_D. As a result of the band bending the energy distribution of electrons in the conduction band as well as donor levels are changed. The electron density n in the surface layer is increased. At high fields the band bending can be large enough and the Fermi level can be above the bottom of the conduction band at the interface of the body with the vacuum (Figure 4.3b). As a result the nondegenerated electron gas in the conduction band of the semiconductor bulk can become degenerated in the surface layer.

The flow of large emission currents through the semiconductor tip, especially in case of its large resistance, is accompanied by the appearance of the voltage drop in the emitter that is mainly concentrated at the apex of the tip. If the internal field caused by these voltage drops is great, it can lead to overheating of the electron gas, that is, to transition from the thermodynamic equilibrium state in the nonequilibrium one, as well as the generation of electrons into the conduction band by impact ionization of valence band electrons or by ionization of electrons of the localized levels (high-field effects). The voltage drop at the tip during the electronic field emission current can also lead to heating of the tip by Joule heat.

The presence of surface states in semiconductors are usually accompanied by the appearance of the internal retarding field (and therefore the additional potential barrier at the surface), which is somewhat reduced at the external field (Figure 4.4). Generally speaking, the electric field due to surface states can also be the opposite sign. The existence of internal barriers should reduce the EFE mainly due to the decrease of the density of the surface electrons.

The complexity of the phenomenon of field electron emission from the semiconductors is also connected with the fact that in contrast to metals, in which electron emission is only from one conduction band, in semiconductors the electrons emission may occur also from the valence band, surface states, and defects within the band gap, and will depend on the type of doping (n- or p-type) used [34]. The presence of defects within the band gap can contribute to the current, especially for low electric fields [35, 36]. Hole current must be accounted for. The effective mass in a semiconductor is not, in general, equal to the electron rest mass in vacuum. The emitted current from wide band gap semiconductors can be strongly influenced or dictated by transport through the Schottky barrier characterizing the back contact, and transport through the wide band gap material has numerous complications of its own [37–41]. Surface

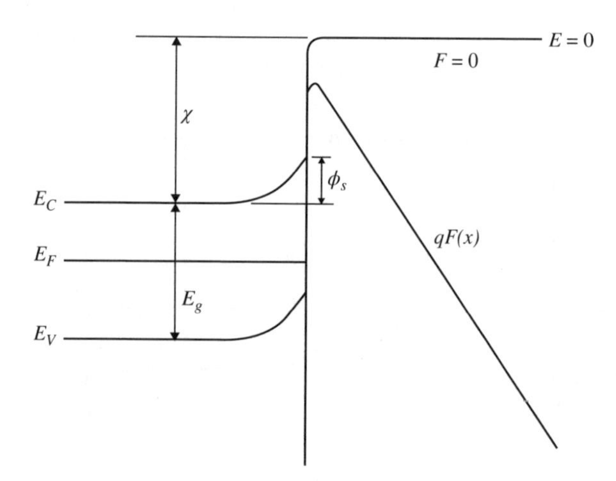

Figure 4.4 Energy band diagrams at electron field emission from semiconductors at existence of the retarding band bending

layers or adatoms [42], and for analogous reasons, layered semiconductor structures which can give rise to resonant tunneling effect [43–45], complicate matters by introducing modifications to either the potential barrier or the supply function. Band bending at the surface is complicated by the fact that the triangular well-like potential has discrete energy states (given approximately by zeros of the Airy function) rather than a continuum of states [46, 47]. Oxides or adsorbents may exist on the surface which may have charged inclusions [48] that preclude a simple Fowler–Nordheim-like relationship, and which may contribute to the large fluctuations observed in field emission from silicon tips [49, 50].

Finally, it is necessary to mention about one more particularity of EFE from semiconductors. In some semiconductor the barrier of electron affinity χ is small, and it can be completely removed by an external field. For example, for electron affinity $\chi = 1$ eV it becomes at achievable in the experiments field of $F = 7 \times 10^6$ V/cm. If this has happened the peculiarities of electron emission from semiconductors will be different.

At present there is no complete physical theory of EFE from semiconductor, taking into account all the marked peculiarities of the phenomenon. Existing theories [10, 19, 51–55] referring to electronic semiconductors, reflect only some aspects of the phenomenon and, therefore, applicable only in the interpretation of experimental data in some special cases.

The complexity of the EFE from semiconductors precludes its full adequate treatment in any analysis.

Let's consider the initial assumptions of existing theories and some conclusions from them.

In Ref. [53] it was assumed that electron interacts with the semiconductor surface in accordance with equation

$$F_s = \alpha \frac{q^2}{16\pi\varepsilon_0 x^2}, \tag{4.38}$$

where $(K_s - 1)/(K_s + 1) < \alpha < 1$, K_s is the dielectric constant of semiconductor.

In principle, this assumption is not valid. However, there is qualitative agreement between the conclusions of the number of theories considering both the image force barrier and triangular barrier. This shows that for some features of EFE from semiconductors the shape of the potential barrier has not the paramount importance. Therefore, in some cases, apparently, it is possible for the qualitative consideration of the phenomenon without applying for the accuracy of quantitative relationships to use the formula (4.38). The transparency of the potential barrier $T(E_x,F)$, as in the case of metals, is determined by the WKB method [56], that is, by

$$T(E_x, F) = \exp\left[-\frac{4\pi(2m)^{1/2}}{h}\int_{x_1}^{x_2}(U(x) - E_x)^{1/2}dx\right]. \tag{4.39}$$

It was assumed that the transparency of the barrier is the same for all electrons of conduction band and is equal to the transparency of electrons located at the bottom of the conduction band [7, 53]. This assumption can be justified only for sufficiently large fields determined by the so-called Stratton's criterion. However, the WKB method, in contrast, can be used in the fields not exceeding certain values. The range of fields, which hold the above assumptions, is limited by the following inequality:

$$2k_0\chi^{1/2}k_BT \ll qF \ll \frac{3}{2}k_0\chi^{3/2}, \tag{4.40}$$

where $k_0 = \dfrac{2\pi}{h}(2m)^{1/2}$.

For example, at $T = 300$ K ($k_BT = 2.585 \times 10^{-2}$ eV) and $\chi = 1$ eV, these inequalities give 3×10^6 V/cm $\ll F \ll 3 \times 10^7$ V/cm.

The electron flux density $v(E_x)$ from the conduction band that fall in the potential barrier of the emitter is determined by several approximations.

1. The electron gas in the conduction band is nondegenerated, the penetration of external field in the semiconductor is neglected, then

$$v(E_x) = n_\infty \left(\frac{1}{2\pi m k_B T} \right)^{1/2} \exp\left(-\frac{\chi - E_x}{k_B T} \right). \quad (4.41)$$

For this approximation the following expression for emission current density can be obtained

$$J = qn_\infty \left(\frac{k_B T}{2\pi m} \right)^{1/2} \exp\left[-\frac{4\sqrt{2m}\chi^{3/2}}{3\hbar qF} \theta(y) \left\{ \left(\frac{K_s - 1}{K_s + 1} \right)^{1/2} \frac{q^{3/2}F^{1/2}}{\chi} \right\} \right], \quad (4.42)$$

where $\theta(y)$ is the Nordheim function and K_s is the dielectric constant of semiconductor.

2. The electron gas in the conduction band is nondegenerated, the weak penetration of the field is considered. For this purpose in Equation (4.41) instead of the equilibrium concentration of electrons in the bulk n_∞ the electron density n in the surface layer at the interface with the vacuum is substituted. It is assumed that the current of EFE practically doesn't disturb the equilibrium distribution of electrons in the surface layer. But then the value of n is determined by the Boltzmann formula (Figure 4.3):

$$n = n_\infty \exp\left(\frac{q\Delta V_s}{k_B T} \right) = n_\infty \exp\left(\frac{\phi_s}{k_B T} \right). \quad (4.43)$$

In case of nondegenerated electron gas, that is, at $|\xi| - |\phi_s| \gg k_B T$ (Figure 4.3a) that realized at

$$F \ll K_s^{1/2} 10^6 \text{ V/cm}, \quad (4.44)$$

where $\xi = E_C - E_F$.

The band bending ϕ_s according to Ref. [51] has view:

$$\phi_s = q\Delta V_s = 2k_B T \operatorname{arcsinh}\left[\frac{qF}{2\varepsilon_s k_B T} \left(\frac{K_s k_B T}{8\pi q^2 n} \right)^{1/2} \right]. \quad (4.45)$$

For the considered approximation the expression for J has the form Equation (4.42), where instead of n_∞ is n.

3. The electron gas in the conduction band due to the strong penetration of the field is degenerated, that is, $|\phi_s| - |\xi| \gg k_B T$ (Figure 4.3b). This takes place at fields

$$F \gg \varepsilon_s^{1/2} 10^6 \text{ V/cm}, \quad (4.46)$$

Then, as it is shown in [52], band bending is

$$\phi_s = |\xi| + bF^{4/5}, \tag{4.47}$$

where b is the constant that depends on the properties of the semiconductor. The flow of $v(E_x)$ is defined in the same way as for the metal. However, since in this case the near-surface electron concentration with increasing field is continuously increasing due to increased band bending, instead of the work function of the semiconductor according to [52] the work function χ' has to use

$$\chi' = \chi - b(qF)^{4/5}, \tag{4.48}$$

where b is the same as in formula (4.47). In this approximation at $b(qF)^{4/5} < \chi$, we have for J

$$J = \frac{q}{8\pi h}\frac{(qF)^2}{\chi}\exp\left[-\frac{4\sqrt{2m}\chi^{3/2}}{\hbar qF}\theta(y)\right] \times \exp\left[-2\frac{k_0 b\chi^{3/2}}{qF^{1/5}} - \left\{1 + \frac{2k_0 b\chi^{3/2}}{qF^{1/5}}\right\}\right]. \tag{4.49}$$

4. In Ref. [52] the formula for the current density at EFE that takes into account the surface states has been obtained.

There are the following conclusions from the Stratton's theory [7, 53]. First, the current-voltage characteristics of the current density of the EFE $lnJ - f(1/F)$ from n-type semiconductor in all the approximations are represented by straight lines, except for the case where the surface states are taken into account. The reason of bending of the current-voltage characteristics that appears in the presence of surface states is the following. At low fields EFE is understated due to an internal inhibitory barrier. At compensation of this barrier by the penetrating external field the increased growth of emissions begins. It is caused by the simultaneous increase of the transparency of the potential barrier and the electron concentration in the surface layer with increasing the external field.

Further, the current density of EFE, if the electron gas of the conduction band in the surface layer is not degenerated, according to Equation (4.42) is proportional to the concentration of the electrons in this layer, that is, $J \sim n$. Without taking into account the penetration of the field $n = n_\infty$, while taking into account field penetration n in more complicated way Equation (4.43) depends on n_∞.

According to Equation (4.43) the current of EFE is also dependent on temperature (mainly because of the dependence $n = f(T)$). For the degenerated electron gas Equation (4.49) J is clearly independent on the temperature. The study of the temperature dependence of EFE from semiconductors showed that with increasing temperature the emission current, as a rule, increases exponentially. At the same time for different temperature ranges the exponent value and, consequently, the activation energy of the carriers to be determined by this value may be different. It should be noted that at measuring of the EFE temperature dependence the great difficulties to the measurement of the temperature of the tip top. Current-voltage (I-V) characteristics shifted to higher currents with the increase in temperature. Moreover, their slopes are not changed or diminished. The constancy of the characteristics slopes is observed in the most cases at low temperatures, whereas a gradual decrease in slope, on the contrary, mainly at higher temperatures. Since according to Equation (4.42) $J \sim n$, the parallel shift of the *I-V* characteristics with increasing T is attributed only to the increase in the concentration of electrons in the conduction band. The changes of slope are usually attributed to the changing of the energy distribution of electrons in the conduction band with growth of the temperature. Temperature growth leads to the increase of average energy of the electrons, and, hence, the increase of transparency of the potential barrier [54, 57].

For description of EFE from semiconductors the *zero emitted current approximation* (ZECA) has been proposed and developed in Refs [10, 19]. As it was shown in Ref. [19], ZECA approximation for field emission from semiconductors, in its simplest form and for $\beta\mu$ sufficiently large, allows using the Equations (4.32) and (4.35) obtained for emission from metal with some modification: (i) the chemical potential μ replaced with $\mu_s = \mu_0 + \phi_s$, (ii) the "work function" Φ is set equal to $\chi - \mu_s$, and (iii) the image charge term Q replaced throughout the F–N equation with the dielectric modification [58]

$$Q(K_s) = \frac{K_s - 1}{K_s + 1}\frac{\alpha \hbar c}{4}, \tag{4.50}$$

where α is the fine structure constant, K_s is the dielectric constant, c is the speed of light, and $\hbar = h/2\pi$ is the reduced Planck's constant.

All the considered theories of EFE originate from thermodynamic equilibrium of electron gas with the lattice (i.e., it is assumed that $E_0 = const$) and, therefore, the effects of the strong field have been neglected. Overheating of the electron gas in the semiconductor in the presence of field in it that appears at the flow through the emitter of conduction current and the increase in the number of carriers in the conduction band with the growth of this field has been taken into account in Ref. [55]. It has been suggested that at the strong internal fields the electron gas in the surface layer has Maxwell's distribution, but it is characterized by higher than the lattice temperature T_e. It is assumed that the electron density n increases exponentially with increasing of the field. According to [55] lnJ increases with $1/V$ faster than linearly.

Electron transport through semiconductors as a rule is described by the Boltzmann Transport Equation (BTE), which, in one dimension [59]

$$-\partial_t f(x,k,t) = \frac{\hbar k}{m}(\partial_x f) - \frac{1}{\hbar}(\partial_x U)(\partial_k f) - \partial_c f\,|_{col}\,. \tag{4.51}$$

The last "collision" term governs scattering events and may be represented by the relaxation time approximation [60]. The scattering time τ is related to the distance electrons travel between scattering events, and for typical silicon parameters, it is of the order of several hundred femtoseconds [61]. The distribution function $f(x, k, t)$ relaxes to thermal Fermi-Dirac distribution

$$f(k) = \frac{m}{\pi\beta\hbar^2}\ln[e^{\beta(\mu - E(k))} + 1]. \tag{4.52}$$

The exception is that in the presence of a varying potential $U(x)$, the chemical potential μ must be replaced by the electrochemical potential $\mu(x)$ [62]. Under time-independent equilibrium conditions, the collision term drops out. Using the thermal Fermi-Dirac distribution function from Equations (4.51) and (4.52) it is then shown that $\mu(x) = \mu_0 + \phi(x)$, where μ_0 is the bulk value and U is replaced by ϕ.

The redistribution of electrons accounted for $\phi(x)$ dictates the extent of band bending under an applied field F_{vac}. At the surface, $\phi(0^-) = \phi_s$, which will be assumed to be measured with respect to the conduction band edge E_c. By virtue of being subject to band bending, the electron affinity χ of a semiconductor is specified rather than work function. ϕ_s may be numerically calculated from Poisson's equation once $F_s = F(0^-) = F_{vac}/K_s$ is specified, where F_{vac} is the externally applied field.

The simplified ZECA assumption regarding the electron density profile is not sustainable near the surface (or interface), as it suggests that the electron density increases until the origin [19]. That approximation runs afoul of several issues. The equilibrium distribution is maintained by scattering events. Assuming that the distribution is thermal near the origin is problematic. Even for a completely thermal distribution (vanishing relaxation time), the wave nature of the electron shows that $\rho(x)$ begins to decline in magnitude, rather than increase, at the distance approximately π/k_F prior to the barrier.

The BTE does not include tunneling effects, and so it will be unable to anticipate the dipole term due to electron penetration of the barrier. Tunneling phenomena are intimately intertwined with the quantum mechanical nature of the incident electrons. Thus, the density implicit in [63]

$$\rho = 4\frac{M_c}{\sqrt{\pi}}\left(\frac{m}{2\pi\beta\hbar^2}\right)^{3/2}\int_0^{\infty}\frac{\sqrt{\varepsilon}d\varepsilon}{1+e^{\varepsilon-\beta\mu}} = N_c\frac{2}{\sqrt{\pi}}F_{1/2}(\beta\mu) \tag{4.53}$$

is not the same as the density:

$$\rho(x) = \frac{1}{2\pi}\int_0^{\infty} f(k)\left|\psi_k(x)\right|^2 dk, \tag{4.54}$$

where M_c is the number of equivalent minima in the conduction band (e.g., 1 for metals, 6 for silicon, etc.), $F_{1/2}(x)$ is the Fermi-Dirac integral, $\varepsilon = E - E_C/k_BT$, and m is the effective mass of the electron.

They become equivalent far into the bulk, where the out of phase oscillations of the wave function for various momenta average out and the bulk density is recovered, but near the origin and into the tunneling potential, the density is decreasing and decaying because the wave functions for all k approximately vanish at the origin, the type of behavior is not incorporated into the simplified ZECA approach. The modification of the BTE is, therefore, required to introduce the quantum effects.

Wigner suggested a quantum distribution function $f(x,k)$ based on the Fourier transform of a generalization of Equation (4.54) for the electron density matrix $\rho(x, y)$

$$f(x,k) = \frac{1}{\pi}\int_{-\infty}^{\infty} e^{-2iky}dy\int_{-\infty}^{\infty} dk' f(k')\psi_{k'}^{*}(x+y)\psi_{k'}(x-y). \tag{4.55}$$

To be specific, the Wigner Distribution Function (WDF) $f(x, k)$ is not a probability distribution function as it takes on negative values, but it reduces to the Boltzman distribution in the classical limit [64]. However, the WDF phase space description [65] is remarkably useful: it can be used to calculate averages, suggests a particle trajectory interpretation analogous to the BTE, and has proven itself remarkably adept for modeling the high frequency behavior of tunneling structures (in particular, but not limited to, resonant tunneling structures) [61, 66–70].

When solved in parallel with Poisson's equation, a self-consistent equilibrium $f(x, k, t)$ can be dynamically obtained by letting the system evolve from an initial state, where the relaxation is introduced by a collision term analogous to that used in the BTE. From the resulting distribution function, density, and current are calculated from the first two "moments" of $f(x, k, t)$ by (compare to Equation (4.54)).

$$\rho(x,t) = \frac{1}{2\pi}\int_{-\infty}^{\infty} f(x,k,t)dk,$$

$$j(x,t) = \frac{1}{2\pi}\int_{-\infty}^{\infty} \frac{\hbar k}{m} f(x,k,t)dk. \tag{4.56}$$

Because $f(x, k, t)$ is time dependent, both the density and the current are likewise time dependent, and so the Wigner function method may be used to model dynamic response, for example, after a sudden change in external field [71].

4.4 Current through Double Barrier Structures

4.4.1 Coherent Resonant Tunneling

Resonant tunneling double barrier structure consists of two potential barriers in series, the barriers being formed by thin layers of wide-gap material like AlGaAs or AlGaN sandwiched between layers of material like GaAs or GaN having a smaller gap. Both barriers are thin enough for electrons to tunnel through. It might seem that the current-voltage characteristics of two barriers in series cannot be any more interesting than those of a single barrier. Ohm's law would suggest that we would simply need twice the voltage to get the same current. That is exactly what would happen if the region between the two barriers were microns in length. But if this region is only a few nanometers long (which is a fraction of the de Broglie wavelength), the current-voltage characteristics are qualitatively different from those of single barrier, underlining once more the failure of Ohm's law on a nanometer scale [1].

The current-voltage (I-V) characteristics of a double-barrier structure are easily understood if we note that the region between the barriers acts like a "quantum box" that traps electrons. It is known from quantum mechanics (see Section 2.2) that a particle in a box has discrete energy levels whose spacing increases as the box gets smaller. We assume that the box is small enough that there is only one allowed energy E_n, in the energy range of interest (Figure 4.5). The structure then acts as a filter that only allows the electrons with energy E_n transmit. Applied bias lowers the resonant energy relative to the energy of the incident electrons from the emitter. When the bias exceeds the threshold voltage V_T, the resonant energy falls below the conduction band edge in the emitter and there is a sharp drop in the current. The current-voltage characteristics thus exhibit negative differential resistance (NDR).

The NDR forms the basis for practical applications as a switching device and in high frequency oscillators [72–75].

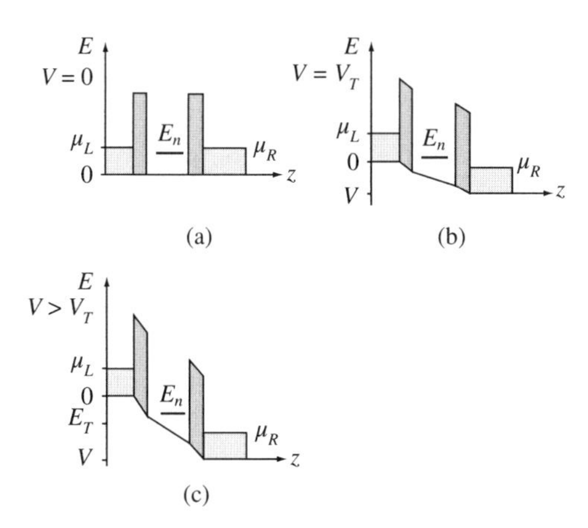

Figure 4.5 Conduction band diagram for a resonant tunneling diode with $V = 0$ (a), $V = V_T$ (b), and $V > V_T$ (c)

In general case the resonant-tunneling current is given by

$$J = \frac{q}{2\pi\hbar}\int g(E)f(E, E_F)T(E)dE = \frac{q}{2\pi\hbar}\int N(E)T(E)dE, \tag{4.57}$$

where the number of available electrons for tunneling (per unit area) from the emitter can be shown to be [76]

$$N(E) = \frac{k_B T m^*}{\pi\hbar^2}\ln\left[1 + \exp\left(\frac{E_F - E}{k_B T}\right)\right]. \tag{4.58}$$

In a resonant-tunneling diode, the variable energy of the incoming electrons is provided by external bias such that the emitter energy is raised with respect to the well and the collector. The incoming energy distribution of tunneling electrons integrated over the sharp resonant-tunneling peaks would seem to predict sharp current peaks and very high peak-to-valley ratios (Figure 4.6), which are not observed in real devices, even at low temperatures. The reason for

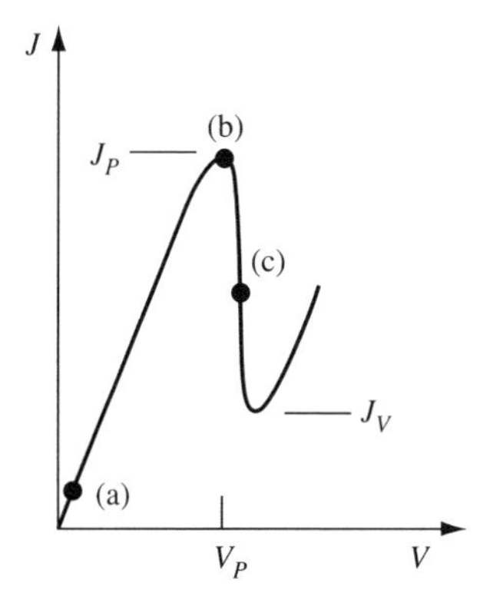

Figure 4.6 I-V characteristic of resonant-tunneling diode with current peak and valley

this is twofold. First, the resonant transmission peaks are exponentially narrow on the order of $\Delta E = \hbar/\tau$, where τ is the lifetime of an electron in the subband E_n with respect to tunneling out and ΔE is the broadening of the energy level E_n [75]. Additionally there exist nonideal effects such as impurity scattering, inelastic phonon scattering, phonon-assisted tunneling, and thermionic emission over the barrier. These effects lead to much larger valley current that diminishes the peak-to-valley ratio.

Let's analyze the coherent resonant tunneling process in detail. The band diagrams drawn in Figure 4.5 are greatly simplified versions where we have assumed that the applied voltage drops linearly across the device. This would be true if there were no space charge inside the device and the surrounding regions were very highly conducting. For quantitative calculations it is necessary to compute the charge density everywhere and use it in the Poisson equation to obtain the actual band diagram [77].

Now at the first simplifying approach to current-voltage characteristic analysis we consider the idealized energy band diagram without influence of charge, but further we will complicate the model to realize situation in real structures and devices.

To analyze resonant tunneling in double barrier structures it is useful to use the generalized barrier with generalized transmission probability (Figure 4.7). The description of current transport through one barrier was given in detail in Section 4.1.

During the analysis of coherent resonant tunneling here we will neglect all scattering processes. For simplicity we will also assume the temperature to be low enough that the Fermi functions can be approximated by step functions:

$$f_1(E) = \upsilon(E_{FL} - E) \text{ and } f_2(E) = \upsilon(\mathrm{E_{FR}} - \mathrm{E}),$$

$$\upsilon(E_F - E) = 1 \text{ if } E_F > E, \tag{4.59a}$$

$$\upsilon(E_F - E) = 0 \text{ if } E_F < E. \tag{4.59b}$$

Equation (4.11) then simplifies to

$$I = \frac{2q}{h}\int_{\mu_R}^{\mu_L} \overline{T}(E)dE. \tag{4.60}$$

So, to calculate the current we need the transition function $\overline{T}(E)$, that we considered in detail in Chapter 1.

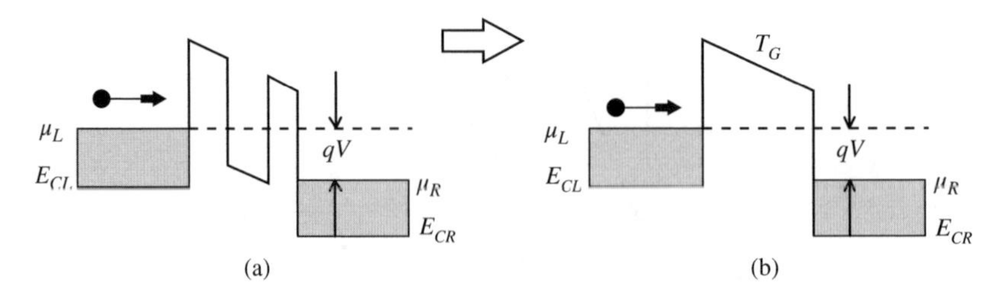

Figure 4.7 Electron tunneling through (a) resonant tunneling structure and (b) generalized barrier

The current is obtained by integrating the transmission function from μ_R to μ_L as indicated in Equation (4.60). Since there is no transmission unless the longitudinal energy E_L is greater than zero (that is, the total energy is greater than ε_m), we can integrate from ε_m to μ_L (see Chapter 1):

$$I = \sum_m I_m, \text{ where} \tag{4.61}$$

$$I_m = \frac{2q}{h}\int_{\varepsilon_m}^{\mu_L} \overline{T}(E)dE = \frac{2q}{h}\frac{\Gamma_1\Gamma_2}{\Gamma_1+\Gamma_2}\int_{\varepsilon_m}^{\mu_L} A(E-E_m)dE. \tag{4.62}$$

Thus, the current carried by a mode depends on the area under its spectral function $A(E-E_m)$ inside the energy window from ε_m to μ_L [1]. If the line shape function for a mode is completely inside the window, then the integral is equal to 2π. It then carries the maximum current I_p that can be carried by a single mode:

$$I_p = \frac{2q}{\hbar}\frac{\Gamma_1\Gamma_2}{\Gamma_1+\Gamma_2}. \tag{4.63}$$

The current carried by a mode m is approximately equal to I_p if it is "resonant," and approximately zero if it is non-resonant. The total current can be obtained approximately by adding up the currents carried by all the resonant modes [1].

4.4.1.1 Current in Three-Dimensional System

Let's consider the three-dimensional system and suppose that the system is described by potential energy that is the function of only z coordinate [2]. That system is invariant in the plane XOY. The energy of the system can be represented as the sum of components. Let k_z is the wave vector of electrons on the left side of the system which is characterized by potential E_L. Three-dimensional wave vector has the following components

$$\vec{K} = (\vec{k}, k_z) \text{ where } \vec{k} = (k_x, k_y). \tag{4.64}$$

Then the wave function has the view

$$\psi_{\vec{k},k_z}(\vec{r}, z) = e^{i\vec{k}\vec{r}}u_{k_z}(z) \tag{4.65}$$

and energy

$$\varepsilon(\vec{k}) = U_L + \frac{\hbar^2\vec{k}^2}{2m^*} + \frac{\hbar^2 k_z^2}{2m^*}, \tag{4.66}$$

where $U_L \equiv E_{CL}$.

That is, the transmission coefficient is only k_z function. Similarly to one-dimensional case, the current density (current through a unit cross section of the plane normal to OZ axis) of electrons emitted from the left side of the device, we write as

$$J_L = 2q\int \frac{d\vec{k}}{(2\pi)^2}\int_0^{\infty} \frac{dk_z}{2\pi} f(\varepsilon(\vec{K}), \mu_L)v_z(\vec{K})T(k_z). \tag{4.67}$$

We write the speed of electrons along the OZ axis as a function of k_z and use Equation (4.66). We obtain

$$J_L = q\int_0^\infty \frac{dk_z}{2\pi}\frac{\hbar k_z}{m^*}T(k_z)\left[2\int \frac{d\vec{k}}{(2\pi)^2}f\left(U_L + \frac{\hbar^2\vec{k}^2}{2m^*} + \frac{\hbar^2 k_z^2}{2m^*}, \mu_L\right)\right]. \tag{4.68}$$

The expression in square brackets, together with factor 2 that represents the degeneracy of spin is nothing more than the density of states of two-dimensional electron gas in the plane *XOY* with bottom energy band that is determined by the energy $U_L + \hbar^2 k_z^2/(2m^*)$. Denote this density of states as $N_{2D}(\mu_L - U_L - \hbar^2 k_z^2/(2m^*))$. Then formula (4.67) can be rewritten as

$$J_L = q\int_0^\infty \frac{dk_z}{2\pi}\frac{\hbar k_z}{m^*}T(k_z)N_{2D}(\mu_L - U_L - \hbar^2 k_z^2/(2m^*)). \tag{4.69}$$

Let's transit from integration over the wave vector to integration over energy of longitudinal movement $E = U_L + \hbar^2 k_z^2/(2m^*)$. For this, note that

$$dE = \frac{\hbar^2 k_z dk_z}{m^*}. \tag{4.70}$$

Then Equation (4.69) yields

$$J_L = \frac{q}{h}\int_{U_L}^\infty N_{2D}(\mu_L - E)T(E)dE. \tag{4.71}$$

Adding here the current associated with emitted electrons from the right side of the device we have the full current

$$J^{(3D)} = \frac{q}{h}\int_{U_L}^\infty \left\{N_{2D}\left(\mu_L - E\right) - N_{2D}(\mu_R - E)\right\} T(E)dE. \tag{4.72}$$

Let's consider the simplifying Equation (4.72) for some limited cases. In the case of *high bias* the states from the right side of the system will not contribute to the full current and it will determine only the states of the left side. If this system is also at very *low temperatures,* the Fermi distribution function can be replaced by step function. Then,

$$N_{2D}(E) = (m^*/\pi\hbar^2)\vartheta(E) \tag{4.73}$$

and the current can be described as

$$J^{(3D)} = \frac{q}{h}\frac{m^*}{\pi\hbar^2}\int_{U_L}^{\mu_L} (\mu_L - E)T(E)dE. \tag{4.74}$$

In the case of *low bias* we obtain the next formula for current density

$$J = \frac{q^2}{h}\frac{m^*V}{\pi\hbar^2}\int_{E_{CL}}^{\infty} f(E,\mu_L)T(E)dE\,|_{T\to 0} = \frac{q^2}{h}\frac{m^*V}{\pi\hbar^2}\int_{E_{CL}}^{\mu_L} T(E)dE, \tag{4.75}$$

where V is the applied voltage.

It should be noted that in this case the current is determined by integral on energies in the range of the E_{CL} to μ_L. It is known that at low temperatures all electrons of semiconductor are degenerated and electronic properties are determined by electrons in a narrow region near the Fermi surface. However, their longitudinal energy can vary over a wide range from E_{CL} to μ_L. In the case of single barrier the electrons with longitudinal energies of the highest value will actually dominate as the probability of transition increases with increasing of the electron energy. Otherwise, the electrons that can be emitted from the left side of the systems have primarily the energy in the region near the Fermi surface. But they fall on the barrier at different angles. A small number of electrons that fall almost normally to the barrier are quasi low-dimensional and may have large values of longitudinal energy. That is, the electrons pass through the barrier with only nonzero k_z wave vector.

4.4.1.2 Current in Two-Dimensional System

Obviously, the current in the case of two-dimensional motion of electrons has the same view excepting that now there is the density of energy states of one-dimensional electron gas in the integral function. Then

$$J^{(2D)} = \frac{q}{h}\int_{U_L}^{\infty} \left\{N_{1D}\left(\mu_L - E\right) - N_{1D}(\mu_R - E)\right\} T(E)dE. \tag{4.76}$$

Note that the views of formulas (4.72) and (4.76) are very similar to the view of formula (4.10). Another characteristic feature of quantum transport is the fact that the current in n-dimensional system is defined by density of states of (n – *1*)-dimensional system.

4.4.2 Sequential Tunneling

It was previously assumed that transport is fully coherent; that is, the electron transmits from the left to the right in a single quantum mechanical process whose probability can be calculated from the Schrodinger equation. This is a reasonably accurate picture if the average time that an electron spends in the resonant state (called the eigenstate lifetime) is much less than the scattering time τ_φ. Otherwise, a significant fraction of the current is due to "sequential tunneling" where an electron first tunnels into the well and then, after losing memory of its phase, tunnels out of the well. The difference between coherent resonant tunneling and sequential resonant tunneling is somewhat like the difference between a two-photon process, and two one-photon processes in optics. Coherent resonant tunneling is like a two-photon process (with a photon energy of zero), where the resonant level E in the device acts as a virtual state, while sequential resonant tunneling is like two one-photon processes where an electron makes a real transition into the resonant level and another real transition out of it [1, 78, 79].

In the sequential-tunneling model, the tunneling from emitter into the well, and that from the well to the collector can be treated as uncorrelated events. In the model of sequential tunneling, tunneling of carriers out of the well to the collector is much less constrained, and tunneling of carriers from the emitter into the well is the determining mechanism for the current flow. This requires available empty states at the same energy level (conservation of energy) and with the same lateral momentum (conservation of momentum) of the available electrons in the emitter as within the well. Since the parallel (to tunneling direction) momentum k_z in a quantum well is quantized this gives rise to quantized level E_n,

$$E_n = \hbar^2 k_z^2/2m^* = h^2 n^2/8m^* w^2, \tag{4.77}$$

the energy of carriers in each subband is a function of the lateral momentum $k_\perp$ only, given by

$$E_w = E_n + \frac{\hbar^2 k_\perp^2}{2m^*}. \tag{4.78}$$

From Equation (4.78) it should be noted that the energy of carriers is quantized only for the bottom of the subband, but the energy above E_n is continuous. The free-electron energy in the emitting electrode is, on the other hand, given by

$$E = E_C + \frac{\hbar^2 k^2}{2m^*} = E_C + \frac{\hbar^2 k_z^2}{2m^*} + \frac{\hbar^2 k_\perp^2}{2m^*}. \tag{4.79}$$

So electrons in the emitter with energy given by Equation (4.79) will tunnel into the energy level given in Equation (4.78). This concept is depicted in Figure 4.8 [72].

We first examine the region of current growth (*a*) of *I-V* curve in Figure 4.6, where the current increases with bias. Figure 4.8a shows that if E_l is above E_F there is little availability of electrons for tunneling. As the bias is increased, E_l is pulled below E_F and toward E_C of the emitter, and tunneling current starts to increase with bias.

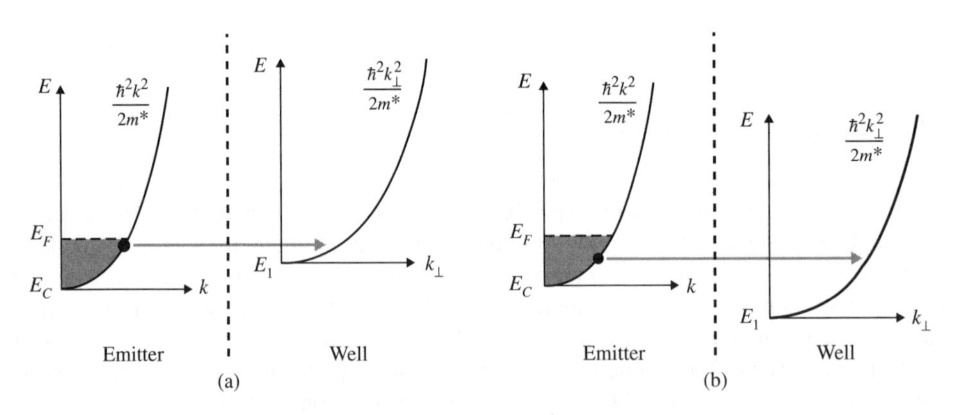

Figure 4.8 Tunneling of electrons from emitter electrode into the well. Note the difference of k vs. $k_\perp$ in abscissas. (a) E_l is higher than E_C and lower than E_F. Resonant tunneling starts to occur. (b) E_l is lower than E_C. Tunneling probability drops significantly [72]. Copyright (2007) by John Wiley & Sons, Inc. This material is reproduced with permission of John Wiley & Sons, Inc.

The decrease of current with bias in region (c) of Figure 4.6 is less trivial. Conservation of lateral momentum requires that the last terms of Equations (4.78) and (4.79) are equal. This, along with the conservation of energy, results in the requirement

$$E_C + \frac{\hbar^2 k_z^2}{2m^*} = E_n. \tag{4.80}$$

From Figure 4.8b, it is seen that $k_\perp$ at the well becomes large. So, for the emitter, since

$$k^2 = k_z^2 + k_\perp^2, \tag{4.81}$$

even with $k_z = 0$, the minimum value of k is $k_\perp$. At low temperature, electrons are within the Fermi sphere of finite momentum. For k_z outside the Fermi sphere, there is no electron available for tunneling. So, the tunneling event in Figure 4.8b is prohibited.

From the above analysis, for maximum tunneling current, E_n should line up between E_F and E_C of the emitter. But at low temperatures, E_n should line up with E_C. With higher bias, the emitter E_C is above E_n and tunneling current is significantly reduced, resulting in NDR. The peak voltage occurs approximately at a voltage

$$V_P = \frac{2(E_n - E_C)}{q}, \tag{4.82}$$

for a symmetrical junction, because half of the bias is developed across each barrier. In practical devices, V_P is larger than that given by Equation (4.82). (Note that E_C for the emitter is different from the well). The electric field can penetrate into both the emitter and collector regions and results in some voltage drop. Second, there is also a voltage drop across each undoped spacer layer. Another effect is caused by the finite charge accumulated within the well under bias. This charge sheet creates the inequity of electric field across two barriers, and extra voltage is required to shift the relative energy between the emitter and the well.

The ratio of local peak current (J_P) to valley current (J_V) is a critical measure of the NDR. In real devices the nonzero valley current is due mainly to thermionic emission over the barriers, and it has large temperature dependence (smaller J_V with lower temperature). Another small but conceivable contribution is due to tunneling of electrons to higher quantized levels. Even though the number of electrons available for tunneling at energy higher than E_F is very small, there is a thermal distribution tail and this number is not zero, especially when the quantized levels are close together.

In the presence of scattering processes, we have the situation where only the fraction of electrons transmit coherently while the remainder gets scattered inside the well and effectively leaks out of the coherent stream [80, 81].

The consideration analysis of current through double barrier structure has shown two important features of transport namely, (1) *resonant tunneling* [1, 80–84] and (2) *single-electron tunneling* [1, 85–89]. Resonant tunneling involves the flow of current through the discrete states formed between the barriers. Single-electron tunneling is observed in double-barrier structures having a small cross-sectional area. If the well is long enough that the energy levels are very close together we do not expect to be able to resolve the resonant tunneling current through the individual levels. But if the capacitance (C) of the structure is small enough, we can still observe tunneling through discrete levels spaced by q^2/C. These discrete levels arise not from the wave nature of electrons (size quantization), but from their particle nature (charge quantization).

4.5 Electron Field Emission from Multilayer Nanostructures and Nanoparticles

In vacuum micro- and nanoelectronics applications it is very important to employ systems with high emission efficiency at sufficiently low applied voltage. The other important requirement is to obtain an electron beam with narrow energy and space distribution. The promising approach to achieve these is to develop new type cathodes with quantum-size effects, the so-called quantum cathode [19, 46, 90–92]. Electron emitters with multilayer structures containing nanoparticles have been actively investigated [44, 47, 93–99]. Coherent resonant electron tunneling [44], sequential electron tunneling [94], and single electron tunneling [95] are the main electron transport mechanisms for emission from quantum emitters.

Recent studies of advanced materials have succeeded in obtaining a whole new class of composites such as nanometer-sized particles [97] and nanotubes [100] with significant potential for electron field-emission (FE) applications. Such materials have generated a great deal of interest due to their unique behavior in FE regimes. While normal emissive materials, such as metals and in most cases semiconductors, exhibit monotonically increasing current-voltage (*I-V*) characteristics, FE from nanostructured materials often produces results that can be attributed to the underlying quantum confinement of the electrons residing in these cathodes. There are two basic types of FE peculiarities reported thus far from these structures: first, resonant tunneling, where electrons with some resonant values of the energy travel ballistically through nanostructures, and second, the sequential or two-step tunneling, where electrons are first supplied in regions of lower potential energies (energy wells) of the nanostructures before tunneling toward the anode. Inelastic processes are usually involved in the two-step tunneling that leads to quasiequilibrium distributions of electron populations in the potential wells below the emitting surface. Both emission mechanisms can appear in real experiments. Resonant tunneling in FE has been thoroughly investigated both from the theoretical [44, 49, 101] and experimental [102, 103] points of view. However, while the existing theoretical models account for the pronounced peaks in the *I-V* characteristics, followed by regions of NDR, experimental evidence is scarce due to the special conditions required for the occurrence of resonant tunneling during FE.

4.5.1 Resonant Tunneling at Electron Field Emission from Nanostructures

The coherent resonant tunneling has been analyzed in detail in Ref. [44]. The field emission of electrons is determined mainly by three factors: (1) the concentration of the source electrons N_n (at the *n*th level); (2) the value of the surface barrier Φ_s and its modification under applied electric field *F;* and (3) the carrier transport through the surface barrier, which is characterized in the first approximation by the tunneling process. The current density of field (cold) emission is expressed as follows:

$$J_{FN} = q\sum_{n,j} N_n^j \omega_n^j \left\langle l_n^j(z) \right\rangle T_n, \tag{4.83}$$

where ω is the collision frequency in the quantum well, $l_n^j(z)$ is the effective width of the quantum well, T_n is the probability of tunneling through the surface barrier and the superscript j denotes the degenerate valleys. The relative role of these factors is influenced by characteristics of the emitting structures and values of the electric field. We shall consider next structures which favor the improvement of field emission. Quantum-sized structures with one-side barrier [44, 47, 98, 99] (the band model is shown in Figure 4.9) and two-side barrier [43, 104, 105] (the band model is shown in Figure 4.10) have been analyzed as field-emitting cathodes with resonance-tunneling electron transport. These two different band models will be explained in the following two sections. The influence of the emitter doping and temperature on their emission characteristics has also been studied.

4.5.1.1 Model of the One-Side Double-Barrier FE Cathode

The potential profile of quantum-size (QS) film cathodes was simulated using a one-side double-barrier structure, in which the input potential barrier was formed by the conduction-band discontinuity at the boundary between the emitter (source of electrons realized using: doped Si, GaAs, GaN, or other semiconductor material) and the QS coating phase. The output barrier is determined by the nonzero QS phase – vacuum affinity χ (Figure 4.9). A quantum well (QW) with triangular potential is formed at the boundary between the ultrathin film (for definition the diamond-like carbon film (DLC)) and the vacuum, due to the slope of DLC conduction band in the presence of an applied external electric field (Figure 4.9). For given DLC layer parameters and electric-field values, the electron-wave interference in the potential well

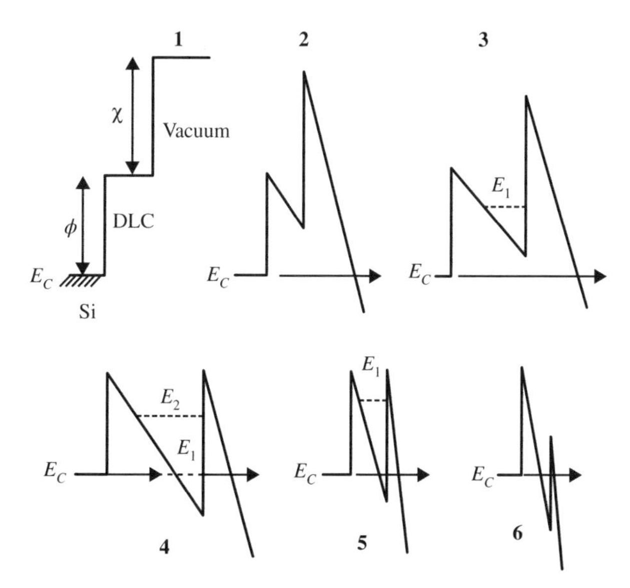

Figure 4.9 Energy band diagrams of Si-DLC film cathodes. (1) – without applied electrical field, (2–6) – with applied field. The electric field and the DLC film thickness are changed: $d(2) < d(3) < d(4)$; $F(5) < F(6)$. E_1 and E_2 are the quantized energy levels in the DLC layer and E_c is the conduction band bottom. Reproduced with permission from Ref. [44]. Copyright (2004), AIP Publishing LLC

causes quantization of electron energy and momentum, and, as a result, the electron emission through the potential profile of cathodes with a DLC film shows a resonance behavior. The energy band diagrams for different DLC-film thickness are shown in Figure 4.9. The number of quantized energy levels in the DLC layer (E_1, E_2) depends on the DLC-film thickness and the applied electric field. For different types of the QS phase (layered, wire, or nanoclusters, that is, 2D, 1D, 0D (QD (quantum dot))) the quantum splitting of energy levels is significantly different, being larger for QDs, but the effect of resonance tunneling transport is qualitatively similar presenting an enhancement of the tunneling current, and peaks of the FE current.

Bending of the emitter conduction band at the boundary between Si and the DLC film creates a second quasitriangular-shaped quantum well before the input barrier, similar to the left-hand side in Figure 4.10, leading to accumulation of 2D electrons in it. The potential profile of the Si-DLC cathode was calculated by solving Poisson's equation. Electric-field continuity was assumed at the cathode boundaries and the 3D electron distribution of the emitter was taken into account. The influence of the emission of 2D electrons from an emitter accumulation layer was also analyzed under conditions of low emitter doping levels.

The current density of the field emission from 3D electron states of the emitter conduction band was calculated as the sum of resonant (J_r) and nonresonant (J_{nr}) components.

$$J_{r,nr} = \sum_{i}^{j} \left(\frac{q m_{3i} kT}{2\pi^2 \hbar^3} \right) \int_{0}^{\infty} T_{r,nr}(F) \ln \left\{ 1 + \exp \left[\frac{E_F - E}{k_B T} \right] \right\} dE, \tag{4.84}$$

where m_{3i} is the electron longitudinal effective mass in the ith valley of the conduction band of silicon; j is the number of the equivalent valleys; q is the electron charge; k_B is the

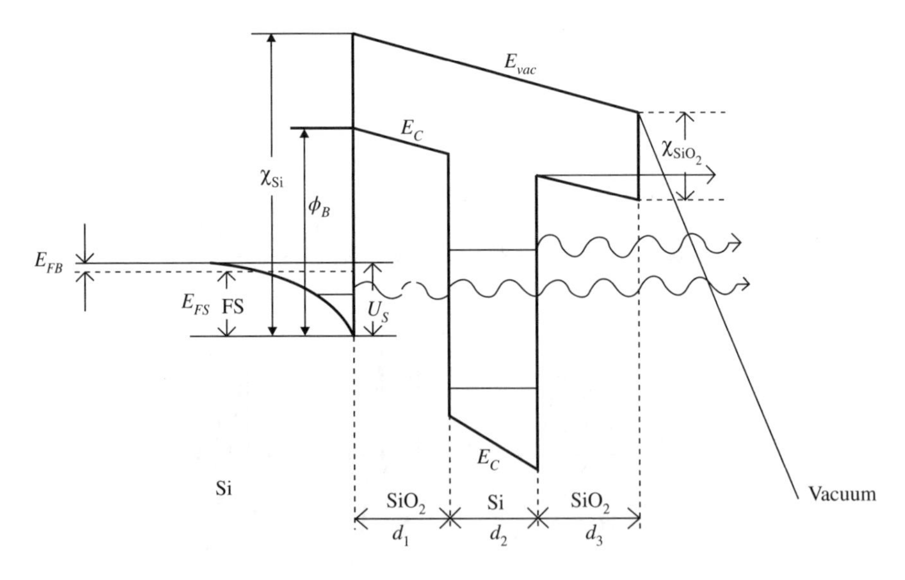

Figure 4.10 Energy band diagram of the layered field-emitting structure (Si-SiO_2-δ-Si-SiO_2). E_{FB} and E_{FS} are the positions of the Fermi level (relative to the position of the conduction band bottom) in bulk and in space charge region, respectively. U_S is the value of band bending due to the field penetration in Si. Reproduced with permission from Ref. [44]. Copyright (2004), AIP Publishing LLC

Boltzman constant, E_F and E are the emitter Fermi energy and the tunneling electron energy, respectively; T_{nr} is the transmission coefficient for nonresonant tunneling of electrons with the three-dimensional (3D) momentum, and T_r is the same parameter for resonant tunneling of electrons.

When calculating the emission of 2D electrons from the accumulation layer, the integration in Equation (4.84) was performed over the width of the accumulation layer by considering the emitting subband. $T_{nr} = T_1 \times T_2$ and T_r could be evaluated by modeling the electron wave interference in the quantum well

$$T_r = \frac{T_1 T_2 \exp\left[-\frac{\Gamma_S}{\Gamma_0}\right]}{\left(\frac{\Gamma}{\Gamma_0}\right)^2} \times \frac{1}{1+4\left(\frac{\Delta E}{\Gamma}\right)^2 \exp\left[\frac{\Gamma_S}{\Gamma_0}\right]}, \tag{4.85}$$

where T_1, T_2 are the transparencies of the input and output barriers, Γ, Γ_S, Γ_0 are the full, relaxation, and intrinsic width of resonant level in the quantum well, respectively. Equation (4.85) is also applied in the case of partial coherence of the electron waves of the carriers interfering in the quantum well. The calculation of T_1 and T_2 was carried out in either a trapezoidal or a triangular approximation for the potential-barrier shape depending on the electric-field strength.

Computer-simulation results of electron-field emission through an ultrathin DLC film are presented in Figure 4.11. The current density dependence *J(E)* on the electric field in vacuum was obtained for cathodes without and with the ultrathin film and for different QW thicknesses.

The characteristics of the DLC film were compared to those obtained experimentally using that structure. The thickness (*d*) of the DLC film has been varied for this purpose from 4 to 110 nm and the barrier height at the Si-DLC interface (Φ) was selected in the range 0.5 – 3 eV. The calculations have been performed for two emitter doping levels: $N_d = 5 \times 10^{15}$ cm^{-3} and $N_d = 5 \times 10^{18}$ cm^{-3} and the current-voltage characteristics of nonresonant current are shown in Figure 4.11.

The analyses show that the current increases as the DLC film thickness increases (for cathodes with identical barrier, Φ). This is in accordance with larger lowering of the output barrier of the coated cathode at increased DLC film thickness values (Figure 4.9). The lower the electric field (F) is, the larger the observed difference is (see curves 2–5 in Figure 4.11a). The current saturates at the same limiting value with increasing F, but the limiting value of J is higher for lower values of potential-barrier height. The resonant maxima of J and of the differential emission conductivity ($G = dJ/dF$) take place at values of the electric field, where the longitudinal energy of emitted electrons coincides with the resonant energy level in the QS subband of the DLC quantum well. An increase of d for constant value of F leads to lowering of the bottom of the DLC quantum well, and resonant levels in the QW are shifted below the emitter conduction band. Therefore, the resonance maxima of J and G are moved to lower electric fields and are narrowed as d is increased. Their largest magnitude is reached at certain optimum values of F which depend on the potential-barrier height and thickness of the DLC coating. This occurred in our case at $d = 5$ nm, $\Phi = 1.5$ eV. A further increase of d leads to the appearance of a second resonance level in the quantum well and, hence, to the additional maximum of J and G at a field near to the threshold value. At a rather large d, the current density at resonance peak to the valley current density, namely, the peak-to-valley ratio (J_{max}/J_{min}), is decreased due to an increase in the difference of the input and output barrier transfer characteristics and

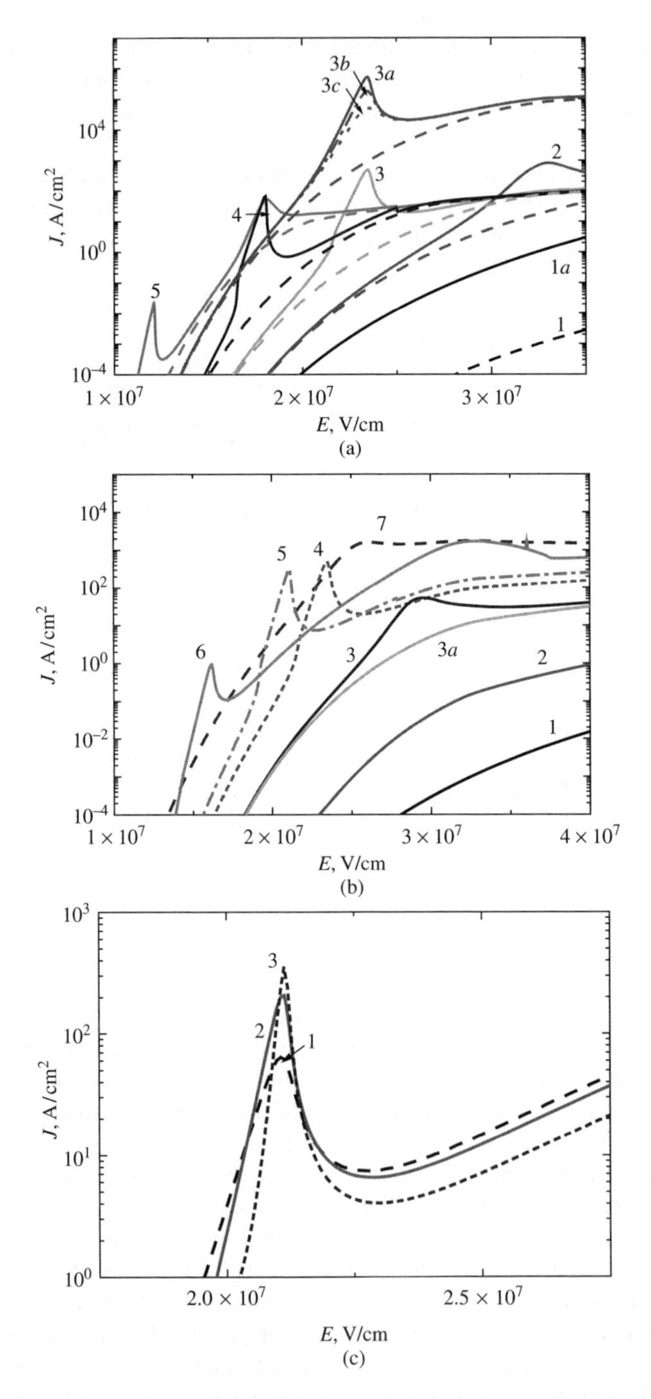

Figure 4.11 Calculated current density vs electric field for Si-DLC-film cathode. (a) $T = 300\,\text{K}$, $\Phi = 1.5\,\text{eV}$: 1–5: $N_e = 1 \times 10^{15}\,\text{cm}^{-3}$ (N_e, emitter doping level); 1a, 3a–c: $N_e = 1 \times 10^{18}\,\text{cm}^{-3}$; 1, 1a: $d = 0$ nm; 2: $d = 4$ nm; 3, 3a–c: $d = 5$ nm; 4: $d = 6$ nm; 5: $d = 8$ nm; 1–5, 3a: $\tau_p = 3 \times 10^{-13}$ s (τ_p, relaxation time); 3b: $\tau_p = 3 \times 10^{-14}$ s; 3c: $\tau_p = 1 \times 10^{-14}$ s; (J_{nr}, dot lines). (b) $T = 300\,\text{K}$, $N_e = 1 \times 10^{15}\,\text{cm}^{-3}$: 1: $d = 0$ nm; 2–7: $d = 5$ nm; 2: $\Phi = 3$ eV; 3: $\Phi = 2$ eV; 4: $\Phi = 1.5$ eV; 5: $\Phi = 1.3$ eV; 6: $\Phi = 0.9$ eV; 7: $\Phi = 0.5$ eV. (c) $N_e = 1 \times 10^{15}\,\text{cm}^{-3}$, $d = 5$ nm, $\Phi = 1.5$ eV: 1: $T = 300$ K; 2: $T = 200$ K; 3: $T = 77$ K. Reproduced with permission from Ref. [44]. Copyright (2004), AIP Publishing LLC

a reduction of electron wave coherence. The resonant emission from DLC coated cathodes is absent when the one-way length for the resonant-tunneling electrons in the quantum well considerably exceeds their De-Broglie wavelength. The J_{max}/J_{min} is also decreased as the electron momentum relaxation time τ_p is reduced (curves 3*a*–*c* in Figure 4.11a).

The influence of the potential-barrier height is illustrated in Figure 4.11b. The increase of the potential-barrier height at constant d leads to reduction of the nonresonant current level and a shift of J and G resonance maxima to higher values of the electric field (curves 3–6 in Figure 4.11b). The resonance maxima of current for the cathode with $d=5$ nm, $\Phi=0.5$ eV at $F=2.5\times10^7$ V/cm (curve 7 in Figure 4.11b) and for the case where $d=8$ nm, $\Phi=1.5$ eV at $F=1.6\times10^7$ V/cm (curve 5 in Figure 4.11a) correspond to the resonant tunneling through the second resonant level in the quantum well, while other maxima (Figure 4.11) are caused by tunneling through the first energy level in the QW.

The temperature dependence of the above characteristics has been analyzed between 300 and 77 K. The decrease of temperature leads to increase and narrowing of resonant current peak and J_{max}/J_{min} ratio due to the increase of the electron momentum relaxation time τ_p (Figure 4.11c). On the other hand, the nonresonant current is reduced due to the decrease of the average energy of emitted electrons.

The time constants τ_1 and τ_2 of electron delay at the first and second resonant levels in the quantum well were calculated for the investigated electric-field values and DLC coating parameters. Their values were found to be between 5.0×10^{-15} s and 3.5×10^{-13} s, and are significantly shorter than the maximum values of τ_n for multilayer Si-SiO_2-δ-Si-SiO_2 cathodes [106] because of the greater transparency of the potential barriers of DLC cathodes in comparison with multilayer structures. The observed negative dynamic conductivity suggests the possibility of generation or amplification of electromagnetic oscillations in diode structures based on DLC-coated silicon cathodes. Oscillations in the frequency range from 1.5 to 100 THz are expected using such structures.

4.5.1.2 Model of Field Emission through QWs with Two-Side Barrier

The band model for a field emitter using QW with two-side barrier design is shown in Figure 4.10. This structure was proposed in the form of a Si field emitter, coated by an ultrathin oxide film with an embedded δ-Si layer. The calculation was performed for degenerated bulk carriers (electrons) and quantized electron states in the surface space-charge region (SCR) using the triangular well approximation on the left-hand side in Figure 4.10. Here the emission current is also determined by the sum of two (but different from the above case) contributions: $J=J_b+J_s$, where J_b is the bulk contribution and J_s is the contribution from the quantum levels in SCR. J_b can be calculated using Equation (4.84), while the following equation can be used for the current from the SCR quantum levels:

$$J_s=\sum_j\sum_{i=1}^{N} n_j\left(\frac{m_{dj}}{\pi\hbar^2}\right)kT\times\ln\left[1+\exp\left(-\frac{E_i^j-E_F}{kT}\right)\right]v_i^jT_j(E_i^j), \qquad (4.86)$$

where v_i^j is the frequency of electron oscillations on the level E_i^j in the triangular SCR well,

$$v_i^j=\frac{qF}{2(2m_{3j}E_i^j)^{1/2}}. \qquad (4.87)$$

The transmission coefficient T_j was calculated by solving numerically Schrodinger's equation for the considered system starting from the vacuum region with a wave function of quasiclassical type

$$\psi = C\{2[W - V(z)]\}^{-1/4} \exp\left[i \int_{z_0}^{z} \sqrt{2(W - V(z)}dz\right] \tag{4.88}$$

and using the boundary conditions $\Psi_+ = \Psi_-$ and $(1/m_+)\Psi_+/dz = (1/m_-)\Psi_-/dz$ at the interfaces to obtain a continuous quantum flux.

Figure 4.12 compares the Fowler–Nordheim (F–N) plots of the current-voltage characteristics of various designs using $N_b = 10^{19}$ cm^{-3}, (100) Si orientation, with QW (curve 1, $d_1 = d_2 = d_3 = 1$ nm) and without QW (curve 2, $d_1 = d_2 = 0$, $d_3 = 2$ nm). In this case the degeneracy factor n_j for valleys with heavy m_{3j} electron mass ($j = 1$) is $n_1 = 2$, while it is $n_2 = 4$ for light m_{3j} mass ($j = 2$); $m_{3(j=1)} = 0.92m_0$, $m_{3(j=2)} = 0.19m_0$. It follows from this figure that the introduction of additional QW layers leads to a substantial increase in emission at defined field values (when effective resonant tunneling occurs) and makes it possible to achieve emission currents higher than those obtained when using the system without such additional layers. The nonmonotonous character in the F–N plot for the system with QW is caused by the presence of resonance tunneling between quantum levels in QW and the levels in SCR at certain field values. The inset of Figure 4.12 shows the ratio of the current from QW with the thickness $d_2 = 1$ nm to that from the system with $d_1 = d_2 = 0$ nm (see Figure 4.10) and demonstrates the appearance of pronounced resonance peaks in the current from the QW system. Increasing of the concentration of carriers in the bulk of Si (from $N_b = 10^{18}$ to 10^{20} cm^{-3}) leads to a remarkable increase of the resonance peaks values.

The influence of the emitter doping level N_e is more clearly shown in Figure 4.13. Lowering the N_e value from 5.0×10^{19} cm^{-3} to 5.6×10^{17} cm^{-3} for the structure with $d_1 = d_3 = 0.5$ nm,

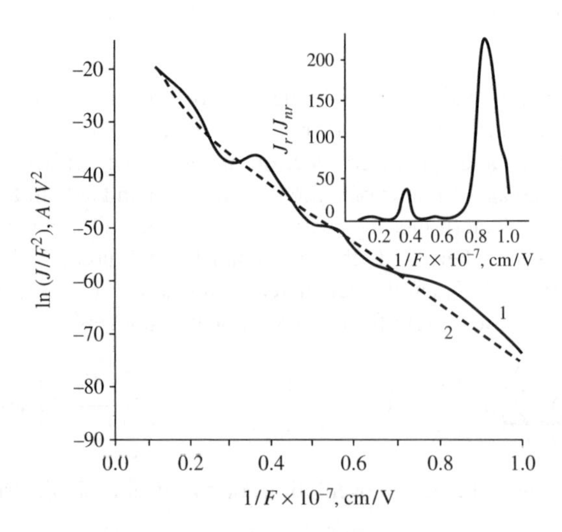

Figure 4.12 Fowler–Nordheim currents from MLC structures: 1: $d_1 = d_2 = d_3 = 1$ nm (j_1) and 2: $d_1 = 0$ nm, $d_2 = 0$ nm, $d_3 = 2$ nm (j_2) (in insert-relationship between currents for two structures). Reproduced with permission from Ref. [44]. Copyright (2004), AIP Publishing LLC

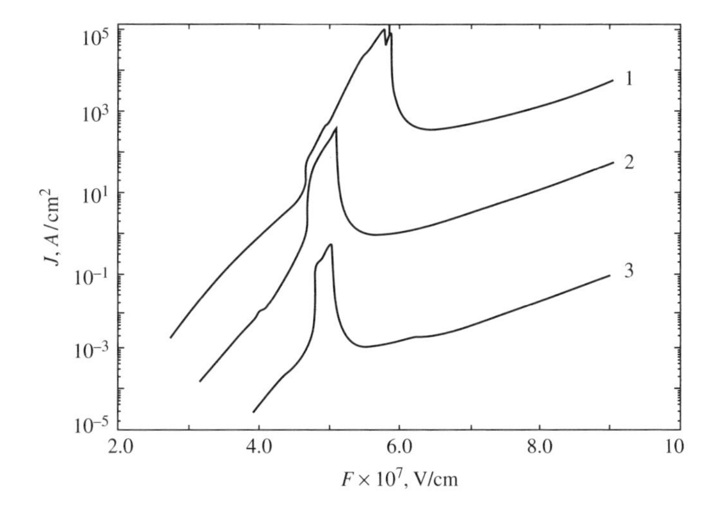

Figure 4.13 Theoretical current density on electric-field dependences: 1: $N_e = 5 \times 10^{19}$ cm^{-3}; 2: $N_e = 5.6 \times 10^{17}$ cm^{-3}; and 3: $N_e = 1.5 \times 10^{15}$ cm^{-3}. $d_2 = 1$ nm, $d_1 = d_3 = 0.5$ nm, $N_w = 5.6 \times 10^{17}$ cm^{-3}, $T = 300$ K. Reproduced with permission from Ref. [44]. Copyright (2004), AIP Publishing LLC

$d_2 = 1$ nm and carrier concentration in the QW of $N_w = 5.6 \times 10^{17}$ cm^{-3} leads to a considerable decrease (about 10^2 times) of resonant and the nonresonant current, a narrowing, and a shift of the current resonant maxima (CRM) to smaller electric fields values (about 7×10^6 V/cm). This happens without a change of J_{max}/J_{min} in the resonance region. Such a result can be explained, in the case of nondegenerated semiconductor, by the smaller width of the electron distribution function and the reduced dependence of the density of states on the electron energy. In this case the CRM occurs at the region of largest overlap between the two-dimensional (2D) electron miniband in QWs and the three-dimensional (3D) electron energy distribution at the emitter. For the degenerate emitter electron distribution the CRM coincides with the overlap of two-dimensional electron miniband in QWs and the bottom of the conduction band in the emitter.

The dependence of the current-voltage characteristic of a multiplayer cathode (MLC) on carrier concentration in the QWs N_w has been calculated. The position of the conduction band bottom in the QW layer is higher than that in the emitter layer. This leads to the increase of the resonant electric field value for every resonant level in the QW. For the MLC with $d_1 = d_3 = 0.5$ nm the resonant level in the QW coincides with the conduction band bottom of emitter at field values of 8.3×10^7 and 5.75×10^7 V/cm for intrinsic and doped to 5.6×10^{17} cm^{-3} QW material, respectively. The CRM is somewhat narrower due to a small electron lifetime constant.

The width of the CRM decreases at lower resonant electric-field values and wider potential barriers. For every MLC parameter there is an optimum number of resonant levels in the QW. Resonant tunneling through them corresponds to the highest CRM and J_{max}/J_{min} ratio values at resonance. The highest values of CRM ($J_m = 10^5 - 10^6$ A/cm^2) and J_{max}/J_{min} ratio at resonance (10^2) take place at temperature $T = 300$ K and field E in the range $4.5 \times 10^7 - 8.5 \times 10^7$ V/cm for MLC with $d_1 = d_3 = 0.5$ nm (Figure 4.14, curves 2–5), and N_e and N_w equal to 1.5×10^{19} and 5.6×10^{17} cm^{-3}, respectively.

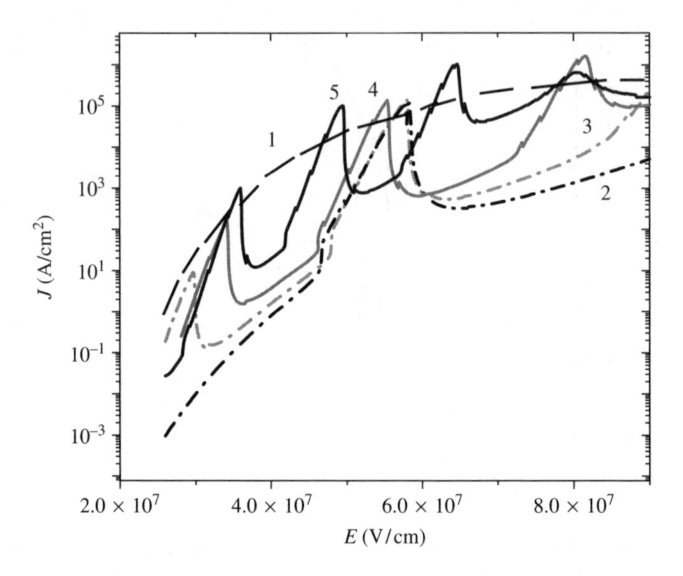

Figure 4.14 Theoretical current density on electric-field dependences: 1: $d_2 = d_1 = 0$, $d_3 = 0.5$ nm and 2–5: $d_1 = d_3 = 0.5$ nm, $d_2 = 1$, 1.5, 2, and 3 nm, respectively; $N_e = 5 \times 10^{19}$ cm^{-3}, $N_w = 5.6 \times 10^{17}$ cm^{-3}, $T = 300$ K. Reproduced with permission from Ref. [44]. Copyright (2004), AIP Publishing LLC

With increasing d_1 and d_2 the J_{max}/J_{min} ratio at resonance and the CRM are decreased. However, for quite thin potential-barrier layers (0.5 nm) and small QW widths (1–3 nm) the resonant current density of the MLC for resonant electric-field values of more than 3.5×10^7 V/cm is a few times higher than the current density in conventional cathodes under the same electric-field values (Figure 4.14, curve 1). These points out the possibility of the effective work function reduction using the MLC with optimum parameters. A remarkable shift of the main peak to lower electric fields is also observed when the thickness of QW layer d_2 is decreased (from 2 down to 0.5 nm).

The resonance emission current was evaluated at different temperatures ($T = 300$, 77, and 40 K) under the assumption that the time constants of electron pulse relaxation in Si are equal to 5.0×10^{-13}, 1.8×10^{-12}, and 2.5×10^{-12} s. The resonant-current temperature dependence is reduced with field. This is explained by the increase of the contribution of resonant-level natural broadening in comparison with that due to relaxation broadening. By lowering the temperature from 300 down to 77 K the nonresonant current level is reduced and the J_{max}/J_{min} ratio at resonance becomes three times higher while CRM is narrowed (1.5 times) because of the electron-mobility increase and redistribution of the electron state density in the emitter conduction band. For $T = 40$ K the CRM is narrower in comparison with $T = 77$ K because of the higher electron mobility.

4.5.2 Two-Step Electron Tunneling through Electronic States in a Nanoparticle

A frequent observation in FE experiments is the steplike *I-V* characteristic [97, 102, 106], which can be associated with a two-step tunneling phenomenon. The explanation behind the peculiar behavior of the *I-V* characteristics resides in the existence of nanometer-sized particles (or layers) on the sample that produces corresponding nonmonotonic spatial variations in the potential energy. The two-step tunneling was described in detail in Ref. [94].

A possible complex cathode configuration is one which contains discrete nanoparticles in its structure. Due to their size, quantization of the electron energy takes place, which has important effects on the field-emission characteristics.

The structure examined is simplified to the metallic substrate, which acts as the cathode, on top of which a wide band-gap (WBG) material has been deposited [94]. One quantum confining structure (QCS) (e.g., metallic nanoparticle) is placed on the top of the dielectric material (Figure 4.15a). The potential energy within the QCS is assumed to be lower than its level in the cathode, so that the dielectric layer plays the role of a barrier which controls the supply of charge from the bulk. Figure 4.15b shows the energy diagram for this idealized system when no

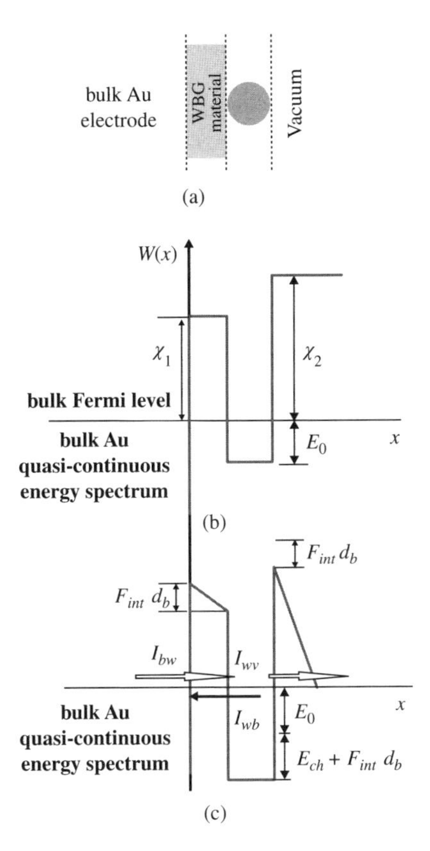

Figure 4.15 (a) Schematics of the idealized system representing a bulk cathode separated from a spherical nanoparticle by a thin WBG material. (b) The potential-energy diagram for the system in (a) (not drawn to scale) at thermodynamic equilibrium with no applied field in vacuum. (c) The potential-energy diagram for the system in (a) after the application of the electric field. The spatial ranges are of the order of few nanometers and the energies extend to a few electronvolt. The origin for the energy scale is set at the Fermi level of the bulk metallic substrate. Reprinted with permission from Ref. [94]. Copyright (2009) by the American Physical Society

electric field is applied in vacuum. To avoid unnecessary technical complications, the present model is based on the common one-dimensional electron-transport approach, which is present in various theoretical studies related to the tunneling phenomenon [44, 49, 95, 101, 103].

More precisely, although the real system consisting of the bulk Au electrode, the WBG material, and the QCS on the top is essentially a three-dimensional (3D) structure, the electron tunneling rates in and out of the QCS will be described using the 1D WKB approximation. Therefore, a simple 1D approach for the energy diagram of that system is more suited. However, not all the aspects of the system under consideration were 1D. For example, the electron current from the bulk inside the well was calculated using the notion of a supply function [72], which uses the "perpendicular" component of the wave vector in the substrate. The emission takes place from the QCS toward the vacuum through a second barrier whose transparency is controlled by the applied electric field. In order to point out the influence of the discrete energy levels on the FE current, a confinement region (nanoparticle) within the heterostructure is modeled as a rectangular asymmetric 1D potential well. The origin of the potential energy is considered to be at the bulk Fermi level. When the system is in thermodynamic equilibrium, there is no electron flow and the Fermi level of the nanoparticle equals that of the bulk. The nanoparticle is assumed to be electrically neutral in this slate. Given a potential well, the energy levels for the thermodynamic equilibrium case, E_l^0 can be found in the usual way and considering a Fermi-Dirac statistics, the initial average number of electrons can be calculated as

$$n_0 = 2\sum_{l=1}^{N_0} f_F(E_l^0, 0), \tag{4.89}$$

where $f_F(E, \mu) = 1/\{1 + \exp[(E - \mu)/(k_B T)]\}$ is the Fermi-Dirac distribution, the factor 2 accounts for the spin degeneracy, N_0 is the number of energy levels at equilibrium, and μ is the local chemical potential, which, in thermodynamic equilibrium, equals the bulk Fermi level. Other symbols in Equation (4.89) are k_B, the Boltzmann constant and the overall absolute temperature T.

When an electric field F is applied, the thermodynamic equilibrium is broken and an electron flow is established through the system. For a given value of the applied electric field it is assumed that the system reaches a steady nonequilibrium state. As the electrons in the nanoparticle are separated from those of the substrate by a consistent potential barrier, it is assumed that the local thermodynamic equilibrium can be considered in each region. With this hypothesis, the local chemical potential can be defined for the electrons contained in the well as a measure of their average number. The electron flow through the system will result in the nanoparticle becoming charged due to excess or depletion of electrons in the potential well. The related charging energy can be described as

$$E_{Ch}(n) = \frac{|n_0 - n|\,(n - n_0)q^2}{2C}, \tag{4.90}$$

where n is the average number of electrons inside the well for a given steady-state flow condition (that is, for a given applied field) and C is the electrical capacitance of the nanoparticle and takes values around the capacitance of a conductive spherical nanoparticle of the same diameter. Some remarks regarding the form of Equation (4.90) are in order. The one-electron Hamiltonian used in the present model has position-independent potential energy in the well.

In a rigorous treatment, any extra charge introduced in the nanoparticle should produce a potential-energy term that is a solution of 3D Poisson equation with some specified boundary conditions. While useful for quantitative comparison with experiments, such an approach would greatly complicate the model and impose a more rigorous 3D approach of the electron transfer at the nanoparticle boundary. Therefore, in order to keep the model as simple as possible, the position-independent charging energy in the well was adopted. However, in order to keep with the rigorous solutions of the Poisson equation, the potential energy resulting from the presence of an extra charge distributed in the well has to be made sensitive to the sign of this addition; an electron depletion in the nanoparticle will give rise to a decrease in the potential-energy term of the electronic Hamiltonian while an electron excess will have the opposite effect. In order to outline this extra-charge sign influence in the Hamiltonian, the product $|n-n_0|(n-n_0)$ has been used in Equation (4.90). As the nanoparticle is assumed electrically neutral in the zero-field situation, n_0 represents the number of positive elementary charges in the well.

The application of the external electric field not only removes the thermodynamic equilibrium of the system, but also generates two important effects in the QCS. The first, and the most common, is an overall decrease in the potential barriers together with the well's bottom (Figure 4.15c). This is due to the electric-field penetration into the WBG material, which is obtained from the Gauss theorem by taking into account the electric charge inside the well. Considering a surface area σ for the QCS, transverse on the direction of the electron flow, the internal field F_{int}, is obtained as a function of the average number of electrons n, inside the QCS and the external applied field F,

$$F_{\text{int}}(n,F) = \frac{1}{\varepsilon}\left(F + \frac{q\left(n_0 - n\right)}{\varepsilon_0 \sigma}\right), \tag{4.91}$$

where ε is the relative permittivity of the WBG material. The external field will produce the overall thinning of the potential barriers in the WBG layer and at the vacuum interface. It will not, however, interfere with the well's energy-level distribution or their number. The second influence of the electric field is the change in the number of electrons residing in the nanoparticle. A consequent local charging will occur, as described by Equation (4.90). The bottom of the potential well will therefore change due to both these combined effects: field penetration and local charging (which also depends on the applied field). As a result, the number and the positions of the energy levels inside the well will change accordingly with the average number of electrons present inside the well and with the applied field F. Unfortunately such dependence cannot be written in any analytical compact form and has to be obtained numerically, as will be described below.

In the hypothesis of local thermodynamic equilibrium inside the potential well, a Fermi-Dirac distribution of electrons over the available energy levels is assumed. The connection between the average number of electrons and the local chemical potential can be obtained from the sum of the average population of each of the $N(n)$ energy levels, $E_l(n,F)$,

$$n = 2\sum_{l=1}^{N(n)} f_F[E_l(n,F),\mu]. \tag{4.92}$$

The above transcendental equation provides the dependence of the chemical potential $\mu(n,F)$ on the average number of electrons inside the potential well and on the applied field F. Due

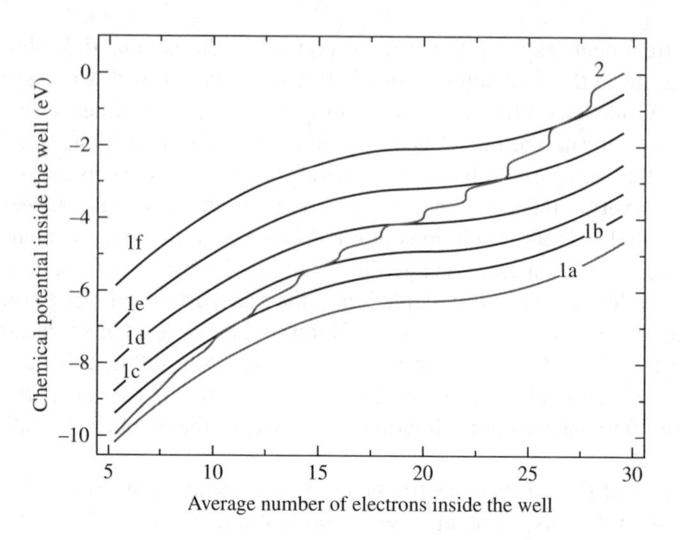

Figure 4.16 The variation in the potential-well bottom energy levels and local chemical potential relative to $\chi_l - F_{int}d_b$ in the nonthermodynamic equilibrium state with the average number of electrons in the well. For convenience, only a limited number of energy levels have been displayed, $\chi_l = 4.2$ eV, $\chi_2 = 4.3$ eV, $E_0 = -1$ eV, $d_b = 3$ nm, $d_w = 3$ nm, $C = 3.3 \times 10^{-17}$ F, and $T = 300$ K. Reprinted with permission from Ref. [94]. Copyright (2009) by the American Physical Society

to discrete structure of the energy spectrum, when n increases there are abrupt shifts in the number of populated energy levels in the well and the obtained n dependence of the chemical potential presents a characteristic staircaselike shape. Figure 4.16 shows the calculated dependence of the chemical potential (relative to $\chi_l - F_{int}d_b$) on the average number of electrons in the well. As the average electron number increases, the well's bottom moves upward followed by the energy levels and the chemical potential is forced to increase every time of the maximum population is attained at each predetermined level corresponding to the size of the particle (well). This peculiar shape of the chemical potential has important consequences for the behavior of the current of electrons tunneling out of the well.

Having found the energy levels and the chemical potential for a given electron content of the well, it is now possible to calculate the electron current. A two-step tunneling electron transfer phenomenon is proposed in which steady-state equilibrium is attained between the electron flow from the substrate into the well and from the well, both into vacuum as FE current, and back into the substrate. To compute the electron supply current from the bulk substrate, $I_{bw,}$ the well-known kinetic formalism of the supply function (SF) is used [107]. Electrons with energies different from the confinement levels can enter the well through inelastic processes. Such processes are assumed to take place in the barrier region and at the interfaces,

$$I_{bw}(n, F) = q\sigma \int f_{supply}(E_\perp) T_{bw}(n, F, E_\perp) dE_\perp, \tag{4.93}$$

where σ is the cross-sectional area of the nanoparticle and $E_\perp$ is the so-called "transverse" part of the incoming electron energy. The symbol $f_{supply}(E_\perp)$ represents the electron supply function defined as [106]

$$f_{supply}(E_{\perp}) = \frac{m_0 k_B T}{2\pi^2 \hbar^3} \ln\left[1 + \exp\left(-\frac{E_{\perp}}{k_B T}\right)\right], \tag{4.94}$$

where m_0 is the free-electron mass. $T_{bw}(n,F,E)$ in Equation (4.93) is the 1D WKB transmission coefficient for the potential barrier between the bulk substrate and the potential well (Figure 4.15c). For simplicity, it will be approximated in computations by the square of the potential-well width d_w. The transmission coefficients have been expressed using the WKB approximation through the generic formula

$$T(E) = \exp\left\{-\int_{x1}^{x2} \sqrt{\frac{8m}{\hbar^2}\left[E_b(x) - E\right]}dx\right\}, \tag{4.95}$$

where $E_b(x)$ is the potential energy in the 1D potential barrier, x_1 and x_2 are the energy-dependent turning points of the potential barrier, and m is the electron effective mass in the barrier region [108]. In order to avoid irrelevant complications by matching Schrodinger's equation solutions across various layers, the effective mass will be held position invariant in the numerical computations. The electron current (or the FE current) from the potential well toward the vacuum can be obtained through the attempt-to-escape frequency formalism [109]

$$I_{wv}(n,F) = 2q\sum_l f[E_l(n,F), \mu(n,F)]T_{wv}[n,F,E_l(n,F)] \times f_{att}[n,F,E_l(n,F)], \tag{4.96}$$

where

$$f_{att}(n,F,E) = \frac{1}{d_w}\sqrt{\frac{2}{m_0}[E - E_0 - E_{Ch}(n,F)]} \tag{4.97}$$

is the attempt-to-escape frequency of an electron with energy E inside the well and T_{wv} is the transmission probability through the vacuum barrier. The current in Equation (4.96) is obtained through the summation over all the energy levels inside the potential well. Finally, the well-to-bulk current takes the form

$$I_{wb}(n,F) = 4q\sum_l \Phi[E_l(n,F), \mu(n,F)]T_{wb}[n,F,E_l(n,F)] \times f_{att}[n,F,E_l(n,F)], \tag{4.98}$$

where

$$\Phi(E,\mu) = f(E,\mu)[1 - f(E,0)] \tag{4.99}$$

is the probability for an electron inside the potential well, initially occupying a given energy level E to find an empty level of the same energy in the bulk.

The essential assumption of the present model is that steady-state equilibrium is attained and the balance occurs between the stationary incoming and outgoing electron fluxes from the QCS. No reverse electron current from vacuum into the well is allowed. The balance equation can be written as a function of the average number of electrons inside the potential well, n, and of the applied external field F as

$$I_{in}(n, F) = I_{out}(n, F), \tag{4.100}$$

where $I_{in}(n, F) = I_{bw}(n, F)$ and $I_{out}(n, F) = I_{wv}(n, F) + I_{wb}(n, F)$ are the well's incoming and outgoing currents, respectively. Equation (4.100) can, in principle, provide a stationary value of the average number of electrons in the QCS for a given external field. Thus, the balance between the incoming and outgoing currents will finally decide the average amount of electrons within the potential well suited to each value of the applied field.

When an electric field is applied to the system, the average electron number in the potential well varies and its structure changes accordingly (Figure 4.16). The applied field thus moves the local chemical potential. For example, at high field strengths, when the electron population decreases, the chemical potential approaches the bottom of the potential well. As a consequence of the electron tight confinement in the QCS, the allowed energy levels are well separated and the local chemical potential, as given by Equation (4.92), experiences regular jumps.

This behavior is further transferred to the currents flowing out of the well. By contrast, the incoming electron flux from the quasicontinuous spectrum of the substrate will be almost unaffected by the well's quantization. At each value of the field, the intersection of $I_{in}(n)$ and $I_{out}(n)$ a steady nonequilibrium state determining the value of n is appeared. Therefore, by varying the applied field, the balance between the smooth $I_{in}(n)$ and the ridged $I_{out}(n)$ resulst in a steplike shape (or at least in some slope discontinuities) of the device's current-field characteristics.

It must be stressed out that although this behavior appears similar to what is described as Coulomb blockade in electron field emission (FECB) [95], the two phenomena differ essentially in their background. In the FECB theory, unlike the classical treatment of the Coulomb blockade [87], the ladder-shaped characteristics are mainly an effect of the discreteness of the electric charge and of the local charging of the particle. In the present model, the confinement quantization of the electron energy in the QCS plays the main role, along with the local variations in the number of electrons produced by the strong external field. For this reason, Coulomb-blockade effects are normally observed at low biases and low temperatures [110, 111], where the local charging energy is not exceeded by the electron thermal energy or by the energy provided by the applied electric field. This effect is typically manifested by blocking the current through the QCS. By contrast, two-step electron tunneling kinetic model is supposed to be less affected by the temperature change (since the involved energy leaps are hundreds of times larger than the electron thermal energy) and will give rise only to alternating parts of high- and less-bias sensitivity in the characteristics. This description of FE current is simplified to the stationary flow of electrons driven by chemical-potential differences between adjacent zones assumed in local thermodynamic equilibrium. The emitted current is controlled by the high field applied to the QCS that penetrates into the substrate barrier and leads also to local charging effects.

The first important result of the sequential tunneling model is the calculated field-emission characteristic as a function of the applied field in vacuum. The steplike shape is obtained for the emission current using the present model. The electric field has been varied from 1 to 3 V/nm and the choice is not arbitrary. For very low values of the applied field the vacuum barrier is too opaque, so that no balance between the incoming and the outgoing currents is possible. On the contrary, for too large value of the field, the QCS is so depleted of electrons that only the very bottom levels in the well (which are much denser) are occupied and the steplike shape of the chemical potential is washed out. As a consequence, the features in the characteristic also disappear.

It is interesting to investigate how the properties of the nanoparticle and the WBG potential barrier influence the current-field characteristics. For example, changing the nanoparticle diameter will automatically affect the number and the positions of the energy levels inside the potential well, which leads to substantial modifications in the current. The results are summarized in Figure 4.17. As can be seen, the structure of the diagram changes as the potential well becomes larger (Figure 4.17a). The number of steplike features increases as the potential-well width increases because wider wells have higher densities of available energy levels. For the same reason, at large values of d_w the chemical potential will be located deeper into the well, electrons will face thicker vacuum barrier. The current is thus expected to be overall smaller from larger particles. However, a different behavior is obtained if the parameter d_w is maintained constant and the WBG thickness is varied (Figure 4.17b). There are again two features to be observed. The first one is the decreasing number of steps as the potential barrier is getting thicker. Thicker barrier toward the substrate means smaller income of elections in the QCB and more advanced electron depletion therein. Consequently, the lower and denser energy levels will be occupied and the stairs are again washed out. Also, for a thicker substrate barrier, the emitted current at small fields is lower. Yet the two cases seem to converge toward similar values as the field is increased. This is due to the fact that, at higher fields, the deformation of both

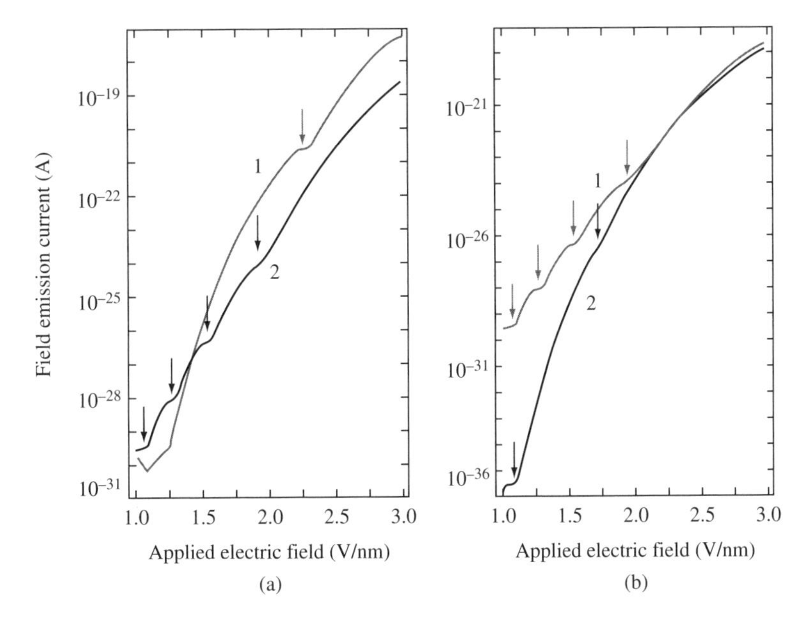

Figure 4.17 (a) Field-emission current as a function of the applied electric field for two values of the potential-well width: 1: $d_w = 3$ nm and 2: $d_w = 6$ nm. The potential-barrier thickness is $d_b = 3$ nm, $\chi_1 = 4.2$ eV, $\chi_2 = 4.3$ eV, $E_0 = -3$ eV, $C = 4.45 \times 10^{-17}$ F, and $T = 300$ K. (b) Field-emission current as a function of the applied electric field for two values of the potential-barrier thickness: 1: $d_b = 3$ nm, and 2: $d_b = 4.5$ nm. $d_w = 6$ nm, $\chi_1 = 4.2$ eV, $\chi_2 = 4.3$ eV, $E_0 = -3$ eV, $C = 4.45 \times 10^{-17}$ F, and $T = 300$ K. Reprinted with permission from Ref. [94]. Copyright (2009) by the American Physical Society

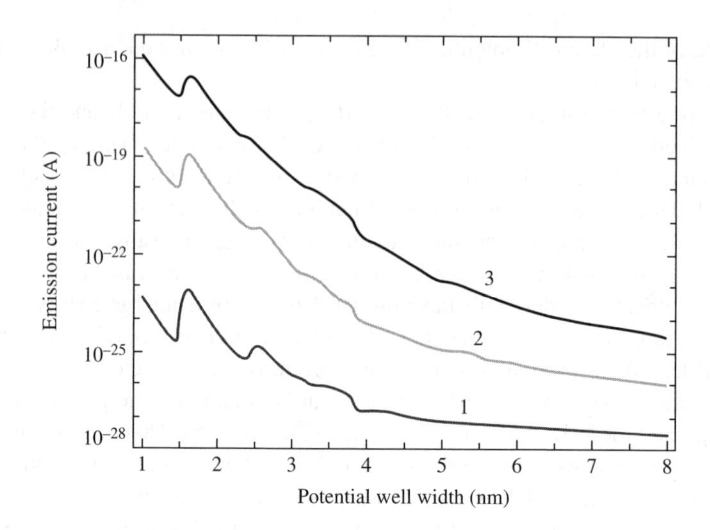

Figure 4.18 Field-emission current as a function of the potential-well width d_w for three values of the applied field: 1: 1.5 V/nm, 2: 2 V/nm, and 3: 2.5 V/nm. $d_b = 3$ nm, $\chi_1 = 4.2$ eV, $\chi_2 = 4.3$ eV, $E_0 = -1$ eV, $C = 4.45 \times 10^{-17}$ F, and $T = 300$ K. Reprinted with permission from Ref. [94]. Copyright (2009) by the American Physical Society

the vacuum barrier and the potential-energy barrier in the WBG material increases the electron injection from the bulk, and thus the chemical potentials in the two cases become comparable.

At this point it is worth noting that the average number of electrons in the QCS n is expected to be less than n_0 (the value for the substrate-well thermodynamic equilibrium) for large extraction fields. The well region will therefore be positively charged in such cases. However, when the applied field is small, the value of n may exceed n_0 as well and the QCS becomes negatively charged in the steady state. Therefore, even if for some value of the applied field the equality $n = n_0$ may occur, this should not be interpreted as an overall thermodynamic equilibrium situation; as the system is under external field, it is open to electron flows and the thermodynamic equilibrium can be approximately conceived only in very limited regions.

In order to better emphasize the influence of the electron confinement in the QCS on the FE current, the dependence of the current versus the potential-well width has been plotted in Figure 4.18. It can be observed that, for narrow wells, for which the confinement is tight and the separation between the energy levels is large, several sharp "oscillations" may appear in the FE current. This is typical for quantum behavior that follows from the (field-induced) shift of the quantized levels in the energy range that is the most favored for emission into the vacuum. The effect is rapidly wiped away by the increase in the well width that would sharply decrease the separation between the energy levels. Another feature to be observed in the data of Figure 4.18 is the abrupt decrease in the FE current with the increase in the well's width. As has already been pointed out, this is a consequence of the variation in the local chemical potential in the well. For wider wells, where the energy levels are denser, the steady-state number of electrons will occupy the states with energies close to the bottom of the well. Consequently, the chemical potential approaches the well's bottom in an energy range, where the vacuum barrier is thicker and this will lead to a severe decrease in the FE current.

The model described above has been developed around the idea that various electron-confinement regions present all the surface of an emitter (e.g., insulated conductive nanoparticles) may interfere in a sequential way with the tunneling transport process to produce the steplike features in the field-emission characteristics. Such circumstances can appear experimentally in various configurations [97, 102, 106] but a dedicated experiment has been needed to sustain the theoretical model presented above. Therefore, using a wet chemistry technique, a structure similar to the one sketched in Figure 4.15a has been obtained as follows [94]. A microscope glass slide has been covered with a layer of sputtered Au, 100 nm thick, which represents the bulk electrode in Figure 4.15a. Following the Au deposition, the so-called layer-by-layer (LbL) deposition technique has been used to realize the WBG layer. This technique is a very flexible deposition method, which is template assisted and develops from a solid substrate provided with an electrostatic charge.

The last stage in the sample preparation was adding the "impurities" which consisted of Au nanoparticles with a nominal diameter between 3.5 and 5.5 nm.

Experimental measurements of field emission *I-V* characteristics showed a clear sequential tunneling characteristic containing the steplike features. One important aspect that has been noted is the difference between the values of theoretical current and the experimental data. This is due to the fact that the theoretical current is obtained from a model which only takes into account only a single nanoparticle, while the experimental data has been obtained from the extended area covered with Au nanoparticles. Another factor explaining the quantitative discrepancy between theory and experiment lies in the lack of accurate correlation between the anode potential and the local field generated at the surface of each emitting nano-particle. While quantitative reproduction of experimental results cannot be expected from such a simplified approach, its main merit is the apparition of the steplike features for the room-temperature field-emission experiment described above [94]. It can be concluded that the experimental data support of the theoretical model presented above and the further refinements of this method could lead to more quantitative comparison with the experimental data.

4.5.3 *Single-Electron Field Emission*

The single-EFE was described in detail in Ref. [95].

The discrete nature of electric charge reveals itself in the transport of electrons through small conductors (nanoparticles or other nanoscale objects) weakly coupled to the source and drain electrodes (current-carrying leads) owing to the Coulomb blockade effect. There are numerous manifestations of the charge quantization in transport properties. The most familiar of them are the Coulomb blockade oscillations of the electric current as a function of the gate voltage and the Coulomb staircase in the current-voltage characteristics [112]. Since the fundamentals of the transport theory in the Coulomb blockade regime have been established [1, 87, 89, 113], the Coulomb blockade-based physics has been applied to various issues of electron transport in mesoscopic systems, and the field of its applications expands in line with the advances in nanotechnology.

Usually, the influence of the Coulomb blockade on the current in two-terminal devices is considered under the assumption that the coupling between the nanoscale object and the leads is not sensitive to the number of electrons N determining the object charge qN. This corresponds to the introduction of ohmic (or nearly ohmic) effective resistances describing this coupling. Though this assumption often works well, it can be violated, for example, in nanomechanical systems [114–116], where charging of the object gives rise to its displacement toward one of the leads thereby changing its tunnel coupling to both leads.

Let's consider structure when small (nanoscale) objects are formed on the source electrode (cathode), the latter is then negatively biased with respect to the drain electrode (anode) in vacuum [95]. The current between the electrodes flows owing to the field emission of electrons from nanoscale objects, because the electric field F at the tips of the objects is higher than in the other places of the device. The field-emission current can be described by the Fowler–Nordheim formula [3].

$$I = ASF^2 \exp\left(-\frac{B}{F}\right), B = \frac{4\sqrt{2m}}{3q\hbar}\Phi^{3/2}, \tag{4.101}$$

where m is the free electron mass, Φ is the work function of the emitting material, S is the effective emitting area, and A is a constant expressed through the work function and Fermi energy E_F of the emitting material

$$A = \frac{q^3\sqrt{E_F/\Phi}}{4\pi^2\hbar(E_F + \Phi)}. \tag{4.102}$$

The effective field F, which describes the tunnel coupling between the nanoscale object and the anode, depends on the object charge, which is induced by the applied voltage $V = V_1 - V_2$, where V_1 and V_2 are the cathode and anode potentials, respectively. Under conditions of Coulomb blockade, that is, when the electric connection between the cathode and the object is weak and the charging energy of the object considerably exceeds the temperature T, the continuous variation of the voltage V leads to discrete changes of the object charge in units of q, and, consequently, to corresponding discrete changes of the field F. Therefore, one may introduce the field F_N, which is a function of the discrete number N and continuous variable V. Next, if the current in the device is limited by the field emission, the single-electron tunneling processes become important. This means that, at a fixed voltage V, the object stays mostly in the states with N and $N - 1$ electrons, the number N is determined by the voltage. In the N-electron state, no electrons can come to the object from the cathode until an electron leaves the object by tunneling through the barrier (Figure 4.19a). Then the object appears in the $N - 1$-electron state and returns to the N-electron state before the next Fowler–Nordheim tunneling event takes place. The field-emission current in these conditions is given by Fowler–Nordheim with $F = F_N$ and can be denoted as I_N. If the bias qV increases, the state with $N + 1$ electrons becomes more favorable, and the current changes into the step-like fashion from I_N to I_{N+1}. This leads to staircaselike current-voltage characteristics, which may look similar to the usual Coulomb staircases [117–119]. However, since the sensitivity of the tunneling to the number of electrons is involved, the staircaselike current-voltage characteristics can exist under rather peculiar conditions, when the source-drain bias is the orders of magnitude larger than the charging energy.

The case of classical (or metallic) Coulomb blockade, in which the electron energy level separation in the nanoscale object can be neglected in comparison with both temperature and charging energy, is considered. Since the object is assumed to be weakly coupled to the cathode, the study of the sequential tunneling process and not of the coherent one is performed. It is convenient to investigate the electron transport by applying the kinetic equation [113] (Master equation) for the distribution function P_N describing the probability for the object to be in

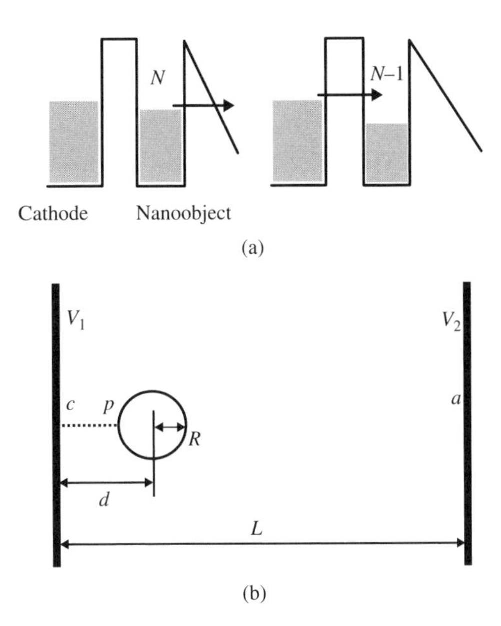

Figure 4.19 (a) The mechanism of single-electron tunneling in the Fowler–Nordheim regime. (b) Schematic representation of the idealized emitter. Reprinted with permission from Ref. [95]. Copyright (2006) by the American Physical Society

the state with N electrons. Assuming that the electric connection between the cathode and the object is characterized by the conductance G, this equation is written as

$$\frac{\partial P_N}{\partial t} = Q_{N+1} - Q_N, \tag{4.103}$$

where

$$Q_N = \frac{G}{q^2} \frac{\Delta E_N}{1 - \exp\left(-\frac{\Delta E_N}{T}\right)} \left[P_N - P_{N-1} \exp\left(-\frac{\Delta E_N}{T}\right)\right] + P_N I_N / q. \tag{4.104}$$

Here $\Delta E_N = \left(-\frac{q^2}{C}\right)[N - 1/2 - C_2 V/q]$ is the difference in Coulomb energies for the objects with N and $N - 1$ electrons, C is the total capacitance, and C_2 is the capacitance of the object with respect to the anode (the capacitance with respect to the cathode is given by $C_1 = C - C_2$). The first term in expression (4.104) has the usual form [113] and corresponds to the current between the object and the cathode. It is written as a difference of the contributions describing the departure of an electron from the object in the N-electron state and arrival of an electron at the object in the $N - 1$-electron state. The second term corresponds to the field-emission current from the object in the N-electron state. Since no electrons come to the object from the anode, this term does not contain a contribution describing arrival of electrons. In the stationary case, Equation (4.103) is reduced to the form $Q_N = const$, where the constant can be chosen equal to zero. After determining P_N from the equation $Q_N = 0$ with the use of the normalization condition $\Sigma_N P_N = 1$, the total current is given by

$$J = \sum_N P_N I_N. \tag{4.105}$$

Under the condition $GT \gg |q|I_N$, which means that the object is in thermal equilibrium with the cathode, the stationary solution of Equation (4.103) is written as $P_N = Z^{-1} exp(-E_N/T)$, where $E_N = (q^2/2C)[N - C_2V/q]^2$ is the Coulomb energy, and $Z = \Sigma_N exp(-E_N/T)$ is the partition function. The current in this case is determined by the expression

$$J = Z^{-1} \sum_N I_N \exp(-E_N/T). \tag{4.106}$$

Let us apply the solution Equation (4.106) to the idealized model of emitter, Figure 4.19b, when the emission takes place from a spherical nanoparticle of radius R, placed at a distance d from the cathode. The distance between the cathode and anode is L. The connection $c - p$ denotes a low-transparent contact (for example, tunnel barrier) between the particle and the cathode, which does not contribute to the field-emission properties and electrostatics of the device. Assuming $d \gg R$, we have $C = R$, $C_2 = Rd/L$, and neglect the charge polarization of the particle because this polarization is small in comparison with the total charge qN induced by the applied voltage. The number of electrons is estimated as $N = C_2V/q = RdF_0/q$, where $F_0 = -V/L$ is the applied electric field. The effective field for the nanoparticle with N electrons is $F_N = qN/R^2$, and the partial currents I_N in these conditions are given by

$$I_N = AS(qN/R^2)^2 \exp(-BR^2/qN), \tag{4.107}$$

where the emitting area S, in the idealized model considered here, can be approximated by the total surface area of the nanoparticle, $S = 4\pi R^2$. In Figure 4.20 we plot the current-voltage characteristics of the idealized emitter, calculated according to Equations (4.106) and (4.107), where A is given by Equation (4.102) with $\Phi = 5.1$ eV and $E_F = 5.5$ eV (taken for Au), and the geometrical parameters are chosen as $R = 5$ nm and $d = 0.5$ μm. The characteristics look like staircases with flat regions (plateaus) between the steps, which are visible even at room temperature. It is possible to estimate the relative heights of the steps by calculating the ratio of the currents I_N and I_{N-1} emitted from the nanoparticle with N and $N - 1$ electrons

$$\frac{I_N}{I_{N-1}} = \exp\left[\frac{BR^2}{qN(N-1)}\right]. \tag{4.108}$$

In spite of the fact that the charged nanoparticle typically contains a large number of electrons, $N \sim 100$, one can always find a regime when the ratio I_N/I_{N-1} is not small in comparison with unity. This necessarily implies a weak Fowler–Nordheim tunneling, when $B/F = BR^2/qN \gg 1$.

In the calculations described above, the applicability of the Fowler–Nordheim formula requires $R \gg \Phi/qF$, which is rewritten as $R \ll q^2N/\Phi$, or, according to $N = RdF_0/q$, as $qF_0 \gg \Phi/d$, independent of the nanoparticle radius. This condition is satisfied at high enough applied voltages. If $qF_0 = qV/L \leq \Phi/d$, the approximation of a triangular potential barrier is not quite good, and one should consider the tunneling through the barrier described by the potential energy $\Phi - q^2N(1/R - 1/r)$ at $r \geq R$, where r is the distance from the center of the spherical nanoparticle; the tunneling through the potential barrier of this form is described in Ref. [120]. Even under the condition $qF_0 \gg \Phi/d$, which is satisfied in the calculations shown in Figure 4.20, the relative change of the current per one step, $I_N/I_{N-1} - 1$, appears to be significant, because the exponent $BR^2/qN(N-1)$ in Equation (4.108) is estimated as $c(\Phi/qF_0d)^2$, where the dimensionless constant $c = 4/3(2mq^4/\hbar^2\Phi)^{1/2}$ is noticeably larger than unity.

If the current is high enough, the field emission cannot remain the bottleneck for the electron transfer from the cathode to the anode, and a finite resistance G^{-1} becomes essential. The nanoparticle in these conditions is no longer in equilibrium with the cathode. This means that the distribution P_N is established kinetically, and several states with different charges coexist at a fixed voltage (see the inset in Figure 4.20). As a consequence, the Coulomb blockade features are washed out. This case requires a numerical solution of the equation $Q_N = 0$. The corresponding current-voltage characteristics of the idealized emitter calculated by using the RC time $C/G = 100$ ps are also shown in Figure 4.20. The degradation of the current steps appears to be stronger with increasing voltage, because the current increases and the nanoparticle-cathode link becomes more important. The shape of the steps in this case resembles the usual Coulomb staircase.

Let's consider more complex cases [95]. The model example discussed above has certain disadvantages. First of all, it is hardly possible to connect a particle placed far from the cathode surface by a link *p-c* (Figure 4.19b), which does not contribute to the electrostatic properties of the device. Second, the model of uniform charging is insufficient: the charge polarization of the

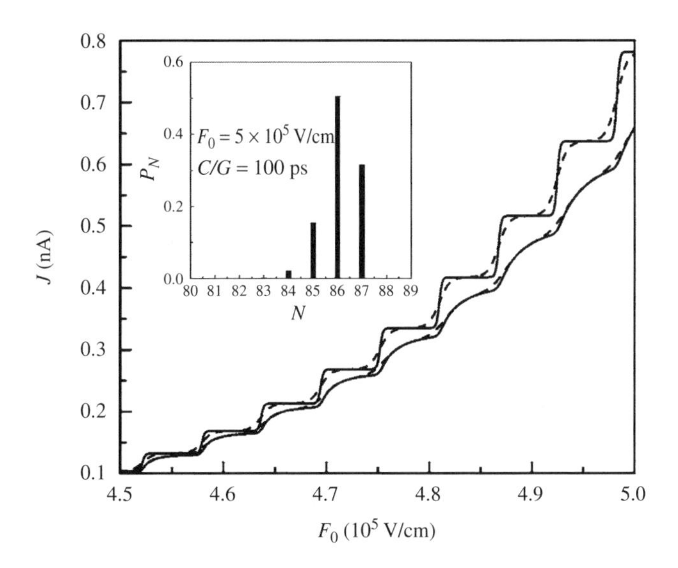

Figure 4.20 Current from the idealized emitter as a function of the applied field $F_0 = -V/L$ for the case of small C/G (nanoparticle in thermal equilibrium with the cathode, upper curves) and for the case of $C/G = 100$ ps (lower curves), at the temperatures $T = 77$ K (solid) and $T = 293$ K (dashed). The inset shows the distribution function P_N at $F_0 = 5 \times 10^5$ V/cm for the second case. Reprinted with permission from Ref. [95]. Copyright (2006) by the American Physical Society

nanoscale object appears to be important and should be always taken into account. Therefore, the model shown in Figure 4.19b is suitable only for the purposes of illustration of the basic physics described by Equations (4.103)–(4.106). The nanoscale objects investigated in experiments on field emission can be roughly divided into two classes: the objects whose dimensions in all directions are comparable (nanoclusters or nanoparticles), and the objects whose length in the direction of the applied field is much larger than their transverse size (nanowires or nanowhiskers). The following consideration is carried out for the cases of nanoclusters, when the electric fields F_N and the capacitances C and C_2 can be determined consistently by solving corresponding electrostatic problems. The current is calculated according to Equation (4.106), under the assumption that the objects are in equilibrium with the cathode.

Let's consider the field emission from a nanocluster modeled by a spherical metallic particle of radius R deposited on the flat cathode surface. To provide a finite capacitance C, one should assume a finite separation d-R between the particle and the metallic cathode plate (for instance, one can imagine that the particle resides on an oxidized surface) (see the inset to Figure 4.21). Besides, this assumption provides electrical isolation of the particle from the cathode, which is a necessary condition for the Coulomb blockade. The electrostatics of the plane-sphere system is known, and the field and charge distributions in this case can be found in the form of rapidly converging infinite series arising from the potentials of image point charges and point dipoles [121].

Such a consideration allows one to present the distribution of the electrostatic potential energy near the particle in the approximate form

$$U(r,\theta) = W + q\{\beta F_0 R[1 + \gamma(\cos\theta - 1)] - \lambda[qN - C_2 V]/C\}\frac{r-R}{r}, \quad (4.109)$$

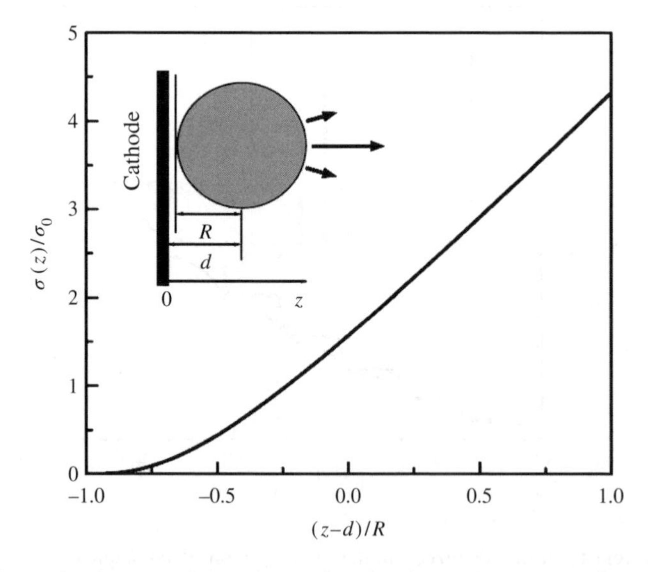

Figure 4.21 Charge density per unit length in z direction for a spherical metallic nanocluster placed at the distance 0.1 R from the metallic cathode. Here $\sigma_0 = F_0R/2$. The inset shows the geometry of the problem and the directions of the field emission (arrows). Reprinted with permission from Ref. [95]. Copyright (2006) by the American Physical Society

where r and θ are the radial and azimuthal coordinates of the spherical coordinate system with the origin at the center of the particle, and β, γ, and λ are the dimensionless constants of the order of unity, which are to be determined from numerical calculations. Such calculations also give us the capacitances C and C_2. Note that if the charge quantization is neglected (so that $N = C_2V/q$, when the particle is in equilibrium with the cathode), β is identified with the field enhancement factor conventionally used in the physics of field emission. The expression (4.109) provides an excellent description of the electrostatic potential at $r - R < R/2$ and at small θ. It allows one to take into account deviations of the potential energy from the linear form $\Phi - qF(r - R)$ and, therefore, to find corrections to the Fowler–Nordheim tunneling exponent. Neglecting such corrections in the prefactor, we obtain the following expression for the partial currents

$$I_N = ASF_N^2 \exp\left[-\frac{B}{F_N} f\left(\frac{qF_N R}{W}\right)\right], \quad f(x) = \frac{3}{2}\left[\frac{x^2}{\sqrt{x-1}}\left(\frac{\pi}{2} - \arctan\sqrt{x-1}\right) - x\right], \tag{4.110}$$

where A is given by Equation (4.102), the dimensionless function $f(x)$ describes the corrections to the tunneling exponent, and the effective emitting area $S = 2\pi R^2(F^2{}_N/\gamma B\beta F_0) = 2\pi R^2(F_N/\gamma B)$ is reduced due to the angular dependence of the radial field described by Equation (4.109). The field F_N is given by

$$F_N = \beta F_0 + \lambda \frac{qN - \overline{C_2}F_0}{CR}, \tag{4.111}$$

where the quantity $\overline{C_2} = C_2 L$ does not depend on the distance L between cathode and anode. Note that, since it is always assumed that L is much larger than any dimension of the nanoscale object, the capacitance C_2 is always proportional to $1/L$, and it is more convenient to replace $C_2\,|V|$ by $\overline{C_2}F_0$. This substitution also allows us to represent the Coulomb energy standing in Equation (4.106) as

$$E_N = \frac{q^2}{2C}[N - \overline{C_2}F_0/q]^2. \tag{4.112}$$

Further calculations are done for the separation $d - R = 0.1R$, when $C = 2.16R$, $\overline{C_2} = 1.74R^2$, $\beta = 4.32$, $\lambda = 1.22$, and $\gamma = 0.66$. Figure 4.21 shows the distribution of negative charges on the surface of the spherical particle staying in equilibrium with the cathode for this case ($|q|N = \overline{C_2}F_0$ is assumed). The distribution of the radial field $F(z)$ at the surface of the particle is given by the same dependence, $F(z)/F_0 = \sigma(z)/\sigma_0$.

The field-emission current from the nanocluster described above has been calculated according to Equations (4.106) and (4.110)–(4.112) at $R = 5$ nm. The results of the calculations shown in Figure 4.22 demonstrate the staircaselike behavior caused by the Coulomb blockade. However, in contrast to the staircases shown in Figure 4.20, the current continues to increase between the steps. This occurs because of electrostatic polarization of the nanoparticle.

According to Equation (4.111), when the particle charge is constant, the increase in the applied field F_0 leads to an increase in the effective field F_N because the factor $\beta - \lambda\overline{C_2}/CR$ is positive. For the chosen particle radius, the steps of the current are clearly visible at liquid nitrogen temperature, but poorly visible at room temperature. Nevertheless, the Coulomb blockade features at room temperature become quite distinct in the plots of the derivative of the current, as shown in the inset to Figure 4.22.

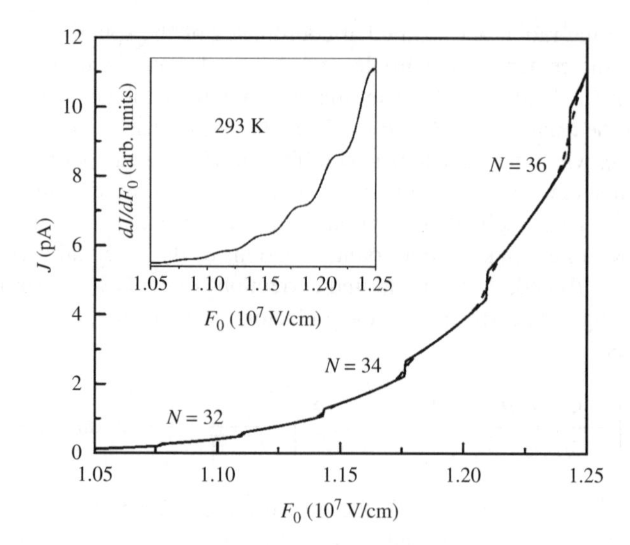

Figure 4.22 Current from the spherical nanocluster of radius $R = 5$ nm as a function of the applied field $F_0 = -V/L$ at $T = 4.2$ K (solid) and 77 K (dashed). The inset shows the derivative of the current at $T = 293$ K. Reprinted with permission from Ref. [95]. Copyright (2006) by the American Physical Society

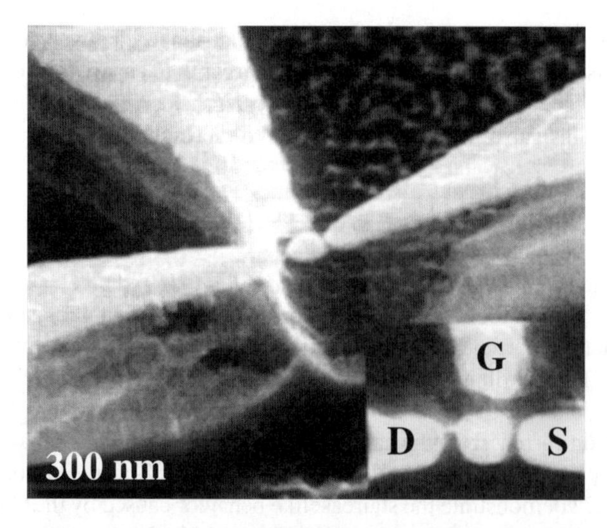

Figure 4.23 A suspended metallic island "glued" between source (S), drain (D), and gating (G) electrodes. The island with dimensions 80 nm × 80 nm × 60 nm (length with height) is attached to the electrodes by a thin layer of CF2 (white). Insert shows top view of the suspended device, indicating the close proximity to the source electrode. The CF2 layer appears as a white film. Copyright (2010) IEEE. Reprinted with permission from Ref. [122]

The experimental evidence of single EFE under Coulomb blockade has been obtained [122]. The experimental structure contains a metallic island between two electrodes freely suspended by a thin insulating layer (Figure 4.23).

The Coulomb-controlled current shows a stepwise increase in the low bias regime at 77 K [96].

References

1. S. Datta. *Electronic Transport in Mesoscopic Systems*, ed. H. Ahmed, M. Pepper, and A. Broers (Cambridge University Press, 1995), pp. 247–75.
2. O.V. Tretyak and V.Z. Lozovskyy. *Foundation of Semiconductor Physics*, Vol. **2** (Kyiv University Press, 2009) (in Ukrainian).
3. R. H. Fowler and L.W. Nordheim, Electron emission in intense electric field, *Proceedings of the Royal Society of London Series A* **119**, 173 (1928).
4. E.L. Murphy and R.H. Good, Thermionic emission, field emission, and the transition region, *Physics Review* **102**, 1464 (1956).
5. S.G. Christov, General theory of electron emission from metals, *Physica Status Solidi B* **17**, 11 (1966).
6. J.W. Gadzuk and E.W. Plummer, Field emission energy distribution (FEED), *Reviews of Modern Physics* **45**, 487 (1973).
7. R. Stratton, Energy distributions of field emitted electrons, *Physics Review* **135**, A794 (1964).
8. G.N. Fursey, Early field emission studies of semiconductors, *Applied Surface Science* **94**, 44 (1996).
9. R. Gomer, *Field Emission and Field Ionization* (American Vacuum Society Classics. American Institute of Physics: New York, 1993).
10. A. Modinos, *Field, Thermionic, and Secondary Electron Emission Spectroscopy* (Plenum: New York, 1984).
11. F.M. Charbonnier, Developing and using the field emitter as a high intensity electron source, *Applied Surface Science* **94/95**, 26 (1996).
12. P.H. Cutler and D. Nagy, The use of a new potential barrier model in the Fowler-Nordheim theory of field emission, *Surface Science* **3**, 71 (1964).
13. H.Q. Nguyen, P.H. Cutler, T.E. Feuchtwang, *et al.*, Investigation of a new numerical method for the exact calculation of one-dimensional transmission coefficients: Application to the study of limitations of the WKB approximation, *Surface Science* **160**, 331 (1985).
14. A. Mayer and J.P. Vigneron, Quantum-mechanical theory of field electron emission under axially symmetric forces, *Journal of Physics: Condensed Matter* **10**, 869 (1998).
15. J. He, P.H. Cutler, N.M. Miskovsky, *et al.*, Derivation of the image interaction for non-planar pointed emitter geometries: application to field emission I–V characteristics, *Surface Science* **246**, 348 (1991).
16. P.H. Cutler, J. He, J. Miller, *et al.*, Theory of electron emission in high fields from atomically sharp emitters: Validity of the Fowler-Nordheim equation, *Progress in Surface Science* **42**, 169 (1993).
17. R.G. Forbes, Field emission: New theory for the derivation of emission area from a Fowler–Nordheim plot, *Journal of Vacuum Science and Technology B* **17**, 526 (1999).
18. R.G. Forbes, Use of a spreadsheet for Fowler–Nordheim equation calculations, *Journal of Vacuum Science and Technology B* **17**, 534 (1999).
19. K.L. Jensen, Theory of field emission in *Vacuum Microelectronics*, ed W. Zhu (John Wiley & Sons, Inc., New York, 2001), pp. 33–104.
20. R.G. Forbes, Refining the application of Fowler–Nordheim theory, *Ultramicroscopy* **79**, 11 (1999).
21. R.G. Forbes and K.L. Jensen, New results in the theory of Fowler–Nordheim plots and the modelling of hemi-ellipsoidal emitters, *Ultramicroscopy* **89**, 17 (2001).
22. D.A. Kirkpatrick, A. Mankofsky, and K.T. Tsang, Analysis of field emission from three-dimensional structures, *Applied Physics Letters* **60**, 2065 (1992).
23. K.L. Jensen, Semianalytical model of electron source potential barriers, *Journal of Vacuum Science and Technology B* **17**, 515 (1999).
24. R. Kubo, H. Ichimura, T. Usui, and N. Hashitsume, *Statistical Mechanics* (North-Holland: New York, p. 241, 1965).
25. M. Ravindra and J. Zhao, Fowler-Nordheim tunneling in thin SiO_2 films, *Smart Materials and Structures* **1**, 197 (1992).

26. S. Horiguchi and H. Yoshino, Evaluation of interface potential barrier heights between ultrathin silicon oxides and silicon, *Journal of Applied Physics* **58**, 1597 (1985).
27. G. Pananakakis, G. Ghibaudo, R. Keis, and G. Papadas, Temperature dependence of the Fowler–Nordheim current in metal-oxide-degenerate semiconductor structures, *Journal of Applied Physics* **78**, 2635 (1995).
28. J. Sune, M. Lanzoni, and P. Olivo, Temperature dependence of Fowler-Nordheim injection from accumulated n-type silicon into silicon dioxide, *IEEE Transactions on Electron Devices* **40**, 1017 (1993).
29. J.J. O'Dwyer, *The Theory of Electrical Conduction and Breakdown in Solids Dielectrics* (Clarendon, Oxford, 1973).
30. I.S. Gradshteyn and I.M. Ryzhik, *Table of Integrals, Series, and Products* (Academic Press: New York, 1980).
31. C.A. Spindt, I. Brodie, L. Humphery, and E.R. Westerberg, Physical properties of thin-film field emission cathodes with molybdenum cones, *Journal of Applied Physics* **47**, 5248 (1976).
32. K.L. Jensen, Exchange-correlation, dipole, and image charge potentials for electron sources: Temperature and field variation of the barrier height *Journal of Applied Physics* **85**, 2667 (1999).
33. L.N. Dobretsov and M.V. Gomoyunova. *Emission Electronics* (Nayka, Moskow, 1966) (in Russian).
34. Z.-H. Huang, P.H. Cutler, N.M. Miskovsky, and T.E. Sullivan, Calculation of electron field emission from diamond surfaces, *Journal of Vacuum Science and Technology B* **13**, 526 (1995).
35. W. Zhu, G.P. Kochanski, S. Jin, and L. Seibles, Defect-enhanced electron field emission from chemical vapor deposited diamond, *Journal of Applied Physics* **78**, 2707 (1995).
36. W. Zhu, G.P. Kochanski, S. Jin, and L. Seibles, Electron field emission from chemical vapor deposited diamond, *Journal of Vacuum Science and Technology B* **14**, 2011 (1996).
37. J. Robertson, Mechanisms of electron field emission from diamond, diamond-like carbon, and nanostructured carbon, *Journal of Vacuum Science and Technology B* **17**, 659 (1999).
38. O. Groning, O.M. Kuttel, P. Groning, and L. Schlapbach, Field emission properties of nanocrystalline chemically vapor deposited-diamond films, *Journal of Vacuum Science and Technology B* **17**, 1970 (1999).
39. M.S. Chung, B.-G. Yoon, J.M. Park, *et al.*, Calculation of bulk states contributions to field emission from GaN, *Journal of Vacuum Science and Technology B* **16**, 906 (1998).
40. P. Lerner, N.M. Miskovsky, and P.H. Cutler, Model calculations of internal field emission and J–V characteristics of a composite n-Si and N–diamond cold cathode source, *Journal of Vacuum Science and Technology B* **16**, 900 (1998).
41. S.R.P. Silva, G.A.J. Amaratunga, and K. Okano, Modeling of the electron field emission process in polycrystalline diamond and diamond-like carbon thin films, *Journal of Vacuum Science and Technology B* **17**, 557 (1999).
42. R. Johnston and A.J. Miller, Field emission from silicon emitters, *Surface Science* **266**, 155 (1992).
43. V.G. Litovchenko, A.A. Evtukh, Yu. M., *et al.*, Observation of the resonance tunneling in field emission structures, *Journal of Vacuum Science and Technology B* **17**, 655 (1999).
44. V. Litovchenko, A. Evtukh, Yu. Kryuchenko, *et al.*, Quantum-size resonance tunneling in the field emission phenomenon, *Journal of Applied Physics* **96**, 867 (2004).
45. A. Evtukh, V. Litovchenko, N. Goncharuk, and H. Mimura, Electron emission Si-based resonant-tunneling diode, *Journal of Vacuum Science and Technology B* **30**, 022207 (2012).
46. V.G. Litovchenko and Yu. V. Kryuchenko, Field emission from structures with quantum wells, *Journal of Vacuum Science and Technology B* **11**, 362 (1993).
47. N.M. Goncharuk, The influence of an emitter accumulation layer on field emission from a multilayer cathode, *Materials Science and Engineering A* **353**, 36 (2003).
48. K.L. Jensen and J.L. Shaw, Simulation of the influence of interface charge on electron emission, Gunn effect in field-emission phenomena, *Materials Research Society Symposium Proceedings* **621**, R3.3 (2000)
49. V. Litovchenkoa, A. Evtukh, O. Yilmazoglu, *et al.*, Gunn effect in field-emission phenomena, *Journal of Applied Physics* **97**, 044911 (2005).
50. A. Evtukh, O. Yilmazoglu, V. Litovchenko, *et al.*, Electron field emission from nanostructured surfaces of GaN and AlGaN, *Physica Status Solidi C* **5**, 425 (2008).
51. N.D. Morgulis, About field emission composite semiconductor cathodes, *Journal of Technical Physics* **17**, 963 (1947).
52. R. Stratton, Field emission from semiconductors, *Proceedings of the Physical Society* **63**, 746 (1955).
53. R. Stratton, Theory of field emission from semiconductors, *Physics Review* **125**, 67 (1962).
54. G.F. Vasiliev, About theory of field emission from semiconductor, *Radiotechniques and Electronics* **3**, 7 (1958).
55. M.I. Elenson, Influence of internal fields in semiconductor on its field emission," Radiotechniques and Electronics, *Radiotechniques and Electronics* **4**, 140 (1959).

56. L. Shiff, *Quantum Mechanics* (McGraw-Hill, Moscow, 1957).
57. M.I. Elenson, F.F. Dobryakova, V.R. Krapivin, *et al.*, About theory of field and thermionic emission from metals and semiconductors, *Radiotechniques and Electronics* **8**, 1342 (1961).
58. L. Eyges, *The Classical Electromagnetic Field*, (Dover: New York, 1972).
59. O. Madelung, *Introduction to Solid State Theory*, 2nd edn, Springer-Verlag: New York, 1978 , See Section 4.2 "The Boltzmann Equation".
60. J.M. Ziman, *Principles of the Theory of Solids*, 2nd edn (Cambridge University Press: Cambridge, 1972).
61. K.L. Jensen, Simulation of time-dependent quantum transport in field emission from semiconductors: Complications due to scattering, surface density, and temperature, *Journal of Vacuum Science and Technology B* **13**, 505 (1995).
62. J.S. Blakemore, *Semiconductor Statistics* (Dover: New York, 1987).
63. S.M. Sze, Physics of Semiconductor Devices, *2nd edn*, John Wiley & Sons, Inc.: New York, 1981.
64. L.E. Reichl, Probability distributions in dynamical systems, in *A Modern Course in Statistical Physics*, Chapter 7 (University of Texas Press, Austin, TX, 1980), pp. 364–80, Sec. E–G.
65. Y.S. Kim and M.E. Noz, *Phase Space Picture of Quantum Mechanics*, (World Scientific: Teaneck, NJ, 1991).
66. J.R. Barker, D.W. Lowe, and S. Murry, in *The Physics of Submicron Structures*, H.L. Grubin, K. Hess, G.J. Iafrate, and D.K. Ferry (eds) (Plenum: New York, p. 277, 1984).
67. aW. Frensley, Wigner-function model of a resonant-tunneling semiconductor device, *Physical Review B* **36**, 1570 (1987);bW. Frensley, Quantum transport modeling of resonant-tunneling devices, *Solid State Electronics* **31**, 739 (1988).
68. N.C. Kluksdahl, A.M. Kriman, C. Ringhofer, and D.K. Ferry, Quantum tunneling properties from a Wigner function study, *Solid State Electronics* **31**, 743 (1988).
69. F.A. Buot and K.L. Jensen, Lattice Weyl-Wigner formulation of exact many-body quantum-transport theory and applications to novel solid-state quantum-based devices, *Physical Review B* **42**, 9429 (1990).
70. K.L. Jensen and A.K. Ganguly, Numerical simulation of field emission from silicon, *Journal of Vacuum Science and Technology B* **11**, 371 (1993).
71. K.L. Jensen, Improved Fowler–Nordheim equation for field emission from semiconductors, *Journal of Vacuum Science and Technology B* **13**, 516 (1995).
72. S.M. Sze and K.K. Ng. *Physics of Semiconductor Devices*, 3rd edn (John Wiley & Sons, Inc. New York 2007).
73. *Physics of Quantum Electron Devices*, ed. F. Capasso (Heidelberg, Springer-Veriag, 1990).
74. L.L. Chang, E.E. Mendez and C. Tejedor (eds), *Resonant Tunneling in Semiconductors: Physics and Applications* (New York, Plenum Press, 1991).
75. H. Mizuta and T. Tanoue, *The Physics and Applications of Resonant Tunnelling Diodes* (Cambridge University Press, 1995), pp. 213–27.
76. S. Luryi and A. Zaslavsky, Quantum-effect and hot-electron devices, in S. M. Sze (ed.), *Modern Semiconductor Device Physics* (John Wiley & Sons, Inc., New York, 1998), pp. 253–341.
77. M. Cahay, M. McLennan, S. Datta, and M.S. Lundstrom, Importance of space-charge effects in resonant tunneling devices, *Applied Physics Letters* **50**, 612 (1987).
78. T. Weil and B. Vinter, Equivalence between resonant tunneling and sequential tunneling in double-barrier diodes, *Applied Physics Letters* **50**, 1281 (1987).
79. S.M. Booker, F.W. Sheard, and G.A. Toombs, Effect of inelastic scattering on resonant tunneling in double-barrier heterostructures, *Semiconductor Science and Technology* **7**, B439 (1992).
80. A.D. Stone and P.A. Lee, Effect of inelastic processes on resonant tunneling in one dimension, *Physical Review Letters* **54**, 1196 (1985).
81. M. Jonson and A. Grincwajg, Effect of inelastic scattering on resonant and sequential tunneling in double barrier heterostructures, *Applied Physics Letters* **51**, 1729 (1987).
82. F. Chevoir and B. Vinter, Scattering-assisted tunneling in double-barrier diodes: Scattering rates and valley current, *Physical Review B* **47**, 7260 (1993).
83. R. Lake, G. Klimeck, M.P. Anantram, and S. Datta, Rate equations for the phonon peak in resonant-tunneling structures, *Physical Review B* **48**, 15132 (1993).
84. P. J. Turley, C.R. Wallis, and S.W. Teitsworth, Tunneling measurements of symmetric-interface phonons in GaAs/AlAs double-barrier structures, *Physical Review B* **47**, 640 (1993).
85. M. A. Kastner, The single-electron transistor, *Review of Modern Physics* **64**, 849 (1992).
86. Y. Meir, N.S. Wingreen and P.A. Lee, Transport through a strongly interacting electron system: Theory of periodic conductance oscillations, *Physical Review Letters* **66**, 3048 (1991).

87. C.W.J. Beenakker, Theory of Coulomb-blockade oscillations in the conductance of a quantum dot, *Physical Review B* **44**, 1646 (1991).
88. G. Klimeck, R. Lake, and S. Datta, Conductance spectroscopy in coupled quantum dots, *Physical Review B* **50**, 2316, 5484 (1994).
89. D.V. Averin and K. Likharev, in *Mesoscopic Phenomena in Solids*, eds. B.L. Altshuler, P.A. Lee and R.A. Webb (Elsevier, Amsterdam, 1991), pp. 173–271.
90. A.A. Evtukh, V.G. Litovchenko, R.I., et al., Parameters of the tip arrays covered by low work function layers, *Journal of Vacuum Science and Technology B* **14**, 2130 (1996).
91. K.L. Jensen and F.A. Buot, The methodology of simulating particle trajectories through tunneling structures using a Wigner distribution approach, *IEEE Transactions on Electron Devices* **ED-38**, 2337 (1991).
92. A.A. Evtukh, V.G. Litovchenko, R.I. Marchenko, and S. Yu. Layered structures with delta-doped layers for enhancement of field emission, Kudzinovski, *Journal of Vacuum Science and Technology B* **15**, 439 (1997).
93. A.A. Evtukh, V.G. Litovchenko, and M.O. Semenenko, Electrical and emission properties of nanocomposite SiO_x(Si) and SiO_2(Si) films, *Journal of Vacuum Science and Technology B* **24**, 945 (2006).
94. L.D. Filip, M. Palumbo, J.D. Carey, and S.R.P. Silva, Two-step electron tunneling from confined electronic states in a nanoparticle, *Physical Review B* **79**, 245429 (2009).
95. O.E. Raichev, Coulomb blockade of field emission from nanoscale conductors, *Physical Review B* **73**, 195328 (2006).
96. C. Kim, H.S. Kim, H. Qin, and R.H. Blick, Coulomb-controlled single electron field emission via a freely suspended metallic island, *Nano Letters* **10**, 615 (2010).
97. W.M. Tsang, V. Stolojan, B.J. Sealy, *et al.*, Electron field emission properties of Co quantum dots in SiO_2 matrix synthesised by ion implantation, *Ultramicroscopy* **107**, 819 (2007).
98. H. Shimawaki, Y. Neo, H. Mimura, *et al.* (2010) Photo-assisted electron field emission from MOS-type cathode based on nanocrystalline silicon, Proceedings of the 23rd International Vacuum Nanoelectronics Conference, Palo Alto, CA, 26–30 July 2010, pp. 74–75.
99. H. Mimura, Y. Abe, J. Ikeda, *et al.*, Resonant Fowler–Nordheim tunneling emission from metal-oxide-semiconductor cathodes, *Journal of Vacuum Science and Technology B* **16**, 803 (1998).
100. R.C. Smith, J.D. Carey, R.J. Murphy, *et al.*, Charge transport effects in field emission from carbon nanotube-polymer composites, *Applied Physics Letters* **87** 263105 (2005).
101. A. Evtukh, H. Hartnagel, V. Litovchenko, and O. Yilmazoglu, Two mechanisms of negative dynamic conductivity and generation of oscillations in field-emissions structures, *Materials Science and Engineering A* **353**, 27 (2003).
102. W. Chen, C.W. Zhou, L.Q. Mai, *et al.*, Field emission from $V_2O_5 \cdot nH_2O$ nanorod arrays, *Journal of Physical Chemistry C* **112**, 2262 (2008).
103. J.C. She, N.S. Xu, S.Z. Deng, *et al.*, Experimental evidence of resonant field emission from ultrathin amorphous diamond thin film, *Surface and Interface Analysis* **36**, 461 (2004).
104. Yu. V. Kryuchenko and V.G. Litovchenko, Computer simulation of the field emission from multilayer cathodes, *Journal of Vacuum Science and Technology B* **14**, 1 (1996).
105. V.G. Litovchenko and Yu. V. Kryuchenko, The dynamic characteristics of the field emission from the structures with quantum wells, *Journal of Physics (Paris), Colloquia* **6**, C5-C141 (1995).
106. H.Y Yang, S.P. Lau, S.F. Yu, *et al.*, Field emission from zinc oxide nanoneedles on plastic substrates, *Nanotechnology* **16**, 1300 (2005).
107. R.H. Good, Jr., and E.W. Muller. Handbuch der Physik (Springer, Berlin, 1956).
108. N.F. Mott and T.N. Sneddon. *Wave Mechanics and its Applications* (Clarendon Press, Oxford, 1948).
109. R.W. Gurney and E.U. Condon, Quantum mechanics and radioactive disintegration, *Physics Review* **33**, 127 (1929).
110. M. J. Kelly, *Low-Dimensional Semiconductors-Materials, Physics, Technology, Devices* (Clarendon Press, Oxford, 1995).
111. V.V. Mitin, V.A. Kochelap, and M.A. Stroscio, *Quantum Heterostructures- Microelectronics and Optoelectronics* (Cambridge University Press. Cambridge, 1999).
112. *Single Charge Tunneling*, ed. H. Grabert and M.H. Devoret, NATO ASI Series B 294 (Plenum Press, New York, 1992).
113. I.O. Kulik and R.I. Shekhter, Kinetic phenomena and charge discreteness effects in granulated media, *Soviet Physics Journal of Experimental and Theoretical Physics* **41**, 308 (1975).
114. L.Y. Gorelik, A. Isacsson, M.V. Voinova, *et al.*, Shuttle mechanism for charge transfer in Coulomb blockade nanostructures, *Physical Review Letters* **80**, 4526 (1998).

115. A. Erbe, C. Weiss, W. Zwerger, and R.H. Blick, Nanomechanical resonator shuttling single electrons at radio frequencies, *Physical Review Letters* **87**, 096106 (2001).
116. D.V. Scheible, C. Weiss, J.P. Kotthaus, and R.H. Blick, Periodic field emission from an isolated nanoscale electron island, *Physical Review Letters* **93**, 186801 (2004).
117. J.B. Barner and S.T. Ruggiero, Observation of the incremental charging of Ag particles by single electrons, *Physical Review Letters* **59**, 807 (1987).
118. K. Mullen, E. Ben-Jacob, R.C. Jaklevic. and Z. Schuss, I-V characteristics of coupled ultrasmall-capacitance normal tunnel junctions, *Physical Review B* **37**, 98 (1988).
119. R. Wilkins, E. Ben-Jacob, and R.C. Jaklevic, Scanning-tunneling-microscope observations of Coulomb blockade and oxide polarization in small metal droplets, *Physical Review Letters* **63**, 801 (1989).
120. L.D. Landau and E.M. Lifshitz, Quamtum Mechanics (Pergamon, Oxford, 1977).
121. W.R. Smythe, *Static and Dynamic Electricity* (McGraw-Hill, New York, 1968).
122. Kim, C., Kim, H.S., Qin, H., and Blick, R.H. (2010) Field electron emission under Coulomb blockade in suspended metallic island. 23 International Vacuum Nanoelectronics Conference, p. 69.

5

Electron Energy Distribution

For many applications it is important to have electron beam with narrow energy distribution of electrons. The electron energy distribution (EED) of emitted electrons is considered in this chapter. The simple theory and experimental set up are presented. The peculiarities of the EED spectra at emission from semiconductors and Spindt-type metal microtips are considered. Special attention is paid to EED spectra of emitted electrons from silicon and nanocrystalline silicon due to their importance for perspective integration of solid-state and vacuum nanoelectronics devices.

5.1 Theory of Electron Energy Distribution

The energy distribution from the conducting surface is the most generally given by [1]

$$\frac{dJ}{dE} \propto N(E)D(E), \tag{5.1}$$

where $N(E)$ is the supply function and $D(E)$ is the barrier transmission probability at energy E.

The assuming free electron theory and the rounded triangular potential barrier, the transmission function D is given by

$$D(E) = \exp\left\{-c + \left[\frac{E - E_F}{d}\right]\right\}, \tag{5.2}$$

where c and d are the functions of electric field F and work function Φ. It is a so-called Jeffreys–Wentzel–Kramers–Brillouin (JWKB) solution of transmission function for tunneling through the triangular barrier modified by the image potential. The exponential form of the solution is also derived for other potential functions, but different potentials do change the terms c and d.

If the emitter is actually 1D, the emitted current entering an analyzer located at normal angle to the surface would come from bulk electrons headed toward the surface at normal incidence.

Vacuum Nanoelectronic Devices: Novel Electron Sources and Applications, First Edition.
Anatoliy Evtukh, Hans Hartnagel, Oktay Yilmazoglu, Hidenori Mimura and Dimitris Pavlidis.

In that case the source function N would be given by

$$N(E,T) = \frac{mk_BT}{2\pi^2h^3}\ln\left[1+\exp(-\frac{E-E_F}{k_BT})\right], \tag{5.3}$$

where m is the electron mass, k_B is Boltzman's constant, h is Plank's constant, and T is the temperature (in Kelvin).

Using Equations (5.2) and (5.3) in Equation (5.1) gives the "normal energy distribution." However, because the emitters are curved, current emitted at many angles with respect to the local surface normal may enter the analyzer. Therefore, the source function includes more bulk electrons. The usual theory is to integrate over all angles, giving the total energy distribution (TED). Here, the source function returns to the Fermi function:

$$N(E) = \frac{mk_BT}{2\pi^2h^3}\left\{1+\exp\left(\frac{E-E_F}{k_BT}\right)\right\}^{-1}. \tag{5.4}$$

For large vacuum barriers, the transmission function is exponential as in Equation (5.2). If the vacuum barrier is not large with respect to the electron energy, the transmission function saturates at high field, as may be modeled with the Fermi function [2]. Thus, these very simple approximations give the form [1]

$$\frac{dJ}{dE} = A\left\{1+\exp\left(\frac{E-E_F}{k_BT}\right)\right\}^{-1}\left\{1+\exp\left(\frac{-c+(E-E_F)}{d}\right)\right\}^{-1}. \tag{5.5}$$

During modeling of the electron field emission (EFE) from carbon nanotubes the next expression for EED has been obtained in Refs. [3, 4]:

$$\frac{dJ}{dE} = \frac{2q\sqrt{2mq}}{\pi\hbar^2}r_0 f(E)D(E,F)\sqrt{E+\mu_0}, \tag{5.6}$$

where r_0 is the radius of carbon nanotube, μ_0 in the energetic scale with the origin at the Fermi level, obviously playing the role of the chemical potential of the conduction electrons and having to be connected to the tube content.

The comparison of theory prediction with experimental results on EED at emission from clean metal surface (Figure 5.1) shows that in the first order, they are in good agreement.

However, this theory (Equation (5.5)) fails to predict some features. On the high-energy side of E_F, there is a low current "tail" that extends several volts above E_F. This tail has been attributed to many body effects [5], that is, energy produced when electrons are emitted from below E_F is transferred to nearby electrons at E_F, creating "hot" electrons. If these hot electrons are created within a few angstroms of the metal's surface, they are transmitted to the surface with little loss and emitted readily. Hot electrons can interact with adsorbed molecules on the surface, stimulating desorption or surface chemistry [6]. This mechanism may be responsible for some of the emission changes that occur over time. The low energy side of the distribution is typically not perfectly exponential. Such nonexponential behavior is prominent in emission spectra from the (100) plane of tungsten (where it has been first observed [7]), and from a variety of other metal surfaces [8]. However, the deviation from exponential behavior is typically less than a factor of 10. Nonexponential behavior may be caused either by peaks in the

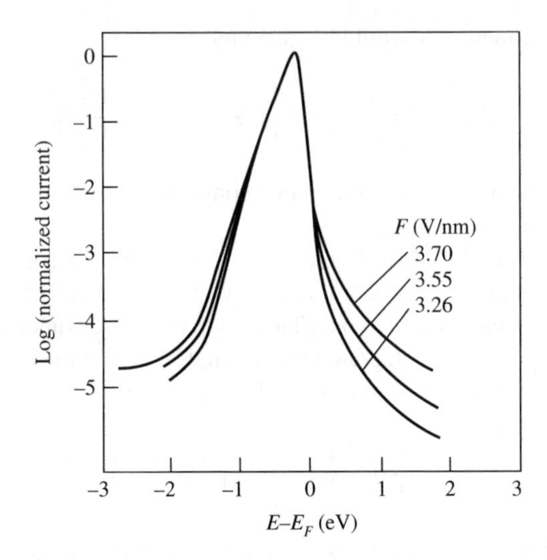

Figure 5.1 Field emission energy spectrum typical for a clean metal surface. Reprinted with permission from Ref. [5]. Copyright (1971) by the American Physical Society

density of states *N(E)*, or by deviations in the transmission function *D(E)* [1]. Structure in the density of states function can be created by surface states and/or bulk band structure.

Structure in the transmission function can be caused by "resonant" tunneling [8] through the atomic like state on the surface (such as might be produced by an adsorbed atom or molecule) (Figure 5.2). Resonance can occur when the surface state exists in close proximity to the substrate free electron gas, but isolated from the substrate by a potential barrier (Figure 5.2, process 1). An electron wave function at the same energy as the surface state can tunnel out with higher probability than the waves with higher or lower energy. Inelastic resonant tunneling occurs when the tunneling electron is scattered into lower energy states in the adjacent potential well (losing energy) before being emitted. This often occurs when the tunneling electron excites vibrations in absorbed atoms or molecules. Field emission spectra from tungsten tips coated with organic molecules has showed large peaks in the energy distribution attributed to inelastic tunneling [9]. Field electron microscope images created by molecules adsorbed on tips are often highly symmetrical, and that is the consequence of scattering [8].

The electron can transit from level E_l in level E_{il} in potential well (Figure 5.2, process 2). If $E_{i1} < E_1$ such transition causes the release of energy equal to $E_l - E_{il}$. This energy can be given as other electron in solid with initial energy E_2 (Auger process, transition 3) and release as a quant of light. The last process has less probability. In case if the energy of exited electron $E = E_2 + (E_1 - E_{i1})$ is higher than the barrier, it can leave the emitter.

The surface potential barrier can be 3D. The field emission patterns showing a single small spot, implying emission from the single atom with a very narrow angle (~4°) has been reported in Refs. [10, 11]. Explanations of this narrow emission angle include the angular dependence of the potential profile [12] and the angular dependence of the resonant surface state [13]. Similar narrow-angle, single spot emission patterns have been reported after the thermal-field build-up treatment [14]. In that work, the energy distribution of the electrons in the narrow spot showed multiple peaks widely separated

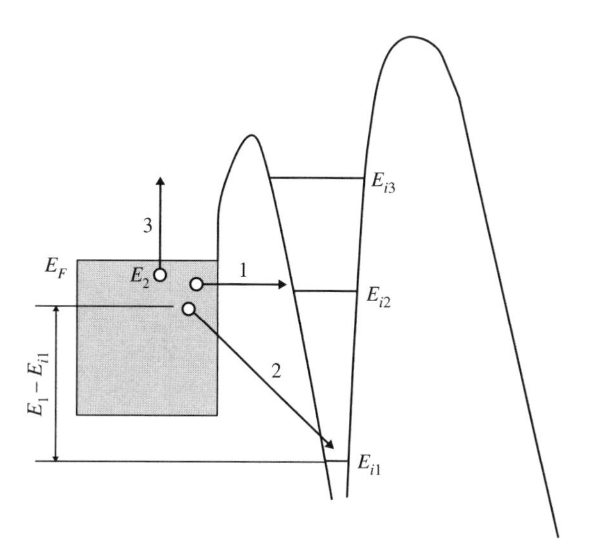

Figure 5.2 Influence of the adsorbed atom on EFE: 1: resonance tunneling process, 2: transition the electron on lower level in atom, and 3: electron excitation

in energy. Other multi-peak emission spectra and narrow angle emission were obtained from tips made from metal carbides and from clean tungsten tips only after carburization [15]. It was postulated that the peaked emission spectra observed were caused by resonant tunneling through states produced by the single atom tip apex [14]. However, it is clear that single atom field emission alone does not necessarily produce multi-peak spectra [16]. It seems more likely that the thermal-field build-up process has added some additional atoms such as C or O near the apex, and that the electronic properties of the contaminated surface have caused the multi-peak emission [17]. Thus, it appears that such contamination can produce narrow angle emission. The narrow emission beams by adsorption of Cs, in which case the width of the energy distribution became narrower, but maintained a similar shape in agreement with free electron theory for a reduced work function were also produced in Ref. [18]. Similar narrow-energy/narrow-angle emission patterns have been obtained from thermal-field build-up tips made on W(111) and Pt(001) [19].

5.2 Experimental Set Up

Energy distributions can be measured with the same energy analysis equipment commonly used for surface analysis such as photoemission spectroscopy. Unlike surface analysis, no external photon or electron source is required, but some means of making contact to the emitters and gates must be provided. An example of the system using a hemispherical analyzer is drawn in Figure 5.3 [1]. The chamber is equipped with probe wires mounted on manipulators, used for contacting the arrays and serving as anodes. The currents from the cathode, gate, and anode are measured independently. The cathode potential can be biased negative or positive with respect to ground to encourage or discourage the beam from hitting the chamber

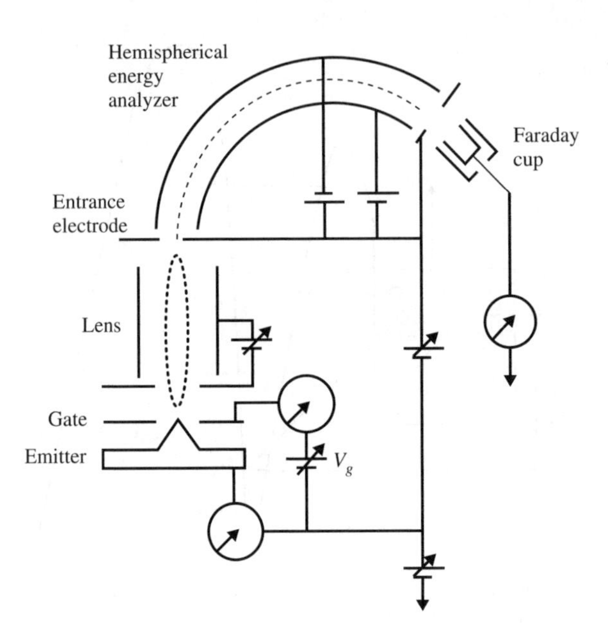

Figure 5.3 Apparatus for simultaneous measuring of the emission current, gate current, and emission energy spectrum of an FEA [1]. Copyright (2001) by John Wiley & Sons, Inc. This material is reproduced with permission of John Wiley & Sons, Inc.

walls. The hemispherical analyzer allows only electrons with specific energy to pass through the circular path to the detector. The potential applied to the slit in front of the hemispheres adds or subtracts energy from the incoming electrons, so that scanning the slit potential while monitoring the detector current produces energy distribution. It is worth noting that the potential seen by the electrons in space is equal to the voltage applied to the slit plus the work function of the slit surface. Thus, the measured energy distribution must be shifted by the work function of the slit when it is plotted with respect to the Fermi level. The work function of the emitter does not shift energy distribution. In this case a Faraday cup is used to detect the analyzer current in order to sample significant fraction of large currents produced by field emission arrays (FEAs) (typical capacitively coupled electron multipliers cannot detect large currents).

The EEDs can also be measured by using a conventional *ac*-retarding-field analyzer consisting of three parallel-plate electrodes [20]. Figure 5.4 shows schematic diagram of the measurement system. G_1 and G_2 are the grids, which have the transmission rate of 50% each, and C is a collector electrode. G_1 is kept at positive potential to obtain constant emission, whereas G_2 is grounded, so that G_2 has the same potential as the Au electrode. The collector electrode C is kept slightly positive with respect to G_2 in order to prevent electrons from backscattering. Retarding voltage V_R, which is the slow *dc* sweeping voltage, is modulated with *ac* signal, which is 0.2–0.5 V_{ps} and 123 Hz, respectively. The corresponding *ac* component of the collector current is detected by a lock-in amplifier and recorded as a function of V_R by a computer. In order to suppress the secondary electron emission, all the electrodes are coated with the lampblack. A variable capacitance C_v is employed to eliminate the capacitance between the Au electrode and the collector electrode.

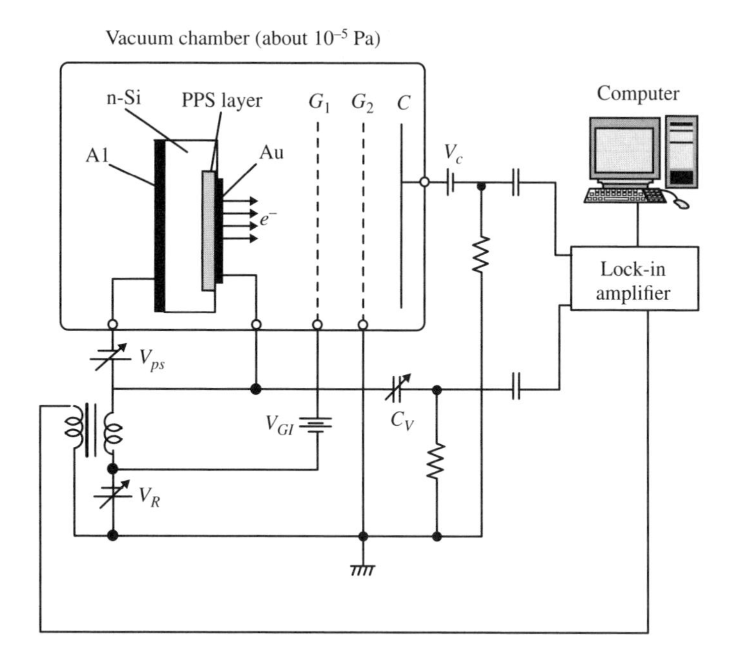

Figure 5.4 Schematic diagram of the measurement system of the electron energy distributions. It is a conventional *ac*-retarding field analyzer consisting of three parallel-plate electrodes. Reproduced with permission from Ref. [20]. Copyright (1999), American Vacuum Society

5.3 Peculiarities of Electron Energy Distribution Spectra at Emission from Semiconductors

Let's consider the issue of the energy spectra of electrons at EFE from semiconductors [21]. The studies were conducted by the retarding field. At EFE of electrons from the conduction band of the semiconductor the value of the potential difference cathode-collector corresponding to saturation of the emission current, that is, U_{sat} can be specify. In this case, U_{sat}, along with a work function of collector and some parameters characterizing the cathode are determined. The most interesting from the possible special cases are shown schematically in Figure 5.5.

As can be seen, in the absence of an electric field in the semiconductor (Figure 5.5a) (low-resistance semiconductor) $U_{sat} = q\varphi_c - E_n$. Taking into account the electric field penetration into semiconductor (Figure 5.5b) $U_{sat} = q\varphi_c - E_n + \Delta\phi_s$. At existence the voltage drop on the emitter ΔV_e (Figure 5.5c) (high resistance semiconductor) $U_{sat} = q\varphi_c - \mathrm{E_n} + \Delta\phi_\mathrm{s} + q\Delta V_e$. For high-resistivity semiconductors $q\varphi_c$ and $q\Delta V_e$ are usually much more than E_n and $\Delta\phi_s$. Therefore, for them it is possible approximately to set $U_{sat} = q\varphi_c + q\Delta V_e$. When at the surface layer of the semiconductor there is an inhibiting barrier $\Delta\phi_s$, associated with the surface states without other complicating the phenomenon factors (Figure 5.5d), $U_{sat} = q\varphi_c - E_n - \Delta\phi_s$. If the source of electrons at EFE is the valence band of the semiconductor, the potential difference of cathode – collector corresponding to the onset of the delay curves, that is, U_{on} will have certain value. In the simplest case (see Figure 5.6) it is $U_{on} = q\varphi_c + E_p$.

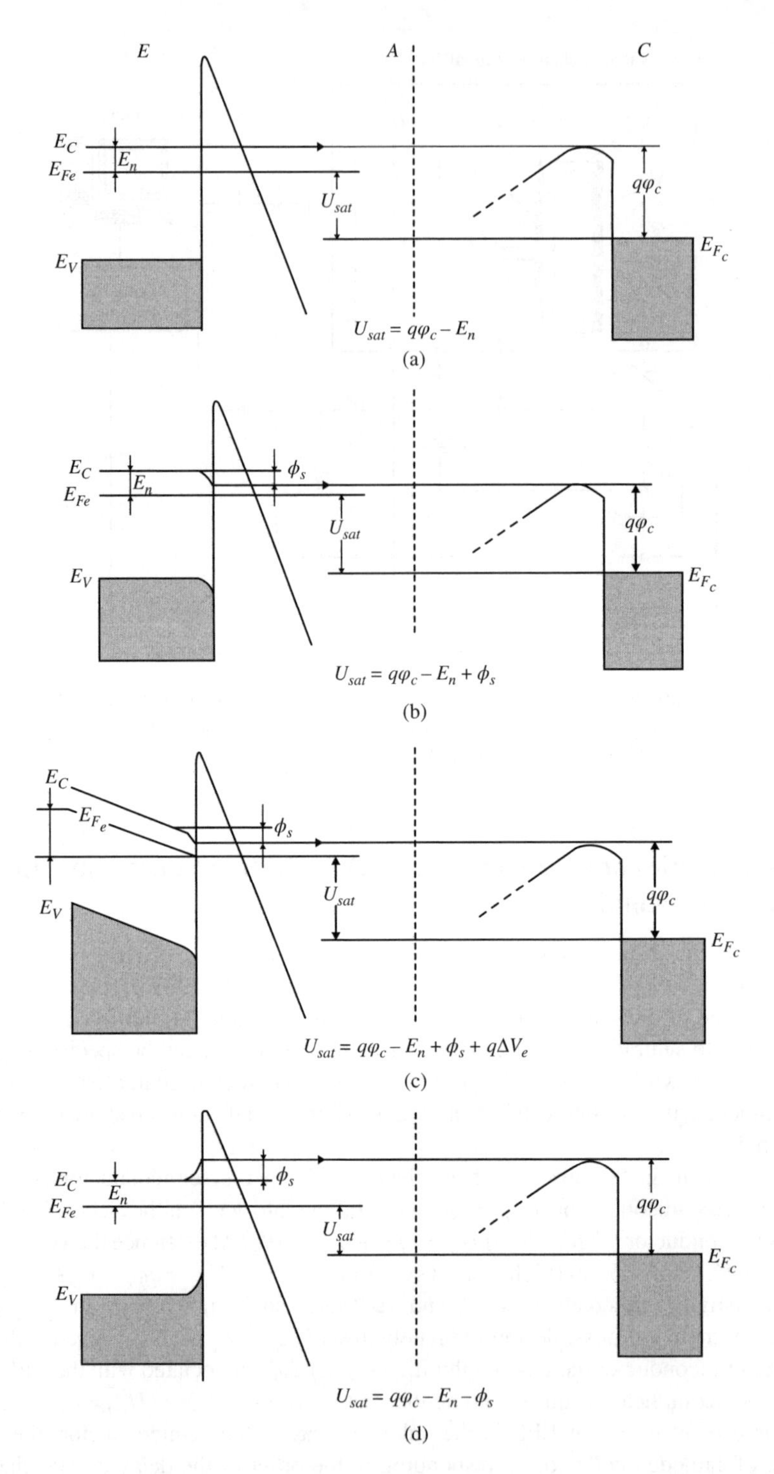

Figure 5.5 Energy band diagrams at measurement of electron energy distribution at emission from semiconductor: (a) without band bending in semiconductor, (b) with band bending in semiconductor, (c) with band bending and voltage drop in semiconductor, and (d) with upward band bending in semiconductor

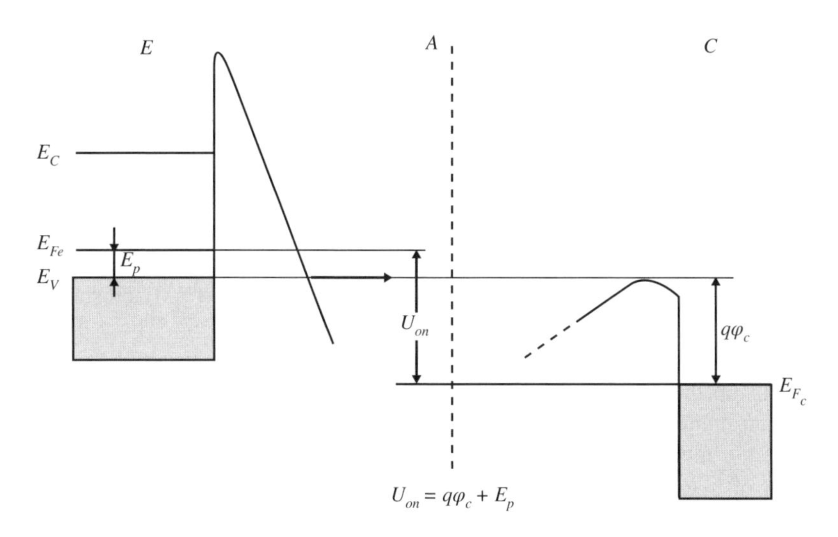

Figure 5.6 Energy band diagrams at measurement of electron energy distribution at emission from semiconductor valence band

The width of the energy spectrum at the condition of equilibrium of the electron gas and the lattice is close to the corresponding spectrum of metals. Significant expansion of the energy distribution should be at overheating of the electron gas. If the electrons at EFE issued simultaneously as from the conduction band and the valence band, the curve of the energy spectrum must obviously have two peaks separated by the distance approximately determined by the width of the band gap of the semiconductor.

Thus, the study of the energy spectra of electrons emitted by semiconductors allows us to estimate the voltage drop across the emitting tip, to detect the overheating of the electron gas in the surface layer of a semiconductor, and under certain conditions, to find out which band of the semiconductor (or both bands at the same time) is the source of electrons at EFE.

5.3.1 Electron Energy Distribution of Electrons Emitted from Semiconductors

The EED has been studied for many semiconductors [22, 23]. It was shown that in the case of high resistive semiconductors, for example, such as CdS, the voltage drop in semiconductor (ΔV_e) is usually equal to tens and even hundreds volts [23]. The quarts doped with carbon have lower resistivity and, in comparison with CdS, ΔV_e has some volts [22]. The EED at existence of voltage drop in the semiconductor is significantly widened due to overheating of the electron gas.

The results presented in work [23] on EFE of tungsten coated by single germanium layer has showed that (1) the coating of tungsten by single germanium layer causes significant decrease of emission current (40–70 times), but emission *I-V* characteristics is parallel to one of clean tungsten (it points out on unchangeable work function) and (2) in EED spectra in addition to one maximum from tungsten there has been revealed second maximum from lower energy electrons. The above mentioned result can be qualitatively explained by the model of adsorbed layer [24, 25].

During the investigations of EFE from diamond field emitter [26, 27] the order of magnitude emission enhancement without spectral broadening, and sensitivity of that structure to the applied electric field have been revealed [28]. The tip radius for diamond emitter has been ~5–10 nm. For the energy spectrum measurements a custom built retardation analyzer which has a resolution function with full width at half maximum (FWHM) of ~0.15 eV has been used [29]. For fixed experimental parameters, the current collected from the diamond emitter exhibits stepwise fluctuations between discrete emission levels. Each stable emission level has a corresponding stable electron energy spectrum (Figure 5.7). These fluctuations in emission and spectrum are due to the adsorption and diffusion of residual gases and other contaminations on the emitter surface. The energy spectrum can develop complex structure due to resonant tunneling through surface states of the adsorbates. Occasionally, the single adsorbate events, such as shown in Figure 5.8, during which the current increases by more than an order of magnitude while the spectral width remains narrow (~0.4 eV), have been observed.

For extended periods of stable emission (~5 minutes) it was the ability to investigate the sensitivity of the emitted spectrum to the applied electric field. The sequence of spectra at various applied electric fields is shown in Figure 5.9. The observable changes in the spectrum with field are completely reversible by returning the field to its previous value.

5.4 Electron Energy Distribution at Emission from Spindt-Type Metal Microtips

Detailed results on EED at emission from Spindt-type metal microtips have been obtained in Ref. [30]. There three metallic (Mo) field emission single integrated microtips (FESIMs) were

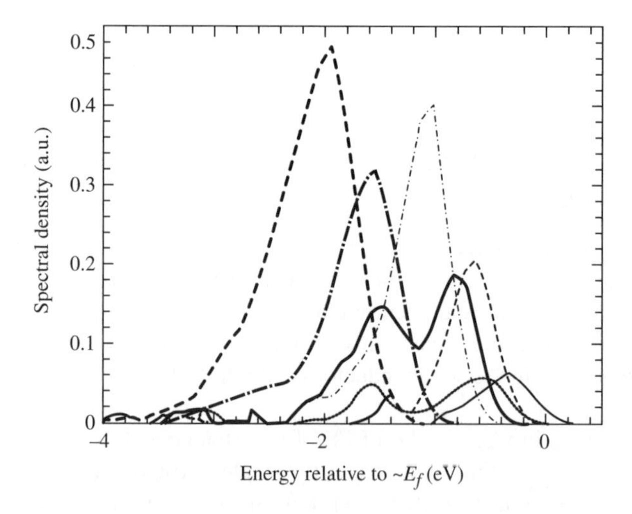

Figure 5.7 Measured spectra for a diamond field emitter for fixed experimental conditions. Copyright (2010) IEEE. Reprinted with permission from Ref. [28]

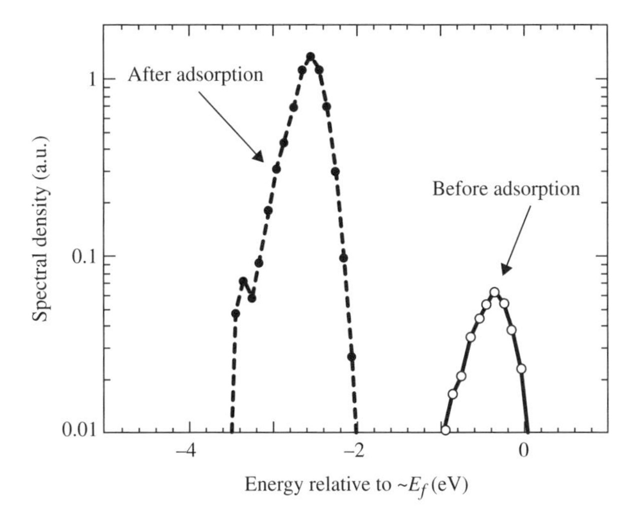

Figure 5.8 Measured energy spectrum before and after a single adsorbate event. The emission current increased by more than an order of magnitude without affecting the spectral width or shape. Copyright (2010) IEEE. Reprinted with permission from Ref. [28]

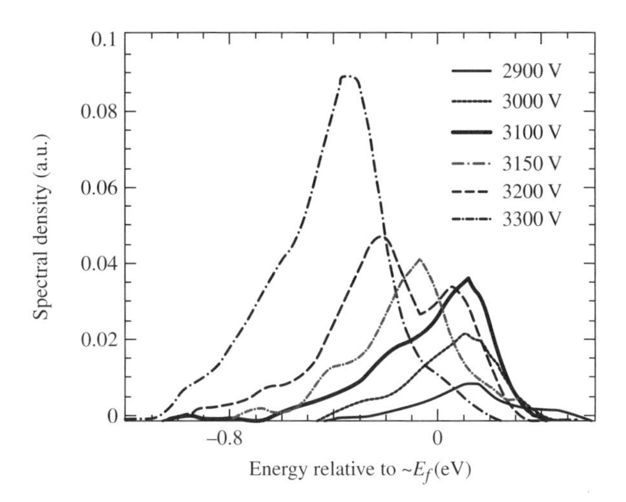

Figure 5.9 The measured spectrum during a period of stable emission for various applied voltages. The anode-cathode gap in this case was ~300 μm. Copyright (2010) IEEE. Reprinted with permission from Ref. [28]

under investigation, that is, monotips, and two-FEAs forming areas of 1 or ~4 mm^2 having a total number of (80 × 80) or (100 × 100) tips. The EEDs were measured with a commercial 135° hemispherical energy analyzer with nominal resolution of 10 meV, positioned behind a probe hole, in which the entrance lenses had been adapted for the field emission electron spectroscopy (FEES) measurements.

The EEDs give the most detailed information on the electron emission and enable deeper understanding of its relation to the seasoning (conditioning) process. During the analysis of the following energy spectra it is necessary to keep in mind that the energy of the emitted electrons is less than V_b (where V_b is the negative voltage applied to cathode) because of two distinct effects: (1) There is a shift due to the presence of the resistive layer under cathode tip which induces a resistive voltage drop V_{IR} between the cathode contact and the metal of the tip. (2) There is a voltage drop V_{FE} intrinsic to the FE (field-emission) mechanism itself localized at the individual emission surfaces.

Characteristic examples of the TED spectra are shown in Figures 5.10 and 5.11 for the FEAs.

1. The EEDs obtained in the early seasoning consisted of a wide, low-energy band that started at an energy level of a few electronvolts and extended to a few tens of electronvolts (Figure 5.10a) [31].
2. After some seasoning, multiple peaks have appeared in the high energy region near V_b as shown in Figure 5.10b,c.
3. After further seasoning, an increasingly larger proportion of high energy electrons were emitted (Figure 5.10c). This could evolve until no low-energy electrons were detected.
4. The overall high energy spread (ΔE_{total}) can reach tens of electronvolts at the highest currents even when there are no low-energy electrons.
5. Certain peaks in the spectrum disappear or modify abruptly in time accompanied with sudden changes in the emission intensity (Figure 5.11b).
6. The multiple peak behavior can occasionally be removed by a large pulsed increase in the FE current at the limit of the emission-induced thermal breakdown.
7. After the flash experiment, upon raising V_b so as to increase the total current to the tens of milliampere range, new localized peaks appeared spontaneously in the EEDs with simultaneous recovery of the previous emission current value.

For explanation of variety of the obtained EED spectra the nanoprotrusion model for metallic microtip emitters has been proposed [30–34]. The nanoprotrusion model is introduced primarily because the presence of localized peaks measured in the EEDs that shift with applied voltage is the signature of the presence of nanoprotrusions ending in atomic scale apexes having the height greater than the critical value [34]. This interpretation does not exclude the role of surface contaminants, however it does not give it the main role as in the former interpretations. An argumentation supporting the five elements of the model and its consequences are detailed further.

The model takes into consideration the evaporation process for the tip fabrication and the absence of high temperature treatment before the field emission. Under these conditions, the microtips are composed of relatively small grains that produce a very corrugated surface of microprotrusions with nanoprotrusions on them [14, 32, 35–37]. The emission can be understood quantitatively with the potential diagram in Figure 5.12. The metal of the base microtip is represented to the left. The formation of a nanometric object creates localized states with narrower energy spreads from which the electrons are emitted through the tunnel barrier. The nanometric object will have a concentration of electric field due to its high curvature and hence

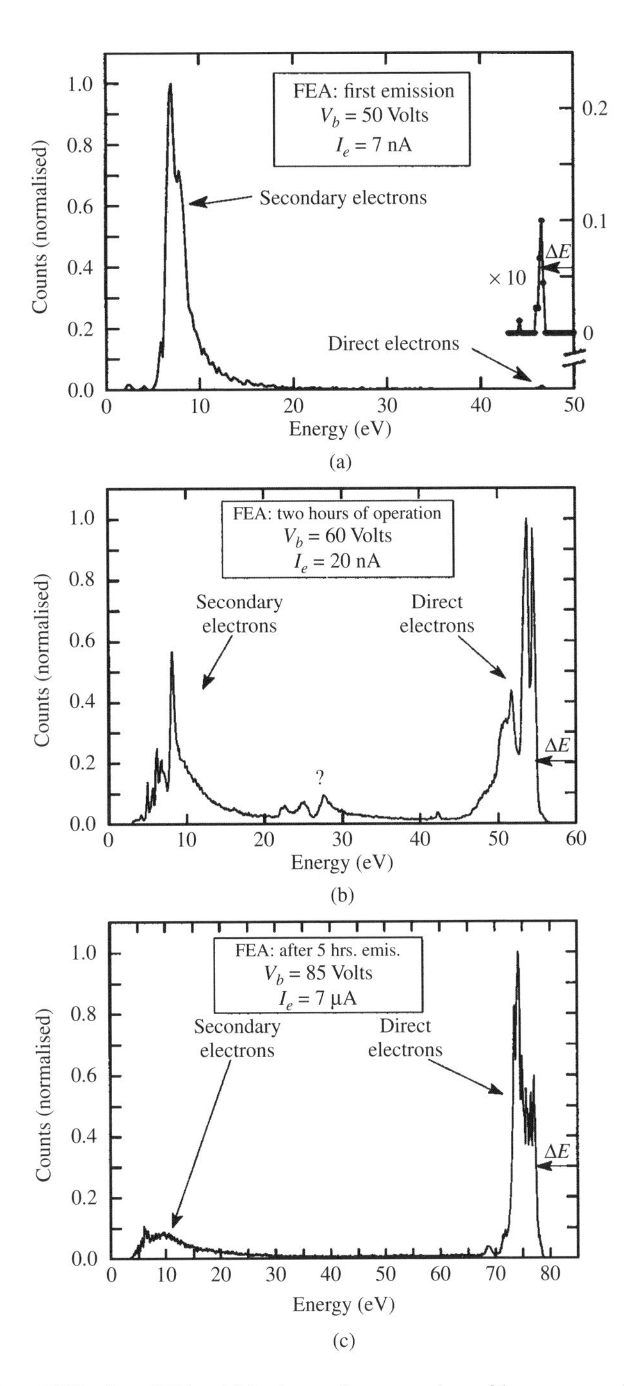

Figure 5.10 Three EEDs for a FEA which show a large number of low-energy electrons. Evident is the relative increase in direct electrons with increasing seasoning. (a) first emission, (b) after 2 hours' operation, (c) after 5 hours' emission. Reproduced with permission from Ref. [30] Copyright (1997), American Vacuum Society

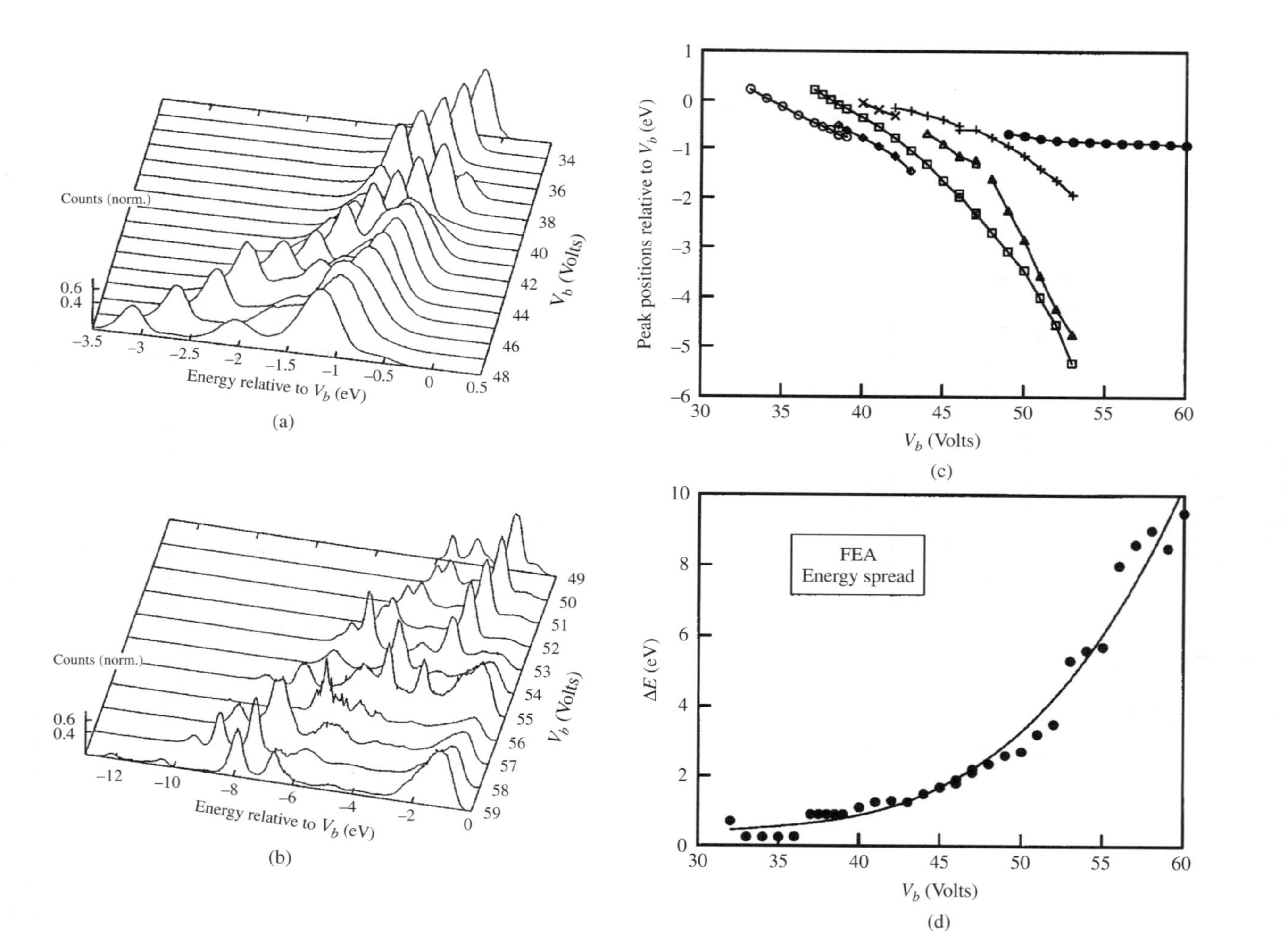

Figure 5.11 (a,b) Series of EEDs for a FEA with minimal low energy electrons. (c) Relative positions of various peaks in (a) and (b). The related shifts of the different peaks show that the total emission is due to multiple protrusions positioned on one or more microtips. (d) Total energy spread of the measured EEDs vs V_b. Reproduced with permission from Ref. [30] Copyright (1997), American Vacuum Society

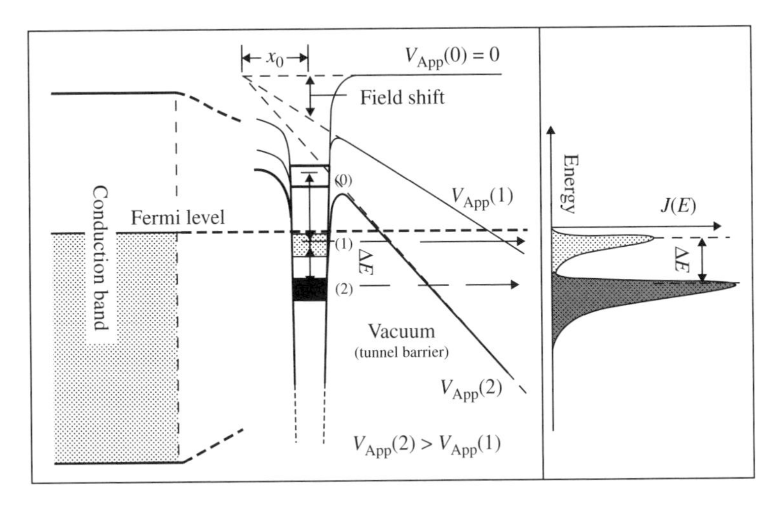

Figure 5.12 Potential diagram for FE when the EEDs consist exclusively of bands. The position of the band for two different values of applied voltage is indicated. Reproduced with permission from Ref. [30] Copyright (1997), American Vacuum Society

an enhanced emission. The band shifting is due to the penetration of the electric field into the physical region of the localized states due to reduced screening by the electrons in this region. Three positions of a local band are depicted in Figure 5.12. In general it has been found that the position of the peaks when extrapolated to zero field lie above EF (position (0)). The bands cannot emit until the band is shifted to E_F and below (positions (1) and (2)).

As the nanoprotrusions were due to the formation of the microtips by an evaporation process (Spindt-type), their density and geometry were not controlled. This induced an erratic and nonuniform emission among these nanoprotrusions and an emission-induced alteration with time. In other words, during the postseasoning of FE the emitting nanoprotrusion distribution was not stable. These alterations are revealed by sudden changes in the emission intensity accompanied by EED modifications. Such modifications are exacerbated by the surface diffusion of contaminants during FE and FE-induced ion bombardment.

5.5 Electron Energy Distribution of Electrons Emitter from Silicon

5.5.1 Electron Energy Distribution of Electrons from Silicon Tips and Arrays

In the case of clean silicon the EED curve with two maximum has been obtained [38]. The distance between the maximum positions is approximately equal to band gap of silicon. It points out as on emission from conduction band and valence band. The EED containing the two peaks was also observed in work [39] during the investigation of the EFE from silicon tip array. The FEAs in size of 7×7 mm^2 with about 3000 tips/mm^2 were investigated by means of an ultra-high vacuum system for photo-induced field emission spectroscopy (PFES) [40], which allows integral current measurements as well as electron spectroscopy. The Si tips are triangularly aligned in a distance of 20 μm and have a typical height of about 2.5 μm and a tip radius of less than 20 nm. The measured spectra are shown in Figure 5.13 and they reveal strong emission from the conduction (right peak) and weak emission from the valence band. The cathode was biased with $V_0 = -83$ V, whereas the Fermi level is located in the spectra at about 63 eV. This difference can be attributed to an electric charging of the emitting tips corresponding to about 20 V, which might be slightly different for the individual tips. This also leads to broadening of both peaks and blurring of the band gap. For the given voltage, the integral counts of the illuminated by green laser spectrum in Figure 5.13 increases for about 26%. This surplus is clearly due to the excited electrons on the high energy side of the conduction band peak. Since both spectra have a very similar shape, they seem to be shifted against each other. This might be attributed to a reduced charging of the illuminated tips due to increased conductivity, or might be a hint for more complex electron excitations.

In general case the characteristic of field emission from silicon emitter, however, is significantly different from metals [41, 42]: there is an obvious peak shift of the EED with external field. The surface states are considerable issue to elucidate the emission mechanism and theoretical calculation shows this issue is also important for the emission from metals [43]. Besides the surface states, the field penetration is another important term for silicon like semiconductor after considering the low conductivity. The field penetration will cause a band-bending and the EED shift to low energy side. Furthermore, the band-bending will made the electron emission from other conduction bands and multi peaks are possible as shown in Figure 5.14 [44].

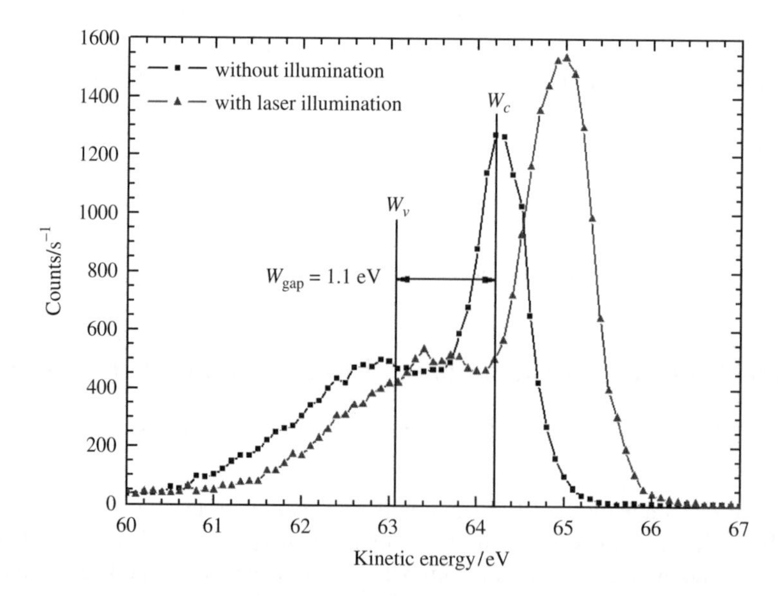

Figure 5.13 Spectra with and without illumination at $F = 5.8$ V/μm. The kinetic energy of the electrons results from the bias voltage V_0, the cathode charging V_C, and the electron binding energy $E_B : E_{kin} = qV_0 - qV_C + E_B$. The band gap of Si is denoted in the spectrum without illumination as the difference between valence and conduction band energies W_V and W_C, respectively. Copyright (2011) IEEE. Reprinted with permission from Ref. [39]

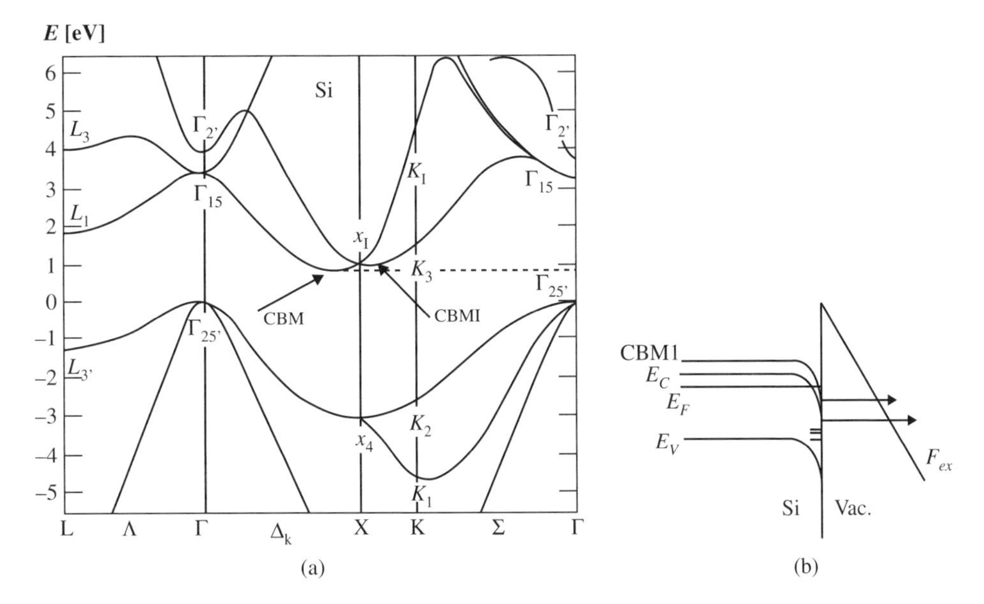

Figure 5.14 (a) The typical band structure of silicon. (b) The schematic energy band of Si emitter during field emission, the CBM1 denotes the other conduction band minimum

Based on free electron approximation, the EED of field emission *I(E)* from conduction bands can be calculated from [45]

$$I(E,F) \propto \frac{1}{\exp\left(\frac{E+0.18-qV_s}{kT_e}\right)+1}\exp\left(-\frac{4\sqrt{2m}}{3\hbar q}\frac{(\phi-E-0.18)^{3/2}}{F_{ex}}\right)$$
$$+\alpha\frac{1}{\exp\left(\frac{E-qV_s}{kT_e}\right)+1}\exp\left(-\frac{4\sqrt{2m}}{3\hbar q}\frac{(\phi-E)^{3/2}}{F_{ex}}\right), \quad (5.7)$$

where V_s is the surface voltage of silicon emitter corresponding to the band bending, F_{ex} is the external electric field, and T_e is the effective temperature.

The EED of Si emitter calculated according to Equation (5.7), where $\alpha = 0.5$ is shown in Figure 5.15.

The field penetration onto the Si-emitter will produce the obvious band bending and distinct EED shift to low energy side. A serious field penetration will generate the electron emission from other conduction bands and multi peaks can be observed.

The EED from silicon FEAs has been considered in detail in Ref. [1]. Energy distributions obtained from silicon FEAs [46, 47] as well as from other semiconductors [17] and molybdenum FEAs [30] show multiple peaks and emission at energies up to several volts below E_F. The peak of the emission distribution typically shifts to lower electron energy linearly with increasing gate voltage. These peaked emission spectra immediately demonstrate that the emission from FEAs does not originate from the bulk band structure of the emitters. The peaks have been interpreted as either direct [46] or resonant [30, 48] tunneling through states

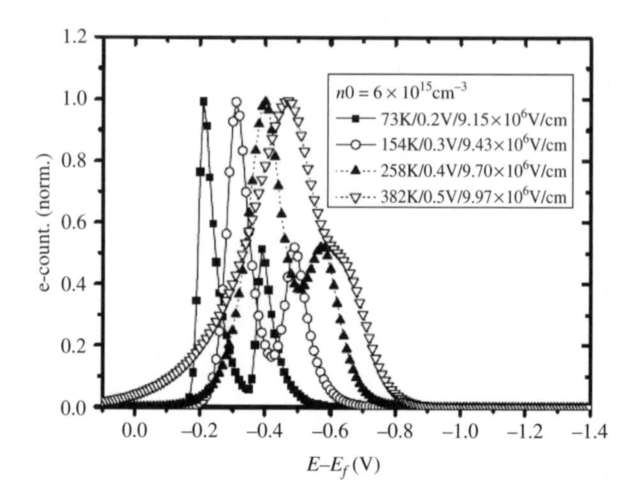

Figure 5.15 EED of Si emitter with low doping content

in a surface dielectric layer. Both models can qualitatively account for the multiple peaks and shifts to lower energy. Field penetration through a dielectric surface layer containing gap states will also shift the states to lower energy in this manner. (Resistive potential drops in the bulk are ruled out, since the peak shifts do not correlate linearly with emitted current.) The surface potential of semiconductors can also shift due to band bending below the oxide interface.

Two band diagrams for an *n*-type semiconductor emitter with a dielectric coating, illustrating different applied fields, are drawn in Figure 5.16. The lateral dimension in the figure is distorted for clarity, but the vertical dimension is drawn to scale. The surface layer is assumed to be SiO_2, having 9 eV band gap, where the conduction band minimum of the oxide occurs about 3.5 eV above E_c (the silicon conduction band minimum) [49]. The native oxide may contain significant amounts of carbon and other species, which can change these properties. Native oxide on flat surfaces is typically 2–3 nm thick. Significant concentrations of allowed states within the SiO_2 band gap are typically present at the Si–SiO_2 interface and within the SiO_2 layer. Two types of surface-related states are drawn in the figure: silicon–oxide interface states, and bulk oxide states. The net charge in interface states causes band bending in the silicon below the interface, and moves the potential of the oxide relative to the Fermi level (dashed line). Charge within the oxide causes band bending in the oxide as well as the silicon, so that the potential of the oxide surface can change with respect to the silicon band edges at the interface. To simplify the figure, oxide states are drawn only at the oxide surface, such that the oxide band edges are straight.

Interface and oxide states, as well as the bulk bands, can serve as initial states in a tunneling process. Emission from the states at the oxide surface will be much more probable than emission from bulk states at the same energy, since electrons in the bulk must tunnel through the oxide as well as the potential barrier in vacuum. However, emission of an electron from an oxide state will leave the state positively charged, and the state must be refilled before emission from the same state can take place again. Thus, most oxide states at energies above the Fermi level will become positively charged. Oxide states near or below the Fermi level may be refilled with charge from the bulk silicon. Transport from the bulk bands into oxide states

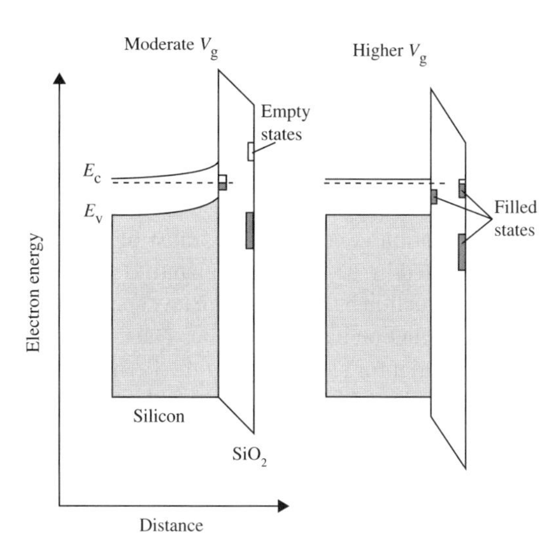

Figure 5.16 Band diagram of an *n*-type semiconductor surface covered with a thin oxide. Examples of filled and empty states at the semiconductor/oxide interface and within the oxide are indicated with filled and empty boxes [1]. Copyright (2001) by John Wiley & Sons, Inc. This material is reproduced with permission of John Wiley & Sons, Inc.

could take place by tunneling through the oxide, either in a single step or in multiple steps via intermediate states. Alternatively, resonant tunneling can occur from an initial state (e.g., in the bulk silicon) through an intermediate oxide state in a single-step process.

In Figure 5.16, the silicon bands are drawn bending upward on the left (moderate V_g case), indicating that the interface has net negative charge. The upward band bending moves electrons away from the surface and exposes positive donors, such that the number of donors is equal to the negative surface charge. Surface depletion such as this often occurs at silicon and other semiconductor surfaces, moving additional negative charge into interface states can screen an electric field outside the surface (such as produced by making the FEA gate positive). The additional negative charge will move the Fermi level closer to the silicon conduction band and reduce the band bending in the silicon. If the density of surface states is large enough, the change in Fermi level position will be small. However, if the gate potential is increased to a point where the density of oxide and interface states is insufficient to terminate the gate field, the bands will flatten (as drawn on the right side of the figure). Electrons that are accumulated in the conduction band near the surface will terminate the additional gate field.

The electric field needed to create a significant tunneling probability may also produce accumulation. Electric fields normally associated with field emission near threshold, 2×10^7 V/cm, correspond to surface charge densities near 10^{13} cm^{-2}. Thus, if the density of interface and oxide charge at the tip apex is below 10^{13} cm^{-2}, the bulk silicon just below the surface in the apex region will accumulate when the electric field adjacent to the surface is strong enough to produce field emission. The charge density at a flat native oxide interface is typically 10^{12} cm^{-2}, well below the density required to screen the electric field applied by the gate. However, higher densities of interface and oxide states are quite possible. High interface state concentrations can be created by impurities (such as metals), ion or electron impact, and the high curvature at the apex. In addition, states can be created when significant energy is transferred from a tunneling electron to the oxide [50, 51]. Thus, it is quite possible for the interface state density near the tip apex to be large enough to prevent accumulation. It is also possible that the interface state density can change due to either environmental factors or emission history, and that such changes could move the surface from accumulation to depletion (or vice versa).

If the interface state concentration is not too high, most of the negative charge needed to screen the gate field is accumulated in the conduction band. Because of the modest density of states near the bottom of the conduction band, the Fermi level position may move several tenths of an electron volt above the conduction band minimum. Thus, accumulation can significantly reduce the emission barrier potential for electrons near the Fermi level at the surface as well as drastically increase the density of initial states. Since the electric field falls off quickly along the tip shank below the apex, the silicon surface along the shank may be depleted even when the apex is accumulated. This result was confirmed via a detailed calculation [52] where the surface state density was assumed to be 3×10^{13} cm^{-2} eV^{-1} (uniform in energy within the band gap, charge neutrality level at midgap).

In dependence on the band bending at semiconductor surface the EED spectra can be significantly different. The energy spectra of the emission from a single n-type (0.02 Ω × cm) silicon tip, measured at three gate voltages are shown in Figure 5.17 [46]. Each of the spectra show emission at energies near the Fermi level (zero), indicating that some initial states near the

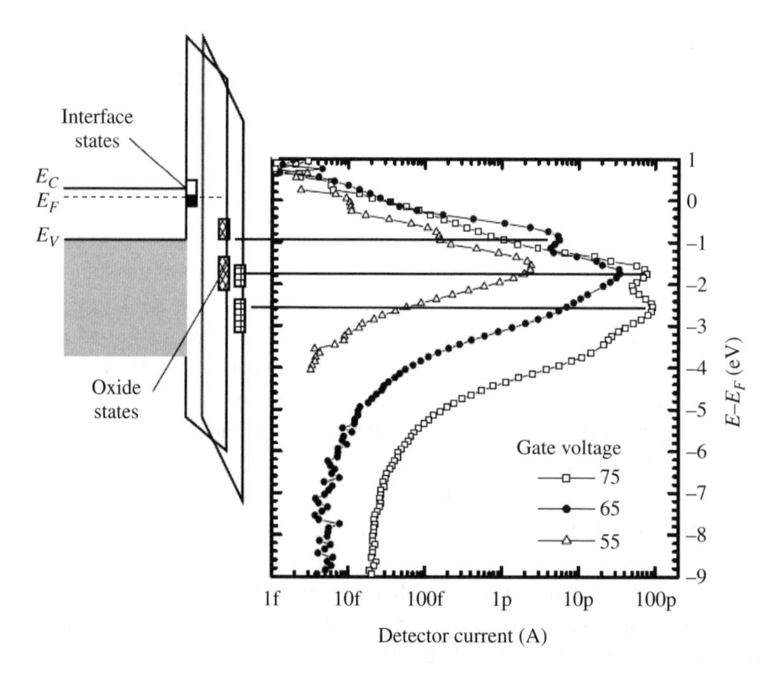

Figure 5.17 Energy spectra of a single silicon field emitter measured at three different gate voltages. A band diagram drawn to the same energy scale is indicated to the left. In the current axis at the bottom, f stands for femto (10^{-15}), and p stands for pico (10^{-12}). Reproduced with permission from Ref. [46]. Copyright (2000). American Vacuum Society

Fermi level are filled. The spectra show peaks, and some of these spectral features shift with gate voltage, indicating that these features come from states in the surface oxide. However, the emission threshold energy does not shift, showing that the band bending in the silicon does not change. This behavior is consistent with flat bands (or accumulation) in the silicon. Two bands of oxide states consistent with the peaks in the spectra are drawn in the figure with and without a field across the oxide layer.

On the other hand, the energy spectra from a second single-tip array, located adjacent to the first, are quite different (Figure 5.18). All the spectra measured at different gate potentials have emission thresholds about 1 eV below E_F. The lack of any emission closer to E_F is consistent with upward band bending near the surface (as shown in the band diagram), since the band bending would prevent electrons in the conduction band from reaching the surface. The consistent threshold energies indicate that the interface state density is high enough to keep the Fermi level position nearly constant. In Figure 5.18, the emission current at energies just below threshold does not increase with gate voltage. Instead, the emission at a given energy saturates with increasing gate potential. Field emission from a free electron gas, with or without a resonant process, would increase indefinitely as the external field is increased. In contrast, the observed current saturation implies that the emission current is limited by transport into the initial states, rather than the tunneling probability. The transport limit can be explained by emission from states on the oxide surface, which must be refilled by transport through the oxide. The oxide states appear to be distributed over a wide energy range. Increasing of the gate potential increases emission from progressively lower energy states, and the states at lower energies emit more current because they are refilled more quickly. (Transport

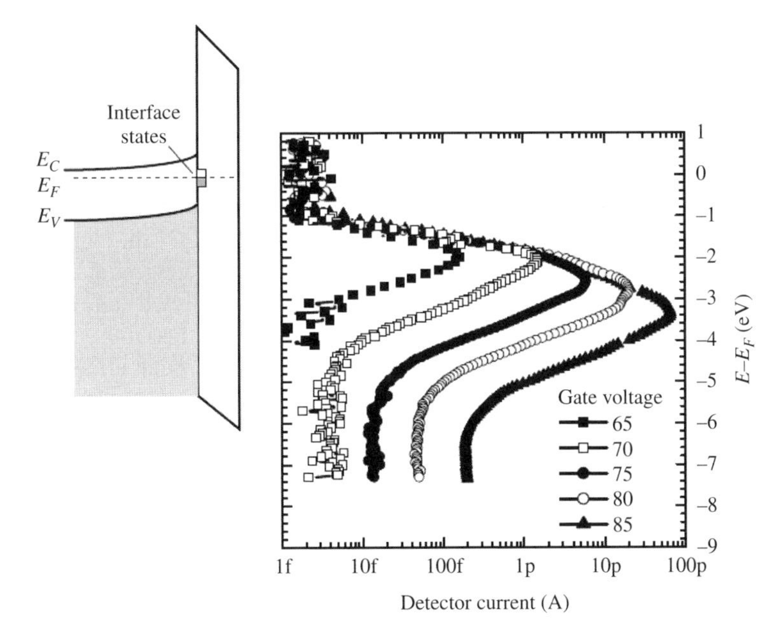

Figure 5.18 Energy spectra of a single silicon field emitter, produced with the gate voltage set at the values indicated. A band diagram with the same energy scale showing surface depletion is indicated to the left. Reproduced with permission from Ref. [46]. Copyright (2000). American Vacuum Society

limits in the silicon below the emission area would shift the entire distribution, and so do not explain the data.)

A hysteresis effect in emission spectra of a third single silicon tip has been observed [1]. Distinct sets of spectra were obtained, depending on whether the gate voltage had been held high or low prior to the measurements. This hysteresis indicates the presence of oxide charge and the change in that charge induced by the gate voltage and/or emission current. The energy spectra have consistent threshold energies below EF, again indicating that the bulk bands are depleted near the surface. The spectra, measured after the gate voltage has been held low for several minutes, are different than the spectra measured after the gate voltage has been held high. The spectra measured after holding the gate low seem to have two emission bands, one being very similar to the spectra made after holding the gate voltage high, plus a second band at higher energy. Thus, holding the gate potential low appears to have "turned on" an additional emission site.

The emission spectra that shift to lower energy with higher gate potential, but shift back up after holding the gate potential high for several minutes have been observed in Ref. [1]. The spectra measured at the lower gate potentials show current saturation at energies just below threshold and the threshold energies shift to more positive potential (lower electron energy) with more positive gate voltage. The spectra measured at 90 V gate potential shifts to higher energy after the gate voltage has been held high for 45 minutes. The shift occurred discretely, suggesting that a single charge or state was responsible.

Based on proposed models [1] taking into account existence of native oxide and interface and oxide states it is possible to explain the peculiarities of EED and their transformation at EFE from silicon. The EED spectra can change under different influences. The density of interface and oxide states can be changed by hot electron induced surface chemistry. Bonds can be broken by the energy released when energetic electrons impact the surface. SiO_2 can be reduced to SiO by bombardment with electrons or photons, and SiO can be thermally desorbed at much lower temperatures than SiO_2. The impacting particles are thought to create ("hot") secondary electrons which interact with the oxide bonds. Energy to create hot electrons can also be generated when electrons are emitted from states well below E_F. Tunneling through the oxide in a silicon-SiO_2-metal (or polysilicon) structure can also create hot electrons. This mechanism is thought to cause "stress induced leakage current" (SILC), the increase in leakage current observed at a low voltage after current is forced through the devices at higher voltages. The additional states increase transport through the oxide, presumably by allowing multi-step tunneling. Field emission from silicon can produce a similar energy loss if the emission occurs at energies well below the Fermi level, and presumably will have similar effects in increasing the density of oxide states and thus transport and emission.

The transformation of emission spectra measured after the tip had been emitting for many hours at V_g between 32 and 50 V (2 V increments) and T = 300 °C was observed in Ref. [46]. The emission current increased exponentially at all energies with very little shift in the peak potential, showing that the emission was not limited by transport to the surface. The peak position was nearly 0.5 eV below E_F, showing that the emission came from surface states rather than the bulk conduction band. The energy spectra extended to energies at least 1 eV above the Fermi level, indicating that electrons were sufficiently concentrated at the emission site to cause significant electronic energy transfer between electrons.

Another experiment showed that after allowing the emitter to rest in Ultra High Vacuum (UHV) (with the gate disconnected) for several days, the minimum gate voltage required to produce emission increased to more than 100 V. Presumably, the change in emission occurred due to oxidation of the clean surface. In contrast, arrays still displaying emission saturation effects can be exposed to air without a significant change in the *I-V* curve (provided that they are not emitting while exposed). Apparently adsorbed molecules, such as O_2 or H_2O or CO_2 do not normally react with the silicon native oxide surface, and are thus free to desorb when the pressure is reduced. However, gas exposure while the emitters are operating does reduce the emission. In this case the energy released during field emission apparently stimulates the oxidation in a manner similar to photo-stimulated oxidation [53, 54]. These reactions can heal unterminated oxide bonds, thus reducing the state density and consequently the emission current. Thus, it appears that surface bonds are concurrently healed and broken while the emitters are operated. The steady state surface condition will depend on the rate balance between bond breaking and reoxidation. The oxidation rate depends on the density of reactive molecules on the surface, the density of dangling bonds, and the flux and energy of emitted electrons. Relatively clean surfaces will oxidize spontaneously (without emission), whereas oxidized surfaces require the energy released by field emission to stimulate reaction. The rate of created oxide states (by breaking oxide bonds) depends on the emission current and emission energy. The range of available energy varies with the condition of the surface and the gate voltage. In general, less energy is released at lower gate voltages. Surfaces with high densities of oxide states behave more like clean metal surfaces and emit at average energies closer to E_F (releasing less energy). SILC measurements [20, 51, 55] indicate that 1.5–2 eV is the minimum energy required to create oxide states. Thus, the surface has come to the equilibrium condition where the energy released by emission is just enough to create oxide states. SILC measurements show that oxide states are created more efficiently when greater energy is released by the electrons; thus a higher rate of bond breaking (for a given emission current) is implied by emission at energies further below E_F. This process explains the initial increase in emission current (as oxide bonds are broken by energy released as electrons are emitted far below E_F) followed by saturation (when reoxidation of a relatively high density of broken oxide states balances bond breaking).

5.5.2 Electron Energy Distribution of Electrons from Nanocrystalline Silicon

The EED was investigated at emission from porous polycrystalline silicon (PPS) metal-oxide-semiconductor (MOS) structures with an ultra-thin oxide layer [20]. The experimental results confirm the mechanism for electron emission from the diode is due to tunneling effect. The porous layer is composed of a number of silicon nanocrystallites. It is assumed that these nanocrystallites are surrounded by a thin SiO_2 layer, which is formed during the photoanodization and subsequent rapid thermal oxidation (RTO) process. When a bias voltage is applied to the diode, the major potential drop is produced in the PPS layer, especially in the region near the PPS surface. Under the positive biased condition, electrons are thermally injected from the *n*-type silicon substrate into the PPS layer and drifted through the PPS layer toward the top Au electrode. Some of them may become hot electrons and excite electrons in silicon nanocrystallites. As the bias voltage is increased further, the field effect becomes dominant. Electrons which are drifted through the PPS layer and produced in the PPS layer by the field induced carrier generation cascade can be emitted into the vacuum through the thin Au electrode as hot electrons via subband of thin oxide layer (Figure 5.19).

The *dc* bias voltage V_{ps} was applied to the top Au electrode with respect to the back electrode (see Figure 5.4). The energy distribution of emission electrons obtained from the sample at different V_{ps} is shown in Figure 5.20. The x-axis indicates the electron energy and the y-axis indicates the number of emitted electrons (arbitrary unit). Unlike the conventional cold cathode device [56], the distributions are not Maxwellian and strongly depend on V_{ps}. The peak energy and maximum energy shift toward the higher energy side in accordance with an increase in V_{ps}. For example, a high peak energy of about 6 eV is shown at $V_{ps} = 18$ V. The original point of the measured energy distribution corresponds to the vacuum level. Taking into account the work function of Au, which can be assumed to be about 5 eV, it can be seen that a significant amount of electrons have the energy of about 50% of the value corresponding to qV_{ps}. The important thing is that the peak of E_p and E_{max} change fairly with V_{ps}.

According to the result of time-of-flight measurements [57], the drift length of the carriers under a strong electric field (about 10^5 V/cm) within the porous silicon (PS) layer reaches 1 μm. This value is a much larger value than the size of the average silicon nanocrystallites. It indicates that conduction electrons can easily become hot electrons. It can be assumed that a major potential drop is produced at the boundaries, such as the surface skin of the nanocrystalline silicon or a thin oxide layer between nanocrystallites, but not at the bulk. This assumption can be applied to the PPS and the drift length within the PPS under a similar electric field is much longer than the grain size (typically 200–300 nm) and, therefore, électrons become hot electrons at the surface area.

The EED was investigated also at emission from nanocrystalline silicon MOS cathodes in dark and under illumination [58]. The energy distribution spectra of emitted electrons from the cathode in the dark and under illumination of a He-Ne laser at the gate voltage of 11 V which

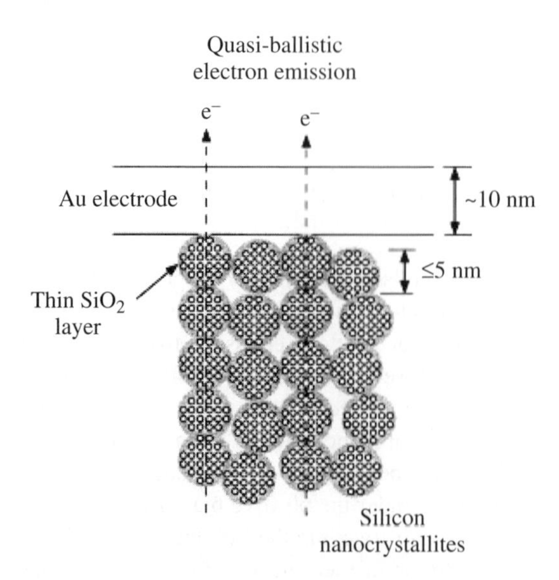

Figure 5.19 Proposed quasiballistic electrons drift model of the PPS surface emitting cold cathode. Reproduced with permission from Ref. [20]. Copyright (1999), American Vacuum Society

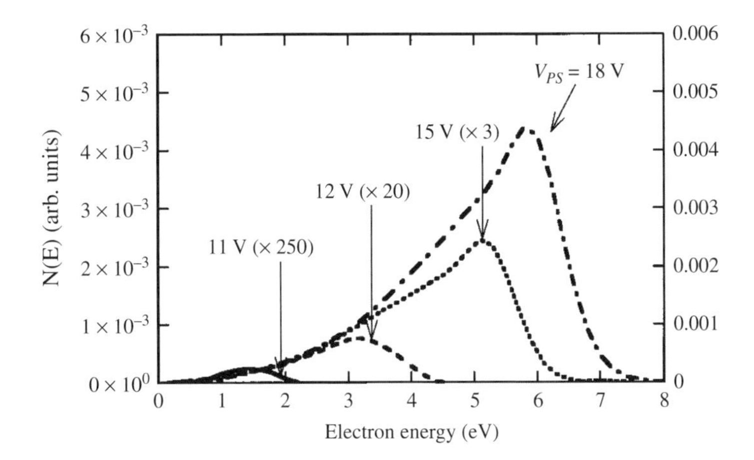

Figure 5.20 Energy distribution of emission electrons obtained from the PPS sample at a different V_{ps}. The distributions are not Maxwellian and strongly depend on V_{ps}. Reproduced with permission from Ref. [20]. Copyright (1999), American Vacuum Society

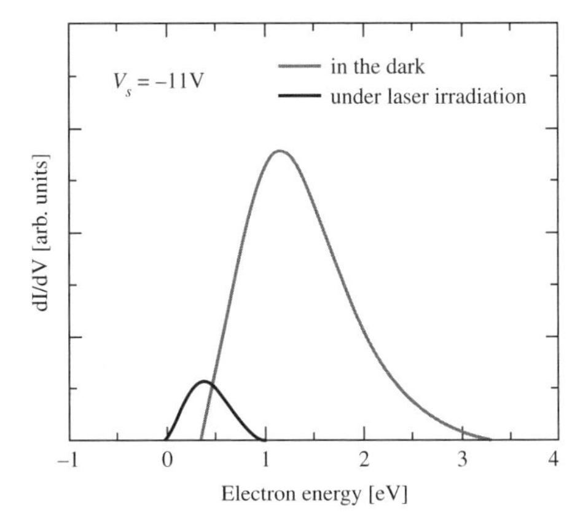

Figure 5.21 Energy spectra of emitted electrons from nc-Si MOS cathode in the dark and under He-Ne laser illumination. Copyright (2011) IEEE. Reprinted with permission from Ref. [58]

is near the threshold voltage in the dark condition are shown in Figure 5.21. The zero energy in the abscissa indicates the vacuum level of the Pt gate electrode. The highest energy of emitted electrons under the illumination is about 2 eV higher than that in the dark; this value is the same as photon energy of the He-Ne laser.

References

1. J. Shaw and J. Itoh, Silicon field emitter arrays, in *Vacuum Microelectronics*, ed. W. Zhu, John Wiley & Sons, Inc., New York, 2001, pp. 187–246.
2. S.C. Miller and R.H. Good, A WKB-type approximation to the Schrodinger equation, *Physical Review* **91**, 174 (1953).

3. V. Filip, D. Nicolaescu, and F. Okuyama, Modeling of the electron field emission from carbon nanotubes, *Journal of Vacuum Science and Technology B* **19**, 1016 (2001).
4. V. Filip, D. Nicolaescu, M. Tanemura, and F. Okuyama, Modeling the electron field emission from carbon nanotube films, *Ultramicroscopy* **89**, 39 (2001).
5. J.W. Gadzuk and E.W. Plummer, Hot-hole-electron cascades in field emission from metals, *Physical Review Letters* **26**, 92 (1971).
6. J.W. Gadzuk, Hot-electron femtochemistry at surfaces: on the role of multiple electron processes in desorption, *Chemical Physics* **251**, 87 (2000).
7. L.W. Swanson and L.C. Crouser, Anomalous total energy distribution for a tungsten field emitter, *Physical Review Letters* **16**, 389 (1966).
8. J.W. Gadzuk and E.W. Plummer, Field emission energy distribution (FEED), *Reviews of Modern Physics* **45**, 487 (1973).
9. 9.L.W. Swanson and L.C. Crouser, The effect of polyatomic absorbates on the total energy distribution of field emitted electrons, *Surface Science* **23**, 1 (1970).
10. H.W. Fink, Point source for ions and electrons, *Physica Scripta* **38**, 260 (1988).
11. H.W. Fink and H. Schmidt, Holography with low-energy electrons, *Physical Review Letters* **65**, 1204 (1990).
12. N.D. Lang, A. Yacoby, and Y. Imry, Theory of a single-atom point source for electrons, *Physical Review Letters* **63**, 1499 (1989).
13. J.W. Gadzuk, Single-atom point source for electrons: Field-emission resonance tunneling in scanning tunneling microscopy, *Physical Review B* **47**, 12832 (1993).
14. V.T. Binh, S.T. Purcell, N. Garcia, and J. Doglioni, Field-emission electron spectroscopy of single-atom tips, *Physical Review Letters* **69**, 2527 (1992).
15. L. Yu. Ming, N.D. Lang, B.W. Hussey, and T.H.P. Chang, New evidence for localized electronic states on atomically sharp field emitters, *Physical Review Letters* **77**, 1636 (1996).
16. N. Ernst, J. Unger, H.W. Fink, *et al.*, Comment on "Field-emission spectroscopy of single-atom tips," *Physical Review Letters* **70**, 2503 (1993).
17. Forbes, R.G. (1996) The electron energy distribution from very sharp field emitters, Proceedings of the 9th International Vacuum Microelectronics Conference, St. *Petersburg, Russia* **1996**, p. 58.
18. R. Morin and H.W. Fink, Highly monochromatic electron point-source beams, *Applied Physics Letters* **65**, 2363 (1994).
19. S.T. Purcell, V.T. Binh, and N. Garcia, 64 meV measured energy dispersion from cold field emission nanotips, *Applied Physics Letters* **67**, 436 (1995).
20. T. Komoda, X. Sheng, and N. Koshida, Mechanism of efficient and stable surface-emitting cold cathode based on porous polycrystalline silicon films, *Journal of Vacuum Science and Technology B* **17**, 1076 (1999).
21. L.N. Dobretsov, M.V. Gomoyunova. *Emission Electronics*, Nayka, Moskow, 1966 (in Russian).
22. A.G. Zdan, M.I. Yelinson, V.B. Sandomirskiy, Investigation of field emission spectra of the electrons emitted from semiconductors, *Radiotekhnika Elektronika* **7**, 670 (1962).
23. N.V. Mileshkina, I.L. Sokolskaya, Field electron emission from thin germanium layers on tungsten, *Physics of the Solid State* **3**, 3389 (1961).
24. R.W. Gurney, Theory of electrical double layers in adsorbed films, *Physical Review* **47**, 479 (1935).
25. L.N. Dobretsov, Electron field emission from metal coated with layer of adatoms, *Physics of the Solid State* **7**, 3200 (1965).
26. W.-P. Kang, J.L. Davidson, *et al.*, Micropatterned polycrystalline diamond field emitter vacuum diode arrays, *Journal of Vacuum Science and Technology B* **14**, 2068 (1996).
27. J.D. Jarvis, H.L. Andrews, C.A. Brau, *et al.*, Uniformity conditioning of diamond field emitter arrays, *Journal of Vacuum Science and Technology B* **27**, 2264 (2009).
28. Jarvis, J.D., Ghosh, N., Jonge, N. *et al.* Resonant tunneling and extreme brightness from diamond field emitters and carbon nanotubes (2010), Proceedings of the 23rd International Vacuum Nanoelectronics Conference, Palo Alto, CA, 2010, pp. 195–196.
29. Y. Cui, Y. Zou, A. Valfells, *et al.*, Design and operation of a retarding field energy analyzer with variable focusing for space-charge-dominated electron beams, *Review of Scientific Instruments* **75**, 2736 (2004).
30. S.T. Purcell, V.T. Binh, and R. Baptist, Nanoprotrusion model for field emission from integrated microtips, *Journal of Vacuum Science and Technology B* **15**, 1666 (1997).
31. R. Baptist, Ghis, A., and Meyer, R., Energetic characterization of field-emission cathodes, *Proceedings of the 2nd International Conference on Vacuum Microelectronics, Bath, UK*, IOP Conference Series vol. 99 (IOP, Bristol, 1989), p. 85–88.

32. P.R. Schwoebel and I. Brodie, Surface-science aspects of vacuum microelectronics, *Journal of Vacuum Science and Technology B* **13**, 1391 (1995).
33. J.D. Levine, R. Meyer, R. Baptist, *et al.*, Field emission from microtip test arrays using resistor stabilization, *Journal of Vacuum Science and Technology B* **13**, 474 (1995).
34. V.T. Binh, N. Garcia, and S.T. Purcell, *Advances in Imaging and Electron Physics*, ed. P. Hawkes (Academic Press, *New York*, 1996), Vol. 95, pp. 63–153.
35. D.J. Rose, On the magnification and resolution of the field emission electron microscope, *Journal of Applied Physics* **27**, 215 (1956).
36. D. Atlan, G. Gardet, V.T. Binh, *et al.*, 3D calculations at atomic scale of the electrostatic potential and field created by a teton tip, *Ultramicroscopy* **42–44**, 154 (1992).
37. V.T. Binh, S.T. Purcell, G. Gardet, and N. Garcia, Local heating of single-atom protrusion tips during field electron emission, *Surface Science* **279**, L197 (1992).
38. A.M. Russel and E. Litov, Observation of the band gap in the energy distribution of electrons obtained from silicon by field emission, *Applied Physics Letters* **2**, 64 (1963).
39. B. Bornmann, S. Mingels, A. Navitski, and G. Müller (2011) Field emission spectroscopy studies on photo-sensitive p-doped Si-tip arrays, *Proceedings of the 24th International Vacuum Nanoelectronics Conference*, Wuppertal, Germany, 2011, pp. 128–129.
40. B. Bornmann, S. Mingels, A. Navitski, and G. Müller (2010) Field emission spectroscopy of carbon nanotube cathodes, *Proceedings of the 23rd International Vacuum Nanoelectronics Conference*, Palo Alto, CA, 2010, p. 20–21.
41. H. Shimawaki, K. Tajima, H. Mimura and K. Yokoo, A monolithic field emitter array with a JFET, *IEEE Transactions on Electron Devices* **49**, 1665 (2002).
42. H. Shimawakia, Y. Suzuki, K. Sagae, *et al.*, Energy distributions of field emission electrons from silicon emitters, *Journal of Vacuum Science and Technology B* **23**, 687 (2005).
43. Q. Huang, M. Qin, B. Zhang and J.K.O. Sin, Field emission from surface states of silicon, *Journal of Applied Physics* **81**, 7589 (1997).
44. G. Yuan, Y. Neo, H. Shimawaki, and H. Mimura (2009) Effect of field penetration on electron energy distribution of field emission from n-Si emitter, *Proceedings of the 22nd International Vacuum Nanoelectronics Conference*, Hamamatsu, Japan, 2009, pp. 113–114.
45. R.D. Young, Theoretical total-energy distribution of field-emitted electrons, *Physical Review* **113**, 110 (1959).
46. J. Shaw, Effects of surface oxides on field emission from silicon, *Journal of Vacuum Science and Technology B* **18**, 1817 (2000).
47. A.J. Miller and R. Johnston, The influence of surface treatment on field emission from silicon microemitters, *Journal of Physics: Condensed Matter* **3**, 231 (1991).
48. R. Johnston, Field emission from silicon through an adsorbate layer, *Journal of Physics: Condensed Matter* **3**, 187 (1991).
49. J.W. Keister, J.E. Rowe, J.J. Kolodziej, *et al.*, Band offsets for ultrathin SiO_2 and Si_3N_4 films on Si(111) and Si(100) from photoemission spectroscopy, *Journal of Vacuum Science and Technology B* **17**, 1831 (1999).
50. K.R. Farmer, C.P. Debauche, A.R. Giordano, *et al.*, Weak fluence dependence of charge generation in ultra-thin oxides on silicon, *Applied Surface Science* **104/105**, 369 (1996).
51. D.J. DiMaria and E. Cartier, Mechanism for stress-induced leakage currents in thin silicon dioxide films, *Journal of Applied Physics* **78**, 3883 (1995).
52. T. Matsukawa, S. Kanemaru, K. Tokunaga, and J. Itoh, Effects of conduction type on field-electron emission from single Si emitter tips with extraction gate, *Journal of Vacuum Science and Technology B* **18**, 1111 (2000).
53. H. Richter, T.E. Orlowski, M. Kelly, and G. Margaritondo, Ultrafast UV-laser-induced oxidation of silicon: Control and characterization of the Si-SiO_2 interface, *Journal of Applied Physics* **56**, 2351 (1984).
54. T.E. Orlowski and D.A. Mantell, Ultraviolet laser-induced oxidation of silicon: The effect of oxygen photodissociation upon oxide growth kinetics, *Journal of Applied Physics* **64**, 4410 (1988).
55. S. Takagi, N. Yasuda, A. Toriumi, A new I-V model for stress-induced leakage current including inelastic tunneling, *IEEE Transactions on Electron Devices* **46**, 348 (1999).
56. C.A. Spindt, C.E. Holland, A. Rosengreen, and I. Brodie, Field-emitter-array development for high-frequency operation, *Journal of Vacuum Science and Technology B* **11**, 468 (1993).
57. R. Sedlacik, F. Karel, J. Oswald, *et al.*, Photoconductivity study of self-supporting porous silicon, *Thin Solid Films* **255**, 269 (1994).
58. H. Shimawaki, Y. Neo, H. Mimura, *et al.* (2011) Electron emission from nanocrystalline silicon based MOS cathode under laser irradiation, *Technical Digest of the 24th IVNC*, 2011, pp. 220–221.

Part Two

Novel Electron Sources with Quantum Effects

6

Si Based Quantum Cathodes

6.1 Introduction

Si tips and field emission arrays (FEAs) are readily fabricated from crystalline silicon using oxidation to create very sharp emitters. The processing steps can be carried out using standard fabrication tools. However, the silicon based emitters have some problems with stability and reliability. As a rule, silicon emitters are contaminated with a layer of molecules that could be adsorbed from the ambient, desorbed from an anode, or segregated from the bulk. The properties of this surface layer can affect the emission properties.

Clean silicon is quite reactive and can be contaminated within hours, even at low pressures. As a rule, Si is coated with grown film of natural oxide SiO_x with uncontrolled properties. To prevent the oxidation of Si in such reactive gases as H_2O, CO, or O_2 it is desirable to cover it by protective film with controlled properties. It can stabilize and, in some cases, increase the electron field emission (EFE). In the case of emitters made from silicon or other semiconductors, the electronic properties of the surface can also affect transport through the semiconductor. A number of semiconductor structures can provide nonlinear current-limiting.

Cold electron emitting devices, as a rule, have a complicated structure and they require a high vacuum and a high supply voltage. A large dispersion of angles of emitted electrons results in a poor resolution. The field emission (FE) devices based on carbon nanotubes addressed some of these problems. However, lifetime is a serious issue for these devices. Planar type electron emitters based on metal-insulator-metal (MIM) or metal-oxide-semiconductor (MOS) structure are very important in overcoming the above-mentioned problems.

In this chapter the novel electron sources based on silicon with quantum effects are considered. Silicon electron sources are important first of all due to the high development of Si-based technologies and the importance of integration of solid-state and vacuum nanoelectronics devices. The peculiarities of EFE from porous silicon (PS), silicon tips with multilayer coating and laser formed silicon tips with thin dielectric layer are described. In all cases the peaks on emission current-voltage characteristics have been revealed and explained in the framework of resonant tunneling theory. EFE from SiO_x(Si) and SiO_2(Si) films containing Si nanocrystals is considered in detail. Special attention is paid to MIM and Metal-Insulator-Semiconductor

Vacuum Nanoelectronic Devices: Novel Electron Sources and Applications, First Edition.
Anatoliy Evtukh, Hans Hartnagel, Oktay Yilmazoglu, Hidenori Mimura and Dimitris Pavlidis.

emitters. They are integrated in solid state and have plain design. The peculiarities of electron transport mechanism and the role of nanoparticles are considered in detail and the importance of further development of Si-based electron field emitters is shown.

6.2 Electron Field Emission from Porous Silicon

There is a continuous search for new emission structures that may have certain advantages over existing emitter [1–6]. The PS layers as on the tips and flat surfaces are actively investigated [7–11].

In Ref. [12] the PS layer was grown both on flat Si surfaces and on Si tips of Si tip array. Arrays of silicon emitter tips were fabricated by wet chemical etching [13–15]. The cathodes were formed on (100) *n*-type Si wafers ($N_d = 10^{15}$ cm^{-3}) by patterning with Si_3N_4 as a masking material. The tip sharpening was performed by oxidation of the as-etched tips at 900 °C in wet oxygen followed by etching in $HF{:}H_2O$ solution. This sharpening technique allowed the production of tips with a curvature radius of 10–20 nm. The height of the silicon tips was 4 μm while the tip density was 2.5×10^5 tips/cm^2. The radii of the tips and their height were estimated from scanning electron microscopy images (Figure 6.1).

PS layers were formed by electrochemical etching of silicon tips as well as on the flat wafer surfaces [16, 17] (Figure 6.1). The electrolyte composition was 1 : 1 of 48% HF and ethanol. The anodization process was held under illumination with the intensity of 30 mW/cm^2 for

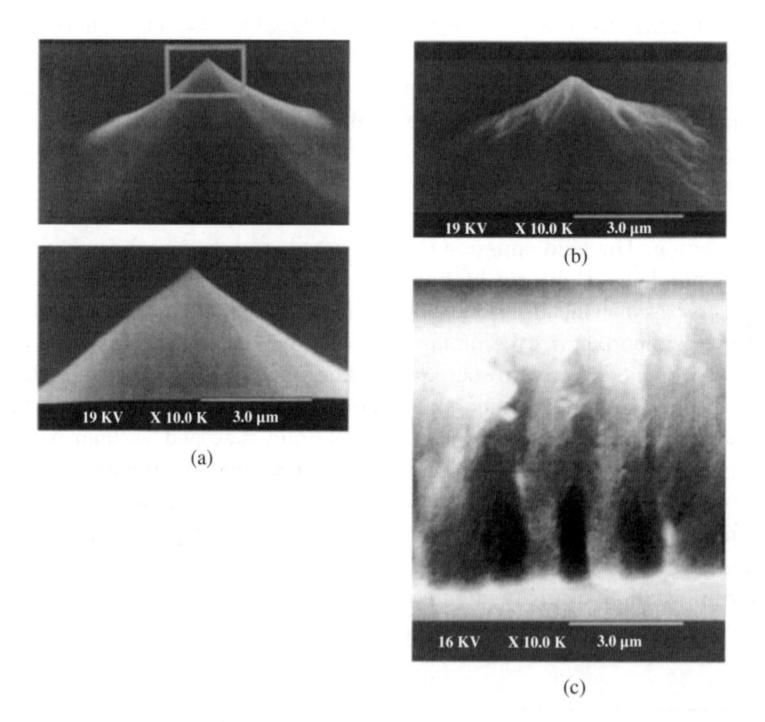

Figure 6.1 SEM micrographs of silicon tips (a) without and (b) with PS layer, and the formed silicon fibrils (c). Reproduced with permission from Ref. [12]. Copyright (2004), AIP Publishing LLC

sufficient hole generation in n-type silicon. The anodization current density and the etching time were varied from 5 to 50 mA/cm^2 and from 15 to 60 seconds, respectively. The thickness of the obtained porous layers and consequently the height of silicon fibrils formed under the investigated conditions increased with the time of anodic etching. The pore width, and, hence, the degree of porosity also increased and the fibril thickness was decreased with the growth of anodization current density. Due to the higher resistivity of PS in comparison with single crystalline Si the PS layer used was thin (<1 μm) to ensure that operation was not current limited.

The measurements of the emission current from samples were performed in a vacuum system that could be pumped out to the stable pressure of 10^{-6}–10^{-7} Torr. The emission current was measured at the ungated cathode-anode diode structure. The test diode structures by sandwiching anode and cathode plates were fabricated. The emitter-anode spacing L was constant and it was chosen to be in the range 7.5–25 μm. Teflon film spacer was used to keep the two plates separated from each other. The investigated emitting structure was used as a cathode and a heavily doped silicon wafer or a molybdenum wire was used as an anode. The emission current-voltage characteristics were obtained with the current sensitivity of 1 nA over the voltage range up to 1500 V. A 0.56–1 MΩ resistor was placed in series with the cathode to provide short-circuit protection.

The I-V characteristics and corresponding Fowler-Nordheim plots of EFE from silicon array with a PS layer are shown in Figure 6.2. The threshold voltage for the silicon tips without PS layer was 530 V. By forming the PS layer the silicon surface becomes rough and therefore an increase of the local electric field and emitting area can be observed in comparison with the case of an untreated silicon tip. An increase of emission area by a factor of 1.5–2 versus anodization time (at fixed etching current density $J = 25$ mA/cm^2) is observed for an etching time of up to 30 seconds. Thinner fibrils are formed on the silicon surface with an increase of the etching time at a constant current density. These cause an enhancement of local electric field and decrease of the threshold voltage of EFE.

Depending on PS preparation conditions two types of I-V characteristics, with smooth features and presence of peaks, are obtained. The peaks on I-V characteristics are caused

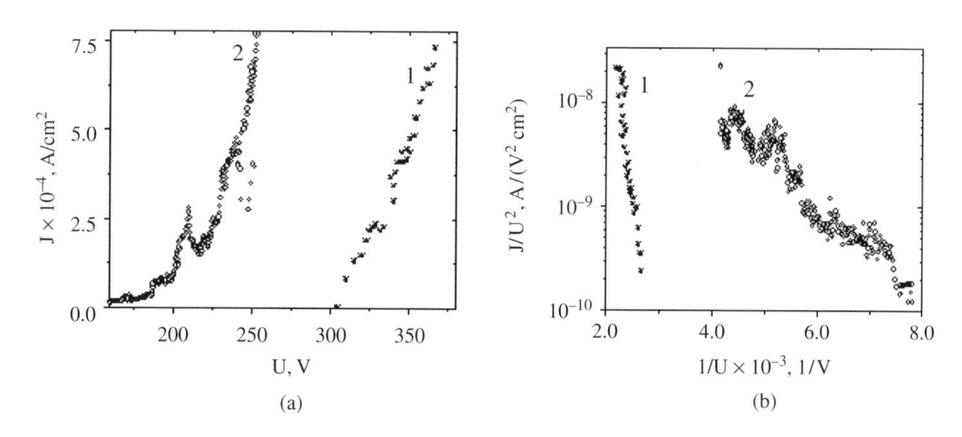

Figure 6.2 (a) Current-voltage and (b) Fowler-Nordheim characteristics of electron field emission from Si tip arrays with porous silicon layer. The two regimes of PS formation are: 20 mA/cm^2 (1) and 50 mA/cm^2 (2), both for 15 seconds. Reproduced with permission from Ref. [12]. Copyright (2004), AIP Publishing LLC

by the resonance-tunneling effect. As a rule, after electrochemical etching and drying in air the surface of PS is coated by an ultrathin layer of SiO_2 (≥2 nm). Therefore, the one-side double-barrier field-emission (FE) cathode is achieved. Ultrathin diamond-like carbon (DLC) film deposition on PS layers does not suppress the peaks.

Further optimization of PS formation by electrochemical etching included the use of technological conditions proposed in Ref. [18]. In this case the PS is formed at low current densities (1–5 mA/cm^2) without applying the voltage [19]. The two-layered PS is formed. The lower layer is microstructured and the upper one is nanostructured. After obtaining the desired thickness of PS the samples were washed with acetone and dried at 100 °C in air or in an atmosphere of acetone. The process of drying was more critical for samples obtained at high current densities.

The emission *I-V* characteristics of structures c-Si/PS are presented in Figure 6.3. As can be seen, the increase of current density at electrochemical etching shifts curves in higher fields region (Figure 6.3, curves 1 and 2). At lower current densities at electrochemical etching the peaks appear on emission *I-V* characteristics. Their shape does not change with repeated measurements (Figure 6.3, curves 3a–c). The presence of emission peaks can be explained as the result of resonant tunneling of electrons under high electric field. Multiple measurements of electron FE confirm the reproduction of the emission peaks pointing out on quite homogeneous size distribution of Si nanowires (Si-NWs) coated with the SiO_x shell. In air atmosphere the diameter of Si-NWs is decreased due to partial oxidation and they can be separated on local Si nanocrystals (Figure 6.4). It restricts the current flow during the electron FE [19]. On removing the partially oxidized layer the threshold voltage at electron FE is shifted to lower values.

There is an interface between micro- and nanostructured layers of PS [20]. After removing the upper nanocrystalline layer in water solution of KOH the electron FE changes dramatically due to changing of the surface morphology. The efficient of EFE can be both increased (for high current densities at electrochemical etching $J = 4$–$5\ \mathrm{mA/cm^2}$) and decreased (for low current densities at electrochemical etching $J = 1\ \mathrm{mA/cm^2}$).

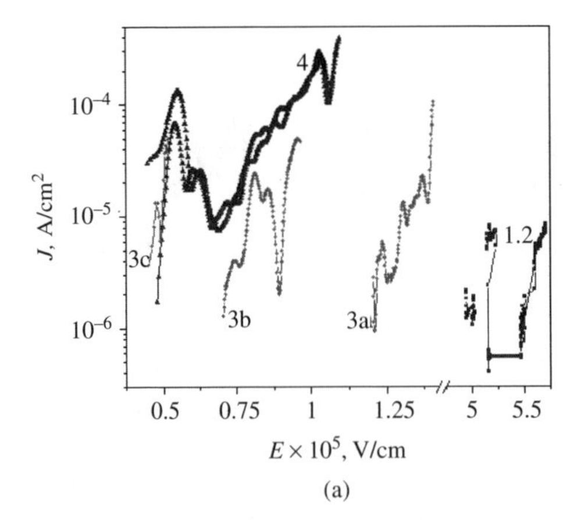

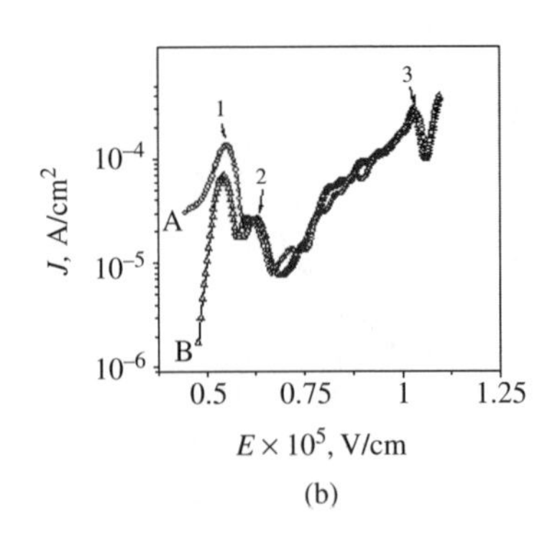

Figure 6.3 (a) *J-E* characteristics of emission current from c-Si/PS with PS layer obtained at different currents densities at electrochemical etching: 1, 2: (4–5) mA/cm^2, 3a–b: 2 mA/cm^2, and 1:1 mA/cm^2 and (b) behavior of curve 4 at repeated measurements

The comparative detailed analysis of emission *I-V* characteristics (Figure 6.3) and photoluminescence (PL) spectra [21] of c-Si/PS samples indicates that the samples that produced at the lowest current density have the best emission and low PL intensity, but in case of high current densities at electrochemical etching the relationship is opposite.

It is obviously caused by the diameters of the Si-NWs. At lower electrochemical etching current the nanowires are thicker and don't restrict the current flow. In the case of the higher electrochemical etching current the nanowires are thin enough and they can be fully oxidized in some thinnest region with creation of the separated nanocrystals (Figure 6.4). That restricts the emission current and enhances the PL intensity.

The curve 4 in Figure 6.3 shows the presence of three peaks at emission *J-F* characteristics of the samples. These peaks are reproduced at repeated measurements (Figure 6.3b). The appearance of the emission peaks can be explained as a result of resonant tunneling through the c-Si/Si-NC/SiO_2/ … /Si-NC/SiO_2/ … /vacuum structure that is formed during the partial oxidation of PS [15, 22].

To explain the peculiarities of electron FE from c-Si/PS the schematic image of coral-like structure of PS and partially oxidized one nanowire have been used (Figure 6.4). The energy band diagram of formed resonant tunneling structure is shown in Figure 6.5.

As a result of size quantization in nanowires there are resonant levels in conduction band of Si NC. The layers of SiO_2 and SiO_2 + vacuum are the energy barriers of resonant tunneling structure. In approximation of infinite high barriers it is possible to estimate the diameter of Si-NC (d) in nanowires [23]:

$$d^2 = \frac{(n_2^2 - n_1^2)\hbar^2}{8m_n^* \Delta E_n}, \tag{6.1}$$

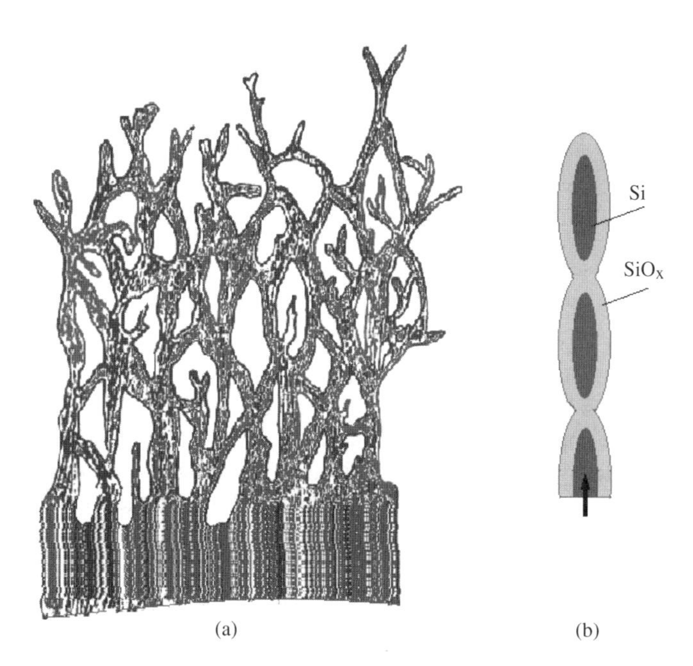

Figure 6.4 (a) Sponge-like structure of porous silicon and (b) view of one nanowire

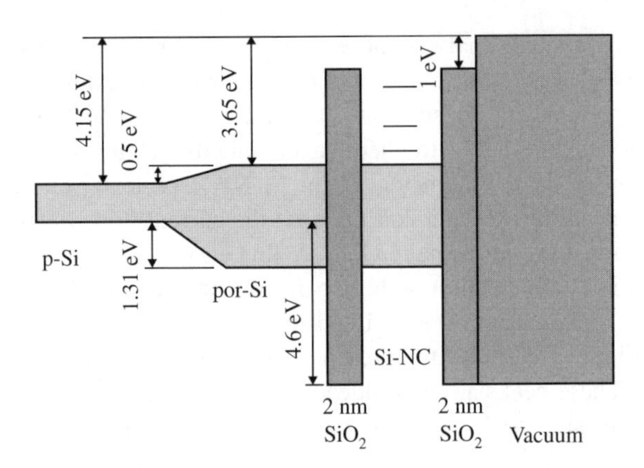

Figure 6.5 Energy band diagram of the c-Si/PS system

where $\hbar$ is the reduced Planck constant, m_n^* is the electron effective mass in Si, ΔE_n is the energy separation between resonance levels, and n_1 and n_2 are the numbers of resonance levels. The approximation of infinite high barriers is quite suitable for high barriers of Si/SiO_2 and Si/SiO_2 + vacuum. The thickness of the first energy barrier SiO_2 in resonance tunneling structure was supposed to be equal to 2 nm (the thickness of native oxide on Si).

To determine the value of the energy separation ΔE_n from the experimental *J-E* characteristics of the emission current it is necessary to calculate the field enhancement coefficient. For this the effective work function has been determined from slope of curves in F-N coordinate.

$$\Phi_{ef} = \frac{\Phi}{\beta^{*\,2/3}}, \tag{6.2}$$

where Φ is the work function and β^* is the electric field enhancement coefficient. The obtained value of effective work was 5.5×10^{-2} eV. The determination of Φ and β^* for case of PS is complicated task. The value of Φ was estimated based on the empiric relationship proposed in Ref. [24]

$$\Phi + E_g \approx 5.5 \text{ eV}. \tag{6.3}$$

In this case it was supposed that $E_g = 2$ eV (in accordance with PL spectra [21]). The obtained value of work function for PS was 3.5 eV and calculated field enhancement coefficient was 545. The diameters of Si-NWs were in the range 1.5–2.8 nm [19].

After removing the upper nanocrystalline layer of por-Si the electron FE comes from the lower microcrystalline layer. In this case the electric field for emission was above four times higher. At repeated measurements the *J-E* characteristics shifted to the region of higher applied fields. The electric field enhancement coefficient calculated from the slope of the curve in the F-N coordinates according to Equation (6.2) was significantly smaller $\beta^* = 65$ and no resonance peaks were observed [19].

6.3 Electron Field Emission from Silicon with Multilayer Coating

The multilayer cathodes (Si-SiO_2-δ-Si-SiO_2) (MLCs) have been formed on silicon tip emitters. The ultrathin SiO_2 and Si layers in MLC structure were deposited using low-pressure chemical-vapor deposition technique, but in some cases the first SiO_2 layer was obtained by thermal oxidation [13, 14]. The layers were deposited on flat silicon wafer and their thickness was measured by an ellipsometer. MLC structures with different thickness of individual layers were obtained by changing the deposition time.

During the electron FE measurements resonant-tunneling effects have been observed for some MLC structures. The measured current-voltage characteristics and the corresponding Fowler-Nordheim plots are shown in Figure 6.6 and compared to the calculated electric field strength. The latter was possible by using the field enhancement coefficient obtained from the Fowler-Nordheim plots according to the procedure described previously [1, 25]. Two separate

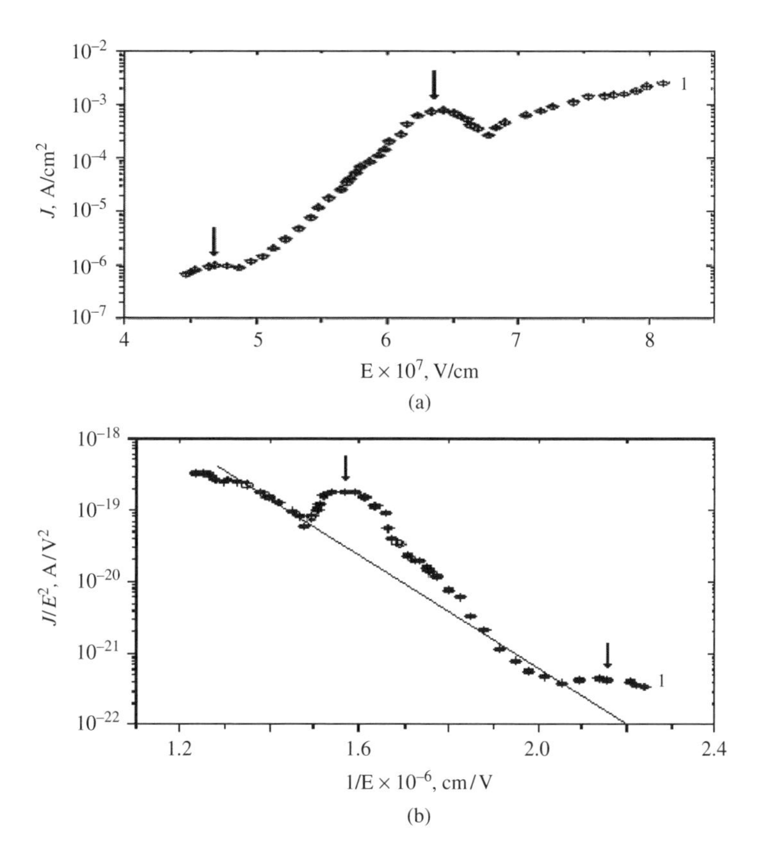

Figure 6.6 Experimental current density dependence on electric field (a) and corresponding Fowler-Nordheim plot (b) ($d_1 = d_2 = d_3 = 2$ nm, $N_e = N_w = 5.6 \times 10^{17} cm^{-3}$, T = 300 K). Reproduced with permission from Ref. [12]. Copyright (2004), AIP Publishing LLC

resonant peaks can be seen in experimental curves (Figure 6.6). The estimated separation between quantum levels is ≈750 meV.

The thickness of δ-doped silicon layer (quantum well) has been determined from the experimental value of the energy separation of two resonant quantum levels according to Equation (6.1). The obtained value of $d_2 = 1.2$ nm. It is lower than given during the structure formation one ($d_2 = 2$ nm). The discrepancy between the experimental and calculated values of d_2 is caused by the reproducibility during the deposition, error in determination of d_2 thickness, and accuracy of the used approximations.

6.4 Peculiarities of Electron Field Emission from Si Nanoparticles

6.4.1 *Electron Field Emission from Nanocomposite $SiO_x(Si)$ and $SiO_2(Si)$ Films*

In Ref. [26] SiO_x(Si) layers were obtained by thermal evaporation of silicon powder in vacuum ((2–3) × 10^{-5} Torr). As shown in Refs. [27, 28], the using of thin SiO_2 coating on silicon tips increases the threshold voltage of FE. The SiO_x film with high excess of Si (low value of x) was chosen in order to ease the electron transport from Si substrate to vacuum. The coatings were deposited as on Si tip arrays and flat silicon wafers. During the deposition process the temperature of substrate was 150 °C. Using quartz oscillator technique with an accuracy of the 3% the film thickness (d) and deposition rate were estimated. For the samples under investigation the deposition rate was ≈(2–3) nm/s. Measurement of the film thickness was performed by microinterferometer MII-4 and profilometer Dektak 3030. In the electrical conductivity and FE experiments the oxides thicknesses were in the region of $d = 10 - 100$ nm. Some samples were annealed in pure argon ambient under the temperature of 1000 °C for 5–40 minutes. As was shown [29], such annealing led to formation of the additional silicon nanoinclusions (nanocrystals) in the oxide film.

The nanorelief of SiO_x(Si) ($x = 0.3$) film surface and its transformation under subsequent treatments are shown in Figure 6.7. As it can be seen, the initial sample surface is characterized

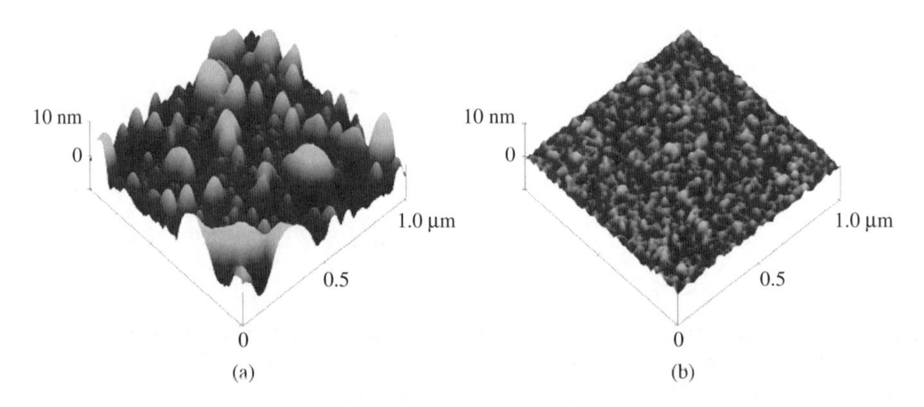

Figure 6.7 AFM images of nanorelief of SiO_x film (a) initial, (b) after annealing (T = 1000 °C, t = 40 min) and following etching in HF solution. Reproduced with permission from Ref. [26]. Copyright (2006), American Vacuum Society

by high roughness connected with large grain nanoprotrusion asperities up to 20 nm. It is caused apparently by the $SiO_x(Si)$ film deposition technology: the film was produced by the deposition of both silicon clusters and SiO_x. In this respect the atomic force microscopy (AFM) results correlate with earlier obtained optical data [30] that demonstrate the composite nature of the as-grown $SiO_x(Si)$ layer structure. As a result of the thermal annealing, the relief character substantially changes. The film surface becomes more uniform and the film surface structure in this case is characterized by nanoprotrusion asperities in the range of 1–3 nm. Thermal annealing of the sample causes decomposition of the initial material onto SiO_2 and Si phases, the latter being in the form of nanoclusters [31]. This fact is revealed, in particular, in the surface morphology of annealed film. Subsequent sample partial etching in $HF{:}H_2O$ solution did not affect on the film surface morphology.

Experimental investigation of EFE from silicon tips coated with $SiO_x(Si)$ ($x = 0.3$) films [26] has allowed to obtain such main results: (1) coating of Si tips with $SiO_x(Si)$ film, as a rule, decreases the electron field emission (FE) efficiency; (2) the efficiency of electron FE decreases with the growth of $SiO_x(Si)$ thickness; (3) partial etching of $SiO_x(Si)$ film increases the FE efficiency; (4) high temperature annealing of $SiO_x(Si)$ film and its transformation in $SiO_2(Si)$ film decreases the electron FE efficiency (sometimes the emission is absent at all investigated regions of electric fields); (5) partial etching of $SiO_2(Si)$ film in all cases allows to significantly increase the electron FE in comparison with $SiO_x(Si)$ film and clean silicon tips; and (6) long etching of $SiO_2(Si)$ film removes it from Si tips and in such a way decreases the FE.

The significant FE was observed from nanocomposite $SiO_x(Si)$ and $SiO_2(Si)$ deposited even on flat silicon surface. The results of EFE from $SiO_x(Si)$ and $SiO_2(Si)$ films in this case are shown in Figure 6.8. As it can be seen, in the case of the initial SiO_x film the emission current has been observed at relatively high voltage (570–770 V) and has been 10^{-7} to 10^{-6} A. The FE from annealed samples was not observed in the investigated electric-field region. But after short etching of the annealed film in the $HF{:}H_2O$ solution the emission is increased in comparison

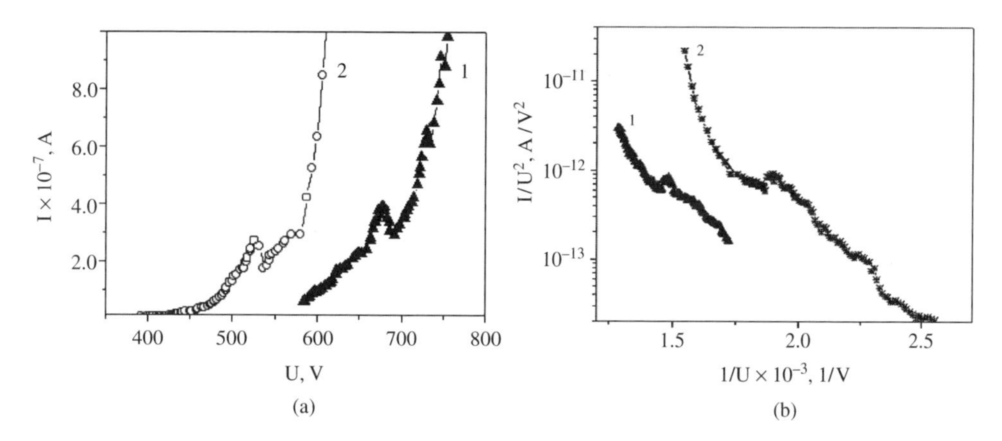

Figure 6.8 Field emission current from $SiO_x(Si)$ films: 1: initial and 2: annealed and etched in $HF{:}H_2O$ solution. Panel (a) shows the current-voltage characteristics and (b) shows the corresponding Fowler-Nordheim plots. Reproduced with permission from Ref. [12]. Copyright (2004), AIP Publishing LLC

with nonannealed samples. The emission began at ≈375 V and the current reached the level of $\approx 10^{-5}$ A. The nonmonotonous region as a peak of emission current is observed at voltages from 430 to 540 V. For an initial SiO_x film, a peak of emission current appeared at voltage of ≈650 V. The existence of current peaks is explained by quantum-size effects in such structures and the appearance at defined electric fields of the additional transport mechanism – resonance tunneling of electrons.

The position of resonant peaks on the voltage axis gives the possibility to estimate the splitting of energy level due to the quantum-size effect by considering the field enhancement coefficient, dielectric permeability, and layer thickness. The estimation of the energy level separation shows that $\Delta E = E_1 - E_2 = 260$ meV.

I-V characteristics of emission currents from Si tips coated with SiO_x(Si) ($x = 0.3$) film of 10 nm thick are shown in Figure 6.9a. The etching of SiO_x film allowed to increase FE

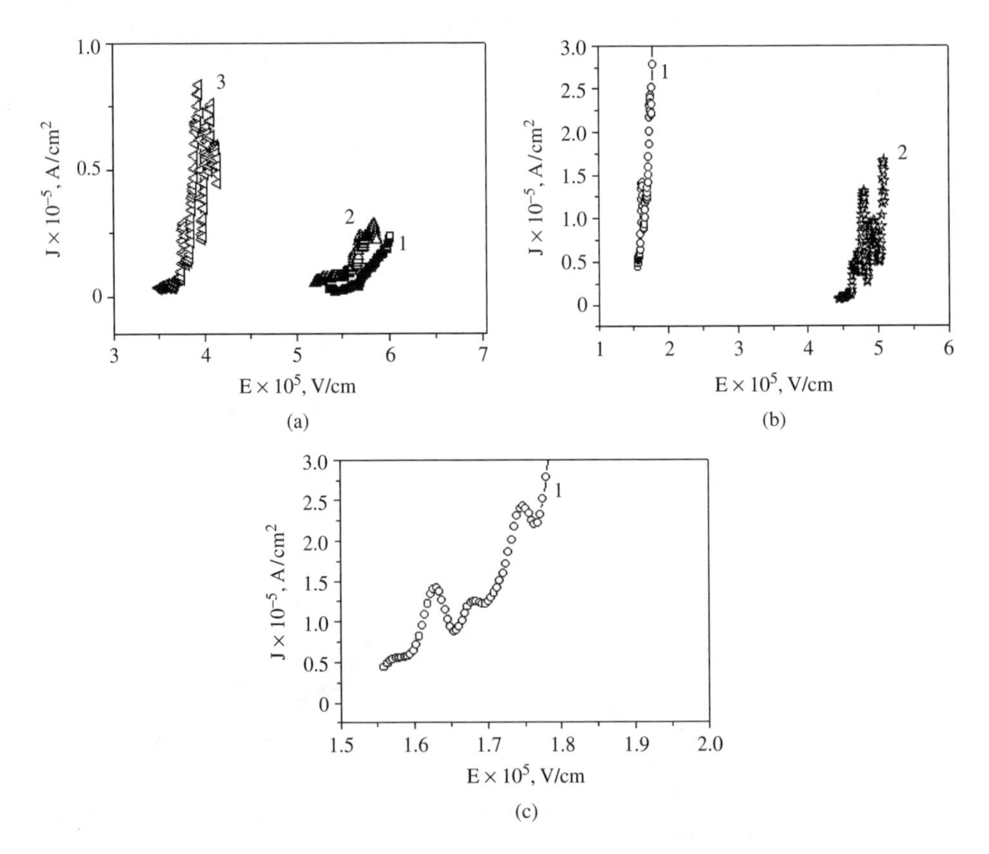

Figure 6.9 Current-voltage dependences of emission current from Si tips coated with nanocomposite SiO_x(Si) film: (a) 1: initial Si + SiO_x(Si) ($d = 10$ nm) system, 2: after following etching $t = 3$ seconds, and 3: after additional following etching $t = 10$ seconds. (b) 1: after high temperature annealing and etching $t = 3$ seconds, 2: after additional following etching $t = 10$ seconds, and (c) curve 1 taken from (b) in other scale. Reproduced with permission from Ref. [26]. Copyright (2006), American Vacuum Society

efficiency. Figure 6.9b shows the *I-V* characteristics of emission from SiO_x(Si) film annealed at T = 1000 °C and in such a way transformed in nanocomposite SiO_2(Si) film including Si nanoclusters embedded into SiO_2 matrix. The FE was not observed without etching of as annealed film. But after etching during $t = 3$ seconds the emission was quite efficient (low electric field, $F < 2 \times 10^5$ V/cm). In this case the two separate peaks were observed on emission *I-V* characteristic (Figure 6.9c). Following additional etching ($t = 10$ seconds) decreases the FE efficiency due to the removing of SiO_2(Si) film (Figure 6.9b).

Experimental results show that in some cases there are peaks on *I-V* characteristics of emission current. The observed peculiarities of the electron FE from SiO_x(Si) and SiO_2(Si) films on semiconductor tips can be explained, taking into account (1) phase composition of the initial and annealed oxide films, (2) amplification of the electric field on the surface silicon clusters, and (3) current carrying mechanisms during FE.

At least a part of relief nanoprotrusions on the film surface observed by AFM are connected with silicon inclusions in SiO_x matrix. These Si clusters due to the local increasing of the electric field promote the EFE and current flow through the film. To leave the film, electrons overcome the combined Si-SiO_x + vacuum barrier at external surface. This barrier is caused by oxide layer coated at the surface of silicon clusters (Figure 6.10a). The breakdown of this layer determines initial sudden change of FE current. Further electron emission passes through Si-vacuum barrier.

Energy band diagram for annealed structure (Figure 6.10b) is similar to initial one. However, in this case the values of the energy barriers at Si-SiO_2 interface are sufficiently higher than at Si-SiO_x interface. Significant barrier height at Si-SiO_2 interface (3.2 eV) is the reason for the absence of FE at applied electric fields for annealed SiO_x(Si) films since external SiO_2 + vacuum barrier has small probability of the electron tunneling. During partial etching in HF:H_2O solution the external SiO_2 layer is removed from Si inclusions on the surface. In this case, to be emitted, electron should be tunneling through only vacuum barrier (Figure 6.10c).

The above-mentioned decrease of the threshold electrical field for FE from annealed and subsequently partially etched SiO_x(Si) films in comparison with initial ones (Figure 6.9) can be connected with the appearance of two factors: (1) increase of the emission center (silicon clusters) concentration that leads to decrease of the distance between them and (2) decrease of their sizes that causes increase of the local electric field in the FE center region. As it can be seen in Figure 6.7, the sizes of Si inclusions for annealed film are smaller, and their density is higher.

The current peak observed on *I-V* characteristics can be explained by the appearance of the electron resonance tunneling transport. The possibility of resonant tunneling mechanism for multilayer Si-SiO_2–δ-Si-SiO_2 structure was theoretically predicted and experimentally proven in work [14]. Due to the similarity of the energy band diagrams of multilayer Si-SiO_2–δ-Si-SiO_2 system and SiO_2(Si) films it is possible to make qualitative suggestion about the resonance tunneling transport at EFE from SiO_2(Si) films. In conduction band of surface Si cluster the discrete resonant energy levels (minibands) may arise due to quantum-size effect that provides the additional resonant tunneling of electrons under certain electric field. The following experiments with varying of the Si cluster size and their density are needed for further support of the resonance tunneling mechanism realization. The optimization of cluster size and nanocomposite film thickness will allow us to obtain the positions of current peaks on *I-V* characteristics in agreement with resonance tunneling theory.

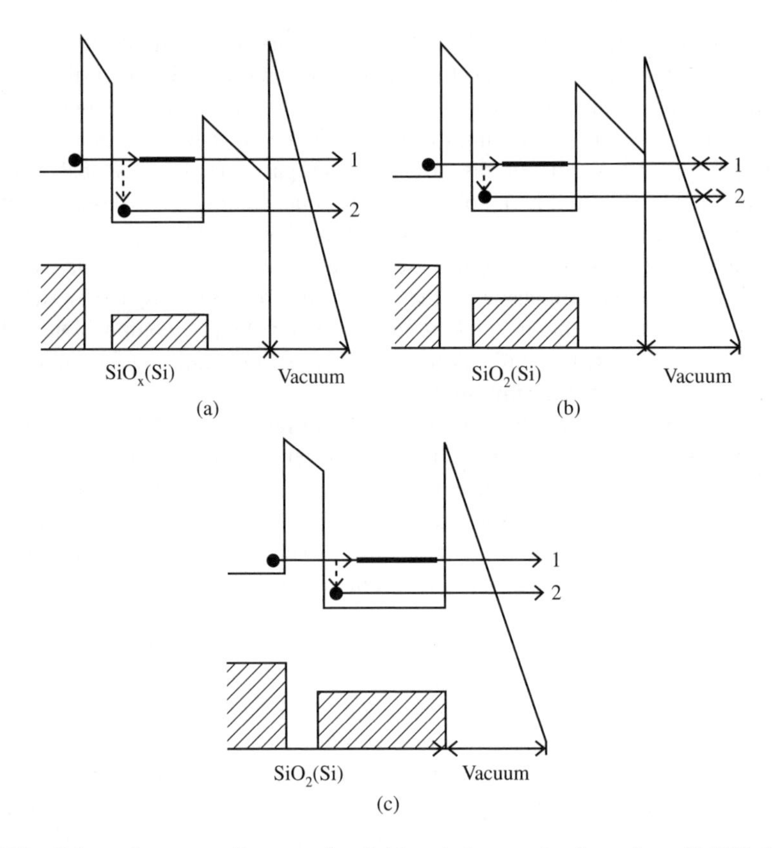

Figure 6.10 Schematic energy diagrams for field emission mechanisms from Si-SiO_x(Si) system: (a) initial, (b) after annealing, and (c) after annealing and processing in HF:H_2O solution (1: resonance tunneling process and 2: Fowler-Nordheim tunneling process). Reproduced with permission from Ref. [26]. Copyright (2006), American Vacuum Society

6.4.2 Electron Field Emission from Si Nanocrystalline Films

SiO_2(Si) films with silicon nanocluster embedded in SiO_2 dielectric matrix is a good candidate for FE cathodes coating [30, 32]. But in order to obtain the efficient emission it is necessary to etch the upper layer of SiO_2. At electron FE from SiO_2(Si) films the effect of resonant tunneling has been observed [30]. It is expected that direct deposition of film of the nanocrystal materials (nc-Si) on a flat surface can be promising for efficient field emitters [33].

In Ref. [34] the results of studies of the EFE from flat nc-Si films formed by laser induced decomposition of silane (SiH_4) [35] have been presented. The silane from gas pipeline flows to the center of the reactor where the CO_2 laser beam has crossed it. The reactor was connected to the molecular-beam analyzer, which included the time of flight mass spectrometer (TOFMS). Reaction products extracted from a conical nozzle into a vacuum and then passed through a sieve in deposition chamber, which preceded the camera TOFMS. Coated with carbon micro rids (sieves) were fixed toward the flow at a distance of 30 cm from the exit nozzle. They served to capture the particles from the cluster beam with or without selection by size using the rotated discontinuous wheel. In typical experiments the nc-Si films were deposited at such technological conditions: the silane flow rate was 40 cm^3/s, the flow rate of helium was 1100 cm^3/s, the total pressure was of 350 mbar, and power of CO_2 laser pulse was 50 mJ.

In this experiment the Si nanoparticles were selected by size. Their diameter was in the range of 3.7–4.2 nm. Si nanoparticles were deposited on silicon substrates. The size of Si nanoparticles is very important to enhance the electric field at electron FE. Morphology of the cathode surface was studied with AFM. AFM image of Si nanoparticles on the surface of the wafer shown in Figure 6.11. As it can be seen, there are a lot of individual nanoparticles, which are grouped in larger clusters. Current-voltage characteristics of emission current and dependence in the Fowler-Nordheim coordinates are shown in Figure 6.12a,b, respectively. Figure 6.12c shows the characteristic in the different scale. The efficiency of FE is quite high for the flat structure (without special formation of tip). There are some nonmonotonous regions of *I-V* characteristics. In the Fowler-Nordheim coordinates these nonmonotonous regions appear as emission current peaks (Figure 6.12b). Repeated measurements have revealed that the low voltage peak 1 in Figure 6.12b is stable (reproducible), but peaks 2 and 3 have disappeared. To explain the results we consider a layer structure of Si nanoparticles. There wasn't special heat treatment for oxidation of Si particles. But the existence of oxygen as impurities in gases and in the reactor cannot be excluded. In Ref. [36] the particle sizes were estimated using the electron microscope with high resolution (HREM). There was revealed the existence of oxide shell around the Si particles. Thickness of the oxide shell decreased linearly from 2.9 to 0.8 nm with decreasing particle size from 33 to 6 nm. In our case, the particle size was 3.7–4.2 nm and thickness of oxide shell was <0.8 nm.

At the FE the electrons from the silicon substrate are injecting into Si nanoparticles layer, pass through it by direct tunneling through the oxide shells between Si nanoparticles and are

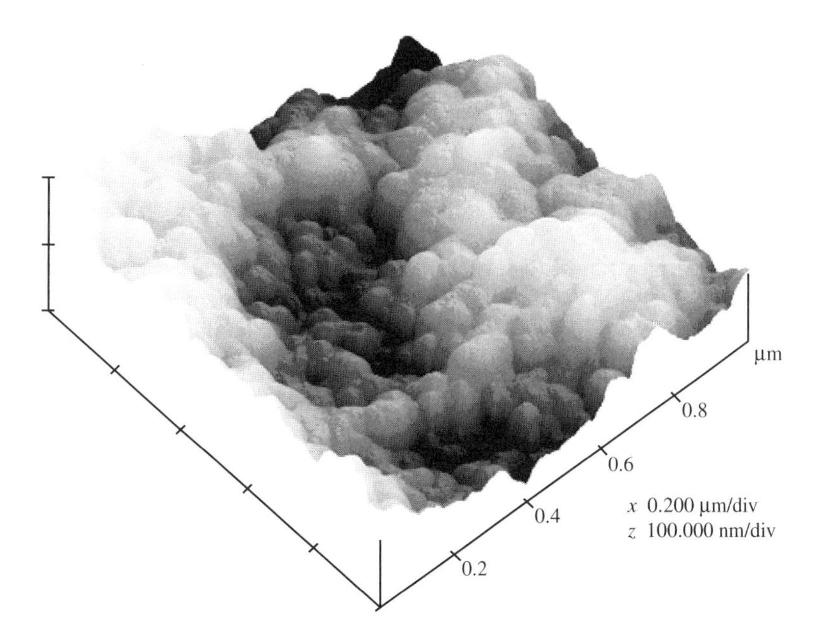

Figure 6.11 AFM image of Si nancrystalline film on the silicon wafer surface

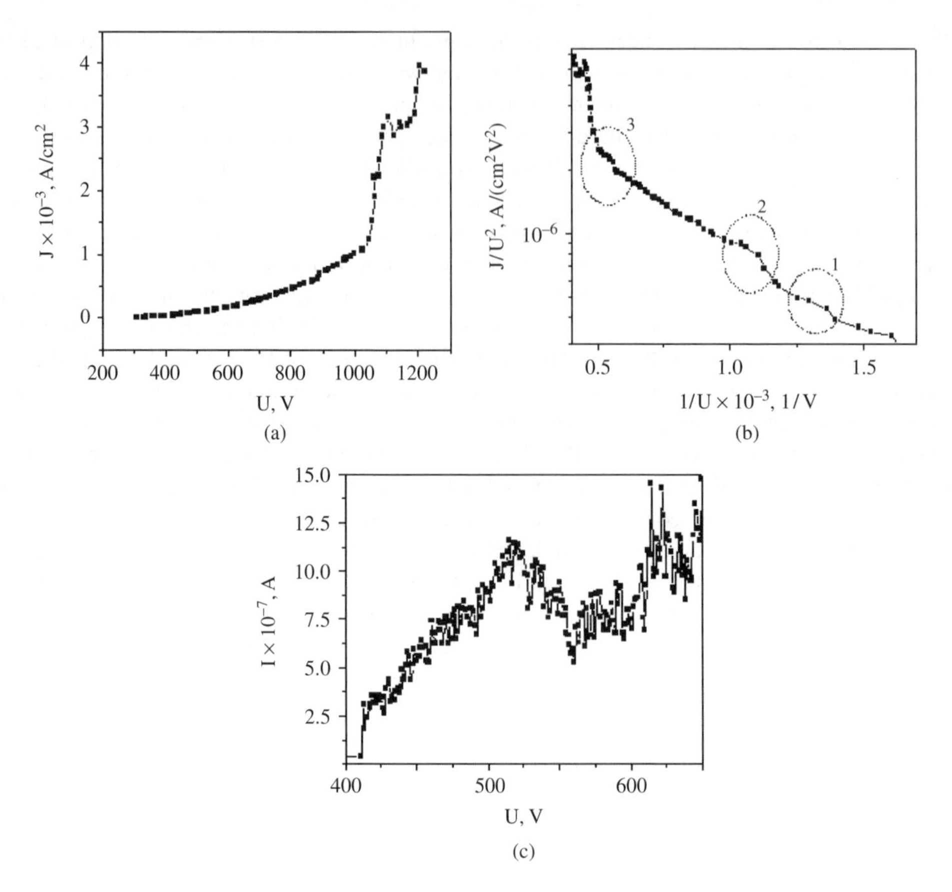

Figure 6.12 *I-V* characteristics (a,c) and the corresponding Fowler-Nordheim plot (b) of emission current from Si nanoparticles ((c) is different scale)

emitted into vacuum. Due to very thin oxide shells the tunneling through the oxide doesn't limit the emission current.

High efficiency of electron FE can be explained by the enhancement of the electric field on Si nanoparticles. In Ref. [37] for calculation of the enhancement coefficient the model of "floating sphere" has been used. Distribution of electric field in this case is

$$E(\theta) = \left(\frac{h}{r}\right) E_0 + 3E_0 \cos\theta, \tag{6.4}$$

where r is the radius of the sphere, h is the altitude of the sphere above the ground (substrate), E_0 is the plane-parallel field of space, and θ is the polar angle. In our case the h is the distance from the substrate surface to the particles, that is, the thickness of the layer of nanoparticles. Accordingly, the electric field near the top of the sphere equals $E_{max} = (h/r + 3)\, E_0$.

Nonmonotonic regions of *I-V* characteristics of emission current are caused by resonant tunneling effect. Because each Si nanoparticle is surrounded by the layer of silicon oxide and it is very small, it can be represented as the quantum well (dot) surrounded by energy barriers. Since Si nanoparticles have small size they show the quantum size effect. As a result, there are resonant levels in the conduction band of Si nanoparticles. At certain values of the electric field the additional mechanism of electron transport, namely, resonance tunneling is realized. Due to this the resonance peaks are observed in current-voltage characteristics of emission current.

6.4.3 *Laser Produced Silicon Tips with $Si_xO_yN_z(Si)$ Nanocomposite Film*

The silicon tip array was formed by a series of single laser pulses which led to the formation of single conical tips [38, 39]. The arrays were relatively uniform. In this process, the silicon substrate (*n*-Si) was heated locally above its melting point by a pulse from a YAG:Nd^{3+} laser. The diameter of the focused laser spot was 20–30 μm. The region exposed to the laser beam was heated to a high temperature and melted leading to strong material evaporation. When the temperature decreased to the crystallization point the conical surface hardened and the arrays of silicon cones could be prepared. Their distance to each other was about 50 μm, the height was in the range 0–100 μm and the curvature radius at the top was ≈1 μm. The laser ablation of *n*-Si took place in air. Due to the pressure of the air (10^5 Pa), most of the silicon particles created by laser heating that had escaped from Si were deposited back on the target and on the conical surface. As the pulse energy was high (0.2 J), there was a broad distribution of particle sizes ranging from several nm to several micrometers. The lighter particles were backscattered further from the plume axis (as far as several millimeters) than the heavier ones. As a result, the conical surface had many protuberances on it and was covered with the nanocomposite film consisting of the nanocrystalline silicon grains in a $Si_xO_yN_z$ matrix (Figure 6.13).

To analyze the content of the surface layer of the silicon tip, Auger profiling was used (Figure 6.14) [38]. Oxygen and nitrogen were observed down to 30 nm from the surface. The real thickness of the nanocomposite film on a silicon surface was smaller and a value of ≈30 nm was obtained due to profiling of a nonflat surface during the Auger measurement.

PL spectra were observed from laser-produced silicon tips (Figure 6.15) [38] and their origin lies in the presence of the nanocomposite film $Si_xO_yN_z(Si)$ on the surface of the silicon tips. Due to small-size nanocrystals in the $Si_xO_yN_z$ matrix, the quantum-size effect causes direct optical transitions. After removing the nanocomposite film $Si_xO_yN_z(Si)$ in a solution of HF in water, no peaks were observed in the PL spectra.

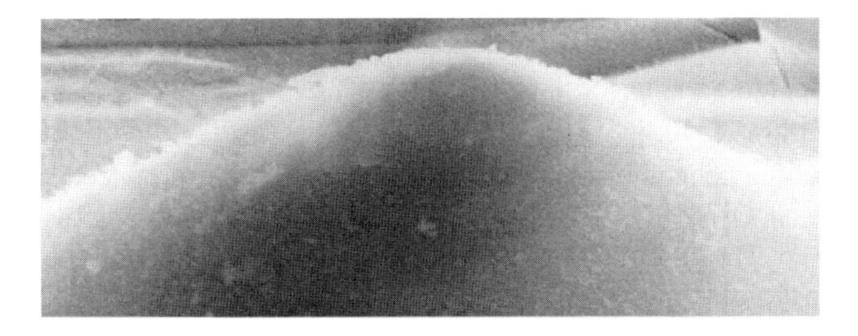

Figure 6.13 SEM image of top of the silicon tip formed by laser. Reprinted from Ref. [38]. Copyright (2002), with permission from Elsevier

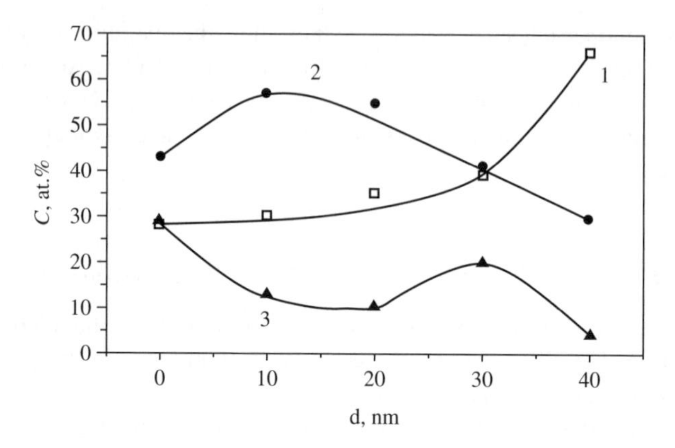

Figure 6.14 Auger profiling spectra of silicon tip array coated with nanocomposite film (1: silicon, 2: oxygen, and 3: nitrogen). Reprinted from Ref. [38]. Copyright (2002), with permission from Elsevier

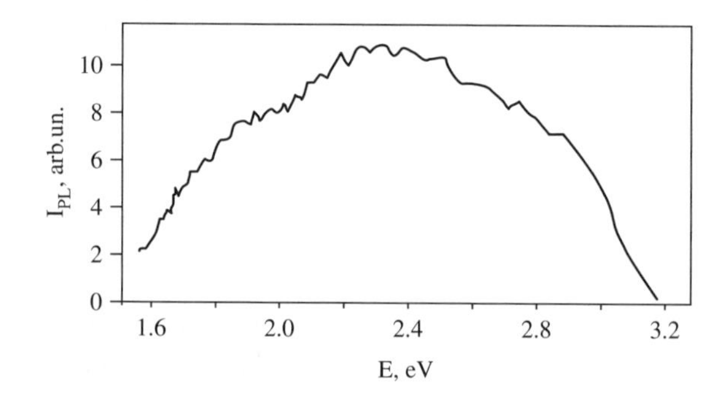

Figure 6.15 Photoluminescent spectrum from Si tips coated by nanocomposite $Si_xO_yN_z$(Si) film. Reprinted from Ref. [38]. Copyright (2002), with permission from Elsevier

The typical current–voltage characteristic of EFE from laser produced silicon tip array corresponds to Fowler–Nordheim law. For the explanation of the features of current–voltage characteristics and their comparison for different tips with nanocomposite film the turn-on (threshold) voltage (V_{th}), emitting areas (α), and local field enhancement coefficient (β^*) have been determined from the Fowler–Nordheim equation according to procedure described in Refs. [1, 25, 40]. For the laser produced silicon tip array relatively high emission parameters, namely effective emission areas around $a = 1.75 \times 10^{-8}$ cm^{-2} and local field enhancement coefficient $\beta^* = 360$ have been obtained.

Resonant-tunneling phenomena have been seen on some laser produced samples that have a nanocomposite $Si_xO_yN_z$(Si) film on their surface. As a rule, one or two resonant peaks have been observed. The current-voltage curves show asymmetry around the maximum in the current, which is typical for electron transport through resonant-tunneling structures. These structures consist of Si quantum-size nanocrystals in the oxynitride matrix on a silicon cone surface. Due to the quantum-size effect, there are some energy levels in the quantum well region that cause an increased tunneling probability under certain values of the electric field.

6.5 Formation of Conducting Channels in SiO_x Coating Film

A simple and low-cost method to form high roughness structures on Si substrates is the chemical or electrochemical etching that produces porous Si (por-Si) [41–45]. The EFEs from emitters based on the por-Si structures have been actively studied and discussed [19, 46–48]. However, por-Si is a complicated material with a wide bandgap due to the quantum-size effect. It contains an ultrathin oxide layer on the surface as well as different kinds of adsorbed atoms such as hydrogen, oxygen, and fluorine. All these factors influence the work function of electrons at FE and cause some modifications of emission current. Deposition of DLC and related films onto the surface of por-Si is known as a good method to decrease the instability of emission current [8, 49–52]. SiO_x(Si) films were used to enhance and stabilize the EFE from semiconductors such as Si [26]. The SiO_x(Si) films were annealed to form nanocomposite SiO_2(Si) films containing Si nanocrystals embedded in the dielectric SiO_2 matrix [26]. Nanocomposite SiO_x(Si) films with different Si compositions can be compared with the por-Si structures regarding their application as EFE emitters. In contrast to Si wires in the por-Si structures, Si nanocrystals are embedded in the nonstoichiometric SiO_x(Si) matrix and covered with protection coating (SiO_2). The protection role of this coating is the same as in the case of DLC films [52, 53]. The size of Si nanocrystals embedded in dielectric matrix and their concentration after annealing depend on the Si concentration in the films before annealing. The initial films with different Si concentrations were grown by thermal evaporation of Si powder [26, 30, 53] or plasma-enhanced chemical vapor deposition (PECVD) from gas mixture [54–56]. The Si nanocrystals embedded in SiO_2 matrix are separated by dielectric layers. Therefore, to obtain conducting Si wires through the SiO_x(Si) films in Ref. [57] the method previously proposed in Ref. [58] has been applied. Electrical conditioning at high current densities through the films leads to a connection of separate Si nanocrystals and in such a way to formation of conducting Si wires.

For investigations described in Ref. [57], different kinds of substrates were used. Glass and flat Si wafers were used for the film thickness and refractive index measurements. Si tip arrays and flat Si substrates were used for current-voltage and EFE current-voltage measurements.

SiO_x(Si) films with $x = 1.2$ were deposited by PECVD method from the gas mixture of $N_2O{:}SiH_4 = 0.19:1$ at the substrate temperature of 75 °C. The SiO_x(Si) films with $x = 0.3$ were deposited by thermal evaporation of Si powder under the residual gas pressure and substrate temperature of $(2.7–4.0) \times 10^{-3}$ Pa and 150 °C, respectively. The values of a stoichiometry index x for SiO_x films were determined by two methods described in Ref. [59]. In the first method, the compositions of SiO_x films were determined from the experimentally measured positions of the edges of the bands of optical absorption spectra. The second method was based on the determination of the positions of the bands of the infrared absorbance spectra of SiO_x films in the range of 1000–1100 cm^{-1} and referring their positions to the film composition. The thicknesses of deposited films on flat Si substrates measured by the Dektak 3030 profilometer were 10–100 nm. All samples were annealed during 5–40 min at 1000 °C under argon flow to induce phase decomposition of nonstoichiometric SiO_x(Si) films and formation of Si nanocrystals in SiO_2 matrix [26, 29, 31]. In order to etch the coating the ammonium fluoride [NH_4]F solution was applied. It etched selectively SiO_2 and opened the Si nanocrystals on the film surface [60]. The etch rate of the SiO_2 film was 150 nm/min. The thicknesses of 7.5 and 25 nm were etched off in 3 and 10 seconds, respectively. Following chemical etching, all samples were rinsed in double de-ionized water.

The procedure used to form the conducting channels in DLC films [58] was applied in Ref. [57] for SiO_x(Si) films. A Si tip array was used as the formation electrode. The mean contact surface area of the Si tip was 3×10^{-9} cm^2. The conducting channels were formed between the flat Si substrates coated with SiO_x(Si) films and the formation electrode. The bias voltage was increased until the current through the film was $I \approx 1 \times 10^{-6}$ A. To create single conducting channels, a gold wire of 80 μm in diameter was used. The tip of the gold wire was etched to obtain a contact surface area of 3×10^{-9} cm^2.

In the described below experiment a 60 nm SiO_x(Si) film with a stoichiometric index of $x \approx 1.2$ was deposited onto a flat Si substrate by PECVD and annealed afterward. The Si inclusion density in the dielectric matrix was small and it was used for conducting channel formation inside the films according to Ref. [58]. The current–voltage characteristics were measured using the Si tip array as the formation electrode and they are shown in Figure 6.16. The current-voltage characteristics after electrical conditioning (curves 1a–4a in Figure 6.16) compared to the nontreated curves (1–4) have much smaller threshold voltage and up to four to five times higher current in the low voltage region. The current transport characteristics through the films tested after electrical conditioning did not change during subsequent measurements. The current transport mechanism during the channel formation was determined with the assistance of a gold electrode. In Figure 6.17, the current-voltage characteristics before (curve 1) and after the electrical conditioning (curve 2) of annealed SiO_2(Si) film are shown. The mechanism of current transport changed after the electrical conditioning (curve 2). At increased applied voltages the electrons tunnel into the conductive band of dielectric and flow according to the Poole–Frenkel (P-F) law (curve 2).

Some explanations of the film conditioning mechanism were presented in Refs. [61–63]. A possible reason for the start of channel formation is the energy release due to the electron transport from the conduction band to the traps in the bandgap or from the conduction band to the Fermi level of the formation electrode. When the electrons flow through the dielectric-metal interface (at the anode contact) they will release energy that can be estimated from the expression $E = Q \times \varphi$ [61–63] (here Q is the charge flow and φ is the contact potential difference at the dielectric-metal interface). For example, the released energy is $E_1 = 0.015$ J/cm^2

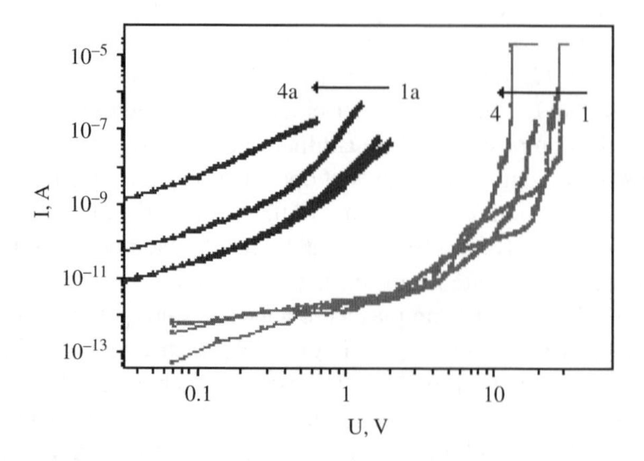

Figure 6.16 Current–voltage characteristics of SiO_x(Si) ($x = 1.2$) films before (curves 1–4) and after electrical conditioning (curves 1a–4a)

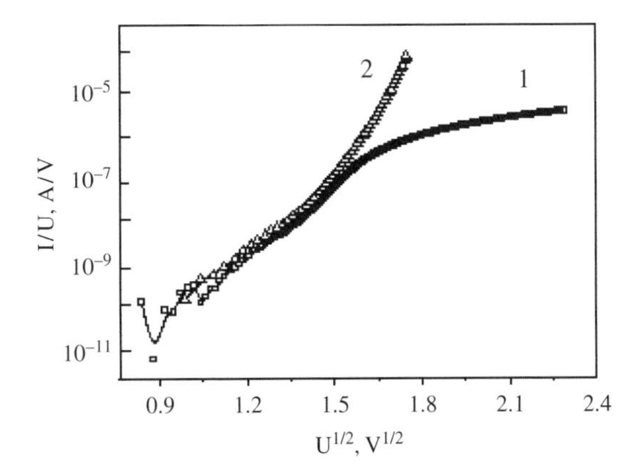

Figure 6.17 Current–voltage characteristics of SiO_x(Si) ($x = 1.2$) film measured by using the single gold electrode before (curve 1) and after electrical conditioning (curve 2) in the P-F coordinates. Reproduced with permission from Ref. [57]. Copyright (2010), AIP Publishing LLC

and $E_2 = 0.02\ \mathrm{J/cm^2}$ at the Al-SiO_2 ($q\varphi = 3.3$ eV) and the Pt-SiO_2 ($q\varphi = 4$ eV) interfaces, respectively. The estimations show that the energy released by electron trapping or falling down to the Fermi level of metal electrode is two to three times higher than the bond breaking energy in SiO_2 (33 kJ/cm^3). The higher is the work function of the formation electrode, the higher is the released energy density. As shown in Ref. [64], the electrical conditioning procedure depends on the anode material. The electrical conditioning and channel formation capability are very high for the anodes with the work functions $\geq$4 eV.

The amorphous SiO_2 films have many Si-H and Si-OH bonds that are the main source of film disordering. During the formation of the films the bonds, mentioned above, act as centers for creation of conducting channels [65]. Break of Si-OH and Si-H bonds produces uncoupled Si atoms during the current transport. These uncoupled Si atoms add energy levels into the bandgap of the energy spectrum of SiO_2. In the course of the current transport through the films, electrons are captured in the traps that lie in the middle of the bandgap and heat up. When the applied voltage is increased, electrons tunnel to the conduction band and then flow according to the P-F law. They are frequently trapped at deep levels in the energy bandgap or fall down to the Fermi level of the anode with some energy release. The released energy is higher than that needed for the bond breaking $SiO_2 \rightarrow SiO + O$ (550.6 kJ/mol). Therefore, the number of Si atoms with uncoupled bonds increases and results in the increase in allowed states in the bandgap of dielectric. Increased concentration of free Si atoms is the main reason for the uncoupled Si bonds to connect to each other. It can result in the formation of conducting Si channels and change in the current transport behavior. As shown in Refs. [61–63], the heat loss was small, and the temperature effect could be neglected during the formation procedure.

To characterize the structure of SiO_x ($x = 1.2$, 60 nm) films the X-ray diffraction has been used [57]. X-ray spectrum of film after annealing, but before conditioning confirms the amorphous structure of SiO_x film with small quantity of Si inclusions [66]. But after conditioning the X-ray diffraction shows the existence of crystalline peaks that confirm the increase in the number of Si inclusions after conditioning treatment. The morphology of the sample surface measured by AFM did not change after the treatment. It can be explained by the fact that no critical breakdown (with melting and cracking of material) occurred in the films under study.

The emission properties of $SiO_x(Si)$ films with various values of stoichiometric index x before and after thermal treatment, and after electrical conditioning were compared. The stoichiometric index influences both the concentration and the size of Si inclusions.

Coating of emitter surface with $SiO_x(Si)$ ($x = 0.3$) films improves the efficiency of EFE. Maximum improvement was achieved when the thickness of coating was in the range of 40–50 nm. The chemical etching and annealing influence were stronger for thicker $SiO_x(Si)$ films. At optimized chemical etching treatment very small Si inclusions remained on the top of films, which resulted in higher field enhancement factors. For thicker $SiO_x(Si)$ films (ranging from 50 to 100 nm), the EFE process began only after short-time etching. The EFE from 100 nm thick film disappeared after annealing. It can be explained by the fact that thick dielectric SiO_2 upper layer is almost nontransparent for FE. The emission curves of the initial silicon tips, after $SiO_x(Si)$ film deposition and following thermal annealing, and after the chemical etching were fitted with linear functions and plotted in Figure 6.18. As was estimated the work function values decreased from 4.15 to 3.16 eV [57].

The F-N characteristics of $SiO_x(Si)$ ($x = 1.2$, 60 nm thick) film deposited onto flat Si substrate by the PECVD method before and after different treatments are shown in Figure 6.19. Curves 1 and 2 have been taken from Figure 6.18. As it can be seen from Figure 6.19, the curve 3 related to the initial $SiO_x(Si)$ ($x = 1.2$) has another slope in comparison with the curves 4 and 5 related to the films after electrical conditioning (two different samples with the same condition). The initial film before thermal annealing (curve 3) contained very small concentration of Si inclusions which could be neglected (x was very high). The slope decrease depends only on the work function. The effective emission area and the field transformation factor can be estimated [1, 25, 40, 67–69]. The field transformation factors are practically the same for curves 3 and 4. The slope change of curve 4 can be explained by the work function increase.

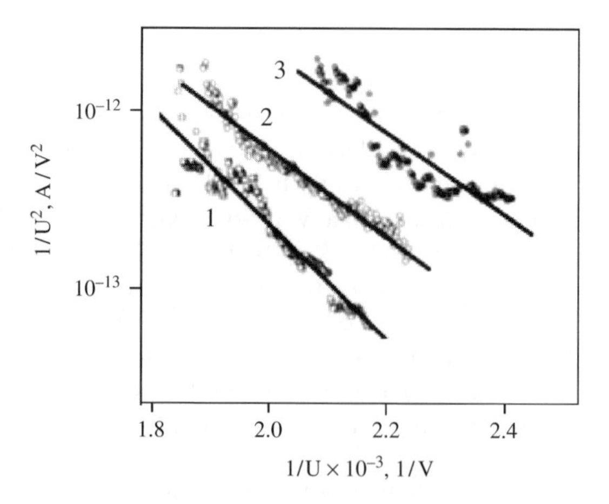

Figure 6.18 F-N characteristics of the Si tip array coated with the $SiO_x(Si)$ ($x = 0.3$) film: (1) initial array of Si tips; (2) Si tips coated with 100 nm thick annealed $SiO_2(Si)$ film; and (3) 10 seconds etched (the thickness is 75 nm). Reproduced with permission from Ref. [57]. Copyright (2010), AIP Publishing LLC

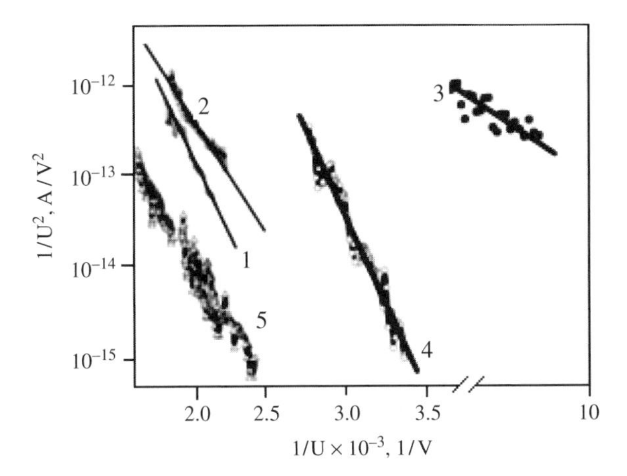

Figure 6.19 F-N characteristics of the SiO_x(Si) deposited onto Si tip array and flat Si substrate: (1) initial Si tips; (2) SiO_x(Si) ($x = 0.3$) film, after annealing (tips); (3) SiO_x(Si) ($x = 1.2$), before treatment (flat); and (4) and (5) series of experiments SiO_x(Si) ($x = 1.2$), after annealing and electrical conditioning (different samples with the same condition, flat). Reproduced with permission from Ref. [57]. Copyright (2010), AIP Publishing LLC

After annealing and further electrical conditioning of SiO_x(Si) ($x = 1.2$) it transforms in SiO_2(Si) with embedded Si inclusions. Therefore, the work function of the created Si conducting channels should be identical to the initial Si tips (curve 1, work function is 4.15 eV). As can be seen from Figure 6.19 the slope of curve 1 is almost identical to the slope of curve 4. This indicates that the EFE process is identical in both cases, which is only possible when the emission spot areas and the corresponding work functions for both emitting structures are equal. The initial tip and the created conducting channel in the SiO_2(Si) film have different ratios for tip height divided by the tip radius. The estimation of the created conducting channel size can be performed according to [26, 37, 58]

$$\beta^* \approx \frac{h}{r}, \tag{6.5}$$

where β^* is the local electric field enhancement coefficient, r is the tip radius of conducting channel, and h is the tip height (film thickness), respectively.

To calculate the conducting channel radius, the effective emission area should be estimated [1, 25, 40, 67–69]. This expression can be determined as the sum of all conducting channels with arithmetic mean of channel radius r in assumption that only the top of each tip provides with the FE as follows:

$$\alpha = \sum_{i=0}^{N} \pi r^2, \tag{6.6}$$

where $N = n \times S$ is a quantity of tips, n is the tip density of the formation array (1.44×10^8 tips/cm^2), and S is the hole area of emitter-anode spacer (7.854×10^{-3} cm^2).

The effective emission area related to curve 2 (see Figure 6.19) is found to be 3.35×10^{-8} cm^2. From Equations (6.5) and (6.6) one finds that the mean values of conducting channel radius and the electric field enhancement coefficient β^* are 0.97 nm and 62, respectively. The diameter of Si inclusion is equal to ≈2 nm. The obtained result is in good agreement with the sizes of Si inclusions measured by AFM (1–3) nm and the calculated ones (2.3 nm) given in Ref. [26].

6.6 Electron Field Emission from Si Nanowires

One-dimensional (1D) structures such as nanotubes, nanobelts, nanorods, and nanowires are ideal candidates for achieving high FE current density at a low electric field because of their high local electric field enhancement by the high aspect ratio [70–74]. Among various 1D nanostructured materials, carbon nanotubes have attracted extensive efforts [70, 71, 73, 74]. However, FE from other materials, including Si, ZnO, SiC, and so on is also interesting and being explored [75–77]. The semiconducting properties of Si-NWs opens distinct possibilities compared to metallic emitters and carbon nanotubes. Si-NWs exhibit a unique sp^3-bonded crystal structure and a low work function (3.6 eV) [75]. Meanwhile Si has been the backbone of the microelectronics industry for decades and it would be desirable to have Si field emitters to be integrated onto Si substrates along with the driving circuitry. Si-NWs can be synthesized by several methods, and FE from various Si nanostructures has been reported [75, 78–82]. For Si-NWs grown on Si substrates in Ref. [83] the current density of 1 mA/cm^2 at such low electric field as 3.4 V/μm has been obtained which is comparable with that of carbon nanotubes [73]. The measurements of EFE from Si-NWs arrays show the linear F-N behavior as for metallic emitters. In addition to strong current saturation in EFE due to the bandgap, their properties could be strongly influenced by surface states because of their large surface-to-volume ratio.

The Si-NWs are usually synthesized from silane by a chemical vapor deposition (CVD) method using Au as catalyst. In Ref. [83] it has been done at a temperature of 480 °C. At the beginning the native SiO_2 layer has been removed from Si substrate in a 5% hydrofluoric acid water solution for 5 minutes, followed by cleaning in alcohol by ultrasonication, then Au as catalyst was deposited onto the Si wafers by immersing the Si wafers into alcohol solution of hydrogen gold tetrachloride ($HAuCl_4 \cdot 3H_2O$, Aldrich). Then the Au-coated Si wafers were loaded into the growth tube furnace for the growth of Si-NWs. After the Si wafers were loaded, the furnace was pumped down to about 1 mTorr, followed by heating up to 480 °C within 20 minutes with flowing H_2 gas and Ar gas. As soon as the temperature reached 480 °C, a helium-diluted silane gas (10%) was introduced into the furnace to build up a pressure to start Si-NW growth. During growth, the pressure was maintained in the range of 590–610 Torr. The typical growth lasts 30 minutes.

The scanning electron microscope (SEM) and the high-resolution transmission electron microscope (HRTEM) images of Si-NW are shown in Figure 6.20. As can be seen, the nanowires are very long (at least 100 μm), the diameter is of about 100 nm and it is fairly uniform. The transmission electron microscopy (TEM) images clearly show that the nanowire has a core and shell structure. The core is about 45 nm in diameter and the shell is about 30 nm thick. The amorphous shell is mainly Si (97%) with a very small amount of oxygen (3%) evidenced by the strong Si peak and weak oxygen peak in the energy dispersive spectroscopy (EDS) spectrum [83].

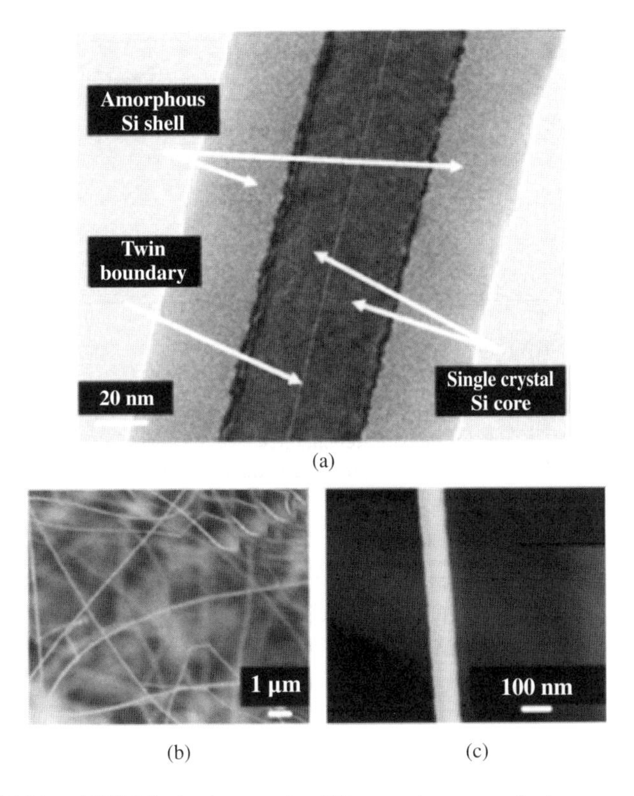

Figure 6.20 TEM (a) and SEM (b,c) micrographs of Si nanowires grown by low-pressure CVD. Reproduced with permission from Ref. [83]. Copyright (2006), AIP Publishing LLC

The *I-V* characteristics of emission current density and corresponding F-N plot are shown in Figure 6.21. A turn-on electric field of 5.5 V/μm was obtained at an emission current density of 0.01 mA/cm^2 for the as-grown Si-NWs, and the highest saturated current density obtained was only 0.03 mA/cm^2 due to probably weak mechanical and electrical contact with the Si wafer. To improve the contact, the same sample was annealed in vacuum at 550 °C for 24 hours and measured again. After annealing the turn-on voltage was drastically reduced to 2.0 V/μm and an emission current density of 1 mA/cm^2 was obtained at 3.4 V/μm.

To assess the size effect of the Si-NWs on FE, samples with different nominal diameters have fabricated by high temperature laser ablation method [84] with varying the synthesis parameters. After that the Si-NWs containing paste has been prepared. The turn-on field of the Si-NW emitters with a nominal diameter ~10, ~20, and ~30 nm are 4.5, 13, and 23 V/μm, respectively [75]. These results clearly show that the diameter of Si-NW emitters plays an important role in the FE properties. The turn-on field and threshold field decrease with decreasing diameter of the Si-NWs. This phenomenon is readily understandable in terms of the field enhancing factor of the Si-NW emitters. As the field enhancing factor is expected to increase with decreasing radius of curvature of the emitting tip, which is in turn related to the diameter of the Si-NWs, it is natural that smaller Si-NWs have larger field enhancing factor and thus smaller turn-on field.

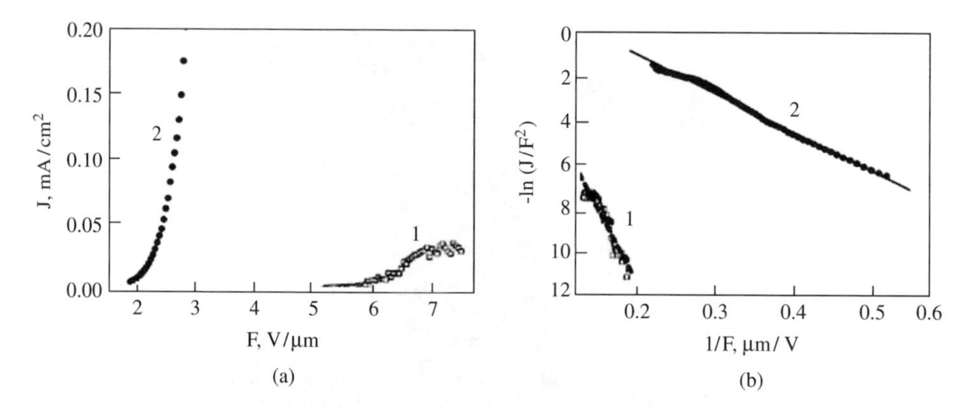

Figure 6.21 Field emission current density dependencies on electric field (a) and Fowler-Nordheim plots (b) 1: as-grown Si NW and 2: annealed during 24 hours. Reproduced with permission from Ref. [83]. Copyright (2006), AIP Publishing LLC

In an attempt to remove the oxide layer and thus enhance the FE of the Si-NWs, the film made with Si-NWs of ~20 nm in diameter was treated with hydrogen (H_2) plasma using electron cyclotron resonance (ECR) CVD [75]. The ECR was operated at a power of 550 W with pure H_2 of 0.01 Torr. The plasma treatment was made for 1 hour. After the H_2 plasma treatment it was found from SEM observations that the Si-NWs became thinner. These observations of EFE indicate that H_2 plasma treatment improved the uniformity of the Si-NW emitters. The uniformity improvement was possibly due to the reduction of the barrier for FE as a result of the thinner oxide shell of Si-NWs.

The study of the EFE from individual Si-NW allows understanding the surface effects and optimizing the EFE characteristics. In Ref. [85] the individual high crystalline Si-NWs with controlled surface passivation were investigated. The Si-NWs were batch-grown by vapor-liquid-solid (VLS) using Au catalyst with no intentional doping [86]. Individual Si-NWs were mounted on standard tungsten tips. The measurements were performed in ultra-high vacuum 2×10^{-10} Torr. Quasi-ideal saturation was obtained accompanied by a strong sensitivity to temperature (current increasing up 2 orders of magnitude) (Figure 6.22). These strong saturation effects must be associated with the high quality and good passivation of the Si-NWs and can be explained by an unintentional *p*-type doping inherent in the Si-NWs growth process. The curves obtained are predicted by the theory of EFE from semiconductor [87]. These are similar to those in *p-n* junctions in reverse bias where the field penetration induces an internal *p-n* junction in the Si-NWs (Figure 6.23).

The role of the surface states was demonstrated by cyclic heating and hydrogen passivation treatments done in situ (Figure 6.24). This treatment allowed determining that the current saturation was linked to the conducting properties of the Si/SiO_2 interface of the Si-NWs. The increasing of the surface states led to suppression of saturation and then to a linear F-N curve (Figure 6.24).

Various types of Si-NW show excellent emission parameters. One-dimensional boron-doped Si nanoparticle chains synthesized in bulk quantity using laser ablating SiO powder

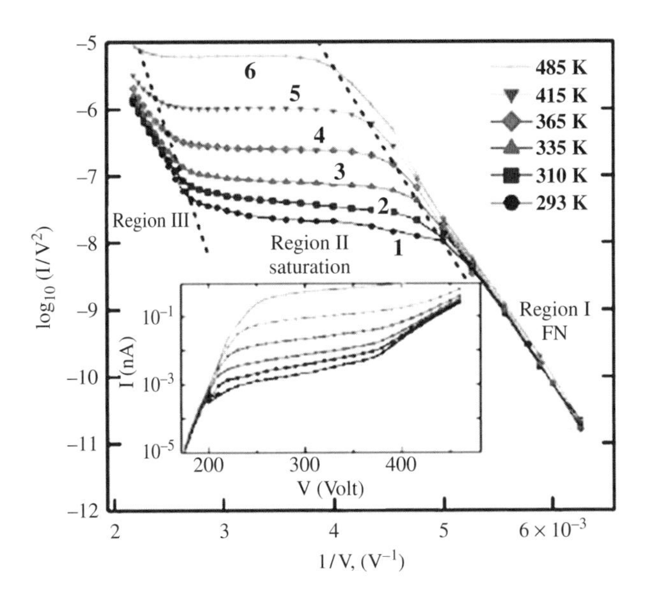

Figure 6.22 Typical Fowler-Nordheim plots for individual Si-NWs with increasing temperature: region I – a standard F-N emission (I < 1 pA), region II – saturation of current with high sensitivity to temperature, and region III – rapid increase of current (1: T = 293 K, 2: T = 310 K, 3: T = 335 K, 4: T = 365 K, 5: T = 415 K, and 6: T = 485 K). Copyright (2012) IEEE. Reprinted with permission from Ref. [85]

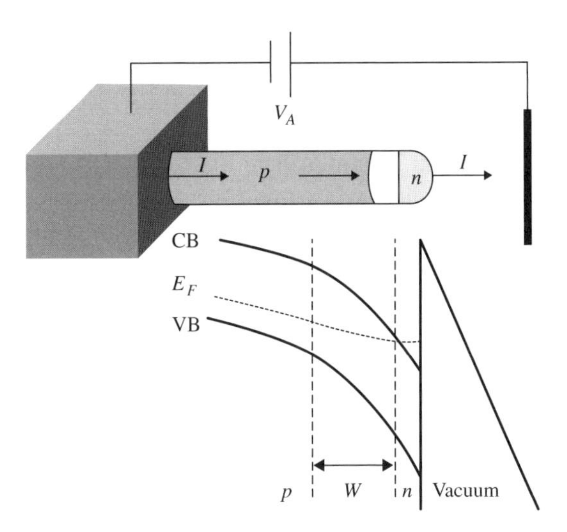

Figure 6.23 Schematics of the emission geometry and the energy band diagram of the internal *p-n* junction formed in *p*-type nanowire during field emission: a depletion region (W) build up behind a pocket of electrons (n) formed in the degenerated conduction band. Copyright (2012) IEEE. Reprinted with permission from Ref. [85]

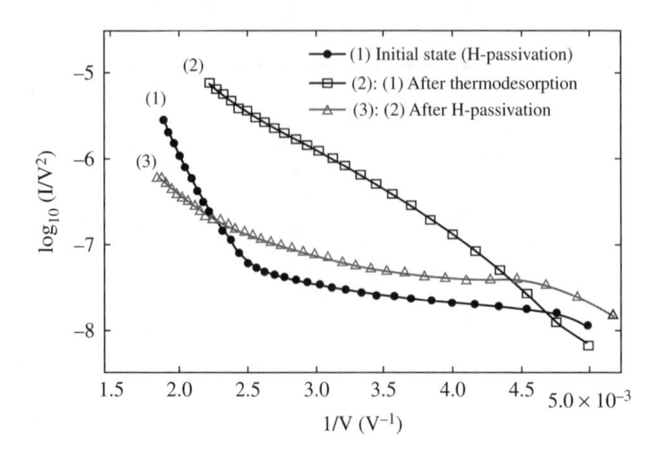

Figure 6.24 F-N plots with heat and hydrogen passivation: the thermodesorption induced a linear F-N plot due to an increase in the surface states (1: initial state (H-passivation), 2: after thermodesorption, and 3: after H-passivation). Copyright (2012) IEEE. Reprinted with permission from Ref. [85]

mixed with B_2O_3 powder have the outer diameters of the nanoparticles in the chains around 15 nm. High-resolution transmission electron microscopy showed that the nanoparticles had perfect lattices with 11 nm crystalline core and 2 nm amorphous oxide outer layer while the distance of the inter particles was 4 nm. Field-emission measurement showed that the turn-on field of the Si nanoparticle chains, which was defined as the electric field leading to a current density of 0.01 mA/cm^2, was 6 V/μm. That was much lower than that of undoped Si-NWs (9 V/μm) [78].

Well-aligned quantum Si-NWs arrays have been synthesized by CVD template method without catalyst. The superior FE behavior (turn-on field for electron emission 14 V/μm) is believed to originate from the oriented growth and the sharp tips of Si-NWs [82].

The taper-like Si-NWs exhibit a turn-on field of (6.3–7.3) V/μm and a threshold field, defined as the electric field leading to a current density of 10 mA/cm^2, of (9–10) V/μm. The excellent FE characteristics are attributed to the taper-like geometry of the crystalline Si-NWs [79].

To improve emission parameter the cesiated Si-NWs grown by the VLS technique have been also prepared and investigated. The average threshold field of cesiated Si NWs was found to be ~7.76 ± 0.55 V/μm and showed a significant improvement over that of as-grown NWs (average threshold field ~11.58 V/μm) [80].

A multistep template replication route was employed to fabricate highly ordered silicon nanotube (Si-NT) arrays, in which annular nanochannel membranes were produced first, and then silicon was deposited into the annular nanochannels by pyrolytic decomposition of silane. Field emission characterization showed that the turn-on field and threshold field for the Si-NT arrays were about 5.1 and 7.3 V/μm, respectively [81].

As was shown, the considered results on the EFE from Si-NWs opened up numerous perspectives for cathodes application, thermal and optical modulation.

6.7 Metal-Insulator-Metal Emitters

Cold electron emitting devices have been studied intensively for their use in flat panel type displays [88], X-ray tubes [89], vacuum microwave devices [90], electron sources in ultrafast electron microscopy [91], electron beam lithography [92], and so on. Spindt-type electron emitters [49, 93] based on FE from Si and metal cones have been studied. High brightness and the ability to operate at high speed were demonstrated. However, these devices have a complicated structure and they require high vacuum and high supply voltage. Large dispersion of angles of emitted electrons results in a poor resolution. The FE devices based on carbon nanotubes have addressed some of these problems [94, 95]. However, lifetime is a serious issue for these devices. Planar type electron emitters based on MIM or MOS structure are very perspective to overcome the above-mentioned problems.

MIM cathodes are attractive for application in vacuum micro- and nanoelectronics because they have some important advantages among others. Significant advantages of MIM emitters are their flat thin film structure. The cathode is less susceptible to surface contamination due to the fact that the emitting material is buried inside, and electrons tunnel through interfacial Schottky barriers instead of surface barriers [96]. Many efforts have been applied to develop this type cathode to practical application [97–102]. The schematic image and energy band diagram of such type of cathode are presented in Figure 6.25. Typically, the device consists of a thin insulating film (e.g., Al_2O_3) sandwiched between two metal electrodes (e.g., Al and Au). The insulating layer is so thin (several nanometers) that an electron can tunnel through it when an electric field is applied across the layer. As the energy diagram in Figure 6.25b shows, the tunneling electrons are injected from the negative electrode (Al, the emitter) through the insulator into the positive electrode (Au, the gate) as hot electrons and are detected as a diode current I_d. A portion of the injected electrons that have kinetic energies larger than the work function of the Au surface can go through Au and emit to vacuum, which is collected as an emission current I_e. However, a majority of the tunneling electrons lose their kinetic energies while they pass through the structure because of scattering events in both the insulator

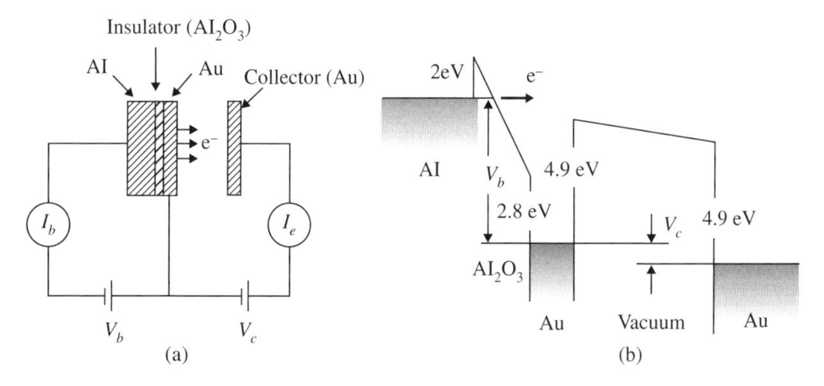

Figure 6.25 Schematic depicting the MIM emission mechanism: (a) MIM structure, (b) energy band diagram of MIM structure under applied voltage. Reproduced with permission from Ref. [102] Copyright (1996), American Vacuum Society

and the gate metal, resulting in a very inefficient emission process with low current transfer ratios ($I_e/I_d < 10^{-3}$). Reducing the thickness of both the insulator and the gate electrode would increase I_e/I_d, but for the insulator, it needs to have a certain thickness to withstand the applied voltage that must be higher than the work function of the gate electrode. The use of low work function materials for the gate electrode would also help to enhance I_e/I_d. In addition, very high current transfer ratio (0.7%) has been reported from the MOS structure, a variant of the MIM cathode [103].

It is important to note that stable and reproducible electron emission from MIM and MOS cathodes requires the fabrication of ultrathin (a few nanometers thick) insulating and metal or semiconductor layers, the structures of which need to be sufficiently controlled on an atomic scale. Roughness at the interfaces or the presence of defects and structural inhomogeneity in these thin layers will cause significant fluctuations in emission current and emission nonuniformity. The emission current density from MIM cathodes is also generally too low (~50 μA/cm^2) for practical applications. Although higher currents can be obtained by increasing the electric field applied across the insulating layer, the insulator would degrade quickly under such high fields, and electroforming of the MIM cathodes can occur [104]. This would further result in unstable emission and poor emission uniformity. Further researches allowed enhancement of the emission current density by 2 orders of magnitude, obtaining the peak current density of 5.8 mA/cm^2 from 30 × 30 MIM cathode array, operating at the gate voltage of 10 V [105].

The perspective application of MOS cathodes has been proposed in Ref. [106]. A MOS tunneling cathode is a promising candidate as a fine electron source, because it has a potential for high current density and pressure insensitivity of emission current and is a flat cathode in contrast to FE cathode. The emission characteristics of MOS cathodes were demonstrated in Ref. [103, 107, 108]. In a MOS cathode, when a voltage exceeding the work function of the gate electrode is applied, electrons travel through the conduction band of the oxide and the gate electrode after tunneling through the potential barrier in the oxide, and some of them are emitted into vacuum. Since electrons are easily scattered by phonons and many types of electron traps during traveling in the oxide, the mean-free path of electrons in the conduction band of oxide is very short (0.4–0.7 nm) [108]. Thus, the phase coherence of electron waves is lost in the oxide for a usual MOS cathode. However, if an oxide layer is sufficiently thin and the interface between SiO_2 and gate electrode is abrupt to keep the coherence, resonant tunneling due to the interference of incident and reflected electron waves at the interface, so-called resonant Fowler–Nordheim (F–N) tunneling, occurs in the triangular potential and resonantly tunneling electrons are emitted into vacuum. In a MOS diode with an ultrathin oxide layer the resonant F–N tunneling is reported for the electrons injected into the SiO_2 layer from the gate metal because the interface of Si-SiO_2 is abrupt in an atomic scale [109]. On the contrary, in Ref. [106] they have achieved the resonant F-N tunneling due to reflection at the interface between SiO_2 and the gate electrode by preparing an extremely abrupt interface with a polycrystalline Si (poly-Si) gate electrode. Using this MOS structure the author confirmed electron emission based on resonant tunneling.

Schematic drawing of MOS cathode and its fabrication process are shown in Figure 6.26a,b, respectively [106]. An n-type Si wafer with the carrier concentration of 1×10^{15} cm^{-3} is oxidized to form thick SiO_2 by wet oxidation. After patterning the gate area, the Si surface is oxidized again in dry O_2 as the gate oxide in the MOS cathode. Although this oxide thickness must be thin enough to satisfy the phase coherence of electron waves, the thickness of 6.3 nm has been employed from consideration of breakdown voltage of SiO_2 because

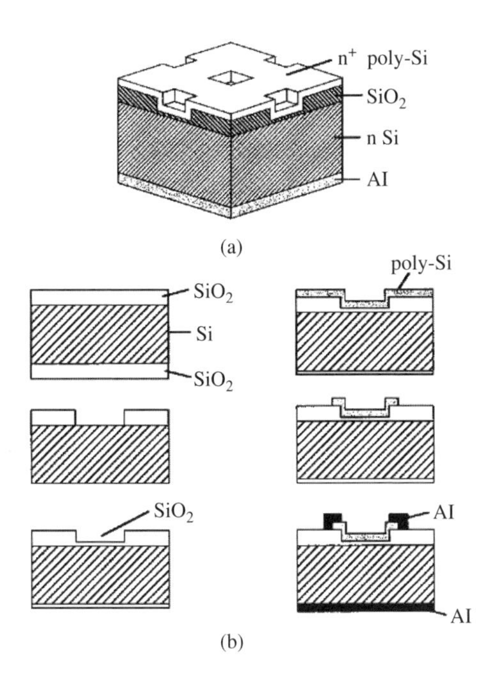

Figure 6.26 Schematic drawing (a) and the fabrication process (b) of the MOS cathode. Reproduced with permission from Ref. [106]. Copyright (1998), American Vacuum Society

the gate voltage larger than the work function of the gate electrode is necessary to extract electrons into vacuum. The phosphorus-doped poly-Si as a gate electrode instead of Al to form a sufficiently abrupt interface in an atomic scale has been used. The phosphorus-doped poly-Si has been prepared by the following processes; amorphous Si is first deposited on the gate oxide by low pressure CVD using Si_2H_6 and PH_3 at a substrate temperature of 550 °C and annealed at 700 °C in dry N_2. The thickness of the poly-Si is chosen to be 20 nm by considering the electron mean-free path of poly-Si of about 4 nm [108] and gate resistance to apply a uniform field on the gate. Finally, an Al electrode is deposited on both sides to reduce the gate resistance and to form an Ohmic contact on the substrate by usual vacuum evaporation. The diode and emission currents have been measured at a pressure of 1.0×10^{-7} Torr. A single cathode with a gate size of 0.25 mm^2 has been used for the experiments (Figure 6.26).

Figure 6.27 shows the band diagram of the MOS cathode, and Figure 6.28 shows the diode and emission currents (a) and the transfer ratio, that is, the ratio of the emission current to the total current (b) of the MOS cathode as a function of the gate voltage. The emission current increases abruptly at 4 V, which corresponds to the work function of the poly-Si gate as shown in Figure 6.27. This indicates that among the electrons tunneling through the potential barrier only electrons with the energy higher than the work function of the poly-Si gate contribute to the emission current and the rest result in the diode current. The transfer ratio has a peak at a gate voltage of about 4.5 V. This means that a considerable number of electrons travel ballistically through the conduction band of the oxide layer below 4.5 V. On the other hand,

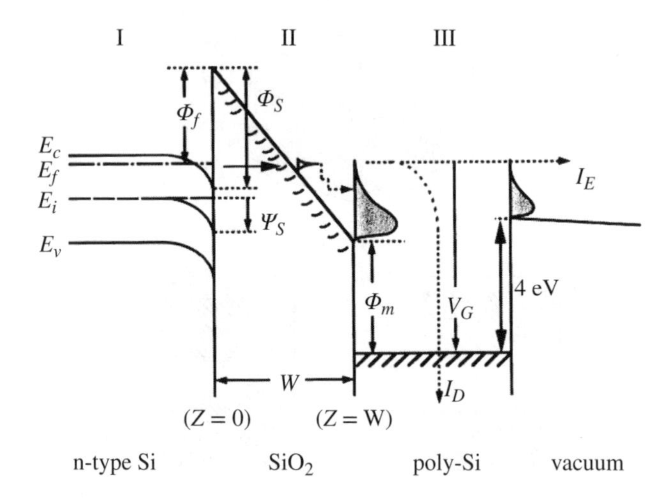

Figure 6.27 Band diagram of the MOS cathode. Reproduced with permission from Ref. [106]. Copyright (1998), American Vacuum Society

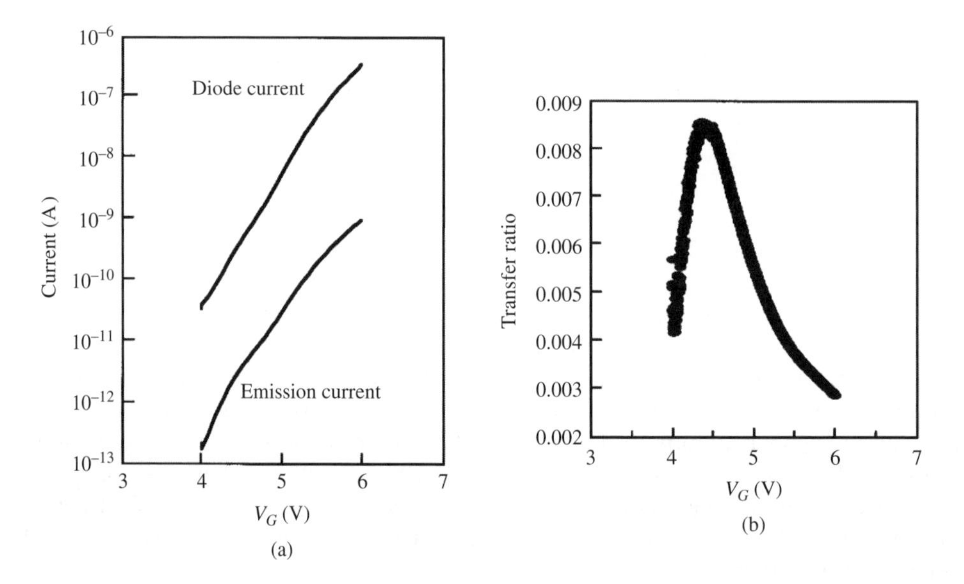

Figure 6.28 (a) Diode and emission currents and (b) the ratio of the emission current to the total current of the MOS cathode as a function of the gate voltage. Reproduced with permission from Ref. [106]. Copyright (1998), American Vacuum Society

electrons lose their energy considerably by scattering in the oxide layer above the gate voltage of 4.5 V, because the traveling distance in the conduction band of SiO_2 becomes longer.

Figure 6.29a shows the F–N plot of the emission current. The F–N fitting using the Wentzel–Kramers–Brillouin (WKB) approximation is also shown by a straight solid line.

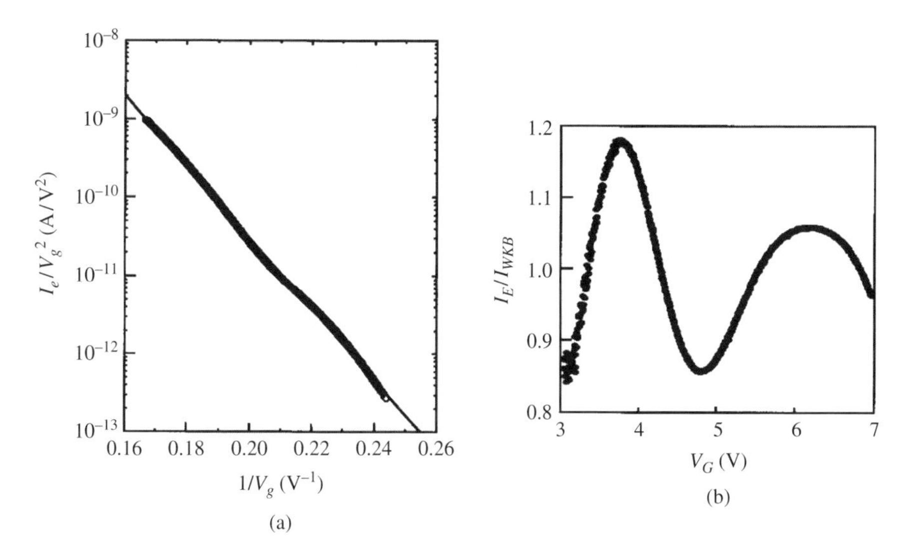

Figure 6.29 (a) F–N plot of the emission current. The F–N fitting using the WKB approximation is also shown by a straight solid line. (b) Ratio of the emission current to the fitting current using WKB approximation as a function of the gate voltage. Reproduced with permission from Ref. [106]. Copyright (1998), American Vacuum Society

The emission current is slightly perturbed from the straight line. To enhance this deviation the ratio of the emission current to the fitting current using the WKB calculation (I_e/I_{WKB}) has been shown in Figure 6.29b. The oscillatory feature in the emission current becomes clearly visible.

In order to confirm that the oscillation is caused by the interference of electron waves reflected at the SiO_2-poly-Si interface the tunneling probability by solving the Schrodinger equation describing electron tunneling through a triangular potential have been calculated and compared it with that of the WKB approximation [106]. The comparison shows that although the magnitude of tunneling probabilities is almost the same, the exact solution exhibits oscillatory behavior which does not appear in the WKB approximation. This oscillation is due to the interference of the incident and reflected electron waves at the interface between the SiO_2 and the poly-Si gate. As ratio of currents I_e/I_{WKB} and ratio of tunneling probabilities T_e/T_{WKB} for exact solution and solution at WKB approximation show the peaks on applied field dependences. The peak positions between experimental and theoretical results agree reasonably, though the peak position at the high field region shifts slightly to high field in the experiment [106]. This disagreement is probably due to the voltage drop by the relatively large tunneling current in the gate electrode. These results indicate that the peaks in the emission current arise from the resonant effect of electrons tunneling. The smaller amplitude of the oscillation in the experiment compared with that of the calculation indicates that the scattering of electrons is still significant in the conduction band of the SiO_2.

The perspective direction of MOS emitter research and development is connected with using of the low-dimensional structures in or instead of the oxide layer.

A planar type electron emitter with narrow beam dispersion formed by PS has been proposed in Refs. [110, 111]. It has been assumed from detailed experimental analyses of the emission characteristics that the observed cold emission is based on the hot-electron tunneling mechanism [112–115]. Under the biased condition, electrons injected into the PS layer from the substrate are drifted toward the outer surface and reach the top electrode as hot electrons, owing to high electric field in the PS layer. Thus, hot electrons are easily emitted into vacuum through tunneling. The emission characteristics of the PS cold cathode strongly depend on the high-field conduction mode in the PS layer. If the carrier transport in PS is appropriately controlled, therefore, further improvement in the emission efficiency and stability should be obtained. To change and control the conductivity of PS the modification of the PS structure during its preparation by electrochemical etching has been performed [110]. Nanocrystalline PS layers has been formed by photoanodizing heavily doped (0.01–0.03 Ω) n^+-type (100) Si wafers in an ethanol solution of 50% HF (HF:ethanols = 1 : 1). Three kinds of PS layers (3–10 μm thick) has been formed under the following different conditions of the anodization current as shown in Figure 6.30: (1) normal: conventional galvanostatic anodization; (2) multilayered: periodic modulation of anodization current; and (3) graded-multilayer: periodic modulation during increase of anodization current. In the case of (2) and (3), the structures of PS layers have been controlled by modulating the anodization current such that the multilayered and graded-band structures with high- and low-porosity layers were formed periodically, as indicated in Figure 6.30b,c. In some cases, the PS layers have been treated by rapid thermal oxidation (RTO). Finally, thin Au films (10 nm thick) have been evaporated onto the PS layers and used as top electrodes. The investigations have been also conducted for nondoped poly-Si films deposited onto n^+-type Si wafers to confirm applicability for large-area devices [111]. The porous poly-Si layer is denoted as PPS.

The introduction of RTO into the normal-structured PS diode is very useful for enhancing the electron emission. The emission efficiency, defined as the ratio I_e/I_d, is improved by employing

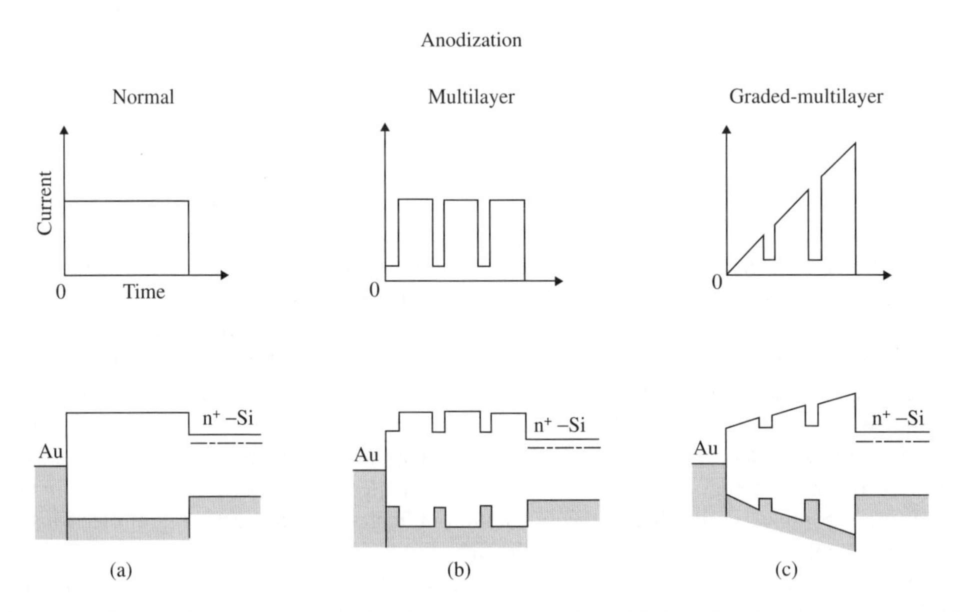

Figure 6.30 Anodization current modulation and corresponding PS diode band structure: (a) normal structure, (b) multilayer, and (c) graded-multilayer. Reprinted from Ref. [110]. Copyright (1999), with permission from Elsevier

the multilayer PS structure and RTO treatment. In this case, emission current comparable to the normal structure case is obtained, while the diode current is significantly suppressed. As a result, the emission efficiency is enhanced up to 12%.

The improvement in the emission efficiency for the multilayered device is presumably caused by a uniform electric field established in the PS layer, since the low-porosity compact layer which has a high electrical conductivity acts as an equipotential plane. It is self-regulated reformation of the electric field distribution in the PS layer, especially in the outer surface of PS close to the top contact. Consequently, hot electrons, which contribute to emission, are efficiently generated there. The introduced low-porosity layers also act as heat sinks owing relatively high thermal conductivity. This contributes to keeping the diode current constant without thermal effects and to stabilizing the electron emission.

Further efficient and stable cold emission was obtained from the PS diode with a graded-multilayer structure [116] and from the RTO-treated PPS diode with a multilayer structure [117]. The electron emission of the graded-multilayer PS diode is quite uniform. There was the strong correlation between the emission stability and output energy distribution. The energy peak and maximum energy E_{max} are shifted toward higher energy fairly in accordance with an increase in volts. These PS results strongly indicate that there is very little serial scattering loss during the drift in the PS layer, and that electrons are emitted quasiballistically. This situation is quite different from that of conventional PS devices, in which the emission fluctuates with spike noises and the corresponding energy distribution curves exhibit broad Maxwellian behavior whose peak energy and width are almost independent of V. It appears that PS in PS diodes with normal or multilayer structure, injected electrons are thermalized after serial scattering losses in PS due to possible potential fluctuations. Thus, some fluctuation is induced in both diode and emission currents.

More efficient quasiballistic emission has been observed in PPS diodes with a multilayer structure. The emission efficiency at $V > 15$ V reaches 1%, and emission current density at $V = 30$ V becomes higher than 200 μA/cm^2. Both the diode and emission currents are quite stable without any signs of spike-like fluctuations. The experimental results of the output electron energy distribution show behavior characteristic of ballistic emission.

The observation of the emission image on the fluorescent screen, the area of the emission shows that the electrons are emitted with a small angular dispersion. Another advantageous feature of this ballistic emitter is the insensitivity of I to ambient pressure. In fact, the emission current of the PPS device, for instance, has showed no changes with Ar gas pressure up to about 1 Pa [110].

For a graded multilayer structure, high electric field exists throughout the PS layer, because the high-porosity wide-gap PS near the substrate owns higher resistivity than that at the outer surface region. In addition, the periodic low-porosity compact layers reform the electric field in the same way as the mentioned above multilayered PS devices. The PS layer consists of a large-number of confined Si nanocrystallites with the same band dispersion as c-Si [118]. The size of Si nanocrystallites in PS is considerably small compared with a drift length of electrons in c-Si. Under a biased condition, major potential drop is produced at the insulating electronic boundaries such as surface skin oxides. In the case of the device with a well-controlled structure such as graded-multilayer, electrons injected into PS can travel for a long distance quasiballistically by multiple tunneling as schematically shown in Figure 6.31. The effective drift length near the outer surface of PS where a high-field of higher than 10^5 V/cm exists possibly reaches about 1 μm. Thus, emitted electrons can easily obtain the energy gain of

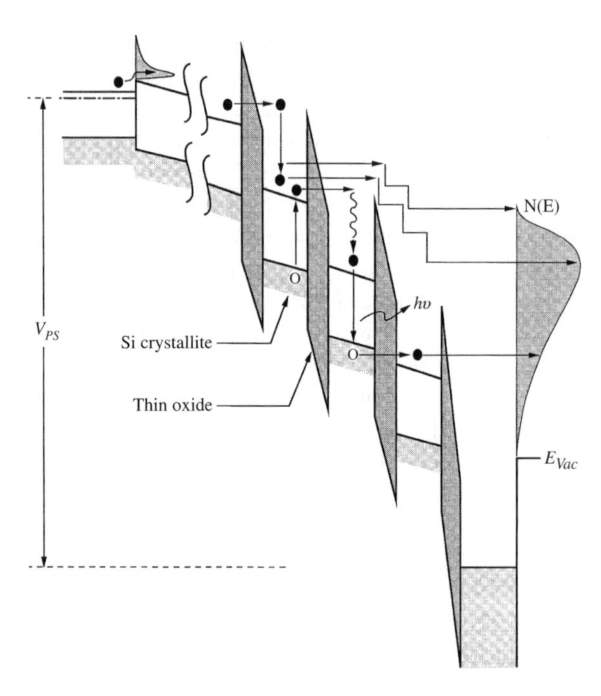

Figure 6.31 Band diagram showing generation and emission processes of quasiballistic electrons in PS diodes under biased condition. Reprinted from Ref. [110]. Copyright (1999), with permission from Elsevier

10 eV. The increased drift length in PS under a high electric field was also suggested in Ref. [119] based on time-of-flight measurements. As a result of significantly reduced scattering and energy dissipation, the diode current is also fully stabilized. The observation of quasiballistic emission provides very important insights for the high-field transport and device physics of PS as a nanocrystalline confined system.

Considered emission device based on PS and PPS does not require the fine control of the position and the size of Si nanostructures in the PS layer because the transport of a large number of electrons is involved. One of the most important mechanisms in these devices is the tunneling process of electrons through a SiO_2 film.

Thermal SiO_2 around nc-Si dots is expected to be more stable than SiO_2 formed by an electrochemical etching process for the PS formation. Thus, an electron emitter using nc-Si dots is expected to deliver higher performance. The attention was focused on nc-Si dots with small spherical shape for a hot electron emission device which behaves like thin film electron emitter devices [97, 120]. The device structure consisted of dots buried in oxide and sandwiched between thin electrode films [121]. It was expected that the enhancement of the electric field around the dots due to the small radius of curvature of the dots should enable a higher yield of hot electrons at the fixed applied voltage. In this device, the problem of size and position distribution of nc-Si dots would not significantly affect the device performance because of macroscopic averaging. An n^+-Si ($0.01\ \Omega \cdot \mathrm{cm}$) wafer was used as the substrate and the electron source. The nc-Si dots were deposited onto the substrate at room temperature by plasma decomposition of SiH_4 [121]. This method enabled to control of the growth time of nc-Si dots.

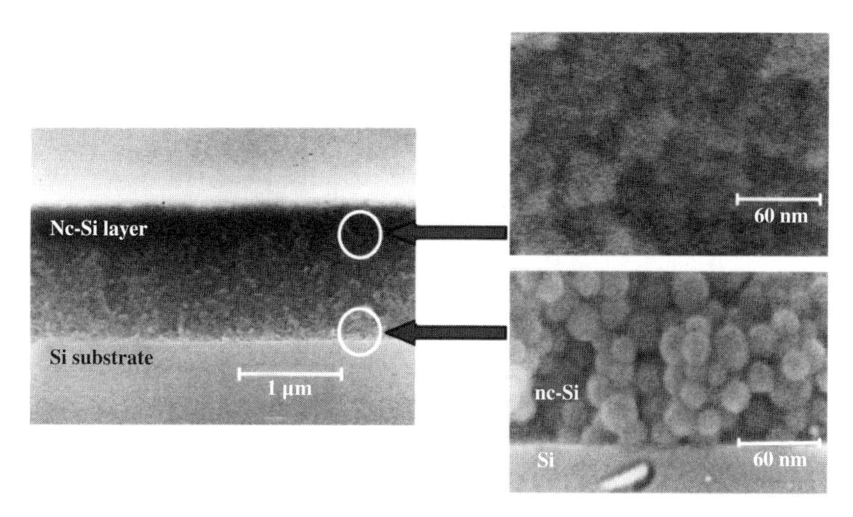

Figure 6.32 Cross-sectional SEM images of the oxidized samples with nc-Si layer. Reproduced with permission from Ref. [121]. Copyright (2002), AIP Publishing LLC

A high-resolution TEM image of nc-Si dot revealed that the crystalline dot was covered with native oxide. The total thickness of the nc-Si layer varied from 0.1 to 1.5 μm. In the next step the samples were oxidized at 700 °C for 1 hour and at 1000 °C for 5 minutes. The low temperature oxidation was performed for covering the dots with SiO_2, and the high temperature oxidation was for forming a thin oxide layer near the top surface. The cross-sectional SEM images of the oxidized sample are shown in Figure 6.32. Near the interface between the Si substrate and the nc-Si layer, many voids were observed between spherically shaped nc-Si dots because nc-Si dots grown in the gas phase had been deposited randomly onto the Si substrate. There was no charge up during the SEM observation because only very thin SiO_2 was formed around each dot. On the other hand, near the top surface, the porosity was small because the molecular volume of SiO_2 was larger than the atomic volume of Si, and the observed image was dark because of charge up effect. This image showed that a thick SiO_2 layer was formed near the top surface. Some samples were planarized by a reflow annealing process. The deep nc-Si layers were not oxidized completely, and nc-Si dots, which were not completely oxidized, remained after the annealing process. For the complete planarization for the thicker nc-Si layer, two optional techniques were added. One was an impurity doping into SiO_2. The other method used for the complete planarization was repeated oxidation and annealing processes.

Finally, an Al ohmic electrode on the backside of Si and a 10-nm-thick Au film on the front surface were formed by electron-beam evaporation. The device structure and the measurement system are shown in Figure 6.33. The current and emission characteristics of the sample with the nc-Si layer thickness of 0.6 μm are shown in Figure 6.34. The emission efficiency in this case was 0.8%. It was observed that the samples with a rough surface had larger average currents. In the sample with smoother surface, the diode current and the electron emission increased with the bias voltage without saturation. The devices planarized by the phosphorus diffusion, which includes the oxidation at 850 °C and following phosphorus diffusion at 1150 °C with P_2O_5 for a few minutes, has showed an emission current of 0.25 $\mu A/cm^2$ and the high efficiency of 10%. The investigation of the effect of the size of nc-Si dots to

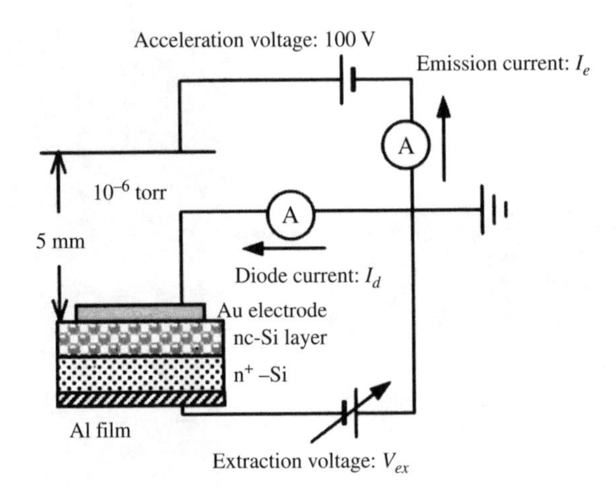

Figure 6.33 Diagram for the measurement of the electron emitter. Reproduced with permission from Ref. [121]. Copyright (2002), AIP Publishing LLC

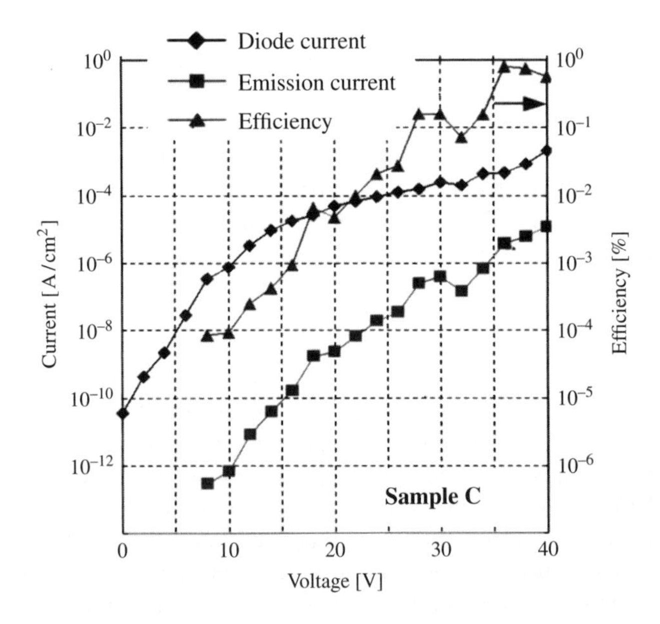

Figure 6.34 Diode and emission current characteristics as a function of the applied voltage. Reproduced with permission from Ref. [121]. Copyright (2002), AIP Publishing LLC

the emission characteristic has showed that the smaller nc-Si dots give higher efficiency (for $d_{nc-Si} = 6$ nm, $\eta = 0.99\%$ and for $d_{nc-Si} = 8$ nm, $\eta = 0.54\%$) [121].

The mechanism for electron emission includes some steps. First, the electrons are injected from the silicon wafer into nc-Si dots without scattering due to the small size of the dots. The electric field is applied mainly within SiO_2 regions covering the dots. Thus the electrons from

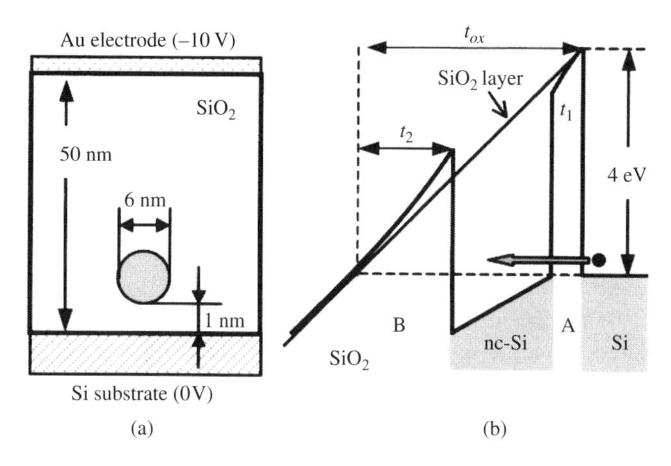

Figure 6.35 (a) Structure of the measured sample, which is estimated from TEM and SEM images, for calculations of the electric field and the potential profile. (b) The band diagram around the nc-Si dot. The straight line marked by the label "SiO_2 layer" indicates the band diagram of the SiO_2 layer without the nc-Si dot. "$t_{1,2}$" and "t_{ox}" indicate the thickness of tunnel barriers for electrons in the conduction band of the Si substrate with and without the nc-Si dot, respectively. The regions labeled as "A" and "B" are the SiO_2 layers between the Si substrate and the nc-Si dot and between the Au film and the nc-Si dot, respectively. Reproduced with permission from Ref. [121]. Copyright (2002), AIP Publishing LLC

nc-Si are accelerated into SiO_2 by the high electric field, allowing ballistic transport through subsequent nc-Si layers. These processes allow electrons to reach the Au electrode with high energy. Electrons with energies larger than the work function of Au can ballistically transport through the Au electrode and reach the collector.

For more detailed investigation, the energy band profile and the electric field in the nc-Si layer were calculated from the Poisson equation under the estimated structure as shown in Figure 6.35a. The dielectric constant of Si and SiO_2 were assumed to be 12 and 3.9, respectively, and the quantum confinement effect in nc-Si dots was ignored. Figure 6.35b shows the energy band profiles of the nc-Si layer with and without a single nc-Si dot. The thickness of the tunnel barrier without dot, t_{ox}, is divided into two thinner SiO_2 layers, t_1 and t_2. Thus, electrons go through the thinner SiO_2 layer labeled as "A" by direct tunneling. Then, accelerated electrons travel through the nc-Si dot with a low scattering rate, followed by a FN tunnel transport through the SiO_2 layer "B" with thinner and lower tunnel barrier than the SiO_2 layer without the nc-Si dot. This means that the electron transport through the SiO_2 layer is enhanced due to the presence of Si dots. Moreover, the electric field around the nc-Si dot is enhanced due to the spherical shape. The nc-Si dots enhance the electron transport density. The calculation of tunnel rates with using the WKB approximation for the 50-nm-thick SiO_2 layer with and without one nc-Si dot showed that the existence of nc-Si increased the tunneling rate by more than 4 orders of magnitude compared with the cases of without a dot.

6.7.1 *Effect of the Top Electrode*

Electrons accelerated in nc-Si dots and the SiO_2 layers are extracted into the vacuum through the top Au electrode. In the Au film, electrons lose their energy due to scattering, and electrons with energy less than the work function of 5 eV are absorbed into the Au electrode without emission. The electron mean free path in a metal film, which is determined by electron–electron interaction and electron-phonon interaction, is shorter than that of Si [122, 123]. Although the mean free path depends on the electron energy, the mean free path of hot electrons with energy of higher than 5 eV is estimated to be 5–7 nm [124]. Thus, a thinner Au electrode is desirable. However, a very thin film, less than 10 nm, formed by an evaporation method can hardly cover the surface completely [125, 126]. Especially, the poor cohesion between Au film and SiO_2 makes the formation of very thin Au film difficult. The effect of the Au film with the thickness of more than 10 nm was investigated [121]. As was revealed, the number of emitted electrons is proportional to $exp(-t/l_m)$, where t is the Au film thickness, l_m the electron mean free path. The mean free path estimated from the slope of the emission efficiency – Au film thickness characteristic is 5.8 nm. When the slope is extrapolated, the interception at zero thickness is not 100% but 10%. This means that 90% of all electrons reaching the Au film are either reflected quantum mechanically at Au surface or having an energy of less than the work function of the Au film. While electrons go through the conduction band of SiO_2, electrons lose their energy. Under this consideration, polycrystalline (poly-Si) or amorphous (a-Si) Si may be good choices of the electrode films because the work function of 4 eV is smaller and the mean free path is longer than those of metal films. Moreover, a thinner film formed by CVD can be obtained easily, and cohesion is also improved. In Ref. [107] the comparison of the Al top electrode with the a-Si top electrode in the electron emitter has been composed of a few-nanometer-thick SiO_2 sandwiched between the top Al or a thin a-Si electrode and a-Si substrate has been performed. They indicated that the a-Si top electrode enhanced the electron efficiency by more than 1 order of magnitude. So, replacing the Au top electrode with an a-Si film may enhance the efficiency of the electron emitter using nc-Si dots.

The MOS emitters with nc-Si dots have been used in Refs. [127, 128] in an investigation of the photoresponse of such cathodes. It is important for the formation of pre-modulated electron beam (a train of periodic electron bunches).

Nc-Si MOS cathode arrays having emission area of 50 μm circles on a 500 μm-diameter active area were fabricated on heavily doped p-type silicon with the resistivity of 0.3–0.5 Ω·cm, as shown in Figure 6.36 [128, 129]. Nanocrystalline silicon was deposited by pulse laser ablation (PLA) and surfaces of nc-Si particles were oxidized by an oxygen radical beam exposure during deposition. The Pt gate electrode of 3 nm thicknesses and nanocrystalline silicon layer of about 300 nm were formed. Thick metal for contact pad was deposited on a Pt gate around the active area.

The schematic of experimental setup for optically modulation is shown in Figure 6.37. The anode was biased at 200 V. The array was irradiated with a He-Ne laser having an optical power of 1 mW and a beam spot of 1 mm in diameter with oblique incidence of 40° from front side. The laser pulses were produced by an optical chopper at frequency of 10 Hz.

The optical response of emission current with irradiating He-Ne laser pulses is shown in Figure 6.38. The gate voltage is set at 14 V. The pulsed emission current is observed corresponding to the pulse frequency and duty, indicating that a modulated electron beam is generated directly from the cathode by He-Ne laser pulse excitation. In this experiment, minimum delay of 10 μs was needed due to the measurement system. Therefore, the disappointing slow responses in the rising and falling of emission are not due to a diffusion process of electrons photogenerated outside the depletion region, but due to the experiment system [127].

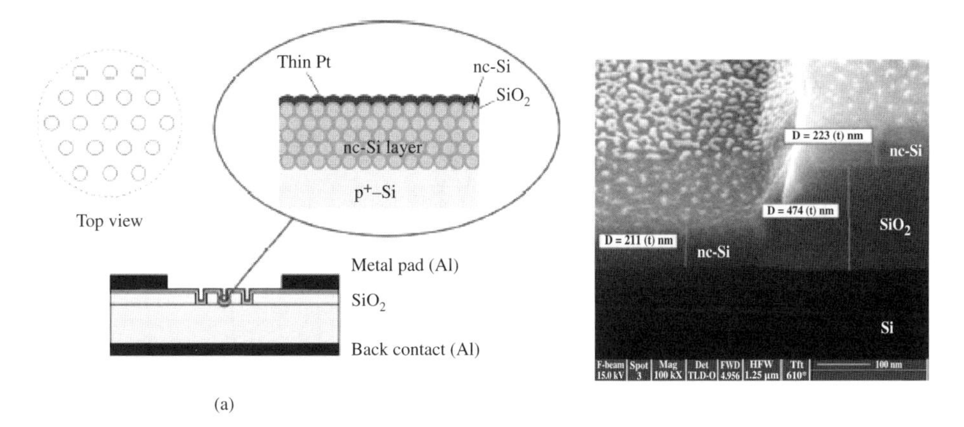

Figure 6.36 Schematic of a *p*-type MOS cathode based on nanocrystalline silicon: (a) schematic view; (b) scanning electron microscopy image. Copyright (2012) IEEE. Reprinted with permission from Ref. [128]

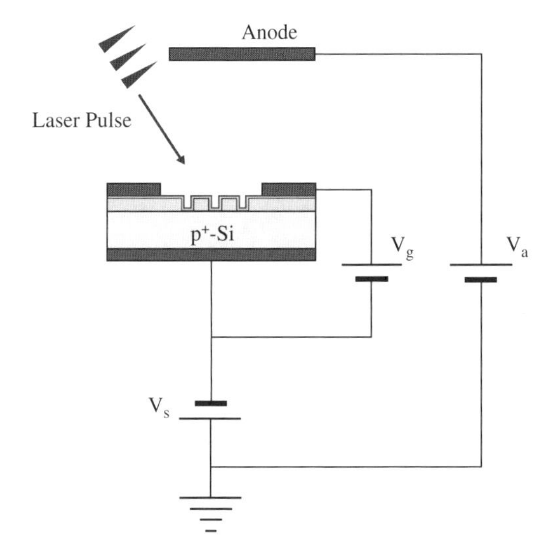

Figure 6.37 Schematic of experimental set up for optical modulation of nc-Si MOS cathode. Copyright (2011) IEEE. Reprinted with permission from Ref. [127]

In another experiment [128] the nc-Si MOS cathode was irradiated with 405 nm pulse laser with the repetition frequency of 10 kHz with oblique incidence of 40° from front side. The original rise and fall times of the laser pulse were less than 1 ns. The photoresponse produced by nc-Si MOS cathode and reference photodetector is shown in Figure 6.39.

The modulated emission current is found to be synchronized with the laser pulse. Both the rise and fall times of the photoresponse in the cathode device were about 4 μs. This indicates that the measured rise and fall times are determined by those of the amplifier used in this experiment and the real rise and fall of the photoresponse in the cathode device are at least less than 4 μs.

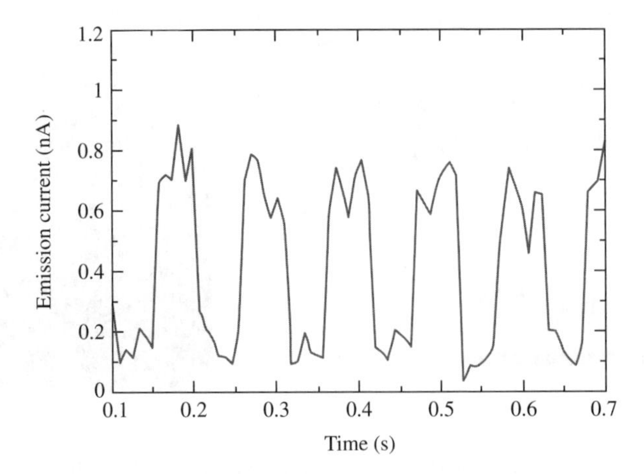

Figure 6.38 Photoresponse of the emission current under He-Ne laser pulse irradiation at the gate voltage of 14 V. Copyright (2011) IEEE. Reprinted with permission from Ref. [127]

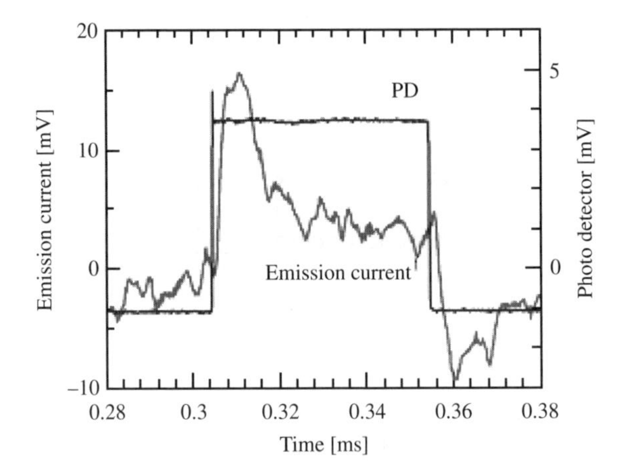

Figure 6.39 Photoresponse of the emission current and photodetector under illumination of a 405 nm pulse laser. Copyright (2012) IEEE. Reprinted with permission from Ref. [128]

6.8 Conclusion

The considered novel Si based cathodes potentially offer the important advantages of high emission parameters, ease in fabrication and low-cost manufacturing. They have unique and interesting FE characteristics. In many experiments the uncontrolled native oxide on the silicon surface is replaced by coatings with determined properties. EFE from SiO_x(Si) and SiO_2(Si) films containing Si nanocrystals was considered in detail. New emission transport mechanisms were revealed in silicon cathodes coated with multilayer films and films with Si nanocrystals. The potential advantages of MIM and Metal-Insulator-Semiconductor emitters were shown. They are integrated in solid state and have plain design. The device applications based on novel Si-based cathodes are being actively researched and explored.

References

1. I. Brodie, C.A. Spindt, Vacuum microelectronics, *Advances in Electronics and Electron Physics* **83**, 1 (1992).
2. K. Derbyshire, Beyond AMLCDs: Field emission displays?, *Solid State Technology* **11**, 55 (1994).
3. J.E. Jaskie, Diamond-based field-emission displays, *MRS Bulletin* **21**, 59 (1996).
4. E.P. Sheshin. *Surface Structure and Electron Field Emission Properties of Carbon Materials* (MFTI, Moskow, 2001).
5. J.-L. Kwo and M. Yokoyama, Numerical indicator field emission display using carbon nanotubes as emitters, *Journal of Vacuum Science and Technology B*. **19**, 1023 (2001).
6. V. Filip, D. Nicolaescu, F. Okuyama, *et al.*, Transport phenomena related to electron field emission from semiconductors through thick oxide layers, *Journal of Vacuum Science and Technology B* **17**, 520 (1999).
7. Yu. A. Skryshevskii and V.A. Skryshevskii, Thermally stimulated luminescence in porous silicon, *Journal of Applied Physics* **89**, 2711 (2001).
8. A.A. Evtukh, V.G. Litovchenko, R.I. Marchenko, *et al.*, Peculiarities of the field emission with porous Si surfaces, covered by ultrathin DLC films, *Journal de Physique IV* **6** (C5) 119-124 (1996).
9. P.R. Wilshaw and E.C. Boswell, Field emission from pyramidal cathodes covered in porous silicon, *Journal of Vacuum Science and Technology B* **12**, 662 (1994).
10. M. Takai, M. Jamashita, H. Wille, *et al.*, Enhanced electron emission from n-type porous Si field emitter arrays, *Applied Physics Letters* **66**, 422 (1995).
11. I. Kleps, D. Nicolaescu, C. Lungu, *et al.*, Porous silicon field emitters for display applications, *Applied Surface Science* **111**, 228 (1997).
12. V. Litovchenko, A. Evtukh, Yu. Kryuchenko, *et al.*, Quantum-size resonance tunneling in the field emission phenomenon, *Journal of Applied Physics* **96**, 867 (2004).
13. A.A. Evtukh, V.G. Litovchenko, R.I. Marchenko, and S. Yu. Kydzinovski, Layered structures with delta-doped layers for enhancement of field emission, *Journal of Vacuum Science and Technology B* **15**, 439 (1997).
14. V.G. Litovchenko, A.A. Evtukh, Yu. M. Litvin, *et al.*, Observation of the resonance tunneling in field emission structures, *Journal of Vacuum Science and Technology B* **17**, 655 (1999).
15. V.G. Litovchenko and A.A. Evtukh, Effects of electron field emission enhancement in structures with quantum well, *Physics of Low-Dimensional Semiconductor Structures* **3/4**, 227 (1999).
16. A.A. Efremov, A.A. Evtukh, D.V. Fedin, *et al.*, Porous silicon as a material for enhancement of electron field emission, *Physics of Low-Dimensional Semiconductor Structures* **1/2**, 65 (2001).
17. A.A. Evtukh, V.G. Litovchenko, Yu. M. Litvin, *et al.*, Porous silicon coated with ultra-thin diamond-like carbon film cathodes, *Materials Research Society Symposia Proceedings* **685E**, D1541 (2001).
18. D.N. Goryachev, L.V. Byelyakov, and O.M. Sreseli, Electrolytic fabrication of porous silicon with the use of internal current source, *Semiconductors* **37**, 477 (2003).
19. A.A. Evtukh, V.G. Litovchenko, M.O. Semenenko, *et al.*, Electron field emission from porous silicon prepared at low anodization currents, *International Journal of Nanotechnology* **3**, 89 (2006).
20. L.V. Byelyakov, T.L. Makarova, V.I. Sakharov, *et al.*, Composition and porosity of multicomponent structures: porous silicon as a three-component system, *Semiconductors* **32**, 1003 (1998).
21. A.A. Evtukh, E.B. Kaganovich, E.G. Manoilov, M.O. Semenenko, A mechanism of charge transport in electroluminescent structures consisting of porous silicon and single-crystal silicon, *Semiconductors* **40**, 175 (2006).
22. A. Evtukh, H. Hartnagel, V. Litovchenko, O. Yilmazoglu, Two mechanisms of negative dynamic conductivity and generation of oscillations in field-emissions structures, *Materials Science and Engineering A* **353**, 27 (2003).
23. A.F. Kravchenko and V.N. Ovsyuk, *Electron Processes in Solid-State Low-Dimensional Structures* (Novosibirsk, 2000) (in Russian).
24. V.G. Litovchenko, Analysis of the band structure of tetrahedral diamondlike crystals with valence bonds: Prediction of materials with superhigh hardness and negative electron affinity, *Physical Review B* **65**, 1531 (2002).
25. D.V. Branston and D. Stephani, Field emission from metal-coated silicon tips, *IEEE Transactions on Electron Devices* **ED-38**, 2329 (1991).
26. A.A. Evtukh, V.G. Litovchenko, and M.O. Semenenko, Electrical and emission properties of nanocomposite SiO_x(Si) and SiO_2(Si) films, *Journal of Vacuum Science and Technology B* **24**, 945 (2006).
27. G. Yang, K.K. Chin, and R.B. Marcus, Electron field emission through a very thin oxide layer, *IEEE Transactions on Electron Devices* **ED-38**, 2373 (1991).
28. A.A. Evtukh, V.G. Litovchenko, Yu. M. *et al.*, Structure with cesium enriched layer for enhancement of electron field emission, *Physics of Low-Dimensional Semiconductor Structures* **1/2**, 21 (2001).

29. V. Ya Bratus, V.A. Yukhimchuk, L.I. Berezhinsky, *et al.*, Structural transformations and silicon nanocrystallite formation in SiO_x films, *Semiconductors* **35**, 821 (2001).
30. A.A. Evtukh, I.Z. Indutnyy, I.P. Lisovskyy, *et al.*, Electron field emission from SiO_x films, *Semiconductor Physics, Quantum Optoelectronics* **6**, 32 (2003).
31. D.J. DiMaria, D.W. Dong, C. Falcony, *et al.*, Charge transport and trapping phenomena in off-stoichiometric silicon dioxide films, *Journal of Applied Physics* **54**, 5801 (1983).
32. V. Ichizli, H. Mimura, and H.L. Hartnagel (1999) Morphology modification of porous (100) GaP for field emitter application, *Proceedings of the 12th International Vacuum Microelectronics Conference, Darmstadt, Germany, July 6–9, 1999*, p. 336.
33. T. Ifuku, M. Otobe, A. Itoh, and S. Oda, Fabrication of nanocrystalline silicon with small spread of particle size by pulsed gas plasma, *Japanese Journal of Applied Physics* **36**, 4031 (1997).
34. A. Evtukh, V. Litovchenko, M. Semenenko, *et al.* (2012) Electron field emission from cathodes with Si and SiGe nanoclusters, *Proceedings of the 25th International Vacuum Microelectronics Conference, Jeju, Korea, July 6–13, 2012*, p. 100.
35. M. Ehbrecht and F. Huisken, Gas-phase characterization of silicon nanoclusters produced by laser pyrolysis of silane, *Physical Review B* **59**, 2975 (1999).
36. H. Hofmeister, F. Huisken, and B. Kohn, Laser production and deposition of light-emitting silicon nanoparticles, *European Physical Journal D: Atomic, Molecular, Optical and Plasma Physics* **9**, 137 (1999).
37. T. Utsumi, Vacuum microelectronics: What's new and exciting, *IEEE Transactions on Electron Devices* **38**, 2276 (1991).
38. A.A. Evtukh, E.B. Kaganovich, V.G. Litovchenko, *et al.*, Silicon tip arrays with nanocomposite films for electron field emission applications, *Materials Science and Engineering C* **19**, 401 (2002).
39. A.A. Evtukh, E.B. Kaganovich, V.G. Litovchenko, *et al.*, Field emission of electrons from laser produced silicon tip arrays, *Semiconductor Physics, Quantum Electronics and Optoelectronics* **3**, 474 (2000).
40. V.G. Litovchenko, A.A. Evtukh, R.I. Marchenko, *et al.*, The enhanced field emission from microtips covered by ultrathin layers, *Journal of Micromechanics and Microengineering* **7**, 1 (1997).
41. A. Uhlir, Electrolytic shaping of germanium and silicon, *Bell System Technical Journal* **35**, 333 (1956).
42. D.R. Turner, Electropolishing silicon in hydrofluoric acid solutions, *Journal of the Electrochemical Society* **105**, 402 (1958).
43. B. Hamilton, Porous silicon, *Semiconductor Science and Technology* **10**, 1187 (1995).
44. A.G. Cullis, L.T. Canham, and P.D.J. Calcott, The structural and luminescence properties of porous silicon, *Journal of Applied Physics* **82**, 909 (1997).
45. O. Bisi, S. Ossicini, and L. Pavesi, Porous silicon: a quantum sponge structure for silicon based optoelectronics, *Surface Science Reports* **38**, 1 (2000).
46. E.C. Boswell, M. Huang, G.D.W. Smith, and P.R. Wilshaw, Characterization of porous silicon field emitter properties, *Journal of Vacuum Science and Technology B* **14**, 1895 (1996).
47. J.R. Jessing, D.L. Parker, and M.H. Weichold, Porous silicon field emission cathode development, *Journal of Vacuum Science and Technology B* **14**, 1899 (1996).
48. M. Tejashree, S.V. Bhave, and S.V. Bhoraskar, Surface work function studies in porous silicon, *Journal of Vacuum Science and Technology B* **16**, 2073 (1998).
49. C.A. Spindt, A thin-film field-emission cathode, *Journal of Applied Physics* **39**, 3504 (1968).
50. A.A. Evtukh, V.G. Litovchenko, R.I. Marchenko, *et al.*, Parameters of the tip arrays covered by low work function layers, *Journal of Vacuum Science and Technology B* **14**, 2130 (1996).
51. J. Robertson, Diamond-like amorphous carbon, *Materials Science and Engineering* **37**, 129 (2002).
52. A.A. Evtukh, H. Hartnagel, V.G. Litovchenko, *et al.*, Enhancement of the electron field emission stability by nitrogen-doped diamond-like carbon film coating, *Semiconductor Science and Technology* **19**, 923 (2004).
53. A. Szekeres, T. Nikolova, A. Paneva, A. *et al.*, Silicon nanoparticles in thermally annealed thin silicon monoxide films, *Journal of Optoelectronics and Advanced Materials* **7**, 1383 (2005).
54. A.A. Evtukh, V. Ya. Rassamakin, and Yu. V. Rassamakin, Chemical vapor deposition of silicon dioxide films in automotive set up with individual wafer treatment, *Materials Science* **10**, 40 (2001).
55. A. Evtukh, O. Bratus', T. Gorbanyuk, and V. Ievtukh, Electrical characterization of SiO_2(Si) films as a medium for charge storage, *Physica Status Solidi C Current Topics in Solid State Physics* **5**, 3663 (2008).
56. O.L. Bratus', A.A. Evtukh, V.A. Ievtukh, and V.G. Litovchenko, Nanocomposite SiO_2(Si) films as a medium for non-volatile memory, *Journal of Non-Crystalline Solids* **354**, 4278 (2008).
57. M.O. Semenenko, A.A. Evtukh, O. Yilmazoglu, *et al.*, A novel method to form conducting channels in SiO_x(Si) films for field emission application, *Journal of Applied Physics* **107**, 013702 (2010).

58. A. Evtukh, V. Litovchenko, M. Semenenko, *et al.*, Formation of conducting nanochannels in diamond-like carbon films, *Semiconductor Science and Technology* **21**, 1326 (2006).
59. V.A. Dan'ko, I.Z. Indutnyy, V.S. Lysenko, *et al.*, Kinetics of structural and phase transformations in thin SiO_x films in the course of a rapid thermal annealing, *Semiconductors* **39**, 1197 (2005).
60. M.S. Dunaevskii, J.J. Grob, A.G. Zabrodskii, *et al.*, Atomic-force-microscopy visualization of Si nanocrystals in SiO_2 thermal oxide using selective etching, *Semiconductors* **38**, 1254 (2004).
61. D.R. Wolters and J.J. Van der Schoot, Dielectric-breakdown in MOS devices: Part I: Defect-related and intrinsic breakdown, *Philips Journal of Research* **40**, 115 (1985).
62. D.R. Wolters and J.J. Van der Schoot, Dielectric breakdown in MOS devices: Part II: Conditions for intrinsic breakdown, *Philips Journal of Research* **40**, 137 (1985).
63. D.R. Wolters and J.J. Van der Schoot, Dielectric breakdown in MOS devices. Part III: The damage leading to breakdown, *Philips Journal of Research* **40**, 164 (1985).
64. G. Dearnaley, A.M. Stoneham, and D.V. Morgan, Electrical phenomena in amorphous oxide films, *Reports on Progress in Physics* **33**, 1129 (1970).
65. G. Revesz, The defect structure of vitreous SiO_2 films on silicon. I. Structure of vitreous SiO_2 and the nature of the Si-O bond, *Physica Status Solidi A Applications and Material Science* **57**, 235 (1980).
66. H. Bouridah, F. Mansour, R. Mahamdi, *et al.*, Effect of thermal annealing and nitrogen content on amorphous silicon thin-film crystallization, *Physica Status Solidi A Applications and Material Science* **204**, 2347 (2007).
67. C.A. Spindt, I. Brodie, L. Humphrey, and E.R. Westerberg, Physical properties of thin-film field emission cathodes with molybdenum cones, *Journal of Applied Physics* **47**, 5248 (1976).
68. V.V. Zhirnov, C. Lizzul-Rinne, G.J. Wojak, *et al.*, "Standardization" of field emission measurements, *Journal of Vacuum Science and Technology B* **19**, 87 (2001).
69. A. Modinos, *Field, Thermionic, and Secondary Electron Emission Spectroscopy* (Plenum Press, New York, 1984, p. 320).
70. W.A. de Heer, A. Chatelain, and D. Ugarte, A carbon nanotube field-emission electron source, *Science* **269**, 1179 (1995).
71. Y.B. Li, Y. Bando, D. Golberg, and K. Kurashima, Field emission from MoO_3 nanobelts, *Applied Physics Letters* **81**, 5048 (2002).
72. Y. Tu, Z.P. Huang, D.Z. Wang, *et al.*, Growth of aligned carbon nanotubes with controlled site density, *Applied Physics Letters* **80**, 4018 (2002).
73. G.Z. Yue, Q. Qiu, B. Gao, *et al.*, Generation of continuous and pulsed diagnostic imaging x-ray radiation using a carbon-nanotube-based field-emission cathode, *Applied Physics Letters* **81**, 355 (2002).
74. C.S. Hsieh, G. Wang, D.S. Tsai, *et al.*, Field emission characteristics of ruthenium dioxide nanorods, *Nanotechnology* **16**, 1885 (2005).
75. F.C.K. Au, K.W. Wong, Y.H. Tang, *et al.*, Electron field emission from silicon nanowires, *Applied Physics Letters* **75**, 1700 (1999).
76. D. Banerjee, S.H. Jo, and Z.F. Ren, Enhanced field emission of ZnO nanowires, *Advanced Materials (Weinheim)* **16**, 2028 (2004).
77. J. Zhou, L. Gong, S.Z. Deng, *et al.*, Growth and field-emission property of tungsten oxide nanotip arrays, *Applied Physics Letters* **87**, 223108 (2005).
78. Y.H. Tang, X.H. Sun, F.C.K. Au, *et al.*, Microstructure and field-emission characteristics of boron-doped Si nanoparticle chains, *Applied Physics Letters* **79**, 1673 (2001).
79. Y.L. Chueh, L.J. Choua, S.L. Cheng, *et al.*, Synthesis of taperlike Si nanowires with strong field emission, *Applied Physics Letters* **86**, 133112 (2005).
80. N.N. Kulkarni, J. Bae, C.K. Shih, *et al.*, Low-threshold field emission from cesiated silicon nanowires, *Applied Physics Letters* **87**, 213115 (2005).
81. C. Mu, Y.X. Yu, W. Liao, *et al.*, Controlling growth and field emission properties of silicon nanotube arrays by multistep template replication and chemical vapor deposition, *Applied Physics Letters* **87**, 113104 (2005).
82. M. Lu, M.K. Li, L.B. Kong, *et al.*, Synthesis and characterization of well-aligned quantum silicon nanowires arrays, *Composites: Part A* **35**, 179 (2004).
83. B. Zeng, G. Xiong, S. Chen, *et al.*, Field emission of silicon nanowires, *Applied Physics Letters* **88**, 213108 (2006).
84. Y.F. Zhang, Y.H. Tang, N. Wang, *et al.*, Silicon nanowires prepared by laser ablation at high temperature, *Applied Physics Letters* **72**, 1835 (1998).
85. M. Choueib, R. Martel, S.C. Cojocaru, *et al.* (2012) Quasi-ideal current saturation in field emission and surface effect studies of individual hydrogen passivated Si nanowires, *Proceedings of IVNC 2012*, pp. 154–155.

86. E. Lefeuvre, K.H. Kim, Z.B. He, *et al.*, Optimization of organized silicon nanowires growth inside porous anodic alumina template using hot wire chemical vapor deposition process, *Thin Solid Films* **519**, 4603 (2011).
87. L.M. Baskin, O.I. Lvov and G.N. Fursey, General features of field emission from semiconductors, *Physica Status Solidi B Basic Solid State Physics* **47**, 49 (1971).
88. M. Nagao, Cathode technologies for field emission displays, *IEEJ Transactions on Electrical and Electronic Engineering* **1**, 171 (2006).
89. A. Haga, S. Senda, Y. Sakai, *et al.*, A miniature x-ray tube, *Applied Physics Letters* **84**, 2208 (2004).
90. Y. Neo, Y. Suzuki, K. Sagae, *et al.*, Smith–Purcell radiation using a single-tip field emitter, *Journal of Vacuum Science and Technology B* **23**, 840 (2005).
91. W.E. King, G.H. Campbell, A. Frank, *et al.*, Ultrafast electron microscopy in materials science, biology, and chemistry, *Journal of Applied Physics* **97**, 111101 (2005).
92. Y. Tanaka, H. Miyashita, E. Tomono, *et al.* (2011) Optically controllable emitter array with pn-junction integrated Si tip, *Technical Digest of 24th IVNC, 2011*, pp. 76–77.
93. I. Brodie and C.A. Spindt, The application of thin film field emission cathodes to electronic tubes, *Applied Surface Science* **2**, 149 (1979).
94. S.C. Lim, H.J. Jeong, Y.S. Park, *et al.*, Field-emission properties of vertically aligned carbon-nanotube array dependent on gas exposures and growth conditions, *Journal of Vacuum Science and Technology A* **19**, 1786 (2001).
95. 95.Y.C. Choi, Y.W. Jin, H.Y. Kim, *et al.*, Electrophoresis deposition of carbon nanotubes for triode-type field emission display, *Applied Physics Letters* **78**, 1547 (2001).
96. W. Zhu, P.K. Baumann, C.A. Bower, Novel cold cathode materials, in *Vacuum Microelectronics*, ed. W. Zhu (John Wiley & Sons, Inc., New York, 2001).
97. C.A. Mead, Operation of tunnel-emission devices, *Journal of Applied Physics* **32**, 646 (1961).
98. Y. Kumagai, K. Kawarada, and Y. Shibata, Energy distribution of electrons tunneling through a metal-insulator-metal sandwich structure, *Japanese Journal of Applied Physics* **6**, 290 (1967).
99. J.G. Simmons, Generalized formula for the electric tunnel effect between similar electrodes separated by a thin insulating film, *Journal of Applied Physics* **34**, 1793 (1963).
100. K. Ohta, J. Nishida, and T. Hayashi, Electron emission pattern of thin-film tunnel cathode, *Japanese Journal of Applied Physics* **7**, 784 (1968).
101. T. Kusunoki, M. Suzuki, S. Sasaki, *et al.*, Fluctuation-free electron emission from non-formed metal-insulator-metal (MIM) cathodes fabricated by low current anodic oxidation, *Japanese Journal of Applied Physics* **32**, L1695 (1993).
102. H. Adachi, Emission characteristics of metal–insulator–metal tunnel cathodes, *Journal of Vacuum Science and Technology B* **14**, 2093 (1996).
103. K. Yokoo, H. Tanaka, S. Sato, *et al.*, Emission characteristics of metal–oxide–semiconductor electron tunneling cathode, *Journal of Vacuum Science and Technology B* **11**, 429 (1993).
104. T.W. Hickmott, Energetic electronic processes and negative resistance in amorphous $Ta\text{-}Ta_2O_5\text{-}Au$ and $Al\text{-}Al_2O_3\text{-}Au$ diodes, *Thin Solid Films* **9**, 431 (1972).
105. T. Kusunoki and M. Suzuki, Increasing emission current from MIM cathodes by using an Ir-Pt-Au multilayer top electrode, *IEEE Transactions on Electron Devices* **47**, 1667 (2000).
106. H. Mimura, Y. Abe, J. Ikeda, *et al.*, Resonant Fowler–Nordheim tunneling emission from metal-oxide-semiconductor cathodes, *Journal of Vacuum Science and Technology B* **16**, 803 (1998).
107. K. Yokoo, S. Sato, G. Koshita, *et al.*, Energy distribution of tunneling emission from Si-gate metal–oxide–semiconductor cathode, *Journal of Vacuum Science and Technology B* **12**, 801 (1994).
108. K. Yokoo, G. Koshita, S. Hanzawa, *et al.*, Experiments of highly emissive metal–oxide–semiconductor electron tunneling cathode, *Journal of Vacuum Science and Technology B* **14**, 2096 (1996).
109. J. Maserjian and N. Zamani, Behavior of the Si/SiO_2 interface observed by Fowler-Nordheim tunneling, *Journal of Applied Physics* **53**, 559 (1982).
110. N. Koshida, X. Sheng, and T. Komoda, Quasiballistic electron emission from porous silicon diodes, *Applied Surface Science* **146**, 371 (1999).
111. T. Komoda, X. Sheng, and N. Koshida, Mechanism of efficient and stable surface emitting cold cathode based on porous polycrystalline silicon films, *Journal of Vacuum Science and Technology B* **17**, 1076 (1999).
112. N. Koshida, T. Ozaki, X. Sheng, and H. Koyama, Cold electron emission from electroluminescent porous silicon diodes, *Japanese Journal of Applied Physics Part 2* **34**, L705 (1995).
113. X. Sheng, T. Ozaki, H. Koyama, *et al.*, Operation of electroluminescent porous silicon diodes as surface-emitting cold cathodes, *Thin Solid Films* **297**, 314 (1997).

114. X. Sheng, H. Koyama, N. Koshida, *et al.*, Improved cold electron emission characteristics of electroluminescent porous silicon diodes, *Journal of Vacuum Science and Technology B* **15**,1661 (1997).
115. X. Sheng, H. Koyama, and N. Koshida, Efficient surface-emitting cold cathodes based on electroluminescent porous silicon diodes, *Journal of Vacuum Science and Technology B* **16**, 793 (1998).
116. X. Sheng and N. Koshida, Quasi-ballistic stable electron emission from porous silicon cold cathodes, *Materials Research Society Symposia Proceedings* **509**, 193, (1998).
117. T. Komoda, X. Sheng, and N. Koshida, Characteristics of surface-emitting cold cathode based on porous polysilicon, *Materials Research Society Symposia Proceedings* **509**, 187 (1998).
118. Y. Suda, K. Obata, and N. Koshida, Band dispersions in photoluminescent porous Si, *Physical Review Letters* **80**, 3559 (1998).
119. O. Klima, P. Hlinomaz, A. Hospodkova, *et al.*, Transport properties of self-supporting porous silicon, *Journal of Non-Crystalline Solids* **164/166**, 961 (1993).
120. J. Cohen, Tunnel emission into vacuum. II, *Applied Physics Letters* **1**, 61 (1962).
121. K. Nishiguchi, X. Zhao, and S. Oda, Nanocrystalline silicon electron emitter with a high efficiency enhanced by a planarization technique, *Journal of Applied Physics* **92**, 2748 (2002).
122. D.J. Bartelink, J.L. Moll, and N.I. Meyer, Hot-electron emission from shallow p–n junctions is silicon, *Physical Review* **130**, 972 (1963).
123. S.M. Sze, J.L. Moll, and T. Sugano, Range-energy relation of hot electrons in gold, *Solid-State Electronics* **7**, 509 (1964).
124. M. Heiblum, Tunneling hot electron transfer amplifiers (theta): Amplifiers operating up to the infrared, *Solid-State Electronics* **24**, 343 (1981).
125. H.L. Caswell and Y. Budo, Influence of oxygen on the surface mobility of tin atoms in thin films, *Journal of Applied Physics* **35**, 238 (1964).
126. J.L. Robins and J.N. Rhodin, Nucleation of metal crystals on ionic surfaces, *Surface Science* **1**, 346 (1964).
127. H. Shimawaki, Y. Neo, H. Mimura, *et al.* (2011) Electron emission from nanocrystalline silicon based MOS cathode under laser irradiation, *Technical Digest of the 24th IVNC, 2011*, pp. 220–221.
128. H. Shimawaki, Y. Neo, H. Mimura *et al.* (2012) Photoresponse of nanocrystalline silicon based MOS cathodes, *Technical Digest of the 25th IVNC, 2012*, pp. 324–325.
129. H. Shimawaki, Y. Neo, H. Mimura, *et al.* (2008) Emission uniformity of nanocrystalline silicon based metal-oxide-semiconductor cathodes, *Technical Digest of the 21st IVNC, 2008*, pp. 128–129.

7

GaN Based Quantum Cathodes

7.1 Introduction

Future generations of electronics and sensors will rely more and more on the special features and inherent properties of new vacuum devices. Over the last two decades, tremendous advances have been made in vacuum devices ranging from the introduction of new operation concepts to new material choices and improvement in fabrication processes. Processing has also improved, opening the way to submicron devices and combining the well-developed silicon technology with new vacuum devices. Technological improvements in solid state electronics have almost reached their limits and new generations of vacuum devices can enable further advances for high frequency generation. New achievements in high frequency sources have not only practical significance in the scientific context, but also beyond it, in the fields such as pharmacy, chemical recognition, bio-engineering, remote radar-type sensing, and tomography. Materials analysis, package monitoring, security screening, and so on, are other possible applications. It has also become clear that these systems must be compact, inexpensive, and easy-to-use to permit development of the technology, interesting for broader markets outside of science. The majority of application requires in general the availability of inexpensive and moderate technology with room-temperature application.

Novel cold electron emitters (field emitters) fabricated with the help of micro-mechanically-based technology have further versatile applications in measurement and sensor technology (e.g., scanning electron microscopy and X-ray image sensors) [1, 2]. The temperature and radiation resistance of field-emitter components are of particular interest for the fabrication of efficient and robust electron sources which make high-resolution imaging possible. These applications have initiated intensive studies in the area of field emission of cold cathode materials. Field emitters with a small emitter tip radius and long arrays are necessary for high field enhancement factors and thus low operating voltages. The micro-nano-mechanically produced (semiconducting) field emitter tips (radius $\sim$10 nm) have shown high field enhancement factors up to $>$1000.

Wide bandgap semiconductors (WBGSs), such as GaN, are very promising for novel applications of vacuum nanoelectronic devices. The large bandgap of group-III nitrides and the chemical and thermal stability allow new applications. Important are also the positive characteristics such as high saturation drift velocity of electrons of nearly 3×10^7cm/s [3] and

Vacuum Nanoelectronic Devices: Novel Electron Sources and Applications, First Edition.
Anatoliy Evtukh, Hans Hartnagel, Oktay Yilmazoglu, Hidenori Mimura and Dimitris Pavlidis.

extremely high breakdown field strengths of (3–5) × 10^6 V/cm [4]. Furthermore, these materials find application for various sensor concepts because of their large piezoelectric coefficients and robustness in harsh environments.

New vacuum nanoelectronic devices can help to overcome the frequency and power limitations of solid state devices [5, 6]. Field emission based vacuum micro-nano-devices (e.g., miniaturized tubes) are promising for amplification and generation of high frequency electromagnetic waves [5, 7, 8]. In such devices, electron transport is performed through vacuum without scattering, as is in case of solid state components, setting up therefore the basis for attaining ultrahigh frequency operation. Efficient electron field emission (EFE) cathodes are here of major importance. New developments in field emitter arrays (FEAs) are therefore discussed and experimentally investigated. FEAs with high current densities are considered as promising sources of cold electrons in miniature tubes for millimeter-wave generation.

Functional field emitter based on nitride materials show new effects, which can be used for miniaturized vacuum devices as well as new sensors. They open the possibility for density modulation of the electron beam with a gate electrode or photo-modulation instead of velocity modulation after electron emission in the vacuum tube. The advantages obtained by pre-bunching the beam are high efficiency and significant reduction in the required radio frequency (RF) interaction length in the tube, thereby simplifying beam transport magnetics and reducing the weight. Devices with high-power and high-efficiency can be obtained [9, 10].

A field emission set-up was built based on such approaches for analyzing the material properties of semiconducting emitters as GaN, AlGaN, and ZnO. A photo-assisted field emission spectroscopy method was developed for characterization of the energy bands of wide bandgap materials used in solid-state and vacuum nanoelectronic devices.

7.2 Electron Sources with Wide Bandgap Semiconductor Films

Many field emission (FE) experiments and applications involve complex architectures in the form of various conductive, semiconductive, or dielectric materials deposited one on top of another. From the point of view of field-emission applications, the governing process is the tunneling of electrons through potential-energy barriers usually created by dielectric layers and vacuum interface.

WBGS materials are attractive for application in vacuum micro- and nanoelectronics first of all due to the low electron affinity (LEA) and as expected low work function, lower threshold voltage, and improved stability. For application of wide bandgap materials it is necessary to solve the problem of efficient electron injection from the back contact.

Since the Fermi level must lie below the conduction band at the surface, the bands must be bent considerably to bring the conduction band close to the Fermi level at the contact interface. Methods of producing this band bending which require n-type doping often cannot be accomplished in wide bandgap low affinity materials, since the energy level of dopant atoms is typically well below the conduction band bottom. Furthermore, impossibly high doping levels would be required when the band bending is large. The energy band diagram with WBGS is shown schematically in Figure 7.1. As the energy barrier at the surface (the electron affinity) is reduced, there is a corresponding increase in the energy barrier at the interface between the conductive substrate and the LEA material (the Schottky barrier or conduction band offset). Thus, LEA surface implies a large conduction band offset or Schottky barrier at the substrate interface.

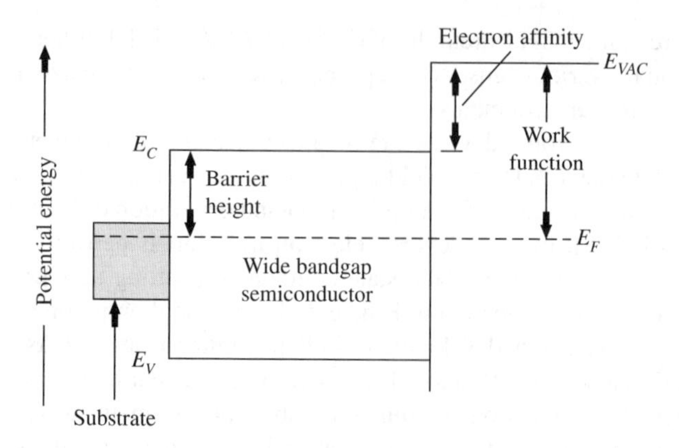

Figure 7.1 Energy band diagram of the emitter with thin wide bandgap semiconductor. Reproduced with permission from Ref. [11]. Copyright (1996), American Vacuum Society

Another problem is connected with stability of emission current [11]. To improve the stability it is necessary first of all to reduce the gas adsorption/desorption and/or to reduce the effects of adsorbates on field emission. Surface layers of adsorbed or reacted gases can drastically reduce the field emission current [12]. The area distribution of surface ad-atoms on field emitter tips can vary with time to cause noisy and nonuniform emission [13]. Several studies have shown that cleaning the emitting surfaces or overcoating them with various metals can substantially reduce the voltage needed to obtain the given current [12, 14–17]. However, the improved properties obtained in this way do not persist when the field emitter arrays (FEAs) are exposed to ambient gas in significant quantity (time or pressure). The improved properties are thus almost certainly due to the creation of a relatively clean metal surface; the subsequent emission decay is due to gas adsorption. Since the reactivity of a surface is at least partially associated with the presence of free charge, any metallic surface that results in low field-emission extraction fields is also very likely to be reactive with polar or electronegative species such as O-H, O, Cl, and so on. Hence, to be useful, such metallic surfaces have to be produced *in situ,* and it requires an exceptionally clean environment to remain clean.

Adsorbed gases or reacted surface layers might also be partially responsible for cathode arcs, which often destroy local areas of FEAs. Arcs occur in FEAs with increasing probability when the emission current is increased, when the ambient pressure is increased, and when the arrays are operated in a constant current mode rather than in a pulsed current mode [18]. Arcs appear to be initiated when sufficient quantities of molecules become ionized near the emitter. A sufficient number of neutral molecules are not present in the gas phase at hard vacuum pressures, but may be adsorbed over time on the gate or other electrodes, then abruptly desorbed due to ion or electron bombardment. Sudden increase in the current intercepted by the gate typically precedes an arc. Changes in gate interception are probably related to changes in the area distribution of adsorbates on the emitter apex. For example, single ion hitting an emitter to one side of center may create a clean spot or asperity, increasing the emission from that location and consequently increasing the gate current. The extreme sensitivity of the field emission process to the condition of the surface allows such large changes in the emission current occurring due to small effects such as the motion of a few atoms. Thus, if the emission could be made less sensitive to the surface electronic properties and small geometrical variations, gate interception, and consequent arcing might be avoided.

7.2.1 AlGaN Based Electron Sources

The interest in electron field emitters based on WBGSs such as diamond and AlGaN has been increasing because the electron affinity of these materials is expected to be very small or even negative [19]. The ternary alloys of the $Al_xGa_{1-x}N$ ($0 < x < 1$) family are very promising materials for field emission device applications. These alloys have an electron affinity χ which ranges between small positive and negative values and can also be n-doped with Si. AlN films are essentially insulators but they exhibit a negative χ. In materials with negative electron affinity (NEA), electrons can be freely extracted from the surface to the vacuum. In contrast, GaN exhibits a positive χ in the range of ~2.1–4.1 eV and can be easily n doped [20]. Thus, by varying x in the $Al_xGa_{1-x}N$ composition, the electron affinity and doping concentration can be simultaneously optimized to maximize field emission. It is known that the alloy $Al_xGa_{1-x}N$ exhibits a NEA for values of $x > 0.75$ [21, 22]. It is difficult to obtain high n-doping for values of x greater than 0.4 [23]. These wide bandgap III-nitride materials also possess several other useful physical properties, such as high thermal conductivity, high bond strengths, and high melting temperatures, making them very attractive for field emission applications. Most FE researches on AlGaN have been focused on GaN [19, 24–26], and there are few studies on AlN [27, 28]. AlN has a wider bandgap (6.2 eV) than diamond (5.5 eV), GaN (3.4 eV), and Si (1.1 eV), and the electron affinity of AlN has been reported as negative [21] and sometimes positive [29, 22, 30, 31]. EFE from sharp tips of GaN crystals was observed [21, 22]. Field emission from patterned flat AlN and AlGaN thin film has also been investigated [27, 28]. On the other hand, for AlN, a less costly SiC substrate compared with GaN is available, because AlN and SiC have almost equal lattice constants and thermal expansions along the a axis. However, the reported FE current from AlN was as low as 50 nA and the threshold electric field was as high as 60 V/μm [27], probably because the AlN samples in the study were not intentionally doped and oxidized very easily.

Theoretical calculations [32] support the possibility of using these wide bandgap materials as electron emitters. However, very little work has been done on optimizing the field emission properties of these materials for applications. Since the electronic structure of these alloys depends upon the stoichiometry, an accurate treatment of field emission from these materials requires a theoretical calculation procedure that goes beyond the conventional or classical Fowler-Nordheim model which assumes only free electron properties of the material.

Using the kinetic theory formalism, there was assumed an independent particle electron model with the current flow only in the $+z$ direction [33]. For a particular energy band, the tunneling current density within this formulation is given by

$$J(F,T) = \frac{2q}{(2\pi)^3\hbar}\int f(E,T)T(E-E_t,F)dk_x dk_y dE, \tag{7.1}$$

where $E = E(k)$ is the total energy of the electron, E_t is the transverse energy parallel to the emission surface, $f(E,T)$ the Fermi distribution function, $T(E_n,F)$ the transmission coefficient with normal energy E_n and external field F, $\bar{k} = (k_x, k_y, k_z)$ is the electron wave vector, q is the electron charge, and $\hbar$ is the normalized Planck constant.

For the wide bandgap ternary alloy semiconductors at solving Equation (7.1), it is important to include the band bending. If there are no surface states, the band bending is due to the applied field. The electrostatic field is obtained by solving the Poisson equation to obtain the potential energy in the space charge region with the appropriate boundary conditions. Details of how the potential is calculated in the one-dimensional model have been presented in Ref. [34]. The total potential of an electron should also include the image interaction. It is possible to choose the classical image potential with a simple quantum correction, that is, replacing $1/z$ by $1/(z + z_0)$, where z_0 is the position of the energy barrier peak [35]. The surface barrier is then constructed. The calculation of $T(E_n, F)$ is performed using the method, where the solution can be expressed as a linear combination of the Airy functions, [36] and the tunneling barrier is divided into N-line segments. The Airy functions (which are exact solutions of the Schrödinger equation for a linear potential) are then matched across segment boundaries. The integration over k_x, k_y in Equation (7.1) yields the energy distribution, $dJ(E,F,T)/dE$. This involves integrating over the states obtained from the projection of the constant energy band in the emission direction. A final integration of $dJ(E,F,T)/dE$ over E yields the field emission current density $J(F,T)$ at a given field F [33].

As a first approximation, the material parameters for $Al_xGa_{1-x}N$ are obtained by assuming that the alloy is a linear combination of AlN and GaN. That is, the values of the longitudinal and transverse effective masses m_l and m_t, respectively, the electron affinity χ, and the dielectric constant ε of $Al_xGa_{1-x}N$ are the weighted averages of these parameters for AlN and GaN [20, 37, 38]. It should be noted, however, that there is some ambiguity in the value of the resulting electron affinity χ_{av}. For GaN, χ ranges from 2.1 to 4.1 eV depending on the sample surface dipoles, so that χ_{av} can vary from 2.7 to 3.3. AlN has been reported to exhibit NEA [20]. It has been suggested that $Al_xGa_{1-x}N$ has a "true" NEA only for $x > 0.75$ [39]. The electron affinity is assumed to depend on the composition (x) through the linear equation $\chi = 3.0–4.0x$. For the calculation of the bandgap E_g, a nonlinearity was included due to field penetration through the bandgap bowing parameter b: $E_g = xE_{g,AlN} + (1 - x)E_{g,GaN} + bx(1 - x)$.

Using the calculated $Al_xGa_{1-x}N$ parameters and Equation (7.1), the field emission current density from $Al_xGa_{1-x}N$ was determined as a function of the stoichiometric composition ($0.1 \leq x \leq 0.4$) [33]. In the range from $x = 0$ to $x = 0.4$, the resistivity increases by several orders of magnitude [20]. The increased resistivity for large x has been attributed to compensation effects (doping efficiency decreases due to compensation by the structural defects) [40]. The J-F (i.e., I-V) characteristics in Fowler-Nordheim (F-N) coordinates are plotted in Figure 7.2 for $x = 0.1$, 0.2, 0.3, and 0.4. These results were first obtained for a fixed carrier concentration, $n = 10^{17}/\text{cm}^3$ [33]. As it can be seen, there is a strong dependence of the J-F characteristics on the composition which reflects the dependence of χ on x. The fields required for $J = 1$ mA/cm^2 are $F = 510$ V/μm at $x = 0.3$ and $F = 350$ V/μm at $x = 0.4$, respectively. These calculated field values are large in comparison with experimental values [27, 39]. However, this is not inconsistent for the following reason. The calculated fields depend on the local tip geometry and are, in general, going to be larger than the measured macroscopic field by one or more orders of magnitude. This effect will obviously be more pronounced as the tip radius decreases and the height of the protrusion (i.e., tip) increases. Experimentally, one usually determines the macroscopic field as the ratio of the applied voltage difference and the tip-anode spacing – the quantity independent of the local tip curvature.

The J-F plots in Figure 7.2 are not linear suggesting anomalous features in the tunneling behavior. Each curve in Figure 7.2 can be approximated by two straight lines of different slope

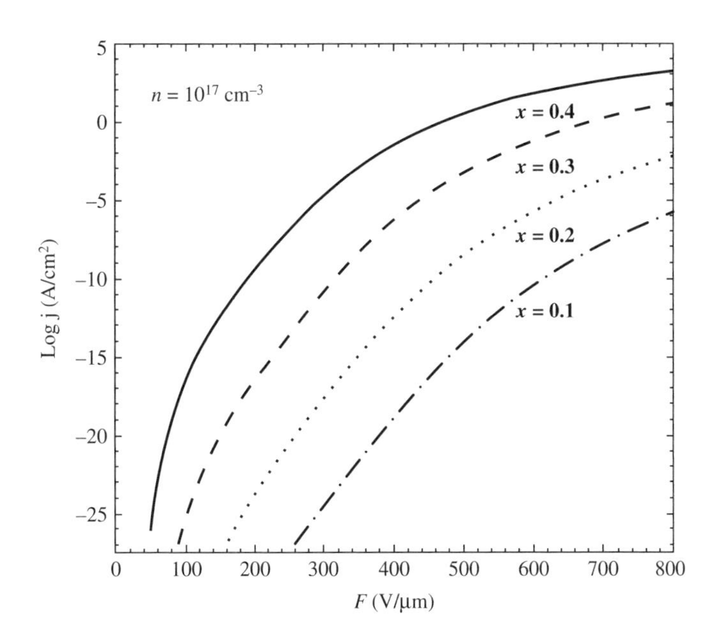

Figure 7.2 *J-F* plots for $Al_xGa_{1-x}N$ for selected values of x. The carrier concentration was $n = 10^{17} cm^{-3}$. Reproduced with permission from Ref. [33]. Copyright (2000), AIP Publishing LLC

above and below a local field of about 300 V/μm, implying different emission mechanisms in the two regimes above and below this value. In Figures 7.3 and 7.4, calculated field electron energy distributions (FEEDs) are presented for values in the high and low field regimes. In Figure 7.3 in which $F = 500\,V/\mu m$, the energy states in the FEED are lower in energy than those of the FEED curve in Figure 7.4 for $F = 250\,V/\mu m$. Qualitatively the two curves differ in the full width at half maximum (FWHM) and peak locations. When F is large, as in Figure 7.3, the potential barrier is relatively thin and the transmission probability does not change significantly over the energy states involved in tunneling. Thus, the emission current density *J(F,T)* is determined mainly by the occupancy of electron states (i.e., electron supply to the barrier). These factors produce the peak in the emitted distribution near the bottom of the conduction band, E_c. The shift of FEED with respect to E_c (in the bulk) is due to the field penetration and resulting band bending at the surface. If F is small, as in Figure 7.4, so that the barrier is thick, then the transmission probability becomes nearly negligible for low energies. Therefore, in this range of fields, *J(F)* is determined by both the transmission coefficient as well as the occupancy factor. The resultant FEED distribution peak is determined by the combination of these two factors and results in a shift toward higher energy where the transmission probability is correspondingly higher and dominates the occupancy factor.

The *j(F,T)* dependences were calculated as a function of the carrier concentrations, $n = 10^{15}$, 10^{17}, $10^{19}\,cm^{-3}$ [33]. It is found in the low field region that there is a weak dependence of the tunneling current on the electron concentration in the conduction band. At higher fields (≥500 V//Am), as n increases from 10^{15} to 10^{19}, the current increases by nearly 4 orders of magnitude. Thus, in the high field regime, there is approximately a linear relationship between the current density and n. The screening effect (i.e., field penetration) is more important at lower values of n, whereas the tunneling contribution dominates at larger values of n.

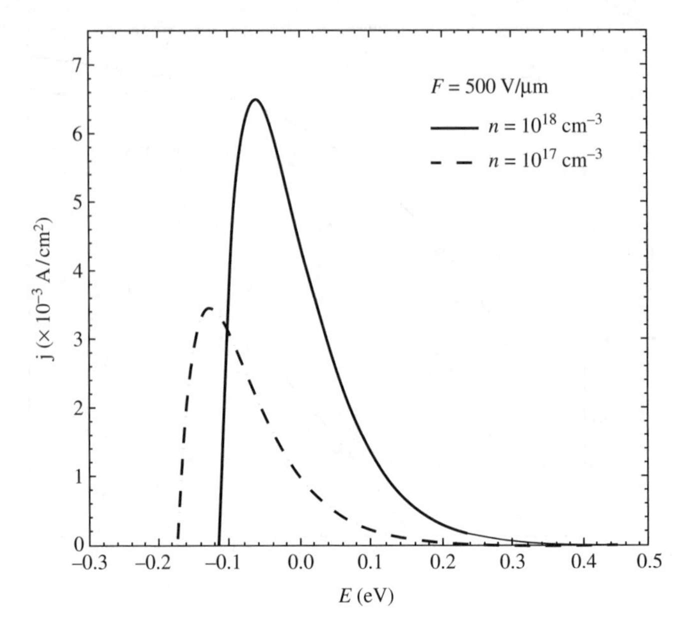

Figure 7.3 Field electron energy distribution (FEED) for $Al_xGa_{1-x}N$ at a vacuum field of $F =$ 500 V/μm. The carrier concentration dependence is seen in both the shift of the peak and the overall width of the FEED distribution. Reproduced with permission from Ref. [33]. Copyright (2000), AIP Publishing LLC

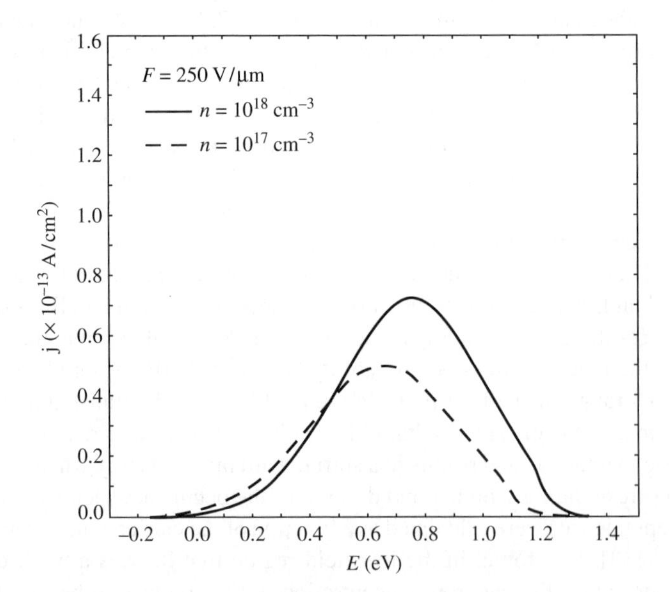

Figure 7.4 Field electron energy distribution (FEED) for $Al_xGa_{1-x}N$ at a vacuum field of $F =$ 250 V/μm. The carrier concentration dependence is seen in both the shift of the peak and the overall width of the FEED distribution. Reproduced with permission from Ref. [33]. Copyright (2000), AIP Publishing LLC

This implies that for large F, the effect of supply is dominant compared to that of screening. The n dependence is also seen in both the shift and the width of the FEEDs in Figures 7.3 and 7.4.

It is also seen from Figure 7.2 that for fixed n, with increasing x, there is a monotonic increase in the current density. Similarly, an increase in the carrier concentration n also increases J. From these observations, it may seem desirable to increase J by increasing the value of x in $Al_xGa_{1-x}N$. However, it is known that the resistivity of these films increases rapidly for $x > 0.4$ [40].

Detailed investigations of electron FE from AlN films have been performed in Ref. [28]. In their case the heavily Si-doped 0.2 μm-thick AlN was grown at 1100 °C by low-pressure (300 Torr) Metal Organic Vapor Phase Epitaxy MOVPE [28]. An n-type (1×10^{18}cm^{-3}) 6H-SiC (0001) substrate was chosen because it had high electric conductivity as well as almost the same lattice constants (mismatch: 1%) and thermal expansions along the a axis as AlN. The sources were trimethylaluminum, trimethylgallium, and ammonia (NH_3). The Si-dopant gas was silane (SiH_4), the most widely used in GaN MOVPE growth. In GaN, Si atoms form a shallow donor level. In Si-doped $Al_xGa_{1-x}N$ ($0 \leq x \leq 0.33$), the resistivity increases exponentially as the Al content x increases [41]. It was confirmed that the resistivity of the AlN with the highest Si dopant density (2.5×10^{20}cm^{-3}) was still too high to measure the free electron density at room temperature (RT) by Hall measurement [28]. The surface roughness of the heavily Si-doped AlN surfaces was less than 50 nm. The heavily Si-doped AlN surfaces were free from cracks. The Si dopant density was measured using secondary ion mass spectrography. For the GaN sample used as a reference, a non-doped 0.1 μm thick AlN layer and 1 μm thick Si-doped (1×10^{19} cm^{-3}) GaN were grown at 1100 and 1010 °C, respectively. The highest Si dopant density in GaN at which a smooth surface could be maintained was 1×10^{19} cm^{-3}.

Figure 7.5a indicates the FE current for different Si-dopant densities N_{Si} in AlN [28]. As the Si-dopant density increased, the threshold voltage decreased and the FE current increased

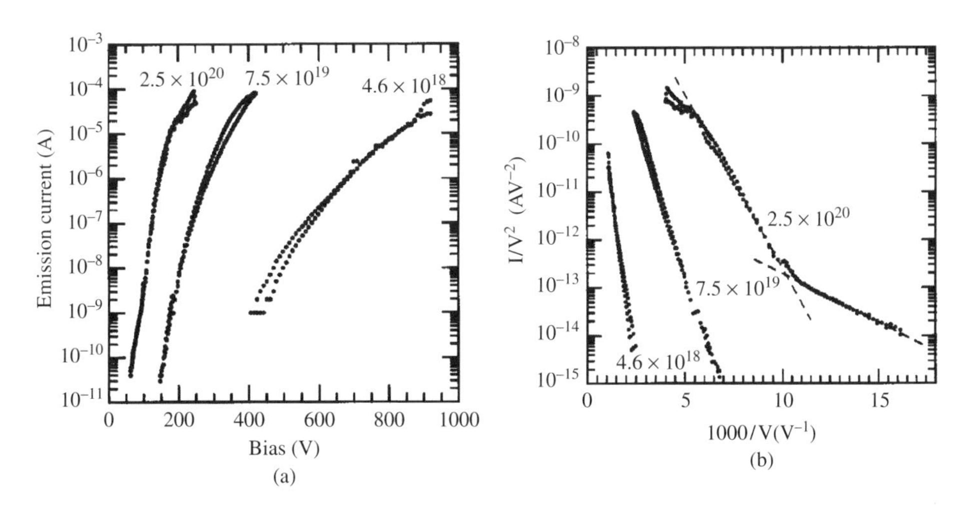

Figure 7.5 Field emission current as a function of (a) the applied bias voltage and (b) its F-N plots of AlN with different Si dopant densities (N_{Si}, cm^{-3}). The sample-probe distance was 1.8 μm. Reproduced with permission from Ref. [28]. Copyright (2000), AIP Publishing LLC

drastically. The threshold voltages defined at 1 nA were 90, 180, and 410 V for $N_{Si} = 2.5 \times 10^{20}$, 7.5×10^{19}, and 4.6×10^{18} cm^{-3} respectively. For $N_{Si} = 2.5 \times 10^{20}$ cm^{-3}, the threshold voltage defined at 0.05 nA was 62 V. At the sample-probe distance of 1.8 μm, the threshold electric field was as low as 34 V/μm. Figure 7.5b shows the Fowler-Nordheim plots (I/V^2 vs. $1000/V$) for AlN with different Si-dopant densities. The negative slope, which corresponds to the energy barrier, decreased with increasing the Si-dopant density. Due to the expected small voltage drop at the back contact and in the sample compared to the applied bias, it is supposed that the obtained energy barrier is mainly the surface energy barrier. Thus, it is speculated that electrons are emitted from Si impurity level over the surface energy barrier by the F-N tunneling [28]. As it is seen in Figure 7.5b, the F-N plots for AlN ($N_{Si} = 2.5 \times 10^{20}$ cm^{-3}) have two slopes, which correspond to two kinds of energy barrier.

Figure 7.6 shows FE *I-V* characteristics of heavily Si-doped AlN ($N_{Si} = 2.5 \times 10^{20}$ cm^{-3}) and those of Si-doped GaN (($N_{Si} = 1 \times 10^{19}$ cm^{-3}), the highest value at which a smooth surface could be maintained). The threshold voltages defined at 0.1 nA were 68 V for AlN and 170 V for GaN. The maximum FE currents were 150 μA for AlN ($N_{Si} = 2.5 \times 10^{20}$ cm^{-3}) and 19 μA for GaN ($N_{Si} = 1 \times 10^{19}$ cm^{-3}). Assuming the probe area of 3.1×10^{-2} cm^2, the maximum FE current densities were estimated to be 4.8 and 0.6 mA/cm^2, respectively. The maximum FE current density is limited by the field evaporation of the surface atoms and subsequent microarc discharge. The reason that AlN has higher maximum current density than GaN is probably that AlN has stronger bonds, and it is more resistant to field evaporation of the surface atoms.

The resistivity of the heavily Si-doped ($N_{Si} = 2.5 \times 10^{20}$ cm^{-3}) AlN was very high. However, the hopping conduction of the electron in the Si impurity level in highly doped materials

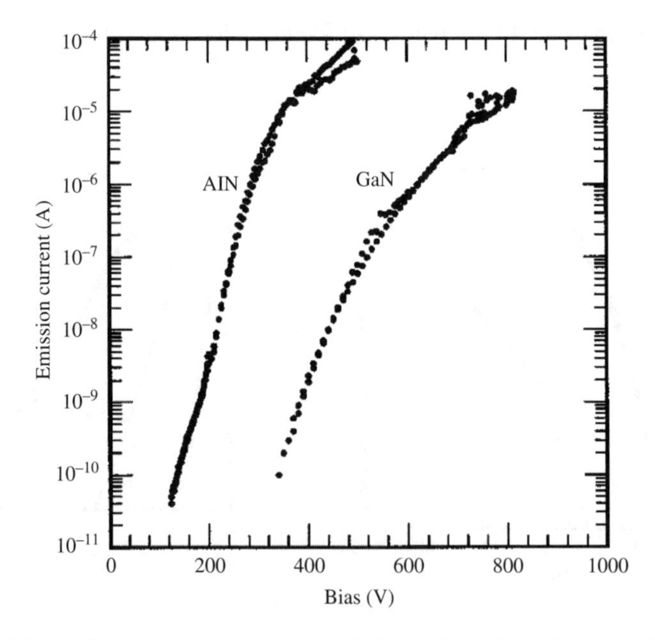

Figure 7.6 Field emission current as a function of the applied bias voltage for AlN ($N_{Si} = 2.5 \times 10^{20}$cm^{-3}) and GaN ($N_{Si} = 1 \times 10^{19}$cm^{-3}). The sample-probe distance was 1.8 μm. Reproduced with permission from Ref. [28]. Copyright (2000), AIP Publishing LLC

and direct tunneling from the Si impurity level to the vacuum are possible. It has been assumed that in heavily Si-doped AlN electrons are supplied from the back contact to the surface by the hopping conduction in Si impurity level and then they are emitted from the Si impurity level to the vacuum without transition to the conduction band.

7.2.2 *Solid-State Field Controlled Emitter*

To overcome the above mentioned problems of $Al_xGa_{1-x}N$ cathodes a new type of electron source with thin WBGS film, namely solid-state field controlled emitter (SSE) has been proposed and investigated [42–44]. The basic structure of SSE is an ultrathin semiconductor cathode (UTSC) layer deposited on a metallic surface. These cold cathodes emit stable electron currents with operating field F_{th} having a threshold value in the range of ~50 V/μm, in a poor vacuum environment (~10^{-7} Torr), and for different cathode geometries as plane, hairpin, or conical tip. The main experimental results are (1) appearance of the emission current for a low threshold value of F_{th} without the need of conditioning; (2) uniform and stable emission over the whole flat surface of the SSE; and (3) emission characteristics that cannot be interpreted by the conventional thermionic or field emission mechanisms. Figure 7.7 is an example of the experimental current density J versus applied field F_{app} characteristics of SSE planar cathodes. The qualitative model to explain these results has been proposed in Ref. [42]. Here, the electron emission from SSE cathodes has a serial two-step mechanism (Figure 7.8). The first step is the injection of electrons at the solid-state Schottky junction from the metal into the UTSC medium, followed by a second step which is the electron emission from the UTSC surface into vacuum. The effective electron emitter becomes, under the control of an applied field F_{app}, a surface with LEA (LEA situation for $\Delta\Phi_E \leq 2$ eV) or with NEA (NEA situation for $\Delta\Phi_E \leq 0$)

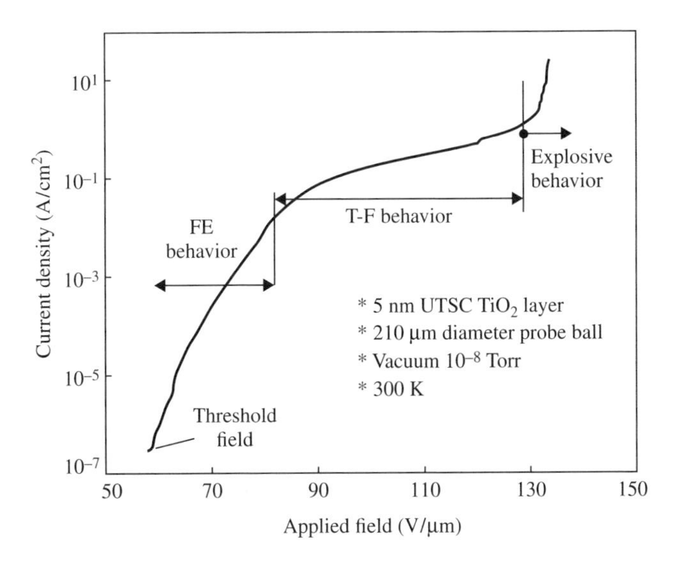

Figure 7.7 Experimental current density J vs. applied field F_{app} characteristics of a SSE (T-F is the thermo-field emission). Reprinted with permission from Ref. [43]. Copyright (2000) by the American Physical Society

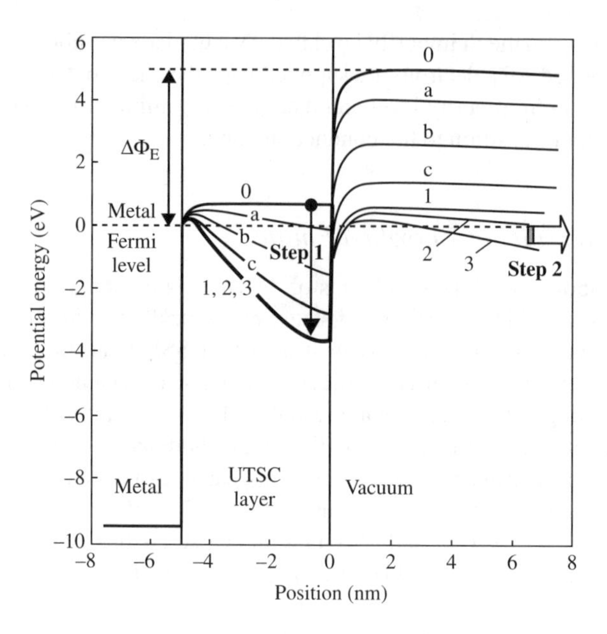

Figure 7.8 Calculated energy-band diagram change representing the serial two-step mechanism for electron emission from SSE cathodes. Step 1: electron injection from metal to UTSV and Step 2: electron emission from UTSC to vacuum. Reprinted with permission from Ref. [43]. Copyright (2000) by the American Physical Society

where $\Delta\Phi_E$ is the potential difference between the metal Fermi level and the vacuum level. The injected charges induce a large band bending in the UTSC layer lower than the emission barrier $\Delta\Phi_E$, which can become low enough to allow electrons emission through or over it.

The model developed in Ref. [43] consists of (1) basic SSE structure with an UTSC layer in the range of 2–10 nm thick TiO_2, for example, with $\varepsilon = 35$; electron affinity = 4.5 eV; bandgap = 3 eV; and a doping level at −0.2 eV under the conduction band (CB), deposited on a metallic surface (Pt, for example, with work function = 5.3 eV); (2) Schottky junction between the metal and the UTSC with the conventional energy band relation [45]; (3) triangular representation of the vacuum barrier including the image potential at the surface of the semiconductor; (4) numerical integration of the Poisson's equation to evaluate the equilibrium space charge distribution Q_{SC} inside the UTSC; and (5) calculation of J by the resolution of the one-body Schrödinger equation using a Green's formalism [46] based on the numerical resolution of the self-consistent Lippmann-Schwinger (LS) equation [47]. The calculations are one-dimensional, as confirmed by experimental measurements showing a very uniform emission over the whole SSE flat surface [42–44]; and the charge densities for electrons in the CB, as well as for holes in the valence band, are assumed to be given by the Fermi-Dirac statistics. The boundary conditions are: (1) at equilibrium and at the Schottky interface, the two Fermi levels are accorded; and (2) the field at the UTSC surface with vacuum is F_{app}/ε, with ε is the dielectric constant.

Q_{sc} is given by the resolution of the Poisson's equation,

$$\frac{d^2V}{dz^2} = \frac{q}{\varepsilon}[n(z) - n_0 - p(z) + p_0], \tag{7.2}$$

within a quasiequilibrium condition, which means within a zero emission current approximation (ZECA) ($n(z)$ is the conduction electronic density, $p(z)$ is the hole density, and n_0 and p_0 are the intrinsic carrier densities of n and p, respectively [45]). This implicitly assumes that the electrons are in thermal equilibrium among themselves and ZECA is valid as long as J is small relative to the electron supply function [48]. The potential distribution $V(z)$ is calculated by starting from the metallic boundary with an initial value corresponding to the Schottky barrier height and propagating it by using a finite differences method based on the second order series development

$$V(z+h) = V(z) + h\frac{dV}{dz} + \frac{h^2}{2}\frac{d^2V}{dz^2}, \tag{7.3}$$

until the UTSC surface with vacuum, where dV/dz is the field inside the UTSC [49] and h is the discretization step. The propagation is done by modifying the polarization of the UTSC in an iterative process until the field at its surface reaches F_{app}/ε.

The quantum transport through this one-dimensional potential barrier is analyzed using a transmission coefficient, $T(E)$, approach [50]. Current density between the electron reservoir (metal) and the vacuum through the UTSC layer is given by [51]

$$J = \frac{qm}{4\pi^2\hbar^3}k_BT\int T(E)[\ln(1+\exp(E_F - E)/k_BT)]dE, \tag{7.4}$$

where m is the mass of electron, k_B is the Boltzmann constant, T is the temperature, E_F is the Fermi level in the metal, and E is the energy of the electron. $T(E)$ is numerically calculated by means of the LS self-consistent equation, which allows us to introduce in a reference system, having analytical solutions, a perturbation corresponding to the UTSC layer [52, 53]. The reference system corresponds to the one-dimensional system constituted by a metal-metal polarized junction. The LS equation is

$$\Psi(z) = \Psi_0(z) + \int dzG_0(z,z',E)V(z)\Psi(z), \tag{7.5}$$

where Ψ is the wave function of the electron in the whole system, Ψ_0 and $G_0(z,z',E)$ are the wave function and the Green's function of the reference system, respectively. The solutions of the corresponding one-body Schrödinger equation are expressed by means of Airy functions.

It has been shown experimentally and theoretically that electrons are emitted from a cathode having its emission barrier controlled by the space charge created inside a deposited UTSC layer. Stable currents are obtained at room temperature, in poor vacuum environment, and with operating fields 2–3 orders of magnitude less than for field emission.

7.2.3 Polarization Field Emission Enhancement Model

To explain the efficient emission from Si cathodes coated by thin GaN film the polarization field emission enhancement model has been proposed [54]. The polarization FE enhancement mechanism with ballistic electron transport comprises two effects (Figure 7.9).

The first effect is the polarization effect [55]. III-V nitrides possess special properties and compared to other III-V compounds, the piezoelectric constants of nitrides are nearly 10 times larger and spontaneous polarization is very large [56–59]. When the growth direction is along

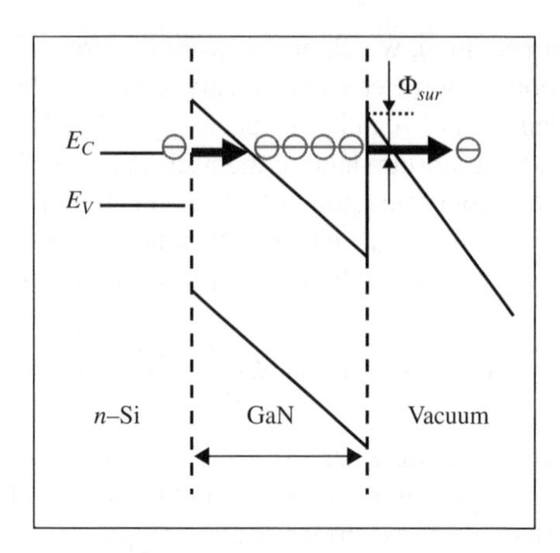

Figure 7.9 Polarization FE enhancement mechanism incorporating ballistic electron transport in the oriented nanostructured GaN films. Reproduced with permission from Ref. [54] Copyright (2010), AIP Publishing LLC

the polar *c*-axis of the wurtzite structure, piezoelectric, and spontaneous polarization can result in a strong polarization field in the range of a few megavolt per centimeters [60, 61]. Polarization field can be formed in the GaN film on Si due to the residual stress along the *c*-axis. When the polarization field is formed, the CB of the GaN is dramatically bended.

The second effect arises from the nanostructure [62, 63]. When the thickness of the GaN layer is less than the mean free path of electrons, electrons are accumulated near the GaN/vacuum interface due to ballistic electron tunneling through the GaN layer. Hence, electrons are emitted not from the CB levels of the GaN near the GaN/vacuum interface, but from the CB levels of the Si substrate. In this case, Φ_{sur} can be defined as the potential difference between CB level of Si and vacuum level. Coupling of the polarization and nanostructure effects leads to greatly reduced Φ_{sur} and consequently, the FE properties are enhanced significantly.

7.2.4 *Emission from Nanocrystalline GaN Films*

Sapphire and SiC are commonly used as the substrates to fabricate GaN [64, 65] and Si is usually not recommended as a substrate for III-V nitrides because of the large lattice mismatch that affects the microstructure (crystallization and orientation) of the deposited layer. However, the use of Si as substrate for FE devices can save fabrication costs and make the integration of field emitters into Si-based integrated circuits possible. Therefore, it is crucial to study the microstructure effects of FE GaN devices produced on a Si substrate.

Detailed investigation of the EFE from nanocrystalline film was performed in Ref. [66]. The series of nanocrystalline GaN films was prepared on Si by pulsed laser deposition (PLD). The microstructure, including crystallization and orientation, could be controlled by the deposition temperature. The obtained films contain GaN nanoclusters embedded into matrix of amorphous GaN (a-GaN). The GaN films were deposited at different substrate temperatures of 1000, 900, 800, 700, 600, and 500 °C (samples designated as *A*-*F*, respectively). The thickness of the GaN films ranged from 220 to 260 nm.

X-ray diffraction (XRD) has revealed that the GaN film deposited at below 500 °C exhibits a typical amorphous pattern. When the temperature reaches 600 °C, the characteristic diffraction peaks of GaN begin to appear as a broad peak around 35°. As the temperature is increased to 800 °C the development of a crystalline structure is observed. When the temperature is further increased to 1000 °C the GaN film is preferentially oriented in the *c*-axis direction. The nanocrystalline GaN content improves and the nanocrystalline GaN grain size increases at a high substrate temperature. The atomic force microscopy AFM images of the GaN surfaces shows a large quantity of small nanoscale protrusions on the surface. As the temperature is increased, the diameter of the surface protrusions decreases from 500 to 100 nm. XRD and AFM reveal that the deposition temperature dramatically affects the microstructure and surface morphology of the GaN films.

J–E emission characteristics are shown in Figure 7.10a, and the corresponding Fowler – Nordheim plots are given in Figure 7.10b. All samples exhibit a clear FE behavior. The F-N plots of all samples in Figure 7.10b exhibit a linear relationship in the high-field region, suggesting that the emission current should originate from a quantum mechanical tunneling process. As can be seen from the insert in Figure 7.10a, the turn-on field E_{on} (E_{on} and threshold field E_{th} are defined at an emission current density of 1 μA/cm^2 and 0.1 mA/cm^2, respectively) of the microstructure nanocrystalline GaN film (sample *b*) is smaller than that of the conventional film GaN materials [67–69] and even comparable to those of 1D GaN materials [70–78] indicating efficient FE from the GaN films with optimal microstructure.

Detailed information concerning the microstructure of the GaN films is crucial to the understanding of the FE mechanism. On the samples prepared at a relatively low temperature (<800 °C), the scanning electron microscopy (SEM) images reveal a number of nanosized surface protrusions consisting of lots of grains on the surface. Furthermore, the diameter of these grains increases from <10 nm (500 °C) to nearly 100 nm (1000 °C), indicating the modification in the crystal structure. Such GaN nanocrystals play an important role in electron transport in the film and electrons emission from the surface; a conducting channel model is proposed to explain electron transport in the nanocrystalline GaN films [66].

In the interior of the GaN film (Figure 7.11), the GaN nanocrystals are embedded in an a-GaN matrix. In the samples fabricated at low temperature, as a result of the small grain size and content, the GaN nanocrystals cannot connect to each other, as shown in Figure 7.11a. As the temperature is increased, both the grain size and content increase considerably, and the connection of separate GaN nanocrystals forms a conducting GaN channel (Figure 7.11b), which determines the electron supply for field electron emission.

The electron supply can be improved due to the effective conducting pathway. On the basis of the aforementioned analysis it can be concluded that the conducting channels depend closely on the nanocrystalline GaN grain size and content. Moreover, these conducting channels may be caused by two different electron transport mechanisms that are crucial to understanding of the principles governing the FE enhancement and improvement of the FE performance. The first one is the electron transport through the interior of nanocrystalline materials.

It has been suggested that the connected GaN nanocrystals embedded in the a-GaN matrix act as the effective conduction pathway in which electrons can traverse across the GaN films from the interior of the conduction channels. However, with regard to sample *a*, XRD shows that increasing temperature further improves the crystallinity and leads to oriented growth in the *c*-axis direction, thus implying that the nanocrystalline GaN content and grain size can be dramatically enhanced. But there were nonmonotonic temperature dependences of E_{on} and E_{th}

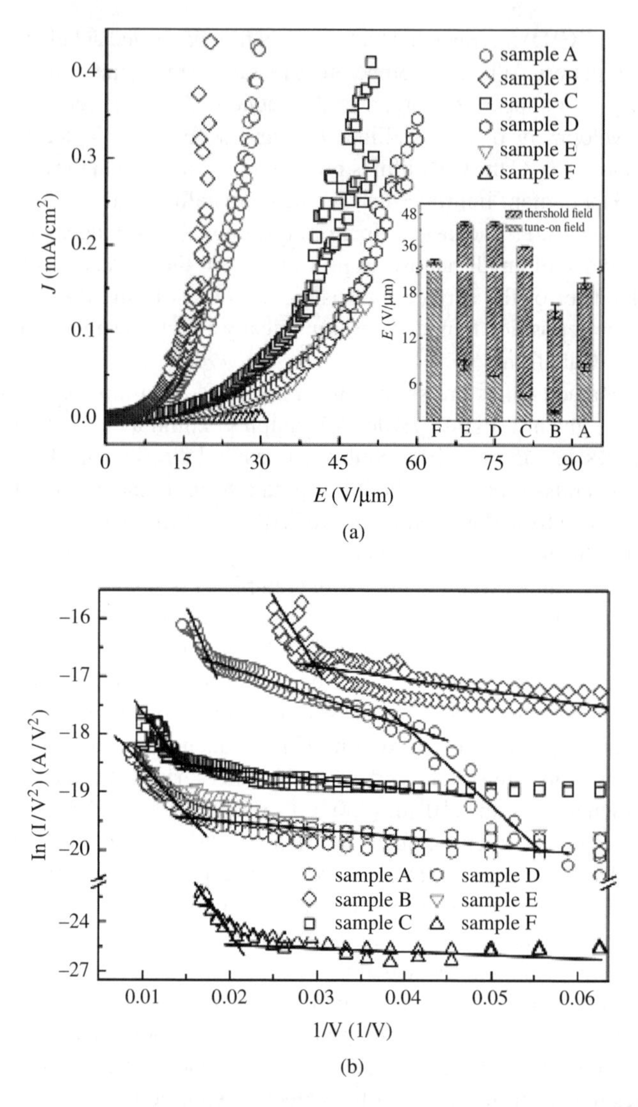

Figure 7.10 FE characteristics of samples grown at different temperature (1000–500 °C); designated as (*A*–*F*): (a) FE current density as a function of the applied electric field (*J*–*E*). The inset shows the variation in E_{on} and E_{th} observed from the GaN films. (b) Corresponding F-N plots of $ln(I/V^2)$ versus 1/*V*. Reprinted with permission from Ref. [66]. Copyright (2013), American Chemical Society

(see insert in Figure 7.10). It confirms that the electron transport does not originate from the interior of the connected nanocrystalline materials.

The second conduction mechanism arises from electron transport through the grain boundary. A recent study suggested that the grain boundary defects introduced localized defect states [79, 80]. The formation of these defect states is thus proposed to be responsible for the high conduction as well as for the emission performance. Herein, sample *a* has compared to sample *b* an increased nanocrystalline grain size and content, whereas the grain boundaries and

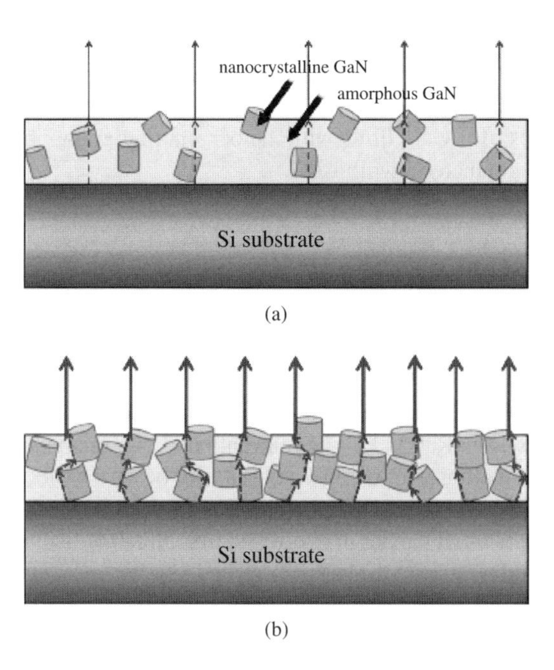

Figure 7.11 Models of electron emission from nanocrystalline GaN films with (a) small and (b) increased grain size and content. Reprinted with permission from Ref. [66]. Copyright (2013), American Chemical Society

corresponding defects are dramatically decreased. Hence, grain boundary conduction should be used to explain the origin of the FE phenomenon, as shown in Figure 7.11b. In addition, the grain size and content can result in a constructive or destructive superposition, thus leading to an additive or subtractive global effect, respectively. The FE properties of GaN can be further enhanced by the suppression of the grain size and improving the nanocrystalline content.

Good emission properties of a-GaN films containing nanoclusters were also observed in Refs. [81, 82]. The GaN thin film containing nanocrystalline grains showed a turn-on field of 5 V/μm and a maximum current density of 500 μA/cm^2 [81]. The emission I-V characteristic in F-N coordinates was nonlinear [81]. It was observed from experimental measurements that the current density J did not increase at low electric fields, but it sharply increased for $E > 20\,\text{V}/\mu\text{m}$. Based on this, the origin of the nonlinear F-N curves for a-GaN can be explained as follows. At the low electric field region, the band bending is not obvious and the electrons entering the conductive band of the semiconductor are mainly from the thermal electrons overcoming the potential barrier. The energy needed for the thermal electron to overcome the barrier is a certain value and so the current density J almost does not change with the change of E at the low field region. In addition, the number of electrons having enough energy to overcome the barrier is very small and so the current density is also very low. On the other hand, at the high field region, the band bending is large and the potential barrier is low enough for the electron to tunnel through. The higher the electric field, the larger is the band bending, the lower is the potential barrier, and hence, the stronger is the tunneling. The current density J increases sharply with the increase of the electric field E. Therefore, the current comes mainly from thermal emission at the low field region and tunneling emission at the high field region.

Moreover, the emission curves at the low field region have been approximately straight lines in the Schottky coordinates. This supports the conclusion that the current comes mainly from the thermal emission at the low field region. There are two possible reasons for the good field emission characteristics of *a*-GaN films with nanocrystalline grains: high tip density on the surface and wide bandgap material with LEA.

7.2.5 Graded Electron Affinity Electron Source

One effective method of injecting electrons into the LEA conduction band is to tailor the composition of a semiconductor such that the energy of the conduction band changes smoothly from the value close to the Fermi level at the backside contact to the value close to the vacuum level at the surface. The use of the LEA materials with field enhancement structures as electron emitter has been proposed in Ref. [11]. The relevant band diagram is shown schematically in Figure 7.12. The electron affinity at the surface needs not be zero or negative to produce emission when an electric field is present at the surface. The graded electronic structure changes the nature of the emission potential barrier in two ways: first, much of the potential energy change occurs inside the solid rather than at the surface; second, the energy change is distributed rather than abrupt. Depending on the specific implementation, this structure may have several benefits: (1) any changes in the surface properties of a graded composition emitter (due to adsorption and desorption) may result in relatively small changes in the *I-V* characteristic; (2) the extraction field needed for emission (for a given geometry) may be drastically reduced without the need of a (reactive) surface dipole; (3) the emission properties may be varied over different areas of the emitter surface so that the emission angle can be controlled; and (4) the emission current can be limited by appropriately adjusting the composition profile of the graded layer.

The progress in epitaxial growth technology allows the consideration of group III-nitride alloys as a potential material for the fabrication of the graded layer. All nitrides have been grown using techniques such as vapor-phase epitaxy (VPE) and molecular beam epitaxy (MBE) which lends themselves to controlled variation of the composition of compound

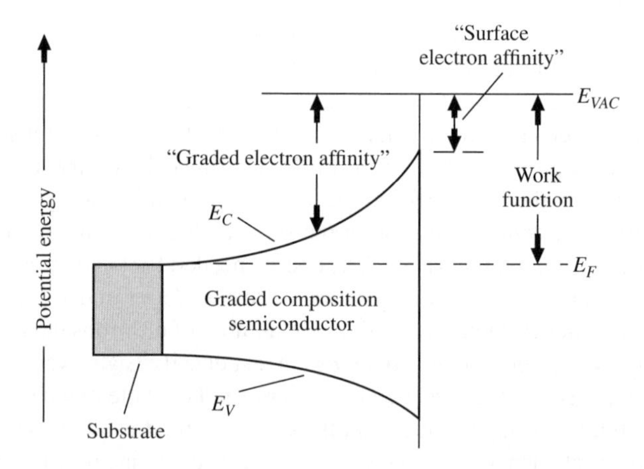

Figure 7.12 Energy band diagram of the emitter with a graded electron affinity semiconductor. Reproduced with permission from Ref. [11]. Copyright (1996), American Vacuum Society

semiconductors. The proposed application does not require any doping levels which remain a challenge for growers [11]. The graded composition layer need not be more than 10–100 nm thick, hence high growth rates are not required and significant strain may be tolerable. The large bond strength of the nitrides may also promote the formation of well-defined crystal planes, which might allow better shape control of the emitter structures.

To analyze the proposed structure, the Poisson equation for several cases in which the electron affinity has been assumed to be a linear function of position have been solved [11]. A linear relationship is not necessarily optimal, but it is simple to visualize and may be relatively simple to fabricate. In the calculations the authors of [11] used the following simplifying assumptions: the system is one dimensional, the energy distribution near the Fermi level associated with finite temperature is ignored ($T = 0$ K), there is no potential drop due to current flow (low or zero current density), and the density of states is described by an ideal parabolic band structure having a relative effective mass of $m^* = 0.2$. While several important aspects of behavior of the proposed system can be discovered based on this analysis, a two-dimensional solution including electron transport will be required to realistically simulate the emission characteristics.

It was proposed to realize a graded electron affinity electron source based on $Ga_xAl_{1-x}N$ alloys deposited on GaAs or GaN pyramids.

Depending on the field values at the surface and in the graded affinity layer, electron emission may take place either by tunneling through a relatively small potential barrier, or by classical thermal emission, or a combination of both. The properties of the cathode suggest that a low voltage emission could be obtained from relatively blunt emitter structures with little sensitivity to surface adsorbates or nanometer-scale geometric variations.

7.3 Resonant Tunneling of Field Emitted Electrons through Nanostructured Cathodes

7.3.1 Resonant-Tunneling $Al_xGa_{1-x}N$-GaN Structures

$Al_xGa_{1-x}N$-GaN emitter can be related to a resonant tunneling diode (RTD), in which the output solid-state barrier is replaced by vacuum. There has been a lot of theoretical and experimental investigations on RTD, and the theory has become quite sophisticated [83]. A matrix method is successfully used for the calculation of the transmission coefficient and the energy levels in RTDs [84]. A similar analysis is quite suitable for the description of resonant tunneling in field emission structures.

To describe the process at EFE from GaN cathodes with multilayer resonant tunneling structures it is possible to use a simple two-step mechanism [85]. In order to explain the change in the surface barrier, the FE process can be divided into two steps. The first step is the electron resonant tunneling into the surface layer, and the second step is an effective lowering of the surface barrier due to accumulated electrons.

But it is possible to combine the two FE processes into a single effect by a self-consistent quantum model [86]. For electron injection from only one boundary and electron movement in one direction (x) the formula for FE current density from a semiconductor film can be written as [87]

$$J = \frac{qm_t}{4\pi^2\hbar^3}k_BT\int T(E_x)[\ln(1+\exp(E_F - E_x)/k_BT)]dE_x = J_0\int J(E_x)dE_x = J_0J_T, \quad (7.6)$$

where

$$J_0 = \frac{qm}{4\pi^2\hbar^3}k_BT \tag{7.7}$$

is the electron supply, J_T is defined by the tunneling factor of the FE structure, q is the unit charge, m_t is the electron transverse mass, k_B is Boltzmann's constant, $E_x = P^2{}_x/2m$ is the normal energy, T is the temperature, h is the Planck's constant, and E_F is the Fermi energy. To compute the FE current, the transmission coefficient $T(E_x)$ is the most important parameter, which can be calculated by the quantum transfer matrix (TM) method based on analytical solution of Schrödinger equation with a linear potential, and the solution can be express as a linear combination of the Airy function or other wave functions. In this method, an arbitrary potential barrier can be divided into square segments which could be treated as linear barriers. Compared with conventional Wentzel-Kramers-Brillouin (WKB) method or LS self-consistent equation [43], the TM method is based on the accurate solution of the Schrödinger equation, so the results are much closer to the realistic tunneling process during field emission. To compute $T(E_x)$ by TM methods, the FE structure must be provided with the stable potential distribution. The invalidity of the previous two-step FE mechanism in multilayer semiconductor films makes it impossible to investigate the FE process by quantum TM methods. Using a self-consistent band bending model and integrating the two-step FE mechanism to the structural enhancement mechanism, the potential barrier of the FE structure will be stable when the external field is applied. Then it becomes simple to investigate the structure effect for controlled field emission from multilayer semiconductor films by the quantum TM method.

In order to compare with experimental results, the following GaN based emitter structure is used (Figure 7.13).

For two WBGSs layers the interface effect between the two layers must be also considered. As shown in Figure 7.13, when a high field is applied, the distribution of space charges can be described by Poisson's equation in the upper GaN layer,

$$\frac{d^2\phi_2(x)}{dx^2} = -\frac{q}{\varepsilon_2\varepsilon_0}\rho_2(x). \tag{7.8}$$

Here $\phi_2(x)$ and $\rho_2(x)$ give, respectively, the potential energy and the total volume charge density at distance x from the GaN-vacuum interface, and ε_0 and ε_2 are the vacuum permittivity and the dielectric constant of GaN, respectively. When a high field is applied, the distribution of space charge can also be described by Poisson's equation in the AlGaN layer

$$\frac{d^2\phi_1(x)}{dx^2} = -\frac{q}{\varepsilon_1\varepsilon_0}\rho_1(x). \tag{7.9}$$

Here $\phi_1(x)$ and $\rho_1(x)$ is the potential energy and the total volume charge density, respectively, at distance x from the AlGaN-vacuum interface, and ε_1 is the dielectric constant of AlGaN.

At the AlGaN-GaN interface, the potential distribution is the subject to the Gauss law and the following equation can be obtained

$$\varepsilon_1\frac{d^2\phi_1(x)}{dx^2} = \varepsilon_2\frac{d^2\phi_2(x)}{dx^2}. \tag{7.10}$$

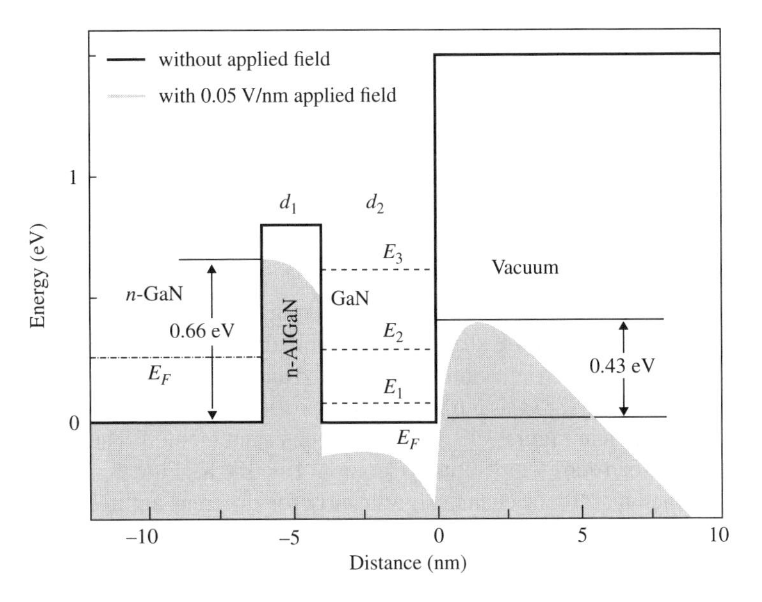

Figure 7.13 Energy band diagram of the n-GaN/$Al_{0.5}Ga_{0.5}N$-GaN/vacuum FE structure. Here n-GaN is the substrate, n-$Al_{0.5}Ga_{0.5}N$ is the first barrier with thickness $d_1 = 2\,\text{nm}$, GaN is the well with $d_2 = 4\,\text{nm}$. The vacuum barrier thickness is 10 nm at 0.05 V/μm. Reprinted with permission from Ref. [86]. Copyright (2005) by the American Physical Society

In the GaN layer, the potential distribution can be written as

$$\int_0^{-x_1} \frac{dx}{\delta} = \int_{\varphi_s}^{\varphi} \frac{d\varphi}{f(\varphi, \varphi_B)}, \tag{7.11}$$

where x_1 is the distance from the GaN-vacuum interface, δ is known as the Debye screening length, and the function $f(\varphi, \varphi_B)$ is [88]

$$f(\varphi, \varphi_B) = \{a[\exp(\varphi) - \exp(\varphi_B)] - b[\exp(-\varphi_B) - \exp(-\varphi)] - 2\sinh(\varphi_B)(\varphi - \varphi_B)\}^{1/2}, \tag{7.12}$$

where

$$a = \left(\frac{m_p^*}{m_n^*}\right)^{3/4} \exp\left(\frac{E_V - E_F}{kT_B}\right), \quad b = \left(\frac{m_n^*}{m_p^*}\right)^{3/4} \exp\left(-\frac{E_C - E_F}{kT_B}\right) \tag{7.13}$$

and φ_s and φ_B are the parameters correlation with the potential energy at the GaN-vacuum interface. $\varphi_S = (\phi_S - \phi_B)/kT$, ϕ_S is the band bending at the interface.

In the AlGaN layer potential distribution can also be written as

$$\int_{-d_{GaN}}^{-x_2} \frac{dx}{\delta} = \int_{\varphi_{s1}}^{\varphi} \frac{d\varphi}{f(\varphi, \varphi_G)}, \tag{7.14}$$

where x_2 is the distance from the AlGaN-GaN interface, d_{GaN} is the thickness of the GaN well, $\varphi_G = (E_F - E_{Fi})/(k_B T)$, E_F and E_{Fi} are Fermi and quasi-Fermi energies of AlGaN, respectively. $\varphi_{s1} = (\phi_{s1} - \phi_B)/(k_BT)$, and φ_{s1} is the band bending at the AlGaN-GaN interface ($x = -d_{GaN}$).

The electric field E_s at the AlGaN-GaN interface can also be obtained from

$$E_S = -\frac{d\phi}{qdx} = -\frac{k_BT}{q\delta}\frac{\delta d\varphi}{dx}\bigg|_{\varphi_S} = -\frac{k_BT}{q\delta}f(\varphi_S, \varphi_B). \tag{7.15}$$

In the present model it is supposed that the surface barrier bending by electron accumulating occurs directly when the field is applied to the FE structure. So the two WBGS layers can then be incorporated into a single layer by combining Equations (7.8)–(7.15), and the overall space charge distribution of the whole FE structure can be obtained.

To deal with the vacuum barrier, a more complicated and realistic image potential involving the image potential shifting [88] was introduced. The potential barrier of the whole FE structure was calculated, as shown in Figure 7.13, by the self-consistent method. The temperature effect could be neglected for room temperature operation T = 300 K. Figure 7.14 shows that the transmission coefficients do not change significantly for different applied voltages 2.2, 4.5, and 9 V. The position of the quantum level remains almost unchanged, and the tunneling factor of the FE structure J_T reaches its maximum.

There are already several experimental investigations and results for GaN based resonant tunneling field emission structures [85, 89, 90].

7.3.2 Multilayer Planar Nanostructured Solid-State Field-Controlled Emitter

The creation of the multilayer structure gives more degree of freedom for the control of the SSE [85], such as the bulk interfacial barrier in addition to the surface barrier and the presence of quantized subbands due to the confined thickness of the surface layer. Specific electron emission has been measured from these cathodes. A model based on a dual barrier mechanism for

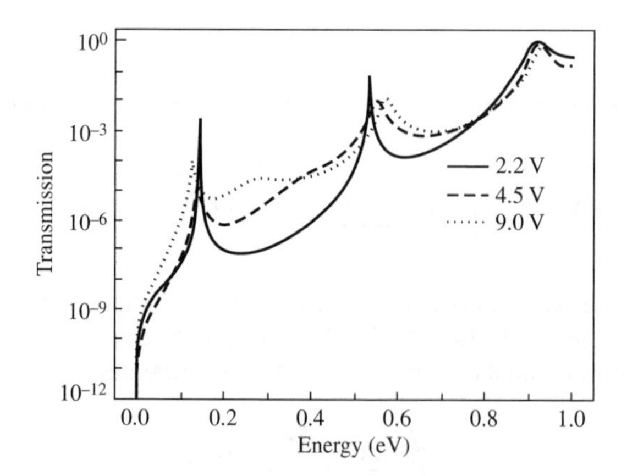

Figure 7.14 Resonant transmission in the n-GaN/$Ga_{0.5}Al_{0.5}N$/GaN/vacuum structure for different applied voltages. (Solid line: 2.2 V; dashed line: 4.5 V; and dotted line: 9 V.) The layer thicknesses are $d_{AlGaN} = 3$ nm and $d_{GaN} = 3$ nm, whereas the vacuum barrier thickness is $d_{vac} = 10$ nm. Reprinted with permission from Ref. [86]. Copyright (2005) by the American Physical Society

electron emission through the nanostructured layers has been proposed from the interpretation of the experimental data [85].

Cathode structure was deposited on n-SiC using low-pressure metalorganic chemical vapor deposition (CVD). Trimethylaluminum, trimethylgallium, silane, and NH_3 were used as the precursors. First, a 0.15 μm thick Si-doped AlGaN layer with Al-content graded from 40% to 15% was deposited on the SiC substrate. It served as the conducting buffer layer. This was followed by 0.25 μm n-GaN, 2 nm n-$Al_{0.5}Ga_{0.5}N$, and 4 nm undoped GaN. The Si-doping level in all the doped layers was about $2 \times 10^{18}\ cm^{-3}$, and the finished surface was characterized to be atomically smooth by atomic force microscope over the whole surface of the cathode having a total area of about $2\ cm^2$.

The band-edge diagram of this multilayer planar nanostructured cathode (Figure 7.15) has a confined layer of 4 nm limited by a first barrier $\Phi_{b1} = 0.8$ eV at the interface of GaN with $Al_{0.5}Ga_{0.5}N$ and a second barrier $\Phi_{b2} = 1.5$ eV at the cathode surface of GaN with vacuum.

The electron emission *I-V* measurements were performed with a piezo-driven scanning anode field emission microscope (SAFEM) [44]. The whole set of data *(I-V)* obtained for different values of *z* was analyzed using electron optics numerical simulations [44] in order to have, for each location at the surface cathode and a given temperature, a unique *J-F* plot. *J* is then the measured emission current density and *F* is the related electric field at the surface of the planar cathode. The measured *J-F* plots were compared directly with theoretical values obtained with different analytical approaches. The measurements were done for different locations across the cathode and for different temperatures. Emission characteristics presented in the following were common to all these measurements and could be considered as specific to this nanostructured SSE.

1. For a given temperature, the *J-F* characteristics presented the following three main characteristics: (i) the threshold field for $J_{mes} = 10^{-4} A/cm^2$ was in the range of 5×10^5–$1 \times 10^6 V/cm$ at room temperature. This value is about 100 to 50 times less than the usual field needed for field emission from a metallic surface with a work function of about 4 eV. (ii) The $ln(J/F^2)$ vs. $(1/F)$ plots of the data were straight lines, that is, the increase of the FE current could be described on the basis of electron tunneling through a barrier deformed by high external electric field known as the Fowler-Nordheim theory [87, 91]. The calculated experimental values of the tunneling barrier height Φ_{FN} were in the range of 0.25–0.5 eV for different locations. There was no dependence of Φ_{FN} on temperature.

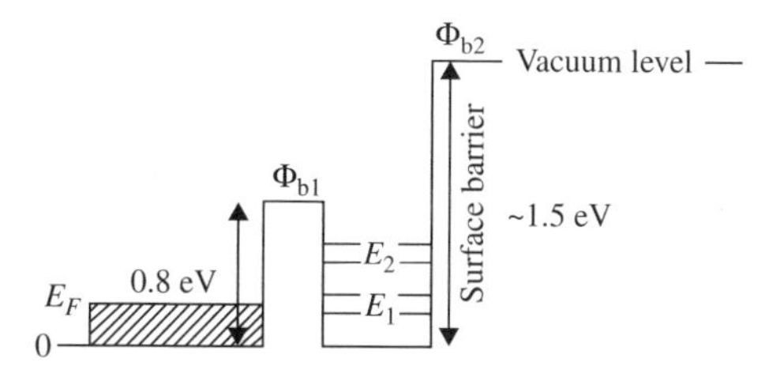

Figure 7.15 Nanostructured layers of the cold cathode. Band-edge diagram in the absence of an external electric field ($V_{app} = 0$): Φ_{b1} and Φ_{b2} are, respectively, the first and second barriers and E_1, E_2 are the energy levels of subbands inside the quantum well. Reproduced with permission from Ref. [85]. Copyright (2004), AIP Publishing LLC

(iii) Sharp increase of the emitted current was obtained in the high field region and for $J > 3 \times 10^{-2} \mathrm{A/cm^2}$. This sharp increase resulted in most cases in a destruction of the surface cathode due to an arc formation between the cathode and the anode.

2. Emission currents were very unstable at the beginning of the emission process. These instabilities slowly decreased after a conditioning period of more than 1/2 hour of continuous emission of the cathode at high currents and the emission currents became very stable. These instabilities were obtained for all cathode temperatures up to 500 K. It seemed that they were intrinsic to the emission mechanism and were not the result of a surface absorption-desorption process that would modify the emission.
3. For a given emission area, the emission currents increased with temperature from 300 to 500 K. As this behavior cannot be explained by the F-N theory, the Richardson-Dushman approach [92] has been used to determine the effective thermal activation energy related to these data. The calculated values from the experimental data were in the range of 0.8–0.9 eV and they were not field-dependent.

The following two characteristics of the field emission from the nanostructured SSE cathode surface are important for the analysis: (1) The values of the tunneling barrier height Φ_{FN} have been in the range of 0.25–0.5 eV, this value has to be compared with the electron affinity of GaN $\chi = 1.5\,\mathrm{eV}$. (2) The value Φ_{FN} is much smaller than the thermal activation energy Q (0.8–0.9 eV) obtained from the temperature dependence emission. This means that the effective tunneling barrier is not the thermionic barrier at the surface. Furthermore, Q is very close to the barrier height at the buffer layer/$Al_{0.5}Ga_{0.5}N$ interface.

This specific electron emission from the nanostructured SSE was modeled based on Figure 7.15. In this model the electron emission is obtained through a serial two-step mechanism under the applied field [43]. In the first step the electrons are injected in the GaN layer from the cathode substrate by tunneling through 2 nm $Al_{0.5}Ga_{0.5}N$ layer. They will occupy the subbands that are under the Fermi level, creating a concentration of electrons inside the GaN layer. Due to this electron concentration or space charge formation, there is an upward energy shift, which is schematically represented in Figure 7.16 with the modification from (b) to (d); leading to a relative lowering of the vacuum level compared with the Fermi level of the substrate.

In this model two concomitant mechanisms for the electron emission are described hereafter in (A) and (B):

A. The first mechanism is the tunneling field emission through a lowering work function, that is, the electrons are emitted by a field emission mechanism from the quantized subbands inside the GaN quantum well.

Figure 7.15 shows two quantum well (QW) states E_1 and E_2 at $V_{bias} = 0$. In the conventional resonant tunneling approach [87, 93, 94] with the application of a bias at $V = V_1$, the state E_1 is aligned with the electron energy from the contact layer resulting in the resonant tunneling current J_{RT1} (Figure 7.16). There is also a small lowering of the effective work function. However, this lowering is not enough to allow electron emission by tunneling for fields in the range of 10^5–10^6 V/cm [94] and cannot be the reason for the very low measured tunneling barrier Φ_{FN}.

In the proposed two-step tunneling model the larger lowering of the work function due to space charge in the QW is essential. When the two-dimensional (2D) quantum state is occupied, the space charge in the QW leads to an additional lowering of the effective

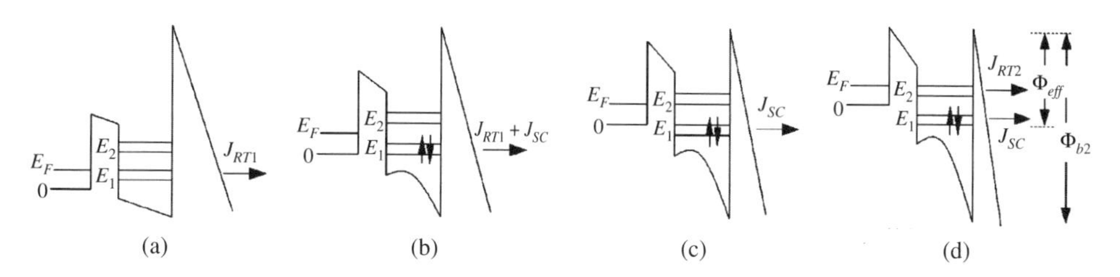

Figure 7.16 Illustration of the different field emission mechanisms by schematic band-edge diagrams of the nanostructured SSE planar cathode with an applied field F and at room temperature: (a) only resonant tunneling mechanism and (b–d) modification with space charge formation inside the GaN layer with, as a consequence, an effective lowering of the surface barrier. In addition to the resonant tunneling, electrons can occupy the quantum state E_1, whenever the level E_1 moves below 0. These electrons can tunnel through the single barrier via the usual FN tunneling, resulting in J_{sc} (c). The total current is $J_{FN} = J_{RT} + J_{sc}$. Reproduced with permission from Ref. [85]. Copyright (2004), AIP Publishing LLC

work function defined by the electron at source and vacuum level. A precise quantitative approach at bias voltage requires the use of the Airy function with self-consistent calculations [43, 48]. Here, a simpler approach was used to estimate the lowering of the work function. After calculating the potential bending due to the space charge inside the 2D quantum well, the usual field emission expression derived from the F-N theory was fitted to the experimentally measured field emission.

The QW with a width w and a perfect confinement has the charge density of $n = q(m^*/\pi\hbar^2) \times (E_2 - E_1)(2/w)sin^2(\pi x/w)$ in the lowest level E_l, where m^* is the effective mass of the electron with charge q. Solving the Poisson's equation the potential energy of the space charge region V_{SC} was obtained

$$V_{SC} = \left(\frac{\rho_0}{4\varepsilon}\right)\left[\left(\frac{w}{\pi}\right)^2 \sin^2\left(\frac{\pi x}{w}\right) + wx - x^2\right] \tag{7.16}$$

for $0 \le x \le w$, where $\rho_0 = q(m^*/\pi\hbar^2) \times (E_2 - E_1)(2/w)$ and ε is the dielectric constant.

The maximum value of V_{SC} is at $x = w/2$, $V_{SC}(w/2) = (0.25\rho_0/\varepsilon)w^2 \times (\pi^{-2} + 0.25)$, and the average of V_{SC} is $V_{SC}(av) = 0.62 \times V_{SC}(w/2)$. Taking the average of the potential at both interfaces $[V(w) - V(0)]$, the total lowering of the work function $\Delta\Phi = \Phi_{b2} - \Phi_{eff}$ (see Figure 7.16d) is

$$\Delta\Phi = V_{SC}(av) + 0.5[V(w) - V(0)] + E_1. \tag{7.17}$$

The effective barrier Φ_{eff} is the actual barrier at the surface after the lowering and it can be determined experimentally from the J-F plots as $\Phi_{eff} = \Phi_{FN}$.

Theoretical estimation of $\Delta\Phi$ was 1.05 eV, with $m^* = 0.22m_0$, $\varepsilon = 8\varepsilon_0$ for GaN, $V_{SC}(w/2) = 0.4$ eV, $0.5 \times [V(w) - V(0)] = 0.62$ eV, and $E_1 = 0.18$ eV. The calculated effective barrier of $\Phi_{eff} = 0.45$ eV is very close to the experimental values Φ_{FN} obtained from the J-F plots, which have been in the range of 0.25–0.53 eV.

Therefore, it was concluded that after the occupation of the quantum level E_1 lying below $E = 0$, the effective tunneling current $J_{FN} = J_{RT} + J_{SC}$ was given by the conventional F-N tunneling through the vacuum barrier, with an effective barrier of only a few tenths of the electron volt (Figure 7.16d). This barrier lowering at the surface controlled the variation of the emitted current J_{FN} with the field.

B. The second mechanism occurs for higher temperatures, that is, $k_BT > 0.8$ eV, when hot electrons can overcome the first barrier located between the conductive substrate and the ultrathin $Al_{0.5}Ga_{0.5}N$ layer. As the second barrier at the surface is lower (less than 0.5 eV due to the space charge), these electrons can emit directly as J_{TH}. The first barrier controls the variation of the emitted current J_{TH} with temperature.

In this dual-barrier model, the measured total emission current J is the sum of both contributions, $J = J_{FN} + J_{TH}$. Numerical simulations have confirmed that an accurate calculation of Φ_{FN} and Q is possible using the plots $ln(J/F^2)$ versus $(1/F)$ and $ln(J/T^2)$ versus $(1/T)$, respectively.

A nanostructured SSE planar cathode gives the possibility to control the effective surface barrier for electron emission by changing the space charge value of an ultrathin layer at the surface. The presence of the first barrier separates the thermionic process from the field emission process, where the layer is controlled by the second barrier at the surface.

7.3.3 Geometric Nanostructured AlGaN/GaN Quantum Emitter

Geometric quantum emitter (GQE) based on a field emission array (FEA) was proposed in Ref. [89]. These quasiperiodical arrays of nanocones were fabricated by one-step nonlithographic method and PLD. The GQEs are more robust than 1D nanostructures and functional under high current density conditions. They are promising candidates for high-performance electron FE.

As shown in Figure 7.17, the ideal GQE structure consists of a conducting buffer layer deposited on a Si substrate together with quantum barrier and quantum well layer. Electrons are confined in the thin surface quantum well layer between the vacuum barrier and quantum barrier layer.

Potential well significantly reduces the tunneling barrier down to 0.2–0.5 eV via resonant energy levels [85, 95] and significantly increases the tunneling probability. In addition, the

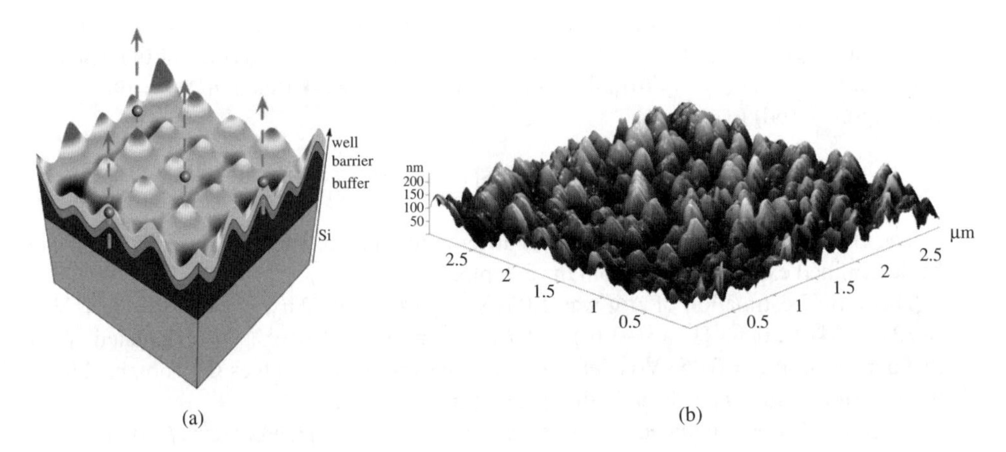

Figure 7.17 Schematic diagram of (a) the GQE structure and (b) the corresponding AFM image of the GaN buffer layer. Reproduced with permission from Ref. [89]. Copyright (2011), AIP Publishing LLC

surface morphology and area can be adjusted by fabricating periodic FEAs. Electrons can be extracted using a smaller field because the protrusions provide some field enhancement. When combining these two effects into a single structure, the original effects are supposed to be magnified exponentially. The overall effects can be estimated by the Fowler-Nordheim theory [96]

$$J = \left(\frac{A\beta^2 F^2}{\Phi}\right) \exp\left(-B\frac{\Phi^{3/2}}{\beta F}\right), \tag{7.18}$$

where J is the current density in A/cm^2, F is the field strength in V/cm, Φ is the surface effective barrier in electron volt, β is the field enhancement factor, and A and B are constants corresponding to 1.54×10^{-6}A (eV) V^{-2} and 6.83×10^7 (eV)$^{-3}$/2V cm^{-1}, respectively.

For comparison, carbon nanotube (CNT) emitters have $\beta = 2400$ and $\Phi = 5.0$ eV [97], conventional multilayer quantum well emitters (MQEs) have $\beta = 10$ and $\Phi = 0.5$ eV [85], GaN coated FEAs have $\beta = 300$ and $\Phi = 3.2$ eV [98], where the GQEs combine the benefits of last two and have $\beta = 300$ and $\Phi = 0.5$ eV.

Although the GaN coated FEAs and MQEs individually do not provide significant efficiency improvement in the FE current, integrating the two devices into a single structure leads to significant enhancement. Moreover, in the low-field region, the FE properties of the GQEs are even better than those of CNTs [89]. As a result, this technique can be applied to ultralow threshold FE applications.

To verify the concept of GQE, the geometric field-enhanced multilayered quantum well structures were fabricated by PLD [54]. A 100 nm thick GaN layer serving as the conducting buffer layer was deposited on (100) n-type Si substrate. Afterwards the GaN conducting buffer layer was coated with a 6 nm thick $Al_{0.45}Ga_{0.55}N$ quantum barrier layer followed by a GaN quantum well layer with varying thicknesses of about 4, 8, 12, 16, and 20 nm (designated as sample no. 1–5, respectively). The AFM image in Figure 7.17b shows that the GaN buffer layer contains a large quantity of quasiperiodical nanocones (10^8–10^9 cm^{-2}) with a diameter of about 200–250 nm. The corresponding root-mean-square surface roughness values of samples 1–5 were 34.5, 30.2, 29.3, 28.4, and 30.9 nm, respectively. These values are larger than those obtained from a roughened surface by post plasma treatment [69, 99]. The results suggest that the geometric field enhancement factor is considerably larger than that of a flat FE cathode [63, 85].

X-ray photoelectron spectroscopy (XPS) was used to determine the composition of the samples. Because of exposure to air, the oxygen O 1s and C 1s core level peaks were always detected besides the Ga 3d and N 1s peaks. The O 1s peak at 531.3 eV with a FWHM of about 2.3 eV is attributed to chemisorbed oxygen [100]. The Ga 3d spectrum can be fitted well at 19.8 eV with the FWHM of about 1.56 eV and can be assigned to nitride [101]. The O 1s and Ga 3d results confirm that the samples are gallium nitride and not oxide.

In order to evaluate the performance of the geometric field-enhanced multilayered quantum well emitter, the FE properties were determined at room temperature using a parallel-plate diode configuration at a pressure of 5×10^{-7}Pa or lower. The substrate to anode distance was 14 μm. In order to remove adsorbates from the cathode surface, the measurements were repeated for several cycles until the J-E characteristics became stable and reproducible. Figure 7.18a compares the FE properties of the GQEs with different quantum well thicknesses. Except for the sample 1, all the samples exhibit clear FE behavior. The turn-on electric field E_{on} which is defined as the field required to obtain current density of 0.1 μA/cm^2 is

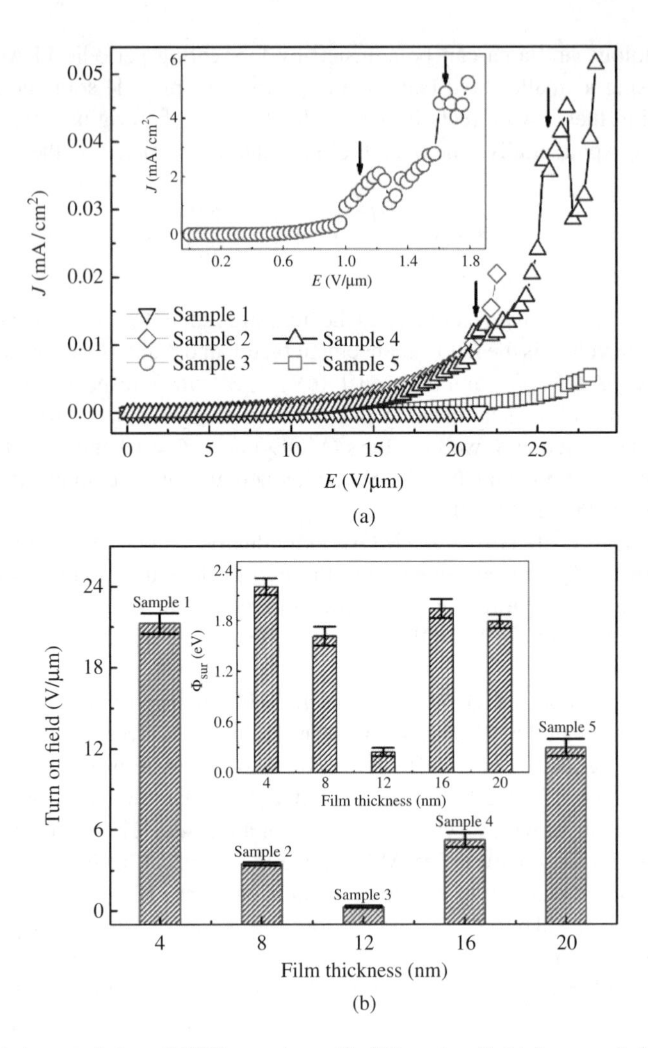

Figure 7.18 FE characteristics of GQE samples with different well thicknesses 1–5: (a) FE current density as a function of the applied electric field (*J–E*) and (b) variation of E_{on}. The inset in (b) shows the variation of Φ_{sur}. Reproduced with permission from Ref. [89]. Copyright (2011), AIP Publishing LLC

shown in Figure 7.18b. Sample 3 exhibits the best FE characteristics with the lowest E_{on} of 0.4 V/cm. Moreover, when the current density reaches 1 mA/cm^2, which is generally required in practice, the electric field is still low with 1.1 V/μm. These values are found to be significantly lower than those reported for GaN coated FEAs and MQEs and even comparable to those of 1D nanostructured CNT emitters. In addition, resonant peaks were observed from samples 3 and 4.

To investigate the electron emission enhancement mechanism, the surface barrier height Φ_{sur} of the samples should be considered. Φ_{sur} can be estimated from the slope of the linearly fitted FN curve [54]. Here, since the surface roughness is quite similar, the impact of the surface morphology should be minimal. Using the reported β value of 150 determined from a roughened GaN film surface (surface roughness was about 17.9 nm) [69], β was estimated to be 300. As shown in the insert in Figure 7.18b, the Φ_{sur} values of samples 1–5 are 2.2, 1.6, 0.3, 1.9, and 1.8 eV respectively. The extraordinary low value observed from sample 3 combined with the resonant peak may be due to two reasons. First of all, the surface electron affinity decreases significantly. In other words, when the emitter surface is coated or chemisorbed with ultralow electron affinity materials, for example, diamond, AlN, Φ_{sur} is dramatically reduced. In some cases, the resonant peak arises from the formation of a quantum confined structure [102, 103]. Second, resonant tunneling [104, 105] occurs through localized subbands in the quantum well structure and the accumulated electrons lower the effective surface barrier.

In order to confirm that the different Φ_{sur} values originate from the quantum well layers rather than surface electron affinity variation, ultraviolet photoemission spectroscopy (UPS) was performed. The electron affinity was evaluated by the procedures described in Ref. [106]. The electron affinity of all the samples is about 3.2 eV which is similar to the reported values of 2.7–3.3 eV [78]. The band diagram of the GQEs can be estimated from the UPS results. Electrons are confined in the quantum well structure forming resonant energy levels. When an electric field is applied, electrons are injected into the quantum well to occupy the subbands causing band bending and lowering the effective surface barrier. Hence, the decrease in Φ_{sur} as well as resonant peak can be attributed to the quantum well structure. By coupling with the geometric field enhancement effects, the current density increases in an exponential fashion while E_{on} can be reduced to few V/μm which is even lower than for some 1D nanostructures.

The FE properties can be further improved by post treatment to sharpen the tips and roughen the surface.

7.3.4 AlN/GaN Multiple-Barrier Resonant-Tunneling Electron Emitter

Nitride semiconductors and quantum wells have unique benefits for new electron emitters. InGaN/GaN and GaN/AlN heterostructure field emitters have shown a strong decrease of the electron affinity on the cesiated surface, owing to the quantum-size effect and strong polarization at the nitride surface [55, 85]. NEA has been reported for cesiated AlN and effective photoelectron emission has also been reported for *p*-type GaN [107, 108]. Cesiated *p*-type GaN has also an effective NEA and a photoemission efficiency of around 20% has been reported [107, 108]. Figure 7.19 shows the band structure of a nitride field emitter [90]. A GaN/AlGaN/GaN or InGaN/GaN quantum-well structure grown in the +*C* direction has a strong polarization field at the surface as shown in Figure 7.19, and the electron affinity is effectively decreased by the quantum-size effect in the triangular surface potential. Thus, these structures are useful as field emitters with small electron affinity.

An AlN/GaN multiple-barrier resonant-tunneling electron emitter has been proposed, utilizing the strong polarization field in an AlN/GaN heterojunction to enable both ballistic electron emission and electron injection into surface accelerating layer through a resonant-tunneling multiple quantum well (MQW). In the AlN/GaN quantum well system, AlN and GaN layers have huge polarization fields due to the spontaneous and piezoelectric polarizations. The polarization field can be effectively used to inject conduction electrons into higher levels. The mid- and near-infrared quantum-cascade structures, where electrons are injected to higher subbands through a few atomic layers of AlN, or through simple resonant tunneling in AlN/GaN layers have been also proposed [109, 110]. The resonant-tunneling voltage of the AlN/GaN MQW is governed mainly by the polarization fields in the quantum

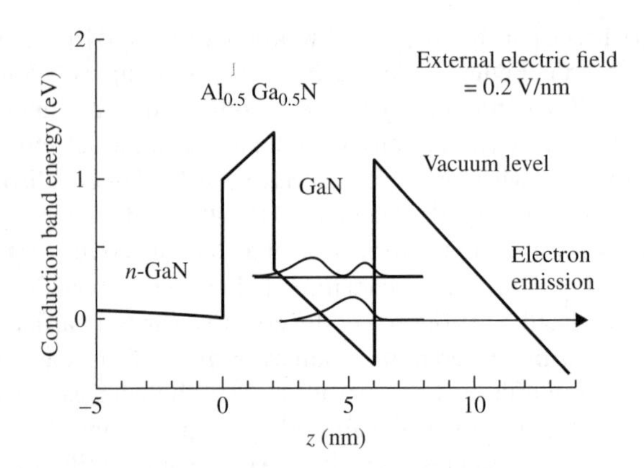

Figure 7.19 Energy band structures of n-GaN/AlGaN/GaN quantum well field emitter. Reproduced with permission from Ref. [90]. Copyright (2005), AIP Publishing LLC

well, and possesses a large value [110, 111]. The large tunneling voltage would be also useful for application in electron emitters because the electric field just beneath the semiconductor surface is controlled by the polarization field, as it is shown in Figure 7.19.

Figure 7.20 shows the schematic device structure of the AlN/GaN resonant-tunneling electron emitter. The band structures of the emitter with and without bias voltage are shown in Figure 7.21. The emitter is grown in the $+C$ direction, and the device is composed of an $(AlN)_{n1}/(GaN)_{n2}$ ($n_1 \sim 5, n_2 \sim 20$ in atomic layers) MQW resonant-tunneling layer on n-type GaN electron-source layer, with n^--GaN accelerating layer with thickness of 50 nm at the emission window. The GaN accelerating layer below the surface electrode ($\sim$100 nm) is thicker than that at the emission window. The carrier concentration at the n^--GaN accelerating layer is less than $1 \times 10^{18} cm^{-3}$, while the resistivity of the surface of the emitter window is kept low, either by highly doping the GaN surface or by thin deposition of metal. Since the AlN layer in the MQW is very thin, the AlN is completely strained and the lattice constant in the

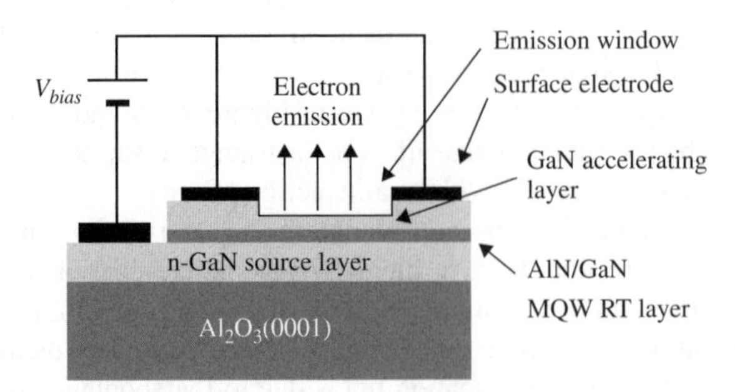

Figure 7.20 Schematic device structure of an AlN/GaN resonant-tunneling electron emitter. The AlN/GaN resonant-tunneling layer is composed of an $(AlN)_3/[(GaN)_{20}/(AlN)_5]_4$ structure, with the thicknesses of the surface GaN accelerating layer being 50 nm at the emission window and 100 nm below the electrode. (The index gives the number of monolayer.) Reproduced with permission from Ref. [90]. Copyright (2005), AIP Publishing LLC

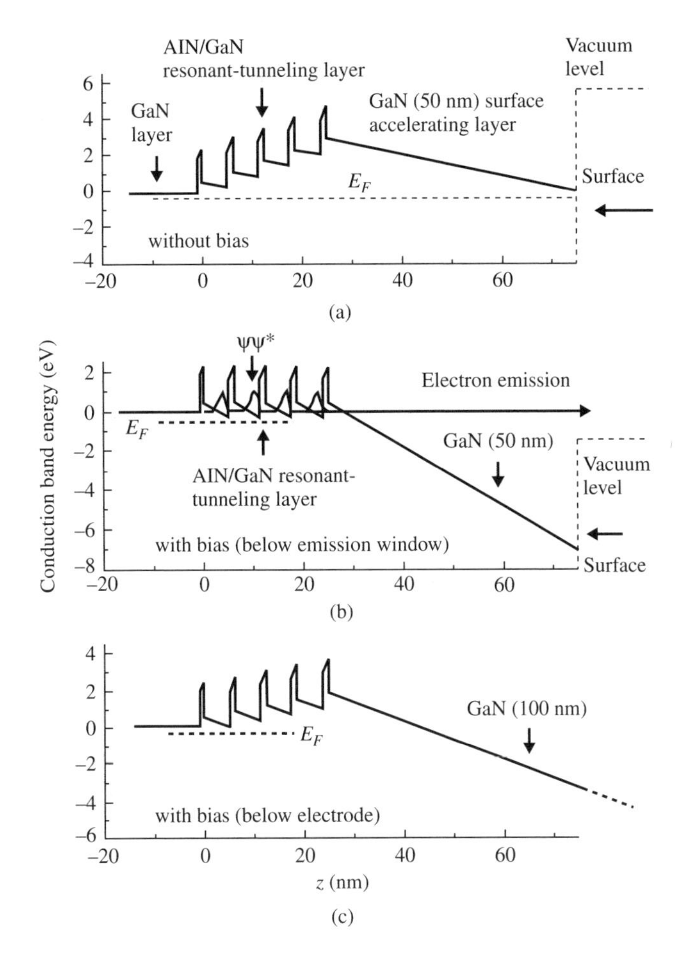

Figure 7.21 Conduction band diagram of resonant-tunneling electron emitter (a) without bias, (b) with bias at the emission window, and (c) with bias below the contact electrode. Reproduced with permission from Ref. [90]. Copyright (2005), AIP Publishing LLC

layer plane matches that of the GaN. As a result, the layer exhibits a high piezoelectric field E_z of 0.56 V/nm, as given by the equation [59]

$$E_z = -\frac{p_z}{\varepsilon} = -[e_{31}(e_{xx} + e_{yy}) + e_{33}e_{zz}]/\varepsilon = 2(e_{33}C_{13}/C_{33} - e_{31})e_{||}/\varepsilon, \tag{7.19}$$

where p_z is the dipole moment due to the piezo effect, ε is the dielectric constant of the material, e_{31} and e_{33} are piezo coefficients, and C_{13} and C_{33} are the elastic stiffness constants of the material. AlN also exhibits high spontaneous polarization due to the deviation from a regular tetrahedral shape. The polarization field is estimated as high as 0.24 V/nm from consideration of the deviation from the regular tetrahedral shape. Thus, total polarization fields E_p in the AlN layer reaches a level of 0.8 V/nm, and the conduction-band edge of the AlN rises to the surface (+C direction), as shown in Figure 7.21a. On the other hand, electron transfer occurs from the surface of GaN accelerating layer to the interface between the MQW and the GaN electron-source layer, so that the Fermi level becomes constant in all regions, and the electric fields in the AlN layer and GaN accelerating layer may be calculated as follows:

$$E_{AlN} = E_p - \frac{\sigma_c}{\varepsilon_{AlN}}, \tag{7.20}$$

$$E_{GaN} = -\frac{\sigma_c}{\varepsilon_{GaN}}, \tag{7.21}$$

where σ_c is the conduction charge density transferred from the interface of the GaN source layer and the MQW layer to the surface of the GaN accelerating layer, and is given by

$$\sigma_c = E_p L_{AlN}/(L_{AlN}/\varepsilon_{AlN} + L_{GaN}/\varepsilon_{GaN}), \tag{7.22}$$

where L_{AlN} and L_{GaN} are the thicknesses of total AlN layer in the MQW layer and the total GaN layer in the MQW and GaN accelerating layer, respectively. The AlN/GaN MQW and GaN accelerating layers are thus depleted, as shown in Figure 7.21a. When an external voltage is applied to the surface electrode the charge density σ_c increases, and electric field at the GaN accelerating layer increases as shown in Figure 7.21b. The resonant condition is simply calculated as follows:

$$E_p W_{AlN} = \sigma_c'[(W_{AlN}/\varepsilon_{AlN}) + (W_{GaN}/\varepsilon_{GaN})], \tag{7.23}$$

where σ_c' is the charge at the surface of the GaN accelerating layer after bias and W_{AlN} and W_{GaN} are total thicknesses of the AlN and the GaN layers in the MQW, respectively. The bias voltage required for resonant-tunneling is given by

$$V_{bias} = E_p\{W_{AlN}[(L_{AlN}/\varepsilon_{AlN}) + (L_{GaN}/\varepsilon_{GaN})]/[(W_{AlN}/\varepsilon_{AlN}) + (W_{GaN}/\varepsilon_{GaN})] - L_{AlN}\}. \tag{7.24}$$

The bias voltage V_{bias} needed to satisfy the resonant-tunneling condition is calculated to be as high as 7 V with these thickness conditions. Thus, accelerated electrons possessing energy higher than the vacuum level are emitted from the emission window. In the $(AlN)_5/(GaN)_{20}$ MQW resonant-tunneling layer, the quantum level is calculated as 0.33 eV from the right bottom of the GaN well by the envelope-function approximation taking into account the polarization fields [112]. As a result, the thickness of the first AlN layer in the MQW has been designed to be three atomic layers in thickness so that the conduction-band edge of GaN source layer has almost the same energy as the quantum level in the MQW layer in resonant-tunneling condition. The tunneling current has been estimated to be as high as 3 kA/cm^2 by WKB approximation if it is assumed that the carrier concentration at the interface between GaN electron-source and MQW layers is 1×10^{18} cm^{-3}.

In a practical device, the loss current or current flow at the surface electrode should be sufficiently low, and the electron affinity of the GaN surface layer should be also lower than the bandgap of the accelerating layer. If the kinetic energy of the conduction electron is higher than the bandgap, impact ionization becomes significant and electron-emission efficiency decreases. Thus, a surface treatment would be necessary to decrease the electron affinity of the GaN surface. The former problem could be overcome by using a thicker GaN surface layer below the electrode than that at the emission window, as shown in Figure 7.20. Resonant-tunneling voltage given by Equation (7.24) depends on the thickness of the GaN accelerating layer, and the resonant-tunneling voltage below the electrode becomes greater than that at the emission window. Thus, selective electron emission is possible merely by varying the thickness of the GaN accelerating layer, as shown in Figure 7.21b,c. Efficiency of electron emission depends on the high-field relaxation time of the conduction electrons. Efficiency comparable to or greater than a *p*-GaN photoelectron emitter is expected for the well-engineered resonant-tunneling device, considering the difference in the operation mechanism between the resonant-tunneling emitter and photoelectron emitter.

7.4 Field Emission from GaN Nanorods and Nanowires

Currently one-dimensional and three-dimensional nanoscale materials are under focus due to their unique electronic, optical, and magnetic properties as well as potential applications in constructing novel nanodevices [113, 114].

Gallium nitride (GaN), as a wide-bandgap semiconductor, has attracted a lot of attention as a material for EFE devices [115]. It has strong chemical and mechanical stability, high melting point (2600 K), and LEA of 2.7–3.3 eV [11, 69, 116–118]. Moreover, FE cathodes based on GaN have longer lifetimes than based on Si or other conventional semiconductors.

There are many researches on FE from GaN films [69, 31] and pyramid arrays [26]. It was reported that the pyramid or rough-surface GaN increased the field enhancement factor (β), lowering the applied voltage for the electron emission. Thus, further enhancement of EFE characteristics is expected by the one-dimensional (1D) GaN structures. Various GaN nanowires and nanorods were fabricated and their field emission properties were investigated [75, 78, 119–121]. In recent years many attempts have been made to synthesize one-dimensional and three-dimensional GaN morphologies. Up to now, one-dimensional and quasi-one-dimensional GaN nanorods [114], nanowires [122], nanotubes [123], nanoribbons [124], nanobelts [125], and prismatic rods and cone nanowires [120] have been synthesized. There are also some reports about three-dimensional GaN nanostructures [126–128].

Such different GaN morphologies were produced by numerous methods such as sol–gel method, simple thermal evaporation method [71, 75], templates method, pulse laser ablation (PLA) [78, 129], and CVD method [71, 74, 121–131]. The 1D GaN structures were obtained either by nanowire (NW) growth or GaN etching, where etching is widely used for GaN surface roughening and nanotip formation [69, 132, 133].

7.4.1 Intervalley Carrier Redistribution at EFE from Nanostructured Semiconductors

EFE offers a new possibility for the study of bandstructure parameters, such as intervalley energy difference, electron affinity from different conduction band valleys, and intervalley free electron redistribution due to heating of carriers by external high electrical fields. The conelike shape and small size of the field emitter cathodes allow reduction in lattice overheating, which is a serious problem for devices operating at high electric fields [67, 134–143]. Important peculiarities of the WBGSs are their large intervalley distance (1–2 eV or higher), as well as charge trapping and piezoelectric phenomena [134, 135, 144, 145]. The conduction band nonparabolicity can also play a significant role leading to the decrease in the electron mobility. Under such circumstances, the drift mobility at high electrical fields decreases even without any carrier redistribution taking place from the central to the upper (satellite) valley [137]. At extremely high electrical fields, additional field emission mechanisms can appear, namely, the emission from charged traps in the wide forbidden gap, localized on the surface or in the bulk, and emission from the valence band. In the case of WBGSs, these effects can be substantially suppressed.

Several EFE measurements and simulations for quantum sized cathodes have shown possible transformation of the energy bands [67, 134–142]. The quantum-size effect decreases considerably the intervalley energy difference for cathodes with tip radii of few nanometers [134, 140–142]. The first theory of field emission from semiconductors was reported in Refs. [146, 147]. It was based on degenerated carrier statistics and took into account different electron masses in semiconductor and vacuum.

Additional important features, such as the complex many-valley structure of the Brillouin zone, hot carrier generation, and intervalley carrier redistribution have been considered in Ref. [148]. Furthermore, the quantum mechanical tunneling process from the backside metal-semiconductor junction is an important mechanism for current transport through thin barriers. The top of the triangular barrier used in the proposed model [148] of the semiconductor-vacuum interface was additionally lowered by the image force potential.

Tunneling probability of electrons from a semiconductor into vacuum has been derived from the time independent Schrödinger equation:

$$-\frac{\hbar^2}{2m_e}\cdot\frac{d^2\Psi}{dx^2}+V(x)\Psi=E\Psi, \tag{7.25}$$

which can be rewritten as

$$\frac{d^2\Psi}{dx^2}=\frac{2m_e(V-E)}{\hbar^2}\Psi, \tag{7.26}$$

where $\hbar$ is the normalized Planck constant, Ψ is the wave function, m_e is the effective electron mass, $V(x)$ is the potential relief on the electron path, E is the energy of electron, and x is the tunneling direction from the surface into vacuum (see inset in Figure 7.22).

Assuming that $V(x)-E$ is independent of position in a section between x and $x+dx$ (WKB approximation), this equation can be solved yielding

$$\Psi(x+dx)=\Psi(x)\cdot e^{-kdx} \tag{7.27}$$

with $k=\dfrac{\sqrt{2m_e[V(x)-E]}}{\hbar}$.

The minus sign is chosen since it is assumed that the particles move from the left to the right. For a slowly varying potential, the amplitude of the wave function at $x=L$ can be related to the wave function at $x=0$ through the following equation [149]:

$$\Psi(L)=\Psi(0)\exp\left\{-\int_0^L\frac{\sqrt{2m_e[V(x)-E]}}{\hbar}dx\right\}, \tag{7.28}$$

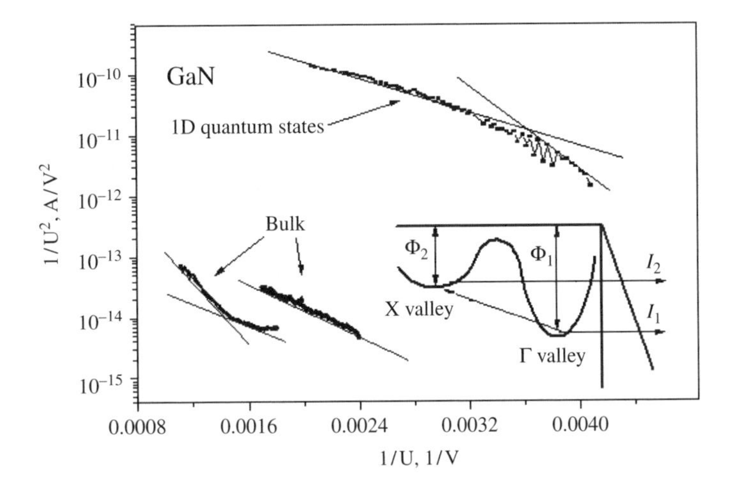

Figure 7.22 Experimental *I-V* characteristics of the field emission current from GaN in F-N coordinates. In the case of bulk material, the difference in two curves is caused by different micrometer radii of curvature for the emitter. The inset shows the schematic energy band diagram of a two-valley semiconductor. Reproduced with permission from Ref. [148]. Copyright (2009), AIP Publishing LLC

where L is the classical turning point (L is the external barrier border that is the last point where an electron is inside of the barrier potential). Equation (7.28) is also referred to WKB approximation.

In general, a triangular barrier is assumed to approximate the semiconductor-vacuum interface. The tunneling probability (Θ) can be calculated in the case of the triangular barrier approximation by the following equation [143]:

$$\Theta = \frac{\Psi(L)\Psi^*(L)}{\Psi(0)\Psi^*(0)} = \exp\left\{ -\int_0^L \frac{\sqrt{2m_e}}{\hbar} \sqrt{q\chi_0\left(1 - \frac{x}{L}\right)}dx \right\}, \quad (7.29)$$

where χ_0 is the barrier height at the cathode-vacuum interface.

The parameter Θ can be evaluated as a function of the electric field F:

$$\Theta_\Delta(F, E) = C \cdot \exp\left(-\frac{4}{3\hbar F} \cdot \sqrt{2m_e(\chi_0 - E)^3} \right), \quad (7.30)$$

where Θ_Δ is the tunneling probability trough the triangular barrier and C is a constant.

The tunneling current $J_i(F)$ from valley i can be obtained from the product of the density of available electrons n, tunneling probability, and effective drift velocity v_D^*. The latter can be different from the ordinary determined bulk drift velocity due to the influence of scattering in the barrier. It has been assumed, however, that the barrier does not considerably disturb the field dependence, which is in any case a much weaker function compared with the other multiplier. The velocity corresponds to the average velocity of the carriers approaching the barrier, while the carrier density equals to the density of available electrons (before the barrier) multiplied with the tunneling probability. Based on the above, *J(F)* can be expressed as:

$$J_{C1}(F) = A_{C1} \int_{E_{C1}}^{E_{C2}} n_1(F,E) \cdot \Theta_1(F,E) \cdot v_{D1}(F,E) dE$$
$$J_{C2}(F) = A_{C2} \int_{E_{C2}}^{\infty} n_2(F,E) \cdot \Theta_2(F,E) \cdot v_{D2}(F,E) dE, \tag{7.31}$$

where $J_{C1}(F)$ is the current from the main valley; $J_{C2}(F)$ is the current from the satellite valley; E_{C1} is the energy position of the main valley; and E_{C2} is the energy position of the satellite valley.

The tunneling current shows consequently an exponential dependence on the 3/2 power of the barrier height, when all other multipliers ($n_1, n_2, v_{D1}, v_{D2}, A_{C1}, A_{C2}$) are constant or differ slightly.

In the well-established F-N approximation, $J_{FN}(F)$ shows only an exponential dependence on the 3/2 power of the barrier height (ϕ) at the cathode-vacuum interface:

$$J_{FN}(F) = C_{FN} \times \frac{F^2}{\varphi} \exp(-B\varphi^{3/2}/F), \tag{7.32}$$

where the multipliers such as $n(F,E)$ and $v_D(F,E)$ (in the degenerated free carriers F-N approximation) are included in the constant C_{FN}, and B is a constant. In considering approximation, $n(F,E)$ can be found from

$$n_{3D}(F,E) = \frac{\sqrt{2} m_e^{\frac{3}{2}}}{\pi^2 \hbar^3} \cdot \frac{\sqrt{E}}{1 + e^{\left(\frac{E-\mu}{k_B T_e(F)}\right)}}$$
$$n_{1D}(F,E) = \frac{m_e^{\frac{1}{2}}}{\sqrt{2}\pi\hbar d^2} \cdot \frac{1}{1 + e^{\left(\frac{E-\mu}{k_B T_e(F)}\right)}} \cdot \sum \frac{1}{\sqrt{E - E_{Ql}}}, \tag{7.33}$$

where $n_{3D}(F,E)$ and $n_{1D}(F,E)$ are the densities of available electrons for the bulk and quantum-size one-dimensional (1D) cases, respectively, E_{Ql} is the energy of the quantum confinement level, k_B is the Boltzmann constant, T_e is the electron temperature, d is the quantum structure diameter, and μ is the Fermi energy.

The distribution functions of density of states (D) and density of available electrons (n) for the 1D and three-dimensional (3D) cases are illustrated in Figure 7.23. Quite different energy behavior of D and n is present for 1D and 3D cases. It is clear that the F-N slope change originates from $n(F,E)$, which is equal to the density of states multiplied by the Maxwell–Boltzmann carrier distribution, since the dependence of $v_D(F,E)$ becomes rather weak (nearly constant) at the high field region when considerable emission takes place [150, 151].

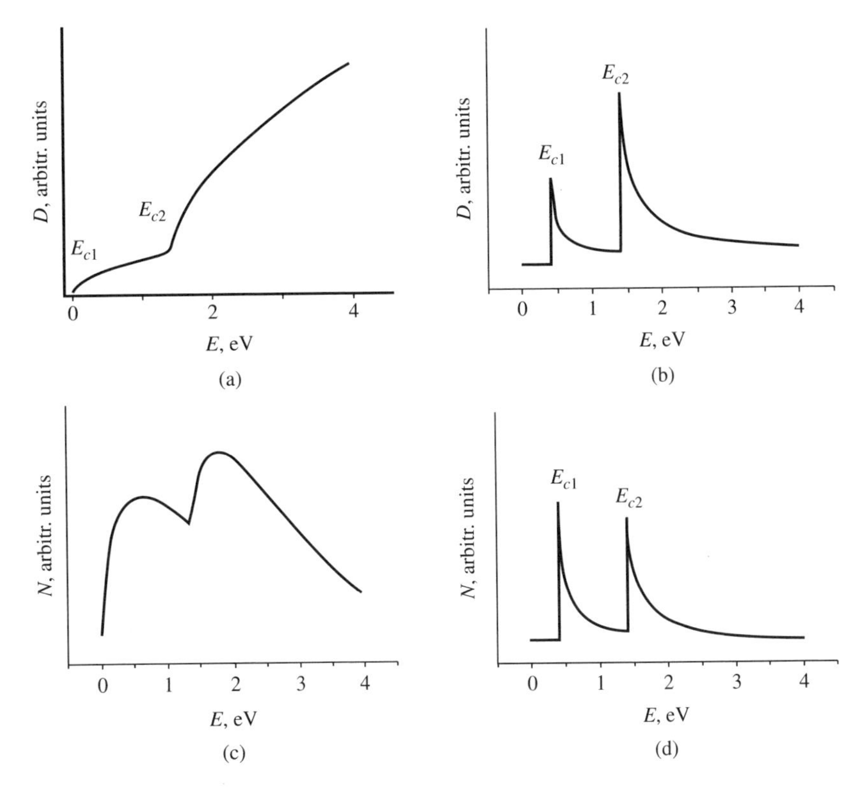

Figure 7.23 Density of states (D) and density of carriers (n) as a function of the band energy for GaN main and satellite valleys in the case of 3D (a) and (c) and 1D (quantum confinement) structures (b) and (d): (a) density of states for bulk semiconductor (3D), (b) density of states for quantum states (1D), (c) density of carriers for bulk semiconductor (3D), and (d) density of carriers for quantum states (1D). (All curves were plotted for the electric field value 5×10^5 V/cm.) Reproduced with permission from Ref. [148]. Copyright (2009), AIP Publishing LLC

In the following, the carrier heating by the applied electric field F is considered, where the hot carrier redistribution in the main and satellite valleys takes place. The effect of quantum confinement on the carrier distribution will also be shown.

First, the distribution of carrier density based on the Maxwell distribution has been analyzed. Contrary to the highly doped (degenerated) case, one can use the shifted Maxwell distribution for strong anisotropic and nondegenerated semiconductors. For low carrier concentration and nonelastic collisions, this assumption is quite reasonable and the following approximation holds:

$$\frac{p_0}{2m_e} \approx \frac{3}{2} k_B T_e(F),$$

where p_0 is the electron momentum, T_e is the effective temperature of carriers, and k_B is the Boltzmann constant.

In this case, the carrier distribution function can be found using the Maxwell approximation for $T_e(F)$ [152–156]:

$$f(p) = C \times \exp\{-(p - p_0)^2/2m_e k_B T_e(F)\}, \tag{7.34}$$

where p is the electron momentum and p_0 is the unheated electron momentum.

When the carrier concentration equals to the ionized impurity concentration, one can use the following form of carrier distribution function [148]:

$$f_0(E) = C \times e\{-E/k_B T_e(F)\}, \tag{7.35}$$

where E is the energy of free electron and F is the electric field.

7.4.1.1 Energy Band Reconstruction

In the case of nanostructures, the energy band of the material under consideration can be reconstructed based on the quantum-size confinement effect. Changing of the energy band parameters (energy bandgap, energy distance between main and satellite valleys, etc.) will influence the EFE significantly. Detailed analysis of energy band reconstruction of GaN due to the quantum-size confinement effect has been presented in Ref. [148]. A quantum oscillator approximation [134, 154] has been used to estimate the possible influence of the quantum-size effects on the increase in the energy bandgap (which degrades the field emission efficiency due to the decreased free carrier concentration) and on the decrease in the electron affinity (which enhances the field emission). GaN has been used as a WBGS material and the impact of the nanostructurization on the electronic properties has been considered. In the oscillator model, the heavy holes are considered as the nonmobile central particles, while electrons are represented as negatively charged spheres. In such a case, the solutions of the Schrödinger equation can be represented as Bessel functions with spherical symmetry. In the effective mass approximation the quantum dot (QD) electron spectra can be obtained from the Schrödinger equation with spherical potential $U(r)$ and finite band barriers (due to offsets of conduction and valence bands at semiconductor-vacuum interface). The eigenvalues of energy are obtained from the Schrödinger equation. The energy difference between the $E_0(X)$ and $E_0(\Gamma)$ conductance band minima is about 1.19 eV [135] for the bulk GaN material and continuously decreases down to 0.6 eV with decreasing QD size d. In Figure 7.24, one can see the calculated curves which illustrate the band transformation due to quantum confinement effects. The increasing of energy bandgaps for main (k-space point Γ) and satellite (k-space point X) valleys has been observed with decreasing nanotip diameter $d < 10$ nm. Due to lower electron effective mass in main valley, the growth of $E_g(\Gamma)$ is higher than $E_g(X)$. As a result, the intervalley energy difference decreases with the decrease in nanotip diameter (Figure 7.24).

The solution of the Schrödinger equation in the case of spherical symmetry corresponds to spherical Bessel functions (J, N):

$$\Psi_1 = BJ(k, r) \tag{7.36}$$

(inside the QD),

$$\Psi_2 = AN(\beta, r) \tag{7.37}$$

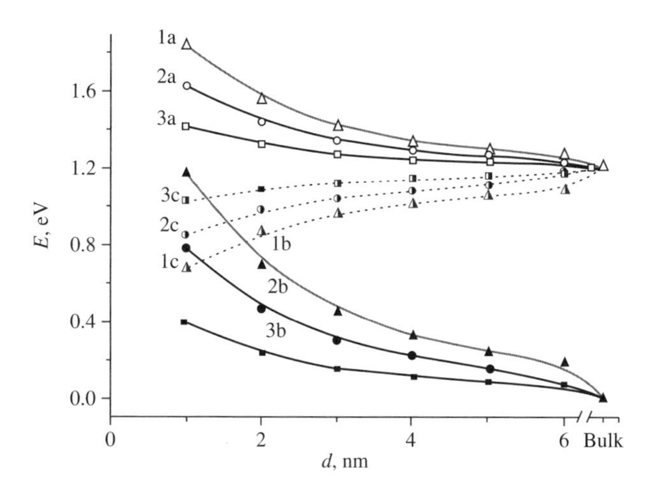

Figure 7.24 Calculated energy shift of (a) satellite valley, (b) main valley, and (c) intervalley energy difference in GaN due to the quantum confinement effect: zero-dimensional case (1a–1c), 1D case (2a–2c), and two-dimensional case (3a–3c). Reproduced with permission from Ref. [148]. Copyright (2009), AIP Publishing LLC

(outside the QD), with

$$\beta = \frac{\sqrt{2m^*_{e2}(V(r) - E)}}{\hbar}, \quad k = \frac{\sqrt{2m^*_{e1}E}}{\hbar}, \tag{7.38}$$

where m^*_{e1} and m^*_{e2} are the effective masses of electrons inside and outside of the QD, respectively, A and B are constants which define the magnitude of wave function, and E is the eigenvalue of energy for ground s-orbital state.

Using boundary and initial conditions such as

$$\Psi_1(d/2) = \Psi_2(d/2), \tag{7.39}$$

$$\frac{1}{m^*_{e,1}} \frac{d\Psi_1(d/2)}{dr} = \frac{1}{m^*_{e,2}} \frac{d\Psi_2(d/2)}{dr}, \tag{7.40}$$

one obtains the following non-algebraic equation:

$$\frac{\frac{d\beta}{2} - 1}{m^*_{e,2}} = \frac{1 - \frac{dk}{2}\cot\left(\frac{kd}{2}\right)}{m^*_{e,1}}. \tag{7.41}$$

The solution of Equation (7.41) provides the electron energy levels. A similar relation takes place for holes (with effective masses of m^*_{h1}, m^*_{h2}). To determine the constants A and B (Equations (7.36) and (7.37)), it is necessary to use the normalization requirement:

$$\int_0^\infty |\Psi|^2 dr = 1,$$

which is typical for single electron approximation.

Calculations have been performed for the energy band reconstruction of nanometer size objects in the presence of the quantum-size confinement effect. The calculation shows a decrease in the energy difference between the main and satellite valleys (Figure 7.24). This enhances the possibility of an intervalley electron transfer effect during field emission.

7.4.1.2 Comparison of the Theory with Experiment Results

Consideration of the above analyzed mechanisms allowed the calculation of the current-voltage characteristics *I-V* of the EFE and the corresponding F-N plot for GaN (Figure 7.25). The F-N curves for nanostructured (1D) and bulk (3D) materials are quite different. Such behavior can be explained by peculiarities of electron intervalley redistribution (Figure 7.26). In contrast to quantum-size cathodes (1D), with almost full intervalley carrier redistribution

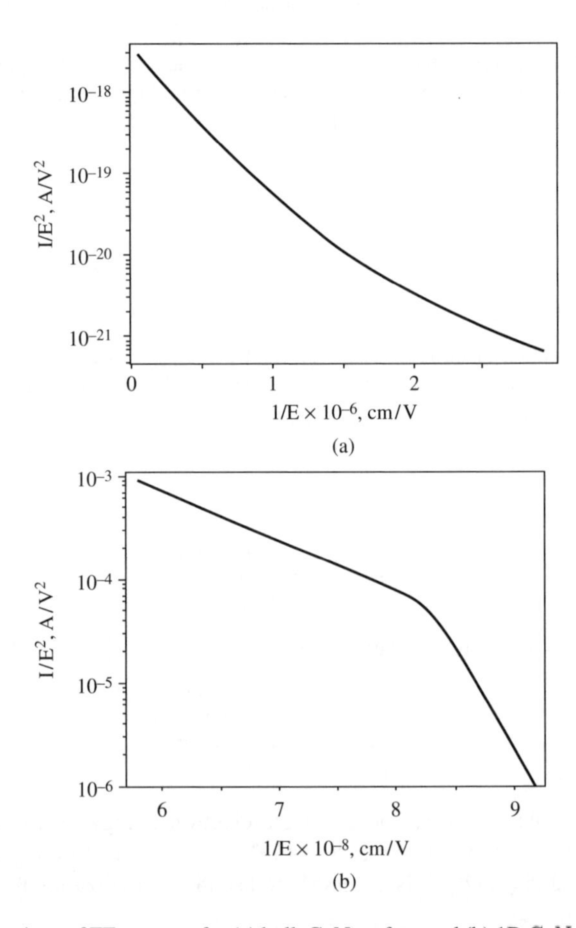

Figure 7.25 Comparison of FE currents for (a) bulk GaN surface and (b) 1D GaN nanotextured surface. Differences in the slope change originate from differences in density of states for bulk and nanotextured many valley semiconductors. Reproduced with permission from Ref. [148]. Copyright (2009), AIP Publishing LLC

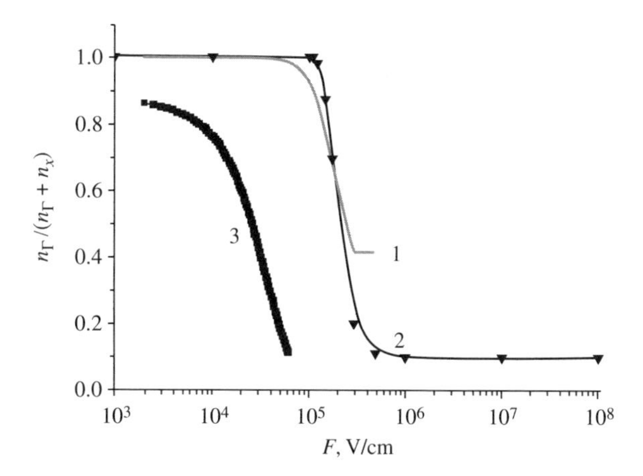

Figure 7.26 Comparison of the intervalley redistribution of electrons for GaN with large intervalley distance (1), GaN with quantum-size confinement (2), and GaAs with small intervalley distance (3). (n_Γ is the electron concentration in the main Γ valley, and n_X is the electron concentration in the satellite valley.) Reproduced with permission from Ref. [148]. Copyright (2009), AIP Publishing LLC

(Figure 7.26) above the threshold field, the bulk (3D) wide bandgap material shows different characteristics. The electron redistribution for GaN ($r = n_\Gamma/(n_\Gamma + n_X)$) has not exceeded 60% (Figure 7.26). Thus the total number of carriers is constant, but when the field is increased the carrier energy increases and electron transfer effect takes place. The values n_Γ and n_X have been calculated according to equations:

$$n_\Gamma(F) = \int_{E_{C1}}^{E_{C2}} N(E,F)dE, \tag{7.42}$$

$$n_X(F) = \int_{E_{C2}}^{\chi_0} N(E,F)dE, \tag{7.43}$$

where n_Γ is the total amount of carriers in the energetic band between E_{C1} and E_{C2} (in the main valley), n_X is the total amount of carriers in energetic band between E_{C2} and vacuum (in the satellite valley), and *N(E,F)* is the carrier density distribution.

Therefore the resulting current consists of two (number of active valleys) parallel currents, which depend on F in a rather different manner. Each of them can be described by the F-N equation for small electric field changes.

The total field emission current is the sum of two currents from the main and satellite valleys $J = J_{C2} + J_{C1}$ (see Equation (7.31)). Therefore, the F-N relation is a very rough approximation in case of multivalley semiconductors, which is only valid for the case of low fields.

The calculated curves in F-N scale (Figure 7.25) were compared with experimental field emission currents (Figure 7.22) over a wide field range. The experimental samples had a nanotextured surface [148]. The experimental curves manifested unique peculiarities (slope change, etc.) but could be described well by the extended modified F-N theory; this theory includes the quantum confinement effects on the nanotextured sample surface and nondegenerated statistics for hot carriers. It is clear that the intervalley redistribution becomes remarkable when the energy of hot carriers reaches the intervalley energy difference of ΔE. Figure 7.26 illustrates the intervalley redistribution of electrons in GaAs with small intervalley difference, GaN with large intervalley difference, and GaN with quantum-size tips. In the case of GaN cathodes, the intervalley redistribution occurs at significantly higher electric fields ($> 1 \times 10^5$ V/cm). An important result is also that for quantum-size GaN emitters, almost all carriers transit into the satellite valley at high electric field, while for bulk GaN emitters, a significant part of carriers is left in the main valley, even at very high fields.

The calculated drift velocity [148] and the electron temperature are shown in Figure 7.27 as functions of the electric field and correspond to the result in Ref. [150]. Indeed the field dependence of electron temperature causes the intervalley electron redistribution [67, 142, 143]. Figure 7.28 shows the general carrier redistribution between both valleys for 1D GaN cathode versus external electric fields.

It should be mentioned that the curve with a smaller slope at higher electric fields was obtained only for quantum-sized cathodes (Figure 7.25). In the case of bulk emitter, a concave curve with larger slope at higher electric fields was observed (Figure 7.25a). Using Figure 7.25a the calculation for bulk emitter has shown saturated carrier redistribution at very high fields with maximum carrier fraction of about 60% in the satellite valley (see Figure 7.26). Whereas in the case of quantum size emitters (Figure 7.25b), sharp and complete carrier redistribution must take place.

The field emission *I*-*V* characteristics in F-N coordinates (Figure 7.25) were calculated according to Equation (7.28), with taking into account the energy (Figure 7.23) and field (Figure 7.26) intervalley electron redistribution for 3D and 1D cases.

Taking into consideration the higher mobility for the main valley, it is clear that for the bulk emitter the current from the upper valley never becomes dominant at high fields. For bulk emitter, the emission starts from the main valley at low fields, with extremely small currents (practically undetectable). At moderate fields, the carrier distribution demonstrates remarkable concentration in the satellite valley. Here, the slope of the F-N characteristic is determined by the satellite valley work function. In the region of the highest electric fields, the main valley begins to play an important role because of the increased probability of tunneling through the

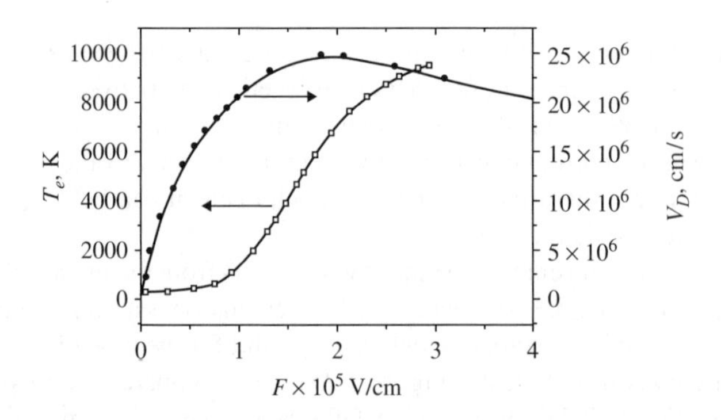

Figure 7.27 Drift velocity and electron temperature in GaN bulk material as functions of the electric field. Reproduced with permission from Ref. [150]. Copyright (1995), AIP Publishing LLC

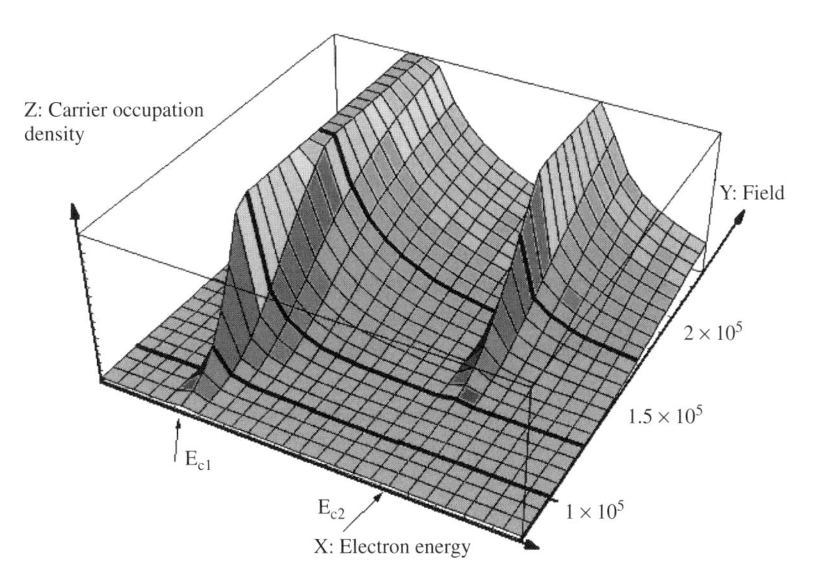

Figure 7.28 Density of carriers for 1D GaN cathode as a function of the band energy and the applied electric field at room temperature. Reproduced with permission from Ref. [148]. Copyright (2009), AIP Publishing LLC

emission barrier and high electron concentration. The slope in the F-N curve increases again (Figure 7.25a).

In the case of quantum-size cathodes, the experimental curve in Figure 7.22 corresponds to electron emission from the main valley (at low electric field; before the slope changing point) and from the satellite valley (at moderate fields; beyond the slope changing point) with two different slopes. The current from the main valley becomes negligible beyond the slope changing point. At moderate fields, most of the carriers are transferred into the satellite valley (Figure 7.26). As a result, the slope of the F-N curve is decreased and determined by the work function of the satellite valley (Figure 7.25b).

The reason for the slope change in GaN compared with GaAs cathodes is the following: in GaAs the carriers are already heated before field emission, demonstrating remarkably high currents. As a result, the slope changes are not seen in the field emission characteristic for a GaAs emitter. The slope change in the case of emission from GaN nanostructured surfaces is a good evidence of the quantum confinement effect on nanosized cathodes.

The calculated and experimental F-N curves are shown together for comparison in Figure 7.29. Different calculated curves in Figure 7.29 correspond to different intervalley distances. The quantum-size effect decreases the intervalley energy differences ΔE (see Figure 7.24). The theoretical and experimental curves are practically coincident for $\Delta E = 0.8\,\text{eV}$ (Figure 7.29, curves 1 and 4). The sharp current increase at low electric fields for a small intervalley distance in GaN (Figure 7.29, curve 2) is caused by easier intervalley carrier redistribution and low work function of the quantum-sized cathode.

The difference in the slope values in the F-N plot can be used for intervalley distance estimation. Large heating of the free electrons and interband electron redistribution take place in the studied GaN field emitter with nanostructured surface.

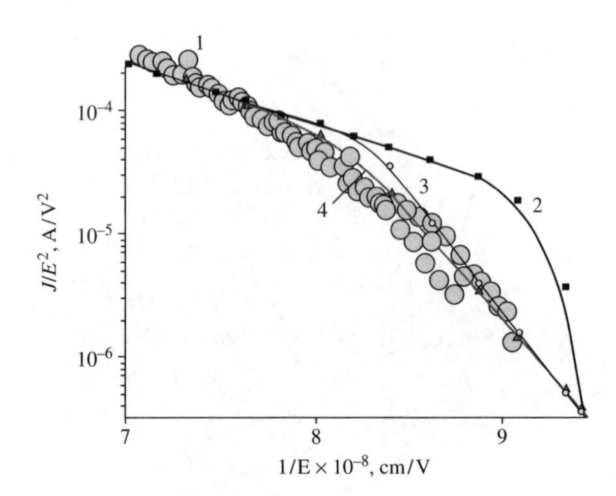

Figure 7.29 *I-V* characteristics of the field emission current (in F-N scale): experimental results (1), calculated curves for the intervalley distance of 0.5 eV (2), 0.7 eV (3), and 0.8 eV (4). Reproduced with permission from Ref. [148]. Copyright (2009), AIP Publishing LLC

7.4.2 Electron Field Emission from GaN Nanowire Film

For growth of grass-like GaN nanostructures the nickel catalyst assisted CVD method was used at 1200 °C [157]. As has been seen from the SEM images a large amount of grass-like nanostructures have been distributed uniformly on the Si substrate. The length/height of the grass-like leaves is in tenth of micrometers and its width/tips have been in hundredths of nanometers. Careful observations of SEM images have shown that the bottom ends of the grass leaves are wide and their tips are very sharp, which are analogous with the natural grass. By a close examination of high resolution SEM image it has been seen that the surface of leaves has been smooth and diameter of the structures has been gradually decreased toward the tip of the leaves having a dimension in nanoscale. A typical vapor–liquid–solid (VLS) growth mechanism has been suggested for the growth of grass-like nanostructured GaN leaves [119].

The EFE properties of such grass-like GaN nanostructures have been investigated [157]. The turn-on field (E_{on}), defined for an emission current density of 0.01 mA cm^{-2}, and the threshold field (E_{th}), defined for an emission current density of 1 mA cm^{-2} [158], have been 7.82 and 8.96 V μm^{-1}, respectively. This small turn-on field value is comparable with many reported values for GaN emitters (Table 7.1).

Excellent field emission properties mainly have been explained by the sharp tips of the non-aligned grass-like nanostructured leaves with small screening effects. Other factors which can affect the field emission properties are tip radius of the emitters, density of emitters, aspect ratio, morphology, intrinsic properties, and dimensions of the emitters, and so on [158]. The linear behavior of I-V characteristics in F-N coordinates at high electric fields shows the electron emission into the vacuum due to quantum tunneling effect [96, 158].

The preparation of highly aligned GaN nanowire arrays has attracted considerable interest, and various approaches have been developed. Vertically aligned and faceted GaN nanorods have been produced in Ref. [159] using a catalyst-free template approach which employed a silicon dioxide mask structured using a porous anodic alumina template. Quasi-aligned GaN

Table 7.1 The comparison of turn-on and threshold field values of 1D GaN nanostructures

#	Type of material	Type of morphology	Turn-on field (V μm^{-1})	Threshold field (V μm^{-1})	References
1	GaN	Nanowires	12	—	[119]
2	GaN	Prismatic sub-micro rods and cone nanowires	9.5	18.0	[120]
3	GaN	Patterned nanowires	8.4	10.8	[78]
4	GaN	Grass-like nanostructures	7.82	8.96	[157]
5	GaN	Needle-like bicrystalline GaN nanowires	7.5	—	[75]
6	GaN	Well-aligned nanocolumns	2.5	4.7	[121]
7	GaN	Well-aligned needle-like nanowires	2.1	4.5	[78]
8	GaN	Dandelion-like GaN	9.65	11.35	[128]

nanowire arrays have also been fabricated via a thermal evaporation of the starting reactants Ga_2O_3/GaN [71]. Furthermore, well-aligned and vertically oriented GaN nanowires have been grown on sapphire by metal organic chemical vapor deposition (MOCVD) [160]. There has also been reported [161] a route to ultrahigh-density and highly aligned single-crystalline GaN nanowires on sapphire by employing ultrathin (submonolayer) Ni catalyst films. However, a catalyst or pattern has been used in the above mentioned technologies. A novel method to fabricate high-density, well-aligned GaN needlelike nanowire arrays over large areas without any catalyst or pattern have been demonstrated in Ref. [76]. Uniformly single-crystal GaN nanowires have been prepared through thermal evaporation of GaN powder with the assistance of HCl gas. The formed nanowires with needlelike tips are vertically aligned without any lateral growth. Furthermore, the GaN nanowires have been extremely dense and uniform (Figure 7.30). They have quite straight morphology and clean surface without any particles. The diameters of the nanowires have been about 80–100 nm, and their lengths have been approximately 1.5 μm.

The sharp, needlelike tips enable such GaN nanowires to exhibit better electron emission properties due to their high aspect ratio and small tip radii of curvature, which could provide a sufficiently high geometric enhancement factor. The turn-on field was evaluated to be about 2.1 V/μm. The emission current density of the GaN nanowires array reached 1 mA/cm^2 at a bias electric field of 4.5 V/μm (Table 7.1). Turn-on field and threshold field were much lower than reported values of GaN nanowires [74, 119, 162]. Electric field enhancement coefficient estimated from the slope of the F-N plots was as high as ~2835. Compared with some other GaN nanostructures [72, 163, 164] the vertically aligned GaN needlelike nanowires showed a relatively larger value.

The excellent field emission properties of the aligned GaN nanowires and nanocolumn arrays grown on Si substrates are mainly attributed to the following three factors: (1) LEA of GaN, (2) high alignment of GaN nanowires and nanocolumns with resulting high field enhancement coefficient, and (3) the easy electron flow across the *p*-type Si-GaN interface under an applied field. The electron affinity of wide bandgap GaN (3.4 eV) is 2.8 eV [165] and that of low bandgap Si (1.1 eV) is 4.01 eV [166]. Thus, heterojunction was formed at the Si-GaN junction. After thermal equilibration a band bending occurs at the junction as shown

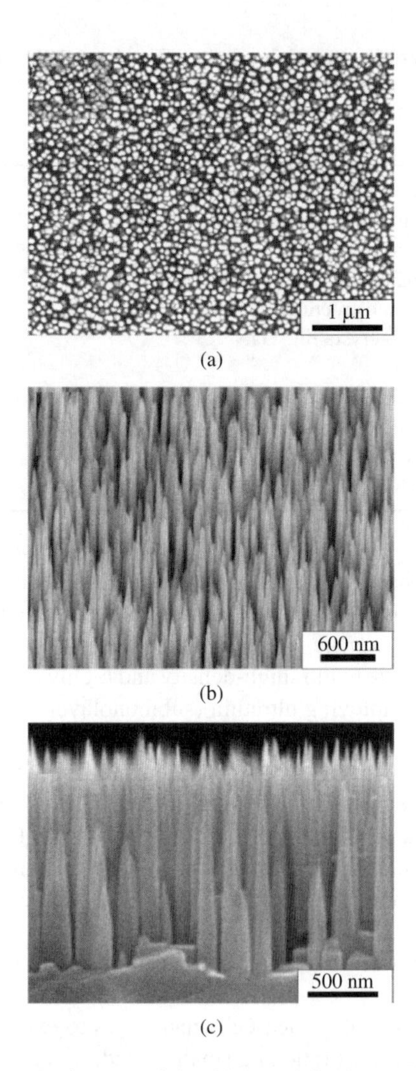

Figure 7.30 Field-emission SEM images of the fabricated GaN nanowire array on a thin GaN film template under (a) top view, (b) 45° side view, and (c) cross-sectional view. Reprinted with permission from Ref. [76]. Copyright (2008), American Chemical Society

in Figure 7.31. Such structure allows easier electron flow across this junction under an applied field [121].

Addition improvement of EFE can be obtained by formation of an AlGaN/GaN heterostructure on the GaN nanorods. The GaN nanorods have been synthesized using a catalyst-free templated approach that employs a silica mask fabricated using a porous anodic alumina template [68, 159]. The nanorods have been vertically aligned, have pointed tip morphologies, and are faceted with an average diameter of 50 ± 5 nm and total height of 100 nm. AlGaN/GaN heterostructure nanorods were grown with a 5 nm thick AlGaN layer to enhance the electron tunneling from the underlying GaN nanorod.

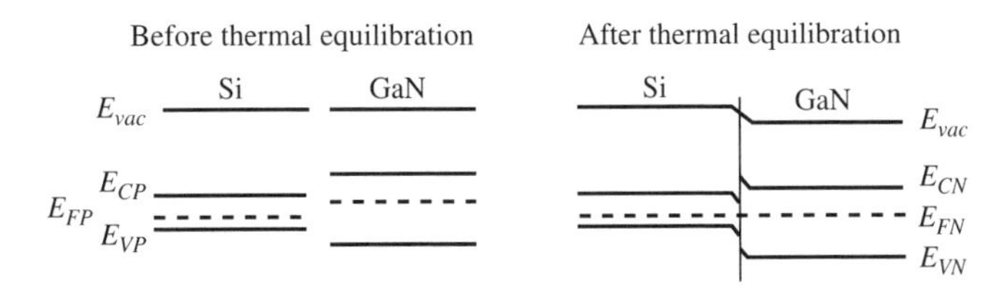

Figure 7.31 Schematic band diagrams of (a) Si and GaN and (b) the *p*-Si-GaN heterojunction. E_{CP}, E_{VP}, E_{FP}, E_{CN}, E_{VN}, and E_{FN} are the conduction band, valence band, and Fermi level of *p*-Si and GaN, respectively. Reprinted with permission from Ref. [121]. Copyright (2009), American Chemical Society

The turn-on fields determined from the field-emission characteristics for GaN and AlGaN/GaN nanorods, were calculated to be 38.71 and 19.33 V/μm, respectively [68]. The energy difference between the conduction band minimum and the Fermi level and the resulting work function for the GaN nanorod emitters were estimated to be approximately 0.1 and 3.3 eV, respectively, assuming an electron concentration of 8×10^{16} cm^{-3} and an electron affinity of 3.2 eV. The field enhancement factor of the GaN nanorod emitters was calculated as 65 by using the F-N and assuming a work function of 3.3 eV. Furthermore, the effective work function of the AlGaN/GaN nanorod emitters was calculated as 2.1 eV. Thus, the large field enhancement factor due to the sharp nanorod tips combined with the lowering of the surface electron affinity by an AlGaN capping layer greatly improves the field-emission properties of GaN-based nanorod emitters. It should be noted here that the samples used in Ref. [68] were unintentionally doped GaN. Increasing the doping level, the AlN mole fraction in the top AlGaN layer and the density of the nanorod emitters could further reduce the turn-on field and increase the field-emission current density [30, 167].

GaN layer was also etched with hydrogen (H_2) plasma and the roughened surface showed an increased field enhancement factor [69]. This technique is applicable to fabricate efficient field emitters. On the other hand, much attention has also been focused on boron nitride (BN) and aluminum nitride (AlN) as one of the promising materials for field emitters with NEA [21, 168, 169]. Several reports on field emission characteristics of BN films [170–172] have demonstrated an electron emission at a considerably low electric field and significantly improved field emission characteristics of BN coated Si tip arrays [173]. The field emission characteristics of H_2 plasma roughened GaN and additional BN coating were studied in Refs. [174, 175]. BN films were synthesized on *n*-type GaN substrates by plasma-assisted chemical vapor deposition (PACVD) with a horizontal quartz reactor [175]. The surface roughness was estimated by AFM to be 13.4–29.3 and 5.3–10.6 nm for GaN and BN/GaN samples, respectively.

EFE measurements allowed estimating the average turn-on electric field between the anode and the sample surface. They were 8.8 and 12.4 V/μm for the BN/GaN and GaN samples, respectively. The F-N plot showed a straight line, suggesting F-N tunneling of electrons.

The electron emission from the GaN surface can be better understood using the band diagram of GaN shown in Figure 7.32. The positive electron affinity of the GaN has been reported to be 2.7–3.3 eV [11, 167]. In the case of *n*-GaN, electrons in the conduction band are emitted due to F-N tunneling through the surface potential barrier. When the surface roughness is formed, the electric field strength is enhanced at the surface under the fixed anode voltage. Assuming the electron affinity of 3.3 eV, the field enhancement factor is estimated to be 150 from the slope of the F-N plot of the GaN substrate. In the case of the as-grown GaN sample (without H_2 plasma etching), on the other hand, no electron emission was detected.

The band diagram of the BN/GaN structure is illustrated in Figure 7.32b to discuss the electron emission process. The Fermi level position of the BN film has been estimated to be 2.8 eV above the valence band edge by ultravoilet photoemission spectroscopy analysis [168]. The energy difference between conduction band edges of the GaN and BN film is assumed to be 3.2 eV because the Fermi level is located at an energy as low as 10–40 meV below the conduction band edge for the *n*-GaN with an electron density of $2 \times 10^{17}\ cm^{-3}$. There exists the NEA at the BN surface.

A possible mechanism of the electron emission is probably due to process (III) consisting of the hopping conduction in the BN film and F-N tunneling from the defect levels to the vacuum at the NEA surface of the BN film. It is reasonable to consider the existence of defects

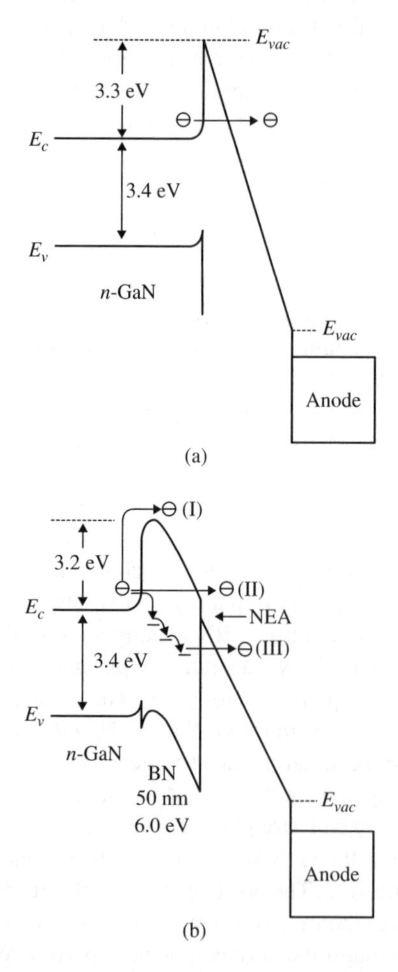

Figure 7.32 Band diagram of (a) GaN and (b) BN/GaN under the application of electric field. Reproduced with permission from Ref. [174]. Copyright (2001), AIP Publishing LLC

in nanocrystalline BN films as mentioned above. Defect states densely distributed near the Fermi level in the BN film make it possible to transport electrons in the conduction band of the GaN to the BN surface due to the hopping conduction. When the defect level is located below the vacuum energy level of the BN film, electrons in the defect level are emitted due to F-N tunneling through the surface potential barrier. Since the surface potential barrier of the BN/GaN sample is lower than that of the GaN sample, a reduction in the turn-on electric field occurs for the BN/GaN sample. Though surface roughness of the GaN sample is estimated to be larger than that of the BN/GaN sample, a reduction in the turn-on electric field occurs. This is because the potential barrier between the Fermi level and the vacuum level at the BN surface is much lower than 3.3 eV, which is the electron affinity of GaN. Moreover, formation of protrusions on the BN surface possibly leads to further reduction in the turn-on anode voltage.

7.4.3 Electron Field Emission from Patterned GaN Nanowire Film

The EFE performance of 1D nanostructured films, though higher than those of bulk films, is dampened by the screening effect between the densely packed nanostructures. Patterning of the nanostructures has been introduced to solve this problem for further FE enhancement [176]. It has been shown that nanotubes arranged in regularly patterned arrays of bundles, with bundle diameter and array spacing of the order of a few micrometers, give much greater FE current densities than dense mats of nanotubes, thin films of nanotubes, or arrayed individual nanotubes [78, 177]. Such arrays appear ideal to be integrated into nanodevices for high-intensity electron beams from FE sources.

The PLA technique was used for growth of patterned GaN nanowire films [78]. PLA can create a highly energetic growth precursor, leading to the formation of nonequilibrium growth conditions. High-quality nanowires can be obtained at a fairly low substrate temperature. GaN wafer was used as the precursor material for growth of GaN nanowires. The laser ablation was conducted with a KrF excimer laser (248 nm, 23 ns at a laser fluence of 5.0 J cm^{-2} for 1 hour at a pulse repetition rate of 3 Hz). Patterned *n*-Si substrates (using a transmission electron microscopy (TEM) Cu grid as a physical mask) coated with 5 nm gold (Au) thin film by an electron-beam evaporator were placed on a substrate holder. It was positioned 3–5 cm perpendicularly opposite to the GaN target. The temperature of the substrates was kept at 700 °C during the synthesis of the nanowires. The special feature of the PLA process is that each laser pulse ablation generates a large quantity of Ga and N species ($\sim 10^6$ atoms/pulse), which are instantaneously deposited onto the Au catalyst surface. The diameters of the grown nanowires ranged from 10 to 40 nm and their lengths range from 1 to 1.5 μm. The tips of NWs were coated with Au/Ga caps of the nucleation site for the catalyzed growth of each nanowire by the vapor-liquid-solid mechanism [129].

I-V characteristics of FE from these patterned single-crystalline GaN nanowires were investigated [78]. The turn-on field of 8.4 V μm^{-1} was obtained based on the definition for the field to produce an emission current density of 0.01 mA cm^{-2}. The emission current density reached 0.96 mA cm^{-2} at an applied field of 10.8 V μm^{-1}. This result is a tremendous improvement from that of continuous GaN nanowire films grown by PLA without patterning under the same experimental conditions. The FE current density from continuous GaN films at 10 V μm^{-1} was 64 times lower. Continuous films suffer from the screening effect due to electrostatic screening between adjacent emitters. Though the field amplification of an emitter is determined by the radius of curvature at the tip as well as the height of the wire over the substrate, the distance between neighboring emitters has a decisive influence due to the screening effects. The electric field near the apex of emitters decreases with decreasing spacing between them. Hence, patterned nanowires prove to have higher and more stable emission current than continuous nanowire films.

The emitted current depends directly on the local electric field at the emitting surface and its work function. Figure 7.33 shows a schematic diagram of the FE magnitude for patterned GaN nanowires compared with continuous GaN nanowire films. The EFE magnitude is greater at regions, where the nanowires are near the edges of the Au catalyst pattern due to reduced screening effect. Therefore, the total current collected during EFE measurement for patterned nanowires is much greater compared with that of continuous nanowire films. Comparing the EFE properties of these patterned GaN nanowires with those of GaN nanowires using other growth methods [71, 74, 75, 119, 130, 178], the turn-on field of 8.4 V μm^{-1} at a current density of 0.01 mA cm^{-2} is reasonably low for good EFE properties of GaN nanowires. Assuming that the work function of GaN is 4.1 eV, the field enhancement factor β was estimated to be 474 from the slope of the F-N plot. This β value reflects the degree of the FE enhancement of the tip shape on a planar surface. It is dependent on the geometry of the nanowires, the crystal structure, and the density of emitting points. Compared with the β values of GaN nanowires already reported [74, 130], the β value of 474 reported here is lower. This lower value may be attributed to the shorter length of GaN nanowires in this study as those with higher β reported have lengths ranging from several micrometers to tens of micrometers. Since the GaN nanowires in Ref. [78] have lengths ranging from 1 to 1.5 μm, the lower β value calculated is considered to be reasonable.

Among various materials that can be used for field emitters, it has been reported that MBE grown GaN nanorods on Si substrate with a native oxide layer exhibit excellent field emission characteristics because GaN has LEA (2.7–3.3 eV) in addition to high chemical and mechanical stability [163]. Moreover, Si substrate has the advantage of economic cost and easy treatment than the sapphire that is commonly used as the substrate for nitride semiconductors such as AlN and GaN.

It is also important to enhance the local field by controlling the morphology and distribution of emitters, in which to determine the barrier through field emitted electrons tunnel. There are many research efforts to control the morphology and distribution of GaN-based field emitters such as patterning nanowires [78].

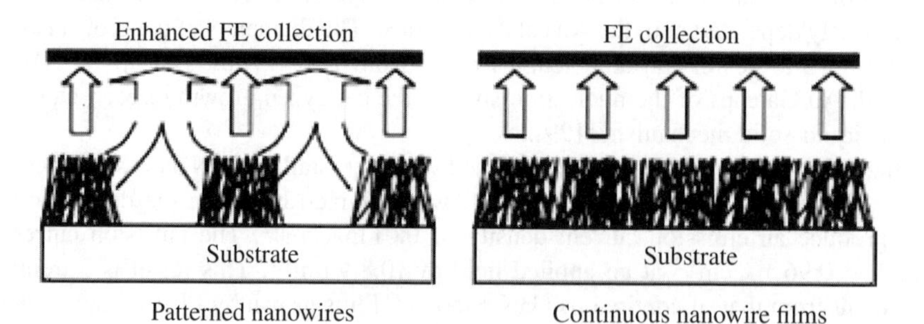

Figure 7.33 Schematic diagram of FE magnitude of (a) patterned GaN nanowires compared with (b) continuous GaN nanowire films [78].

Silicon on insulator (SOI) substrates consisting of 75 nm thick Si (Si_{top}) and 185 nm-thick SiO_2 over Si (Si_{sub}) were used (Figure 7.34) [179]. To prepare patterned SOI substrates, conventional photolithography process was carried out. After forming thin native oxide layers on the Si_{sub} and Si_{top} surfaces, GaN nanorods were grown on the surfaces of both Si_{sub} and Si_{top} areas by rf plasma-assisted MBE (Figure 7.34).

The morphologies of FEAs were analyzed by field emission scanning electron microscope (FESEM). FESEM images for the FEA samples are shown in Figure 7.35. GaN nanorods are uniformly grown both on the Si_{sub} and on the Si_{top} with their *c*-axes being aligned with the surface normal.

The straight lines in emission Fowler-Nordheim plots (J/F^2–$1/F$ relation) indicate that the electron emission is due to the F-N tunneling mechanism. From the slopes of the straight lines, the respective field enhancement factors β were calculated to be 666 and 1685, for the 10×10 and $1 \times 1\ \mu m^2$ window FEA samples, when assuming the electron affinity of 3.3 eV. FEAs fabricated on the $1 \times 1\ \mu m^2$ periodic window substrates exhibited lower threshold electric field of 2.6 V/μm, higher field emission current density of 10^{-3} A/cm^2. The improved field emission characteristics for the FEAs with smaller area periodic window were considered to be due to the edge effect at the patterned area, that is, the electric field at the edge was stronger than that at other areas.

7.4.4 Electron Field Emission Properties of Individual GaN Nanowires

GaN nanowires, which are one-dimensional structures of GaN, are interesting candidates for field-emitters because of their LEA (2.7–3.3 eV) [11, 180] and high-aspect-ratio geometry. Many studies on field-emission properties of GaN nanowires have been focused on collections of these nanowires [73–76, 78, 119]. It has been observed that GaN bundles have lower turn-on field and higher emission current density compared with bulk GaN and relatively longer

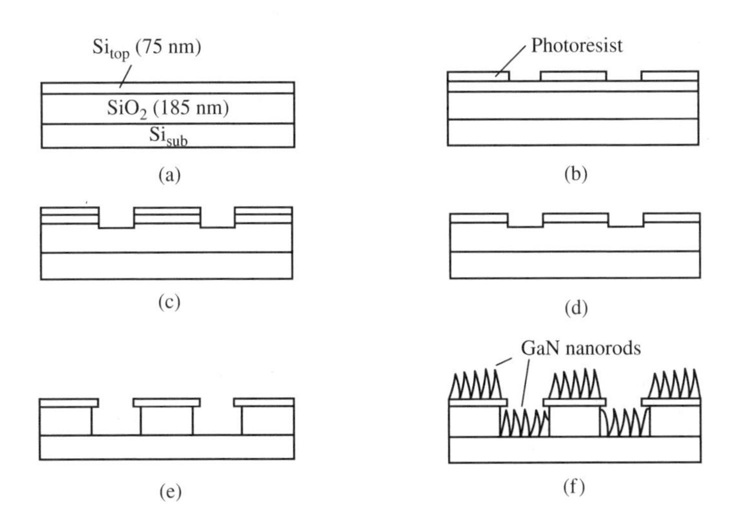

Figure 7.34 Schematic drawing of FEA fabrication process. (a) SOI substrate, (b) photolithography, (c) reactive ion etching (RIE), (d) removal of resist, (e) wet etching, and (f) GaN growth [179]. Copyright (2008) WILEY-VCH Verlag GmbH & Co. KGaA, Weinheim

lifetime and better stability than some of other nanomaterials [74]. However, field-emission properties of individual GaN nanowires are important in order to gain a better understanding of their field-emission behavior and be able to engineer field-emitters based on them. In Ref. [181] GaN nanowires have been synthesized using CVD. Devices containing individual GaN nanowires have been fabricated using the contact printing technique.

A 500-nm-thick SiO_2 layer was thermally grown on a Si (100) substrate and a layer of gold approximately 2 nm thick was deposited as catalyst using electron-beam evaporation. The sample was placed in a quartz-tube furnace and a gallium metal source was placed approximately 3 cm upstream of the sample. The temperature was ramped up to 850 °C, and a flow of 500 sccm of Ar was maintained during heating and subsequent 15 minutes of annealing. For the nucleation of the GaN nanowires, 13 sccm of NH_3 and 300 sccm of H_2 were introduced into the reaction chamber for 5 hours [182, 183].

The GaN nanowires were found to have diameters from 15 to 60 nm and lengths from 1 to 10 μm. Since the as-grown GaN nanowires were highly dense and randomly oriented, they needed to be partially transferred and aligned to make devices containing individual GaN nanowires. For this, the contact printing method was used [184] (Figure 7.36). Because of the high density of nanowires on the as-grown substrate, multiple printing steps were needed to progressively reduce the density of nanowires on the target substrate. Most of the nanowires were transferred onto dummy substrates by the first few printing steps (Figure 7.36b). Eventually, the as-grown sample was left with a suitably low density for transfer to the final target substrate (Figure 7.36c).

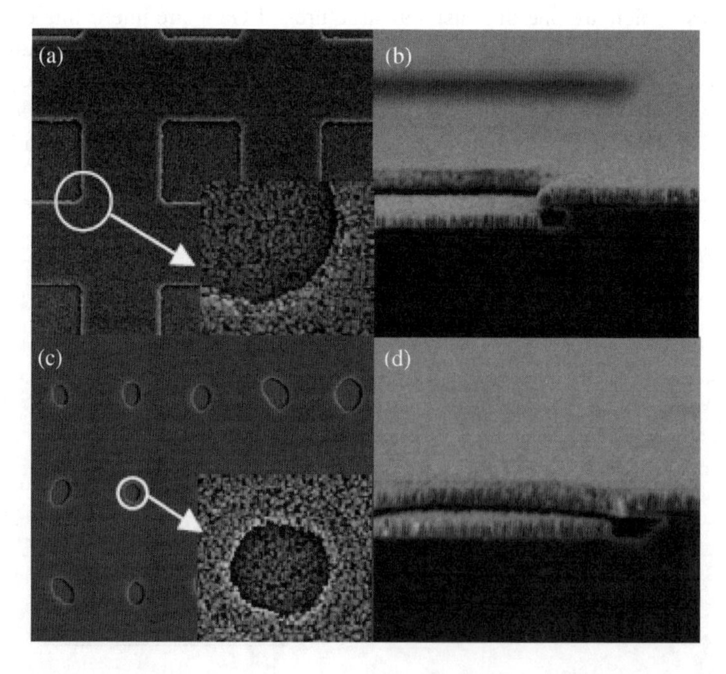

Figure 7.35 FESEM surface (a,c) and cross section (b,d) images for the $10 \times 10\ \mu m^2$ (a,b) and $1 \times 1\ \mu m^2$ (c,d) window FEA samples. The insets in the surface images show the enlarged image of areas indicated by the white circles [179]. Copyright (2008) WILEY-VCH Verlag GmbH & Co. KGaA, Weinheim

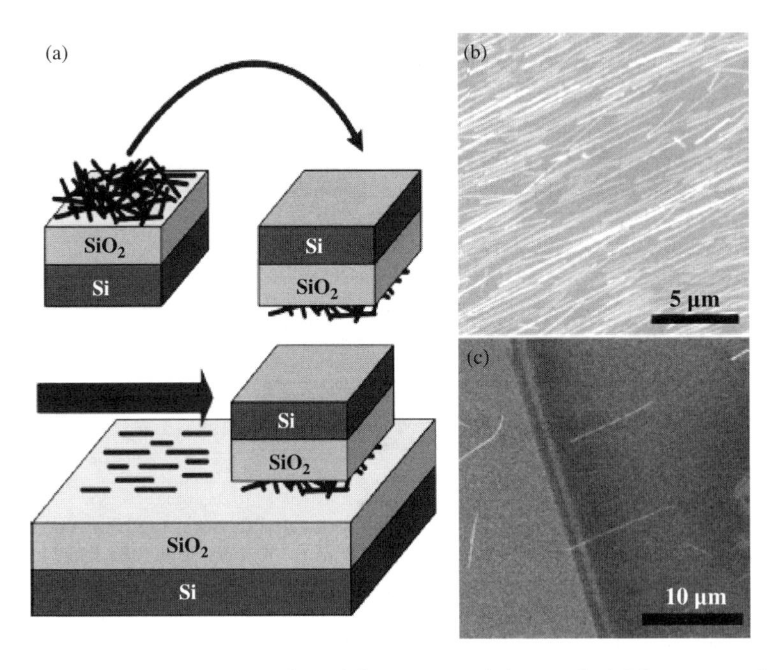

Figure 7.36 (a) Schematic representation of the contact printing method. The source substrate containing GaN nanowires was moved to the target substrate with constant speed of ~20 μm/s and constant force (7 g weight on the back side of source substrate). (b,c) SEM images showing the transferred and well-aligned GaN nanowires on substrates with high and low density, respectively. Reproduced with permission from Ref. [181]. Copyright (2012), AIP Publishing LLC

Two main types of devices were fabricated to investigate the effect of the proximity of the nanowire to substrate: those where the nanowires rest on the oxide surface or are very close to it, Type-1 (Figure 7.37a), electrode over nanowire and Type-3 (Figure 7.37c), nanowire over electrode, and those with the nanowires suspended over a trench, forming a cantilever-like structure, Type-2 (Figure 7.37b). The patterned electrode opposing the nanowire served as anode in the experiments, eliminating the need for an external anode and allowing to know the cathode-anode distance precisely in each case. Ohmic contact was achieved by depositing four metal layers from bottom to top: Ti (20 nm), Al (80 nm), Pt (40 nm), and Au (100 nm) (both devices Type-1 and Type-2 were made with Ohmic contacts) [185]. Type-3 devices were made with two different electrode materials, Mo (50 nm) or a double-layer of Cr (20 nm)/Pd (50 nm).

Maximum emission current obtained from each nanowire was a few tens of nanoamperes before the device was permanently damaged under an applied electric field greater than 35 V/μm. The field enhancement factor, β, can be extracted from the experimental data using the $ln\ (J/V^2)$ versus $1/V$ relationship. Using a work function value of 4.1 eV for GaN, the calculated field enhancement factors of the individual GaN nanowires are 170 and 167 for Type-2 and Type-1 device, respectively. The turn-on field for a collection of GaN nanowires has typically been defined as the field required for generating emission current densities of 0.01 mA/cm^2 [73, 75, 78, 119]. For an individual nanowire with a diameter of a few tens of nanometers, these would correspond to a current on the order of 10^{-16} A, which is below the resolution of the measurement circuitry. In devices under investigation, the first noticeable

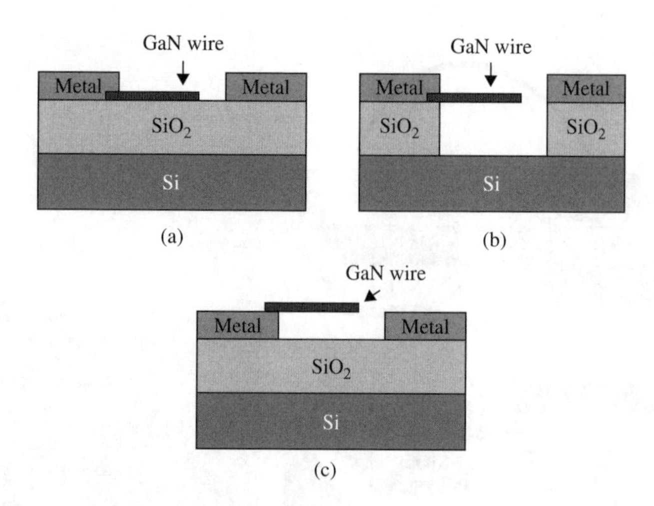

Figure 7.37 Cross-sectional schematic views of the fabricated devices. (a) Electrode on top of the end of nanowire lying on substrate, Type-1. (b) Similar to (a), but oxide etched from underneath the nanowire to form a free-standing, cantilever-like structure, Type-2. (c) Nanowire end on top of the electrode, Type-3. In Type-3 devices, depending on its length, the nanowire either has or has not touched the oxide surface. Reproduced with permission from Ref. [181]. Copyright (2012), AIP Publishing LLC

sign of turn-on was observed at an applied field of ~21 V/μm (at which the current was ~100 pA). This field was greater than that for collections of GaN nanowires (which was typically less than 10 V/μm).

Field-emission theory for semiconducting materials predicts a nonlinear behavior for the emission current [186]. At low electric fields, the emission current is determined by the tunneling probability and the field-emission behavior of the semiconductor is similar to that of metals. At medium electric fields, the emission current becomes saturated due to insufficient carrier supply. In this regime, the emission current is greatly affected by the energy band structure of the semiconducting material. In addition, as the applied field and the resulting emission current increase, the resistance of the emitter (including bulk and contact resistance) and space charge effects among the emitted electrons also contribute to current saturation. At high electric fields, the emission current rapidly increases due to the field penetration into the depletion region, which results in a greatly increased number of carriers. The saturation and subsequent rapid increase of the emission current have not been observed in Ref. [182] at emission from individual GaN nanowire, in contrast with what has been observed for other semiconducting nanomaterials [187–190]. In fact, increasing the field further has led to the destruction of the devices and there has been no possibility to observe a possible saturation regime or beyond.

The Schottky-contact devices (Type-3) were also investigated. Interestingly, no emission current was observed from either Mo- or Cr/Pd-electrode devices. Investigated GaN nanowires were N-type, likely due to O_2 impurities and/or N vacancies, with an energy bandgap of approximately 3.4 eV [183]. Electrons in the metal electrode have to overcome the Schottky barrier, Φ_{bn} (~0.8 eV for Mo electrode and ~1.2 eV for Cr/Pd electrode), in order to be injected into GaN [191, 192]. It is expected that the major part of the externally applied voltage drops over a region around the nanowire tip, and thus the region around the metal-nanowire contact only experiences a small reverse bias (Figure 7.38).

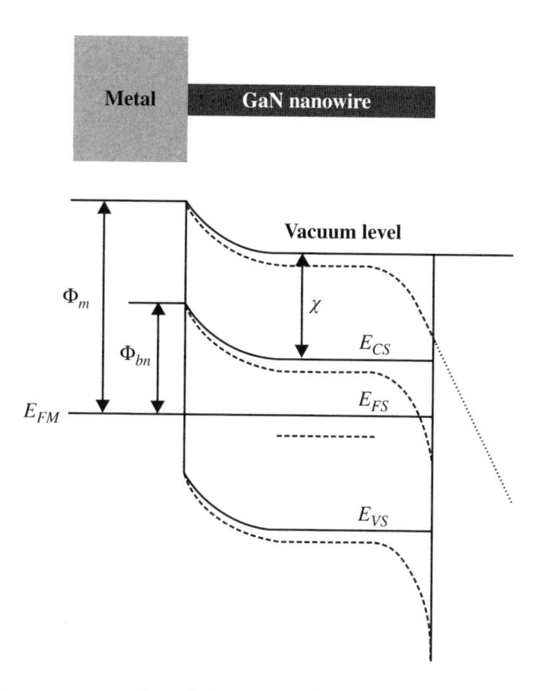

Figure 7.38 Qualitative representation of the energy band diagram of GaN nanowire connected to a metal electrode and forming a Schottky contact. E_{FM} is the Fermi level of the metal, Φ_m is the work function of the metal, and Φ_{bn} is the Schottky barrier height. E_{CS}, E_{FS}, E_{VS}, and χ are the conduction band edge, Fermi level, valence band edge, and electron affinity of the GaN nanowire, respectively. The dotted lines are E_{CS}, E_{FS}, E_{VS}, and vacuum level under field. Under the applied bias, most of the potential drops around the nanowire tip and only a small reverse bias appears over the Schottky junction. Reproduced with permission from Ref. [181]. Copyright (2012), AIP Publishing LLC

Under these conditions, the field-emission current is limited by the Schottky contact barrier as insufficient electrons are transferred to the GaN nanowire. The obtained results show that the contact between the GaN nanowire and metal electrode plays an important role in field-emission behavior, and good Ohmic contact is desirable for sufficient electron supply and high emission current from the nanowire.

7.4.5 Photon-Assisted Field Emission from GaN Nanorods

GaN is a promising material for high-frequency, high-power, and high-temperature electrical, and optical ultraviolet devices due to its unique material parameters [193, 194]. It is very important to known the precise band structure of GaN. Many groups have theoretically calculated the band structure for GaN [195, 196] and measured some characteristic values in it, for example, the intervalley energy distance [142, 197, 198, 108]. But until now, there have been different calculated and measured values for the GaN energy-band parameters, including the difference between the main Γ and the satellite valley.

There are several methods for studying of the energy-band structure of materials. The first method to determine the electronic-band structure for semiconductors is the electroreflectance

(ER) modulation spectroscopy due to its high resolution and its sensitivity with small changes in band structure. The ER modulation spectroscopy method is based on electromodulation, in which reflection spectra are studied [199]. Wide application of this method is complicated due to the requirement that the electric field must be uniform at the space-charge region. The second method is inverse photoemission spectroscopy (IPES) [200]. IPES is a surface-science technique used to study the unoccupied electronic states, structure of surfaces, thin films, and adsorbates. In case of IPES investigations, a well-collimated beam of electrons with energy less than 20 eV is directed to the sample. Due to low energy of incident electrons, their penetration depth is only several atomic layers. The studies of a field-emission energy distribution also give information about electronic states that lie within the Fermi energy [201–203]. The ultraviolet photoelectronspectroscopy method allows one to determine the electron affinity of III-nitrides and their alloys [144]. Photofield-emission spectroscopy is also used for the characterization of the energy-band structure [204]. Several authors have discussed the possibility of investigating the electron states between the Fermi and vacuum levels by studying the field emission of optically excited electrons [205, 206]. This method is based on the work function change under photoexcitation of electrons at their tunneling through a barrier according to the Fowler-Nordheim mechanism. A spectroscopy of the GaN band structure will be possible for field-emission measurements with monochromatic illumination of the depleted emitter tips.

The investigation of the photon-assisted field emission from GaN nanorods and determination of some energy band structure parameters have been performed in Refs. [67, 207]. The GaN field-emitter rods were fabricated on a wafer with n-GaN active layer (5 μm) sandwiched between n^+-GaN cap layer (100 nm) and 300 μm thick n^+-GaN substrate. The active layer and top layer were grown with metal-organic chemical vapor deposition using a standard method. A photoresistant mask was used for the argon plasma etching (Ar flow of 20 sccm, background pressure of 50 mTorr, and *rf* power of 300 W) of a circular mesa with 400 nm height. Afterwards, the samples were cleaned in acetone and photoelectrochemically (PEC) etched [208] in KOH. UV illumination generates electron-hole pairs at the n^+-GaN surface, which enhance the oxidation and reduction reactions taking place in the electrochemical cell, formed by immersing the substrate in a 0.1 M KOH stirred solution. This allows one to achieve selectivity in etching n^+-GaN with respect to n-GaN. The GaN sample was PEC etched for 12 minutes in the presence of Hg lamp illumination. The SEM image of GaN nanorods after PEC etching is shown in Figure 7.39. Further etching up to 38 minutes increased the height of the rods by extending them into the n-GaN layer below the n^+-GaN cap. The n-GaN material around the mesa was not affected by this step and remained unetched. The selectivity in etching the area containing the nanorods but not the surrounding area is due to the presence of the thin n^+-GaN layer in the "nanorod area," which promotes PEC etching. An Ohmic-contact cathode electrode was formed on the back side of the n^+-GaN substrate, while an indium tin oxide (ITO) coated quartz glass was used as the anode electrode. The distance between the GaN emitter and ITO anode was controlled by the 7.5 or 20 μm thick Kapton spacer disk with 1 mm hole diameter. The light was focused onto the field emitter in a high-vacuum chamber through the quartz glass from inside or outside the positioned laser or Light Emitted Diode (LED). LEDs were connected to the anode side for illuminating the sample during FE measurements (Figure 7.40). Laser diodes with different wavelengths, namely, infrared (IR) ($\lambda \approx 880$ nm, P = 5 mW), red (R) ($\lambda \approx 650\,\text{nm}, P = 5\,\text{mW}$), green (G) ($\lambda \approx 532\,\text{nm}, P = 5\,\text{mW}$), and ultraviolet (UV) ($\lambda \approx 365\,\text{nm}, P = 110\,\text{mW}$) were used for the photoassisted field emission. The vacuum achieved in the FE setup used for the characterization was $\sim 2 \times 10^{-8}$ mbar.

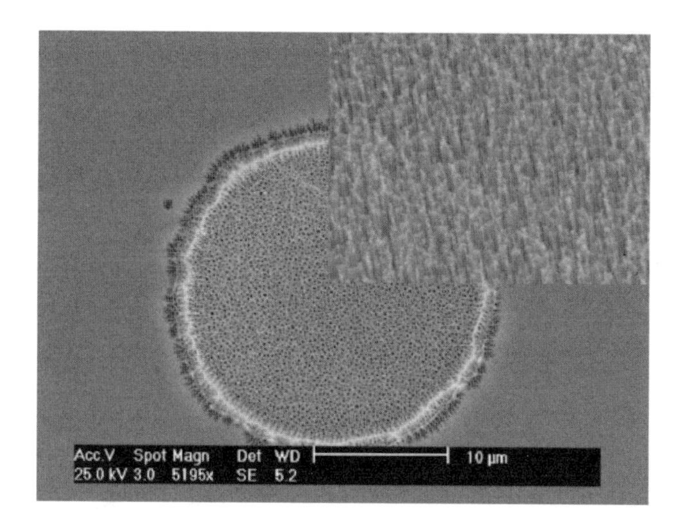

Figure 7.39 SEM image of the argon plasma (20 minutes) and PEC (12 minutes) etched GaN rods. Only n^+-GaN mesa region is affected. On the top right side is the side view of the GaN rods. Reproduced with permission from Ref. [139]. Copyright (2008), AIP Publishing LLC

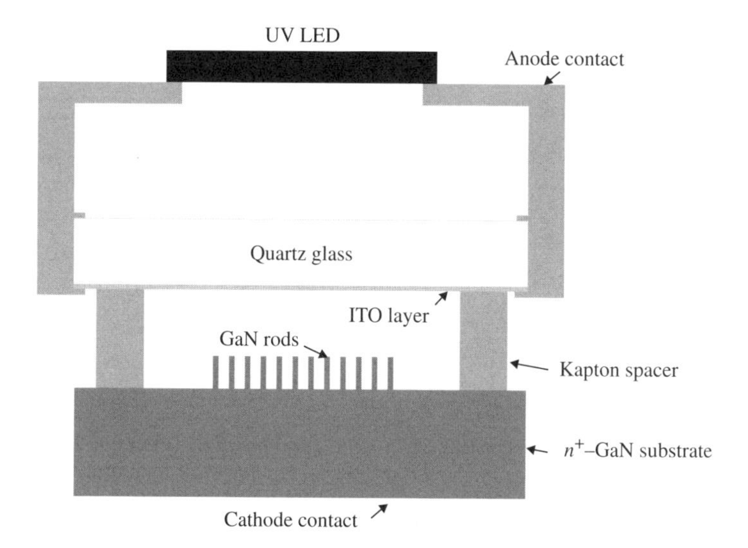

Figure 7.40 Biasing scheme of the GaN rods for photoexcited FE spectroscopy in the vacuum chamber. Reproduced with permission from Ref. [139]. Copyright (2008), AIP Publishing LLC

The experimental electron-field-emission curve in the Fowler–Nordheim coordinates ($lg\ I/V^2 - 1/V$) has two different slopes, namely, a lower slope at low voltages and a higher slope at high voltages (Figure 7.41a). The voltage of the slope change is equal to 670 V, which corresponds to a macroscopic electric field of 8.9×10^5 V/cm. The effective electron affinities

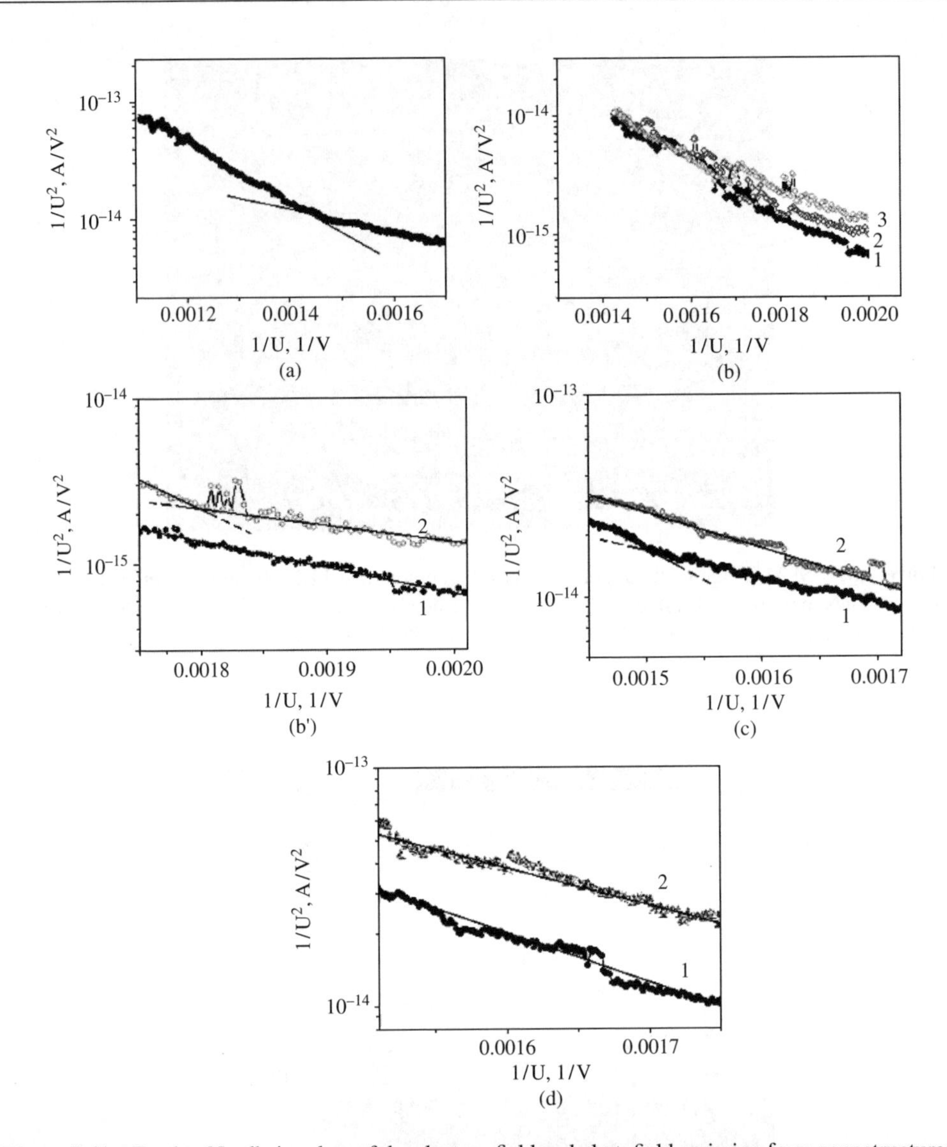

Figure 7.41 Fowler–Nordheim plots of the electron field and photofield emission from nanostructured GaN rods: (a) F–N curve without illumination; (b,b′) without and with IR light illumination (in different scales); (c) without and with G light illumination; and (d) without and with UV light illumination: (1 without illumination; 2 and 3 under illumination). Reproduced with permission from Ref. [207]. Copyright (2010), American Vacuum Society

(barriers for electron tunneling at field emission) have been determined from the slopes of the F-N plot,

$$\chi_{ef} = \chi / \beta^{2/3}, \tag{7.44}$$

where χ is the electron affinity and β is the electric field enhancement coefficient.

The ratio of $\chi_1/\chi_2 = \chi_{1ef}/\chi_{2ef}$ is equal to 2.07. This is in good agreement with the ratio of electron affinities of the GaN Γ valley (3.3 eV) and the X valley (2.21 eV) [144, 198] if we take into account the energy barrier lowering due to Schottky effect ($\Delta\Phi \approx 1.1$ eV).

The electron-field emission from the upper X valley dominates in the low-voltage region. The electric field in GaN is high enough to heat some of the electrons and transfer them from the Γ valley into the X valley. Because the X-valley electrons go through a lower and thinner vacuum barrier (high tunneling probability), their contribution to emission current dominates. But following the growth of voltage and corresponding lowering and thinning of the energy barrier due to the Schottky effect, the influence of Γ-valley electrons is increased. These processes were theoretically described in Ref. [148].

During the investigation of the photofield emission, the influence of illumination on the emission current was observed only in the low-voltage region of the curve. The experimental curves in F-N coordinates without and with illumination were similar in the high-voltage region.

7.4.5.1 IR Irradiation

The experimental field and photofield emission curves at IR illumination are shown in Figures 7.41b,b′. The increase in emission current (1.8 times) and decrease in the curve slope were observed. Under IR illumination, the change of the emission curve slope was at lower voltage ($V = 554$ V, $E_{th} = 7.4 = 10^5$ V/cm). The effective electron affinities were determined from the slopes of the F-N plot measured without (Figure 7.41b′, curve 1) and with IR illumination (Figure 7.41b′, curve 2). The IR light excited the electrons from the bottom of the Γ-valley. Excited electrons gave significant influence on the emission current. The comparison of experimental curve slopes without and with illumination allowed to estimate the energy barrier of IR light-excited tunneling electrons as $\Phi_{IR} = 1.34$ eV (taking into account a Schottky barrier lowering of $\Delta\Phi = 0.56$ eV). The difference of the Γ-valley electron affinity and the energy barrier of IR light-excited electrons was 1.40 eV, which practically coincided with the energy of the IR photon ($h\nu = 1.41$ eV). In this case, the subbandgap transitions (inner band electron transitions) caused by photon adsorption were realized. The adsorption coefficient due to subbandgap transitions in GaN increased with the photon energy [209].

7.4.5.2 Green Light Irradiation

Fowler–Nordheim plots before and under G light illumination are shown in Figure 7.41(c) and Figure 7.43. In the case of green-light illumination ($h\nu_G = 2.25$ eV), the sum of $\Phi_G + h\nu_G$ is noticeably higher (3.8–4.0 eV versus $\Phi_\Gamma = 3.3$ eV), where Φ_G is the energy barrier of green light-excited tunneling electrons determined from experiment. This indicates a partial electron thermalization before the field emission took place.

During electron-field emission from GaN mesa without PEC etching, an increase of the current due to G light illumination was also observed. On the other hand, R light had small influence on emission. This fact could be explained by the lower R light-absorption coefficient in GaN. Some other peculiarities were obtained for photofield emission from the GaN mesa (without PEC etching). One such peculiarity was the sharp growth of the current at a determined voltage (sharp conditioning). The maximum applied voltage was successively increased with initial emission 550 V. The macroscopic-threshold electric field for a cathode-anode distance of 20 μm was 2.75×10^5 V/cm. After the sharp conditioning at 800 V, the G light influence disappeared. Another peculiarity of the photofield emission was the memory effect. The electron-field emission curves were practically coincided in subsequent measurements. At R light illumination, a small growth in current was observed. The subsequent measured curves without illumination showed equal results. Further measurements at G light illumination demonstrated an increase in the emission-current with similar values for the subsequent measurements without illumination.

7.4.5.3 UV Irradiation

The initial FE current was set to several nanoamperes. Upon emitter illumination, the current increased immediately and then fluctuated around a value, which was up to 1 order of magnitude higher than the initial current. Additional electron-hole pairs were generated by absorption of UV photons with energies larger than the bandgap of the GaN nanorods (3.4 eV). Electrons reached the surface quickly due to the small diameter of the GaN nanorods. The field necessary for FE penetrated into the depleted GaN rod (electrical field $E > E_{th} = 150\,\mathrm{kV/cm}$) and enabled electron transfer from the Γ valley to the satellite valleys. The photoexcited electrons contributed to the FE process from the satellite valley due to their smaller electron affinity χ (Figure 7.42). The electron FE current shown in the F-N plot (Figure 7.43 and 7.41(d)) clearly demonstrates two different slopes for nonilluminated and UV illuminated emissions. In the case where the GaN rods are not illuminated, the emitter is almost depleted and emission takes place from the Γ valley ($\chi = 3.3$ eV).

Upon UV illumination, new electrons are generated from the valence band and they occupy the upper valley. The main emission occurs from this valley, which has lower electron affinity. The presence of a high electric field in the emitter tip supports the presence of the electron transfer effect. The electron affinity difference (equal to the energy difference between the valleys) can also be calculated by evaluating the two slopes m of the F-N characteristic.

$$m = -2.84 \times 10^7 (\Phi^{3/2}/\beta). \tag{7.45}$$

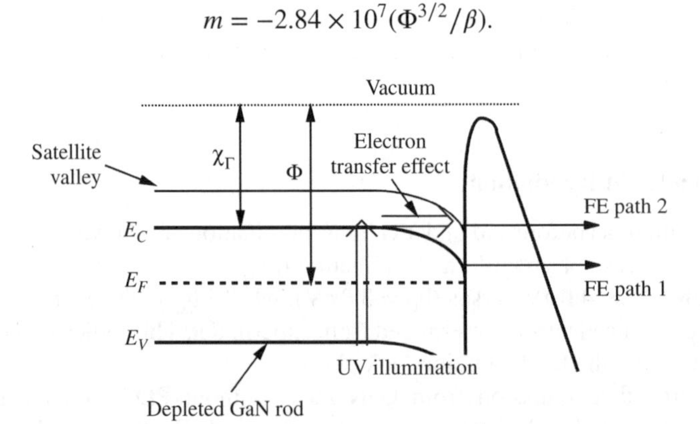

Figure 7.42 Energy band diagram of a depleted GaN rod and FE path from the Γ valley (FE path 1, electron affinity χ_Γ) and from the satellite valley (FE path 2, $\chi_X < \chi_\Gamma$) without and with UV illumination, respectively. The applied field penetrates into the GaN rod with $E > 150$ kV/cm. Reproduced with permission from Ref. [139]. Copyright (2008), AIP Publishing LLC

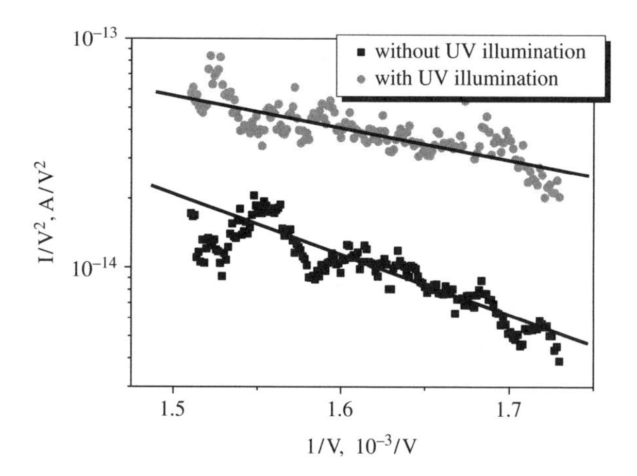

Figure 7.43 F-N plots of the electron FE from nanostructured GaN rods. The slope decreases under UV illumination. Reproduced with permission from Ref. [139]. Copyright (2008), AIP Publishing LLC

Correct estimation of the field enhancement coefficient β is difficult to be made from the SEM image of nanostructured surfaces and so is the calculation of the absolute value of the real work function Φ. The ratio of Φ_1/Φ_2 (work function for Γ valley/ work function for the satellite valley) can, however, be determined from the two emission F-N curves (Figure 7.42) with the same β.

$$\Phi_1/\Phi_2 = \chi_\Gamma/\chi_X = (m_1/m_2)^{2/3} \tag{7.46}$$

The obtained value of $\chi_\Gamma/\chi_X = 1.53$ (GaN rod) can be used to calculate χ_X for given values of χ_Γ, that is, $\chi_\Gamma = 3.3$ eV. The calculated energy difference between the valleys $\Delta E = 1.15$ eV is in good agreement with the values published in Ref. [198] ($\Delta E = 1.1$ eV). The UV illumination and the presence of a high electric field in the emitter tip permit the manifestation of the electron transfer effect in the GaN field emitter structure, leading to the change in the emission characteristics. The current from the upper valley becomes dominant at a rather small electron concentration n in the upper valley due to much smaller electron affinity in this valley, as compared to the Γ valley.

7.5 Conclusions

The fundamental properties of wide bandgap GaN emitters with their low-dimensional (1D and 2D) structures have been investigated. Field emitters of this type have applications for various vacuum nanoelectronic devices and they have showed unique functions with high performance.

The small electron affinity of wide gap structures opened a new way for realizing effective electron field emitters, especially when combined with nano-dimensional features offered, for example by nanowires. The integration of vacuum nano-electronics with laser sources for modulated electron emission allowed an additional degree of functionality involving high-frequency and optical signal interactions. Here the emitted electrons are generated by electron bunching within the cathode and give rise to a signal that can be of very high frequency due to the inherent properties of wide bandgap components, as well as, the fact that signal emission takes place in a different region (vacuum microtube). The use of silicon-based microcavities in conjunction with nanowire emitters will allow the demonstration of single and array type vacuum microtubes (triodes and Traveling Wave Tubes) for high frequency applications.

Photo-assisted field-emission experiments on various WBGS materials allowed a systematic study of the material properties including details of their bandstructure. Nanowires were studied using various semiconductors such as III-N (GaN, AlGaN). This allowed an in-depth study of their energy band diagram such as electron affinities and satellite valley positions. The quantum confinement effect strongly changed the energy band diagram of nanostructures and consequently, the main physical properties of the materials of which they were made. 1D and 0D structures of semiconductors are ideally suited for the realization of devices for high-speed/high-frequency.

References

1. D.-S. Yang, O.F. Mohammed, and A.H. Zewail, Scanning ultrafast electron microscopy, *Proceedings of the National Academy of Sciences of the United States of America* **107**, 14993–14998 (2010).
2. M. Nakagawa, Y. Hanawa, and T. Sakata, CdTe x-ray image sensor using a field emitter array, *Journal of Vacuum Science and Technology B* **27**, 725–728 (2009).
3. B. Gelmont, K. Kim, and M. Shur, Monte Carlo simulation of electron transport in gallium nitride, *Journal of Applied Physics* **74**, 1818 (1993).
4. S.C. Jain, M. Willander, J. Narayan, and R. van Overstraeten, III–nitrides: Growth, characterization, and properties, *Journal of Applied Physics* **87**, 965 (2000).
5. M.-S. Lin, K.-H. Huang, P.-S. Lu, *et al.*, Field-emission based vacuum device for the generation of terahertz waves, *Journal of Vacuum Science and Technology B* **23**, 849 (2005).
6. I. Brodie and C.A. Spindt. "Vacuum microelectronics." *Advances in Electronics and Electron Physics*, Vol. **83**, ed. P.W. Hawkes, Academic Press: New York, 1–106 (1992).
7. K. Yokoo (1999) Functional field emission for high frequency wave application. *Technical digest of IVMC 99, Darmstadt, Germany*, pp. 206–207.
8. O. Yilmazoglu , H. Mimura , K. Mutamba , *et al.* (2001) Generation of a bunched electron beam by field-emitter structures. *ITG Proc. 165, Conference on Displays and Vacuum Electronics, Garmisch-Partenkirchen, Germany, May, 2001*, pp. 263–267.
9. E.G. Zaidman and M.A. Kodis, Emission gated device issues, *IEEE Transactions on Electron Devices* **38**, 2221 (1991).
10. D.R. Whaley, B.M. Gannon, C.R. Smith, *et al.*, Application of field emitter arrays to microwave power amplifiers, *IEEE Transactions on Plasma Science* **28**, 727 (2000).
11. J.L. Shaw, H.F. Gray, K.L. Jensen, and T.M. Jung, Graded electron affinity electron source, *Journal of Vacuum Science and Technology B* **14**, 2072 (1996).
12. R.A. King, R.A.D. Mackenzie, G.D.W. Smith, and N.A. Cade, Atom probe analysis and field emission studies of silicon, *Journal of Vacuum Science and Technology B* **12**, 705 (1994).
13. C. Py and R. Baptist, Stability of the emission of a microtip, *Journal of Vacuum Science and Technology B* **12**, 685 (1994).
14. A.A. Talin, T.E. Fleter, and D.J. Devine, Effects of potassium and lithium metal deposition on the emission characteristics of Spindt-type thin-film field emission microcathode arrays, *Journal of Vacuum Science and Technology B* **13**, 448 (1995).
15. Xie, T., Mackie, W.A., and Davis, P.R. (1995) Field emission from ZrC films on single emitters and emitters arrays, *Proceedings 8th International Vacuum Nanoelectronics Conference (Portland, Oregon, USA, July 30–August 3, 1995)*, pp. 403–407.
16. P.R. Schwoebel, C.A. Spindt, and I. Brodie, Electron emission enhancement by overcoating molybdenum field-emitter arrays with titanium, zirconium, and hafnium, *Journal of Vacuum Science and Technology B* **13**, 338 (1995).
17. P. R. Schwoebel, and C.A. Spindt, Glow discharge processing to enhance field-emitter array performance, *Journal of Vacuum Science and Technology B* **12**, 2414 (1994).

18. S. Meassick, Z. Xia, C. Chen, and J. Browning, Investigation of the operating modes of gated vacuum field emitter arrays to reduce failure rates, *Journal of Vacuum Science and Technology B* **12**, 710 (1994).
19. R.J. Nemanich, P.K. Baumann, M.C. Benjamin, *et al.*, Electron emission properties of crystalline diamond and III-nitride surfaces, *Applied Surface Science* 130–132, 694 (1998).
20. S. Strite and H. Morkoc, GaN, AlN, and InN: A review, *Journal of Vacuum Science and Technology B* **10**, 1237 (1992).
21. M.C. Benjamin, C. Wang. R.F. Davis, and R.J. Nemanich, Observation of a negative electron affinity for heteroepitaxial AlN on α(6H)-SiC(0001), *Applied Physics Letters* **64**. 3288 (1994).
22. C.I. Wu, A. Kahn, E.S. Hellman, and N.E. Buchanan, Electron affinity at aluminum nitride surfaces, *Applied Physics Letters* **73**, 1346 (1998).
23. J.C. She, S.E. Huq, J. Chen, *et al.*, Comparative study of electron emission characteristics of silicon tip arrays with and without amorphous diamond coating, *Journal of Vacuum Science and Technology B* **17**, 592 (1999).
24. R.D. Underwood, S. Keller, U.K. Mishra, *et al.*, GaN field emitter array with integrated anode, *Journal of Vacuum Science and Technology B* **16**, 340 (1997).
25. T. Kozawa, M. Suzuki, Y. Taga, *et al.*, Fabrication of GaN field emitter arrays by selective area growth technique, *Journal of Vacuum Science and Technology B* **16**, 833 (1998).
26. B.L. Ward, O.-H. Nam, J.D. Hartman, *et al.*, Electron emission characteristics of GaN pyramid arrays grown via organometallic vapor phase epitaxy, *Journal of Applied Physics* **84**, 5238 (1998).
27. A.T. Sowers, J.A. Christman, M.D. Bremser, *et al.*, Thin films of aluminum nitride and aluminum gallium nitride for cold cathode applications, *Applied Physics Letters* **71**, 2289 (1997).
28. M. Kasu and N. Kobayashi, Large and stable field-emission current from heavily Si-doped AlN grown by metalorganic vapor phase epitaxy, *Applied Physics Letters* **76**, 2910–2912 (2000).
29. V.M. Bermudez, T.M. Jung, K. Doversoike, and A.E. Wickenden, The growth and properties of Al and AlN films on GaN(0001)–(1×1), *Journal of Applied Physics* **79**, 110 (1996).
30. R.D. Underwood, D. Kapolnek, B.P. Keller, *et al.*, Selective-area regrowth of GaN field emission tips, *Solid-State Electronics* **41**, 243 (1997).
31. I. Berishev, A. Bensaoula, I. Rusakova, *et al.*, Field emission properties of GaN films on Si(111), *Applied Physics Letters* **73**, 1808 (1998).
32. P. Lerner, P.H. Cutler, and N.M. Miskovsky, Theoretical analysis of field emission from a metal diamond cold cathode emitter, *Journal of Vacuum Science and Technology B* **15**, 337 (1997).
33. M.S. Chung, N.M. Miskovsky, P.H. Cutler, and N. Kumar, Band structure calculation of field emission from $Al_xGa_{1-x}N$ as a function of stoichiometry *Applied Physics Letters* **76**, 1143–1145 (2000).
34. T.T. Tsong, Field penetration and band bending near semiconductor surfaces in high electric fields, *Surface Science* **81**, 28 (1979).
35. N.D. Lang, The density-functional formalism and the electronic structure of metal surfaces, *Solid State Physics* **28**, 225 (1973).
36. W.W. Lui and M. Fukuma. Exact solution of the Schrodinger equation across an arbitrary one-dimensional piecewise-linear potential barrier, *Journal of Applied Physics* **60**, 1555 (1986).
37. O. Madelung, *Semiconductor Basic Data*, 2nd edn (Springer, New York, 1996).
38. H. Morkoc, S. Strite, G.B. Gao, *et al.*, Large-band-gap SiC, III-V nitride, and II-VI ZnSe-based semiconductor device technologies, *Journal of Applied Physics* **76**, 1363 (1994).
39. M.C. Benjamin, M.D. Bremser, T.W. Weeks, Jr., *et al.*, UV photoemission study of heteroepitaxial AlGaN films grown on 6H-SiC, *Applied Surface Science* **104**, 455 (1996).
40. C. Stampfl and C.G. Van de Walle, Doping of $Al_xGa_{1-x}N$, *Applied Physics Letters* **72**, 459 (1998).
41. J.M. Redwing, J.S. Flynn, M.A. Tischler, *et al.*, MOVPE growth of high electron mobility AlGaN/GaN heterostructures, *Materials Research Society Symposia Proceedings* **395**, 201 (1996).
42. V.T. Binh, J.P. Dupin, P. Thevenard, *et al.*, Serial process for electron emission from solid-state field controlled emitters, *Journal of Vacuum Science and Technology B* **18**, 956–961 (2000).
43. V.T. Binh and C. Adessi, New mechanism for electron emission from planar cold cathodes: The solid-state field-controlled electron emitter, *Physical Review Letters* **85**, 864–867 (2000).
44. V.T. Binh, V. Semet, J.P. Dupin, and D. Guillot. Recent progress in the characterization of electron emission from solid-state field-controlled emitters, *Journal of Vacuum Science and Technology B* **19**, 1044 (2001).
45. S.M. Sze, *Physics of Semiconductor Devices* (Wiley-Interscience, New York, 1981).
46. E.N. Economou, *Green's Functions in Quantum Physics* (Springer, 2006).
47. B.A. Lippmann and J. Schwinger, Variational principles for scattering processes. I, *Physical Review* **79**, p. 469, (1950)

48. K.L. Jensen and A.K. Ganguly, Time dependent, self-consistent simulations of field emission from silicon using the Wigner distribution function, *Journal of Vacuum Science and Technology B* **12**, 770 (1994).
49. K.L. Jensen, Numerical simulation of field emission from silicon, *Journal of Vacuum Science and Technology B* **11**, 371 (1993).
50. K.L. Jensen and A.K. Ganduly, Numerical simulation of field emission and tunneling: A comparison of the Wigner function and transmission coefficient approaches, *Journal of Applied Physics* **73**, 4409 (1993).
51. C.B. Duke, *Tunneling in Solids* (Academic Press, New York, 1969).
52. N.D. Lang and A.R. Williams, Theory of atomic chemisorption on simple metals, *Physical Review B* **18**, 616 (1978).
53. A.A. Lucas, Morawitz, H., Henry, G.R. *et al.*, Scattering-theoretic approach to elastic one-electron tunneling through localized barriers: Application to scanning tunneling microscopy, *Physical Review B* **37**, 10708 (1988).
54. W. Zhao, R.-Z. Wang, X.-M. Song, *et al.*, Ultralow-threshold field emission from oriented nanostructured GaN films on Si substrate, *Applied Physics Letters* **96**, 092101 (2010).
55. R.D. Underwood, P. Kozodoy, S. Keller, *et al.*, Piezoelectric surface barrier lowering applied to InGaN/GaN field emitter arrays, *Applied Physics Letters* **73**, 405 (1998).
56. J. Simon, V. Protasenko, C. Lian, *et al.*, Polarization-induced hole doping in wide–band-gap uniaxial semiconductor heterostructures, *Science* **327**, 60 (2010).
57. X. Wang, J. Song, F. Zhang, *et al.*, Electricity generation based on one-dimensional group-III Nitride nanomaterials, *Advanced Materials* **22**, 2155 (2010).
58. V. Fiorentini, F. Bernardini, and O. Ambacher, Evidence for nonlinear macroscopic polarization in III-V nitride alloy heterostructures, *Applied Physics Letters* **80**, 1204 (2002).
59. F. Bernardini, V. Fiorentini, and D. Vanderbilt, Spontaneous polarization and piezoelectric constants of III-V nitrides, *Physical Review B* **56**, R10024 (1997).
60. V. Ranjan, G. Allan, C. Priesler, and C. Delerue, Self-consistent calculations of the optical properties of GaN quantum dots, *Physical Review B* **68**, 115305 (2003).
61. F. Bechstedt, U. Grossner, and J. Furthmiiller, Dynamics and polarization of group-III nitride lattices: A first-principles study, *Physical Review B* **62**, 8003 (2000).
62. Z.Q. Duan, R.Z. Wang, R.Y. Yuan, *et al.*, Field emission mechanism from a single-layer ultra-thin semiconductor film cathode, *Journal of Physics D* **40**, 5828 (2007).
63. R.Z. Wang, H. Yan, B. Wang, *et al.*, Field emission enhancement by the quantum structure in an ultrathin multilayer planar cold cathode, *Applied Physics Letters* **92**, 142102 (2008).
64. Z.J. Reitmeier, S. Einfeldt, R.F. Davis, *et al.*, Surface and defect microstructure of GaN and AlN layers grown on hydrogen-etched 6H–SiC(0001) substrates, *Acta Materialia* **58**, 2165 (2010).
65. R.F. Davis, S. Einfeldt, E.A. Preble, A.M. Roskowski, *et al.*, Gallium nitride and related materials: challenges in materials processing, *Acta Materialia* **51**, 5961 (2003).
66. W. Zhao, R.-Z. Wang, Z.-W. Song, *et al.*, Crystallization effects of nanocrystalline GaN films on field emission, *Journal of Physical Chemistry C* **117**, 1518 (2013).
67. A. Evtukh, O. Yilmazoglu, V. Litovchenko, *et al.*, Electron field emission from nanostructured surfaces of GaN and AlGaN, *Physica Status Solidi C* **5**, 425 (2008).
68. P. Deb, T. Westover, H. Kim, *et al.*, Field emission from GaN and (Al,Ga)N/GaN nanorod heterostructures, *Journal of Vacuum Science and Technology B* **25**, L15 (2007).
69. T. Sugino, T. Hori, C. Kimura, and T. Yamamoto, Field emission from GaN surfaces roughened by hydrogen plasma treatment, *Applied Physics Letters* **78**, 3229 (2001).
70. L.T. Fu, Z.G. Chen, D.W. Wang, *et al.*, Wurtzite P-doped GaN triangular microtubes as field emitters, *Journal of Physical Chemistry C* **114**, 9627 (2010).
71. B. Liu, Y. Bando, C. Tang, *et al.*, Quasi-aligned single-crystalline GaN nanowire arrays, *Applied Physics Letters* **87**, 073106 (2005).
72. W. Jang, S. Kim, J. Lee, *et al.*, Triangular GaN–BN core–shell nanocables: Synthesis and field emission, *Chemical Physics Letters* **422**, 41 (2006).
73. D.V. Dinh, S.M. Kang, J.H. Yang, *et al.*, Synthesis and field emission properties of triangular-shaped GaN nanowires on Si(100) substrates, *Journal of Crystal Growth* **311**, 495 (2009).
74. B. Ha, S.H. Seo, J.H. Cho, *et al.*, Optical and field emission properties of thin single-crystalline GaN nanowires, *Journal of Physical Chemistry B* **109**, 11095 (2005).
75. B. Liu, Y. Bando, C. Tang, *et al.*, Needlelike bicrystalline GaN nanowires with excellent field emission properties, *Journal of Physical Chemistry B* **109**, 17082 (2005).

76. C. Lin, G. Yu, X. Wang, *et al.*, Catalyst-free growth of well vertically aligned GaN needlelike nanowire array with low-field electron emission properties, *Journal of Physical Chemistry C* **112**, 18821 (2008).
77. L. Luo, Field emission from GaN nanobelts with herringbone morphology, *Materials Letters* **58**, 2893 (2004).
78. D.K.T. Ng, M.H. Hong, L.S. Tan, *et al.*, Field emission enhancement from patterned gallium nitride nanowires, *Nanotechnology* **18**, 375707 (2007).
79. T. Ikeda and K. Teii, Origin of low threshold field emission from nitrogen-incorporated nanocrystalline diamond films, *Applied Physics Letters* **94**, 143102 (2009).
80. P. Zapol, M. Sternberg, L.A. Curtiss, *et al.*, Tight-binding molecular-dynamics simulation of impurities in ultra-nanocrystalline diamond grain boundaries, *Physical Review B* **65**, 045403 (2001).
81. F. Ye, E.Q. Xie, X.J. Pan, *et al.*, Field emission from amorphous GaN deposited on Si by dc sputtering, *Journal of Vacuum Science and Technology B* **24**, 1358 (2006).
82. D.S. Joag, D.J. Late, and U.D. Lanke, Field emission from a-GaN films deposited on Si (100), *Solid State Communications* **130**, 305 (2004).
83. S. Ganguly, L.F. Register, S. Banerjee, and A.H. MacDonald, Bias-voltage-controlled magnetization switch in ferromagnetic semiconductor resonant tunneling diodes, *Physical Review B* **71**, 245306 (2005).
84. Y. Ando and T. Itoh, Calculation of transmission tunneling current across arbitrary potential barriers, *Journal of Applied Physics* **61**, 1497 (1987).
85. V. Semet, V.T. Binh, J.P. Zhang, *et al.*, Electron emission through a multilayer planar nanostructured solid-state field-controlled emitter, *Applied Physics Letters* **84**, 1937 (2004).
86. R.Z. Wang, X.M. Ding, B. Wang, *et al.*, Structural enhancement mechanism of field emission from multilayer semiconductor films, *Physical Review B* **72**, 125310 (2005).
87. R. Tsu and L. Esaki, Tunneling in a finite superlattice, *Applied Physics Letters* **22**, 562 (1973).
88. R.Z. Wang, X.M. Ding, K. Xue, *et al.*, Multipeak characteristics of field emission energy distribution from semiconductors, *Physical Review B* **70**, 195305 (2004).
89. W. Zhao, R.Z. Wang, X.-M. Song, *et al.*, Electron field emission enhanced by geometric and quantum effects from nanostructured AlGaN/GaN quantum wells, *Applied Physics Letters* **98**, 152110 (2011).
90. A. Ishida, Y. Inoue, and H. Fujiyasu, Resonant-tunneling electron emitter in an AlN/GaN system, *Applied Physics Letters* **86**, 183102 (2005).
91. J.W. Gadzuk and E.W. Plummer, Field emission energy distribution (FEED), *Reviews of Modern Physics* **45**, 487 (1973).
92. A. Reimann, *Thermionic Emission* (John Wiley & Sons, Inc, New York, 1934).
93. L. Esaki, Long journey into tunneling, *Reviews of Modern Physics* **46**, 237 (1974).
94. Y. Yu, R.F. Greene, and R. Tsu, Cooling by inverse Nottingham effect with resonant tunneling. In *Advanced Semiconductor Helerostructures*, ed. M. Dutta and M.A. Stroscio (World Scientific, New Jersey, 2003), Vol. **28**, pp. 145–162.
95. X. Zhang, L. Gong, K. Liu, *et al.*, Tungsten oxide nanowires grown on carbon cloth as a flexible cold cathode, *Advanced Materials* **22**, 5292 (2010).
96. R. H. Fowler and L. Nordheim, Electron emission in intense electric fields, *Proceedings of the Royal Society of London, Series A* **119**, 173 (1928).
97. C. Li, Y. Zhang, M. Mann, *et al.*, High emission current density, vertically aligned carbon nanotube mesh, field emitter array, *Applied Physics Letters* **97**, 113107 (2010).
98. X.L. Tong, D.S. Jiang, Y. Li, *et al.*, Folding field emission from GaN onto polymer microtip array by femtosecond pulsed laser deposition, *Applied Physics Letters* **89**, 061108 (2006).
99. K. Zhu, V. Kuryatkov, B. Borisov, *et al.*, Evolution of surface roughness of AlN and GaN induced by inductively coupled Cl2/Ar plasma etching, *Journal of Applied Physics* **95**, 4635 (2004).
100. D. Li, M. Sumiya, S. Fuke, *et al.*, Selective etching of GaN polar surface in potassium hydroxide solution studied by x-ray photoelectron spectroscopy, *Journal of Applied Physics* **90**, 4219 (2001).
101. N. Shiozaki and T. Hashizume, Improvements of electronic and optical characteristics of n-GaN-based structures by photoelectrochemical oxidation in glycol solution, *Journal of Applied Physics* **105**, 064912 (2009).
102. M. Lyth and S.R.P. Silva, Resonant behavior observed in electron field emission from acid functionalized multiwall carbon nanotubes, *Applied Physics Letters* **94**, 123102 (2009).
103. L.D. Filip, M. Palumbo, J.D. Carey, and S.R.P. Silva, Two-step electron tunneling from confined electronic states in a nanoparticle, *Physical Review B* **79**, 245429 (2009).
104. W.M. Tsang, S.J. Henley, V. Stolojan, and S.R P. Silva, Negative differential conductance observed in electron field emission from band gap modulated amorphous-carbon nanolayers, *Applied Physics Letters* **89**, 193103 (2006).

105. T. Yamada, S.I. Shikata, and C.E. Nebel, Resonant field emission from two-dimensional density of state on hydrogen-terminated intrinsic diamond, *Journal of Applied Physics* **107**, 013705 (2010).
106. H. Yamaguchi, T. Masuzawa, S. Nozue, *et al.*, Electron emission from conduction band of diamond with negative electron affinity, *Physical Review B* **80**, 165321 (2009).
107. C.I. Wu and A. Kahn, Negative electron affinity at the Cs/AlN(0001) surface, *Applied Physics Letters* **74**, 1433 (1999).
108. Z. Liu, F. Machuca, P. Pianetta, *et al.*, Electron scattering study within the depletion region of the GaN(0001) and the GaAs(100) surface, *Applied Physics Letters* **85**, 1541 (2004).
109. A. Ishida, T. Ose, H. Nagasawa, *et al.*, Quantum-cascade structure in AlN/GaN system assisted by piezo-electric effect, *Japanese Journal of Applied Physics Part 2* **41**, L236 (2002).
110. A. Ishida, Y. Inoue, M. Kuwabara, *et al.*, AlN/GaN near-infrared quantum-cascade structures with resonant-tunneling injectors utilizing polarization fields, *Japanese Journal of Applied Physics Part 2* **41**, L1303 (2002).
111. A. Kikuchi, R. Bannai, and K. Kishino, AlGaN resonant tunneling diodes grown by rf-MBE, *Physica Status Solidi A* **188**, 187 (2001).
112. G. Bastard, Superlattice band structure in the envelope-function approximation, *Physical Review B* **24**, 5693 (1981).
113. X. Duan, Y. Huang, Y. Cui, *et al.*, Indium phosphide nanowires as building blocks for nanoscale electronic and optoelectronic devices, *Nature* **409**, 66 (2001).
114. W.Q. Han, S.S. Fan, Q.Q. Li, and Y.D. Hu, Synthesis of gallium nitride nanorods through a carbon nanotube-confined reaction, *Science* **277**, 1287 (1997).
115. H. Yoshida, T. Urushido, H. Miyake, and K. Hiramatsu, Formation of GaN self-organized nanotips by reactive ion etching, *Japanese Journal of Applied Physics* **40,** L1301 (2001).
116. R.J. Nemanich, M.C. Benjamin, S.P. Bozeman, *et al.*, "(Negative) Electron affinity of AlN and AlGaN alloys, *Materials Research Society Symposia Proceedings* **395,** 777 (1996).
117. J.I. Pankove and H. Schade, Photoemission from GaN, *Applied Physics Letters* **25,** 53 (1974).
118. W. Czarczynski, S. Lasisz, M. Moraw, *et al.*, Field emission from GaN deposited on the (100) Si substrate, *Applied Surface Science* **151**, 63 (1999).
119. C.C. Chen, C.C. Yeh, C.H. Chen, *et al.*, Catalytic growth and characterization of gallium nitride nanowires, *Journal of the American Chemical Society* **123**, 2791 (2001).
120. X. Xiang and H. Zhu. One-dimensional gallium nitride micro/nanostructures synthesized by a space-confined growth technique, *Applied Physics A* **87**, 651 (2007).
121. Z. Chen, C.B. Cao, W.S. Li, and C. Surya, Well-aligned single-crystalline GaN nanocolumns and their field emission properties, *Crystal Growth and Design* **9**, 792 (2009).
122. G.S. Cheng, L.D. Zhang, Y. Zhu, *et al.*, Large-scale synthesis of single crystalline gallium nitride nanowires, *Applied Physics Letters* **75**, 2455 (1999).
123. Z. Liliental-Weber, Y. Chen, S. Ruvimov, and J. Washburn, Formation mechanism of nanotubes in GaN, *Physical Review Letters* **79**, 2835 (1997).
124. J.Y. Li, Z.Y. Qiao, X.L. Chen, *et al.*, Morphologies of GaN one-dimensional materials, *Applied Physics A* **71**, 587 (2000).
125. S.Y. Bae, H.W. Seo, J. Park, *et al.*, Single-crystalline gallium nitride nanobelts, *Applied Physics Letters* **81**, 126 (2002).
126. J. Su, G. Cui, M. Gherasimova, *et al.*, Catalytic growth of group III-nitride nanowires and nanostructures by metalorganic chemical vapor deposition, *Applied Physics Letters* **86**, 013105 (2005).
127. S. Gupta, H. Kang, M. Strassburg, *et al.*, A nucleation study of group III-nitride multifunctional nanostructures, *Journal of Crystal Growth* **287**, 596 (2006).
128. G. Nabi, C. Cao, W.S. Khan, *et al.*, Synthesis, characterization, growth mechanism, photoluminescence and field emission properties of novel dandelion-like gallium nitride, *Applied Surface Science* **257**, 10289 (2011).
129. X. F. Duan and C.M. Liber, Laser-assisted catalytic growth of single crystal GaN nanowires, *Journal of the American Chemical Society* **122**, 188 (2000).
130. T.Y. Kim, S.H. Lee, Y.H. Mo, *et al.*, Growth of GaN nanowires on Si substrate using Ni catalyst in vertical chemical vapor deposition reactor, *Journal of Crystal Growth* **257**, 97 (2003).
131. T.Y. Kim, S.H. Lee, Y.H. Mo, *et al.*, Growth of GaN nanowires on Si substrate using Ni catalyst in vertical chemical vapor deposition reactor, *Korean Journal of Chemical Engineering* **21**, 257 (2004).
132. Y. Terada, H. Yoshida, T. Urushido, *et al.*, Field emission from GaN self-organized nanotips, *Japanese Journal of Applied Physics Part 2* **41**, L1194 (2002).

133. P.B. Shah, B.M. Nichols, M.D. Derenge, and K.A. Jones, Sub-100 nm radius of curvature wide-band gap III-nitride vacuum microelectronic field emitter structures created by inductively coupled plasma etching, *Journal of Vacuum Science and Technology A* **22**, 1847 (2004).
134. V. Litovchenko and A. Grigorev, Determination of the electron affinity (work function) of semiconductor nanocrystals, *Ukrainian Journal of Physics* **52**, 897 (2007).
135. U.V. Bhapkar and M.S. Shur, Monte Carlo calculation of velocity-field characteristics of wurtzite GaN, *Journal of Applied Physics* **82**, 1649 (1997).
136. V. Litovchenko, A. Evtukh, Yu. Kruchenko, *et al.*, Quantum size resonance tunneling in the field emission phenomenon, *Journal of Applied Physics* **96**, 867 (2004).
137. V.G. Litovchenko, A.A. Evtukh, Yu. M. Litvin, *et al.*, Peculiarities of the electron field emission from quantum-size structures, *Applied Surface Science* **215**, 160 (2003).
138. O. Yilmazoglu, D. Pavlidis, Yu. M. Litvin, *et al.*, Field emission from quantum size GaN structures, *Applied Surface Science* **220**, 46 (2003).
139. O. Yilmazoglu, D. Pavlidis, H.L. Hartnagel, *et al.*, Evidence of satellite valley position in GaN by photoexcited field emission spectroscopy, *Journal of Applied Physics* **103**, 114511 (2008).
140. V. Litovchenko and A. Evtukh, Vacuum nanoelectronics. In *Handbook of Semiconductor Nanostructures and Nanodevices*, ed. A.A. Balandin and K.L. Wang (American Scientific Publishers, Los Angeles, CA, 2006), pp. 153–234.
141. A. Evtukh, O. Yilmazoglu, V. Litovchenko, *et al.*, Electron field emission from nanostructured surfaces of GaN and AlGaN, *Physica Status Solidi c* **5**, 425 (2008).
142. V. Litovchenko, A. Evtukh, O. Yilmazoglu, *et al.*, Gunn effect in the field-emission phenomena, *Journal of Applied Physics* **97**, 044911 (2005).
143. S. Sze, *Modern Semiconductor Device Physics* (John Wiley & Sons, Inc., New York, 1998).
144. R.J. Nemanich, in *Properties, Processing and Applications of Gallium Nitride and Related Semiconductors*, ed. J.H. Edgar, S. Strite, I. Akasaki, *et al.* (INSPEC, London, 1999), pp. 98–103.
145. V. Bugrov, M. Levinstein, S. Rumyantserv, and A. Zubrilov, *Advanced Semiconductor Materials* (John Wiley & Sons, Inc., New York, 2001), pp. 1–30.
146. N. Morgulis, About field emission composite semiconductor cathodes, *Journal of Technical Physics* **17**, 983 (1947).
147. R. Stratton, Energy distributions of field emitted electrons, *Physical Review* **135**, A794 (1964).
148. V. Litovchenko, A. Grygoriev, A. Evtukh, *et al.*, Electron field emission from wide bandgap semiconductors under intervalley carrier redistribution, *Journal of Applied Physics* **106**, 104511 (2009).
149. Van Zeghbroeck, B. (2007) Principles of Semiconductor Devices (tutorial), http://ecee.colorado.edu/~bart/book/book/index.html (accessed 20 January 2015).
150. J. Kolnik, I.H. Oguzman, and K.F. Brennan, Electronic transport studies of bulk zincblende and wurtzite phases of GaN based on an ensemble Monte Carlo calculation including a full zone band structure, *Journal of Applied Physics* **78**, 1033 (1995).
151. J. M. Barker, D.K. Ferry, D.D. Koleske, and R.J. Shul, Bulk GaN and AlGaN/GaN heterostructure drift velocity measurements and comparison to theoretical models, *Journal of Applied Physics* **97**, 063705 (2005).
152. V. Dienys and J. Pozela, *Hot Electrons* (Mintis, Vilnius, 1971).
153. V.I. Gavrilenko, A.M. Grekhov, D.V. Korbutjak, and V.G. Litovchenko, *Optical Properties of Semiconductors* (Naukova Dumka, Kiev, 1987).
154. Yu. Pozela, K. Pozela, V. Juciene, *et al.*, An increase in the electron mobility in the two-barrier AlGaAs/GaAs/AlGaAs heterostructure as a result of introduction of thin InAs barriers for polar optical phonons into the GaAs quantum well, *Semiconductors* **41**, 1439 (2007).
155. V.G. Mokerov, Yu. K. Pozela, and Yu. V. Fedorov, Electron transport in unipolar heterostructure transistors with quantum dots in strong electric fields, *Semiconductors* **37**, 1217 (2003).
156. J. Pozela, K. Pozela, and V. Juciene, Electron mobility and electron scattering by polar optical phonons in heterostructure quantum wells, *Semiconductors* **34**, 1011 (2000).
157. G. Nabi, C. Cao, W.S. Khan, *et al.*, Preparation of grass-like GaN nanostructures: Its PL and excellent field emission properties, *Materials Letters* **66**, 50 (2012).
158. W.S. Khan, C.B. Cao, Z. Chen, and G. Nabi, Synthesis, growth mechanism, photoluminescence and field emission properties of metal–semiconductor Zn–ZnO core–shell microcactuses, *Materials Chemistry and Physics* **124**, 493 (2010).
159. P. Deb, H. Kim, V. Rawat, *et al.*, Faceted and vertically aligned GaN nanorod arrays fabricated without catalysts or lithography, *Nano Letters* **5**, 1847 (2005).

160. G.T. Wang, A.A. Talin, D.J. Werder, *et al.*, Highly aligned, template-free growth and characterization of vertical GaN nanowires on sapphire by metal–organic chemical vapour deposition, *Nanotechnology* **17**, 5773 (2006).
161. Q. Li and G.T. Wang, Improvement in aligned GaN nanowire growth using submonolayer Ni catalyst films, *Applied Physics Letters* **93**, 043119 (2008).
162. B.D. Liu, Y. Bando, C.C. Tang, *et al.*, Excellent field-emission properties of P-doped GaN nanowires, *Journal of Physical Chemistry B* **109**, 21521 (2005).
163. T. Yamashita, S. Hasegawa, S. Nishida, *et al.*, Electron field emission from GaN nanorod films grown on Si substrates with native silicon oxides, *Applied Physics Letters* **86**, 082109 (2005).
164. K. Lee, C. Shin, I. Chen, and B. Li, The effect of nanoscale protrusions on field-emission properties for GaN nanowires, *Journal of the Electrochemical Society* **154** (10), K87–K91 (2007).
165. K.M. Tracy, W.J. Mecouch, R.F. Davis, and R.J. Nemanich, Preparation and characterization of atomically clean, stoichiometric surfaces of n- and p-type GaN(0001). *Journal of Applied Physics* **94**, 3163 (2003).
166. D.A. Neamen, *Semiconductor Physics and Devices*, 2nd edn, McGraw-Hill: New York, 2003; p 328.
167. O.H. Nam, M.D. Bremser, B.L. Ward, *et al.*, Growth of GaN and $Al_{0.2}Ga_{0.8}N$ on patterened substrates via organometallic vapor phase epitaxy, *Japanese Journal of Applied Physics Part 2* **36**, L532 (1997).
168. K.P. Loh, I. Sakaguchi, M.N. Gamo, *et al.*, Surface conditioning of chemical vapor deposited hexagonal boron nitride film for negative electron affinity, *Applied Physics Letters* **74**, 28 (1999).
169. M.J. Powers, M.C. Benjamin, L.M. Porter, *et al.*, Observation of a negative electron affinity for boron nitride, *Applied Physics Letters* **67**, 3912 (1995).
170. R.W. Pryor, Carbon-doped boron nitride cold cathodes, *Applied Physics Letters* **68**, 1802 (1996).
171. H.H. Busta and R.W. Pryor, Performance of laser ablated, laser annealed BN emitters deposited on polycrystalline diamond, *Journal of Vacuum Science and Technology B* **16**, 1207 (1998).
172. H.H. Busta and R.W. Pryor, Electron emission from a laser ablated and laser annealed BN thin film emitter, *Journal of Applied Physics* **82**, 5148 (1997).
173. T. Sugino, S. Kawasaki, K. Tanioka, and J. Shihafuji, Electron emission from boron nitride coated Si field emitters, *Applied Physics Letters* **71**, 2704 (1997).
174. C. Kimura, T. Yamamoto, T. Hori, and T. Sugino, Field emission characteristics of BN/GaN structure, *Applied Physics Letters* **79**, 4533 (2001).
175. T. Sugino, K. Tanioka, S. Kawasaki, and J. Shirafuji, Characterization and field emission of sulfur-doped boron nitride synthesized by plasma-assisted chemical vapor deposition, *Japanese Journal of Applied Physics Part 2* **36**, L463 (1997).
176. H. Kind, J.M. Bonard, C. Emmenegger, *et al.*, Patterned films of nanotubes using microcontact printing of catalysts, *Advanced Materials* **11,** 1285 (1999).
177. H.M. Manohara, M.J. Bronikowski, M. Hoenk, *et al.*, High-current-density field emitters based on arrays of carbon nanotube bundles, *Journal of Vacuum Science and Technology B* **23,** 157 (2005)
178. H.M. Kim, T.W. Kang, K.S. Chung, *et al.*, Field emission displays of wide-bandgap gallium nitride nanorod arrays grown by hydride vapor phase epitaxy, *Chemical Physics Letters* **377,** 491 (2003).
179. J.U. Seo, S. Hasegawa, and H. Asahi, Molecular-beam epitaxy fabrication and analysis of GaN nanorods on patterned silicon-on-insulator substrate, *Physica Status Solidi C* **5**, 3005 (2008).
180. R.J. Nemanich, P.K. Baumann, M.C. Benjamin, *et al.*, Negative electron affinity surfaces of aluminum nitride and diamond, *Diamond and Related Materials* **5**, 790 (1996).
181. Y. Choi, M. Michan, J.L. Johnson, *et al.*, Field-emission properties of individual GaN nanowires grown by chemical vapor deposition, *Journal of Applied Physics* **111**, 044308 (2012).
182. J.L. Johnson, Y.H. Choi, and A. Ural, GaN nanowire and Ga_2O_3 nanowire and nanoribbon growth from ion implanted iron catalyst, *Journal of Vacuum Science and Technology B* **26**, 1841 (2008).
183. J. Johnson, Y.H. Choi, A. Ural, *et al.*, Growth and characterization of GaN nanowires for hydrogen sensors, *Journal of Electronic Materials* **38**, 490 (2009).
184. Z.Y. Fan, J.C. Ho, Z.A. Jacobson, *et al.*, Wafer-scale assembly of highly ordered semiconductor nanowire arrays by contact printing, *Nano Letters* **8**, 20 (2008).
185. J.S. Wright, W. Lim, B.P. Gila, *et al.*, Hydrogen sensing with Pt-functionalized GaN nanowires, *Sensors and Actuators, B* **140**, 196 (2009).
186. L.M. Baskin, O.I. Lvov, and G.N. Fursey, General features of field emission from semiconductors, *Physica Status Solidi B* **47**, 49 (1971).
187. P. Yaghoobi, K. Walus, and A. Nojeh, First-principles study of quantum tunneling from nanostructures: Current in a single-walled carbon nanotube electron source, *Physical Review B* **80**, 115422 (2009).

188. P. Yaghoobi, M.K. Alam, K. Walus, and A. Nojeh, High subthreshold field-emission current due to hydrogen adsorption in single-walled carbon nanotubes: A first-principles study, *Applied Physics Letters* **95**, 262102 (2009).
189. J.R. Arthur, Photosensitive field emission from p-type germanium, *Journal of Applied Physics* **36**, 3221 (1965).
190. K.X. Liu, C.J. Chiang, and J.P. Heritage, Photoresponse of gated p-silicon field emitter array and correlation with theoretical models, *Journal of Applied Physics* **99**, 034502 (2006).
191. V.R. Reddy, C.K. Ramesh, and C.J. Choi, Structural and electrical properties of Mo/n-GaN Schottky diodes, *Physica Status Solidi A* **203**, 622 (2006).
192. L. Wang, M.I. Nathan, T.H. Lim, *et al.*, High barrier height GaN Schottky diodes: Pt/GaN and Pd/GaN, *Applied Physics Letters* **68**, 1267 (1996).
193. M.S. Shur and M.A. Khan, in *GaN and Related Materials II*, ed. S.J. Pearton (Gordon and Breach, 1999), pp. 47–92.
194. S. Nakamura, in *GaN and Related Materials II*, ed. S.J. Pearton (Gordon and Breach, New York, 1999), pp. 1–45.
195. Y.C. Yeo, T.C. Chong, and M.F. Li, Electronic band structures and effective-mass parameters of wurtzite GaN and InN, *Journal of Applied Physics* **83**, 1429 (1998).
196. C. Bulutay, B.K. Ridley, and N.A. Zakhleniuk, Full-band polar optical phonon scattering analysis and negative differential conductivity in wurtzite GaN, *Physical Review B* **62**, 15754 (2000).
197. C.-K. Sun, Y.-L. Huang, S. Keller, *et al.*, Ultrafast electron dynamics study of GaN, *Physical Review B* **59**, 13535 (1999).
198. M.E. Levinshtein, S.L. Rumyantsev, and M.S. Shur, *Properties of Advanced Semiconductor Materials: GaN, AlN, InN, BN, SiC, SiGe*. John Wiley & Sons, Inc., New York (2001).
199. M. Cardona, K.L. Shaklee, and F.H. Pollak, Electroreflectance at a semiconductor-electrolyte interface, *Physical Review* **154**, 696 (1967).
200. E. Kisker, K. Shroder, M. Campagna, and W. Gudat, Temperature dependence of the exchange splitting of Fe by spin-resolved photoemission spectroscopy with synchrotron radiation, *Physical Review Letters* **52**, 2285 (1984).
201. R.D. Young, Theoretical total-energy distribution of field-emitted elect, *Physical Review* **113**, 110 (1959).
202. R.D. Young and H.E. Clark, Effect of surface patch fields on field-emission work-function determinations, *Physical Review Letters* **17**, 351 (1966).
203. R.D. Young and E.W. Muller, Progress in field-emission work-function measurements of atomically perfect crystal planes, *Journal of Applied Physics* **33**, 91 (1962).
204. T. Radon, Photofield emission spectroscopy, *Progress in Surface Science* **59**, 331 (1998).
205. B.I. Lundqvist, K. Mountfield, and J.W. Wilkins, Photo-field-emission: A new probe of electron states between the Fermi and vacuum levels, *Solid State Communications* **10**, 383 (1972).
206. M.J.G. Lee and R. Reifenberger, Periodic field-dependent photocurrent from a tungsten field emitter, *Surface Science* **70**, 114 (1978).
207. A. Evtukh, O. Yilmazoglu, V. Litovchenko, *et al.*, Peculiarities of the photon-assisted field emissions from GaN nanorods, *Journal of Vacuum Science and Technology B* **28**, C2A72 (2010).
208. C. Youtsey, L.T. Romano, R.J. Molnar, and I. Adesiva, Rapid evaluation of dislocation densities in n-type GaN films using photoenhanced wet etching, *Applied Physics Letters* **74**, 3537 (1999).
209. htpp//www.ioffe.rssi.ru/SVA/NSM/Semicond/

8

Carbon-Based Quantum Cathodes

8.1 Introduction

Carbon materials based field-emission devices have many potential applications, including miniaturized microwave power amplifier tubes, electron sources for microscopes in nanovision systems, miniaturized X-ray tubes, electron beam nano-lithography, flat panel field emission displays, light sources, sensors, and so on. As was experimentally shown, carbon nanomaterials are able to emit electrons at relatively low electric fields, generate useful current densities and they are stable field emitters.

One of the most important properties of diamond is the rather small barrier to the emission of electrons into vacuum. The barrier has been reported to be as low as 0.05 eV [1]. This fact points to a great interest in diamond as a cathode (emitter) material for the vacuum micro- and nanoelectronic devices. Diamond has attracted much attention as a cold cathode also due to its negative electron affinity (NEA) [2]. Diamond is the wide band gap material whose properties (low and negative electron affinity, surface stability, excellent thermal conductivity, etc.) make it ideal for electron field emission (EFE). Additional interesting properties are obtained when diamond is doped with nitrogen that dissolves in diamond with generation of deep donor levels in its band gap [3, 4]. One of the most important achievements in connection with this is a very low threshold for emission [3].

An important advantage of diamond-like carbon (DLC) films compared to diamond films and carbon nanotubes (CNTs) is the simpler process with low-temperature deposition. Hence, DLC films are very promising materials for EFE. The emission properties of silicon tips and tip arrays can be improved by coating with DLC films [5–8]. This process combines both the advantages of silicon tips and those of DLC films. At the beginning, the field emission was mainly characterized in terms of the threshold field defined by a current density equal to $10^{-6}\,A\cdot cm^{-2}$. Different factors, such as the variation of the sp^3 phase content or the nitrogen content provided a low threshold voltage. At the same time, it was shown that a treatment of tetrahedral amorphous carbon (ta-C) type of DLC films in argon, hydrogen, or oxygen plasma resulted in a decrease of the threshold field and an increase of the emission site density (ESD). Extremely low effective work functions of (0.01–0.04) eV were obtained in [9] from the slope of Fowler–Nordheim plots without taking into account the field enhancement coefficient ($\beta = 1$). It is not the real barrier height because for such small barriers there would be

Vacuum Nanoelectronic Devices: Novel Electron Sources and Applications, First Edition.
Anatoliy Evtukh, Hans Hartnagel, Oktay Yilmazoglu, Hidenori Mimura and Dimitris Pavlidis.

temperature-dependent thermionic emission at low temperature. This has never been observed in other experiments.

CNTs have the greatest potential for field emission applications [10–16]. Due to their geometrical structure, the length of several micrometers and diameters down to 1.0 nm nanotubes have been recognized as very promising field emitting structures. Other strong points of nanotube emitters are the possibility of their relatively simple production in very large quantities (in 1 g of pure nanotube material it is expected in the order of 10^{16} nanotubes, each having a field enhancement factor of about 1000) and their chemical inertness [17]. Especially the chemical inertness of carbon field emitters is one of the most important advantages over silicon or metal microtips which suffer from the emission degradation due to sputter erosion and chemical contamination and therefore require high vacuum for operation [18]. In contrast to the emission mechanism of DLC and chemical vapor deposition (CVD) diamond films, field emission from nanotubes, as it seems, is clearly governed by their geometrical field enhancing properties. Yet, assuming that nanotubes behave like metallic needles to describe the field emission properties gives an overly simplified picture of the situation, because we have to deal with a covalently bonded nanoscaled system. Theoretical as well as experimental data show that the electronic properties of single-walled nanotubes (SWNTs) depend strongly on their helical symmetry and diameter. Such nanotubes exhibit semiconducting or metallic electronic properties that can influence their field emission behavior [19–21].

In recent, graphene has emerged as a new material with its unique physical and chemical properties and is attracting considerable attention not only in its basic physics but also in its possible application including field emitters. It has excellent electrical properties and an atomic thin edge. These properties make graphene appropriate for a variety of applications. The unique 2D structure, which includes a single atom layer thickness and a high aspect ratio, assists graphene in providing excellent field emission characteristics, making the materials good candidate emitters for use in field emission devices [22]. With the high specific area and high electrical conductivity, aligned graphene can serve as an ideal efficient edge emitter [23], competitive supercapacitor and other electronic devices. Because electrons are mainly emitted from the graphene edges, the fabrication of the vertically aligned graphene on the substrate with high density is one of the crucial parameters for highly performed graphene emitter. Moreover, low temperature fabrication is prerequisite for the flexible field emission application of the graphene nanosheets.

In this chapter the peculiarities of EFE from various carbon nanostructures, including diamond, DLC films, CNTs, graphene, and nanocarbon are considered and analyzed. The proposed models of electron transport and EFE take into account some features of material properties, such as NEA in the case of diamond, electrically nanostructured heterogeneous (ENH) in the case of DLC films, high electric field enhancement coefficient in the case of CNTs, graphene, and nanocarbons. The perspectives of carbon-based materials for efficient EFE sources are emphasized.

8.2 Diamond and Diamond Film Emitters

8.2.1 Negative Electron Affinity

As has been first reported in [2] and confirmed later by other authors [24–27], the minimum energy of electrons in the conduction band of diamond can be larger than the minimum energy

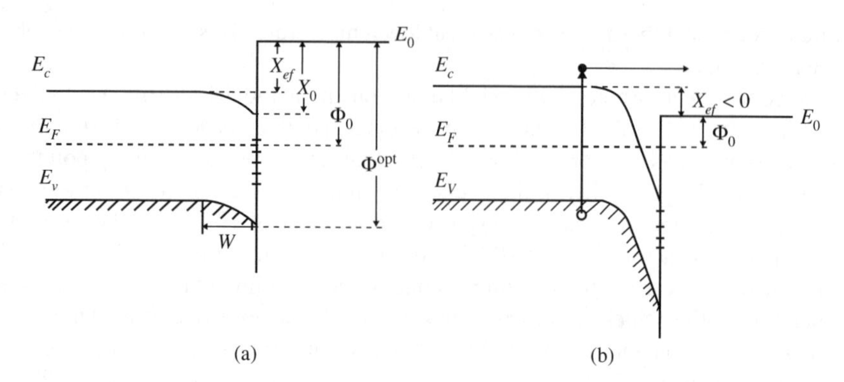

Figure 8.1 Energy band diagrams of semiconductor vacuum system with surface states: (a) positive effective electron affinity and (b) negative effective electron affinity

of electrons in vacuum, that is, diamond can have NEA. This effect is observed when the conduction band edge lies above the vacuum level, so that any electrons in its conduction band can pass into vacuum without energy barrier [27] (Figure 8.1b).

Electron affinity of a semiconductor is defined as the energy required to remove an electron from the conduction band minimum at a macroscopically large distance from the semiconductor (i.e., away from image charge effects). This energy can be shown schematically at the surface as the difference between the vacuum level and the conduction band minimum (Figure 8.1). The electron affinity does not generally depend on the Fermi level of semiconductor. Thus, while doping can change the Fermi level of semiconductor and the work function will change accordingly, the electron affinity is unaffected by these changes. An alternative view is that the electron affinity is a measure of the heterojunction band offset between the vacuum and the conduction band of the semiconductor of interest. For most semiconductors, the conduction band minimum is below the vacuum level, and electrons in the conduction band are bound to the semiconductor by energy equal to the electron affinity. In some cases, the surface conditions can be obtained in which the conduction band minimum is above the vacuum level. In that case the first conduction electron would not be bound to the sample but could escape with a kinetic energy equal to the difference in energy of the conduction band minimum and the vacuum level. This situation is termed a negative electron affinity. (Note that the electron is still bound to the vicinity of the sample by Coulomb forces.)

The electron affinity or work function of a material is usually ascribed to two aspects of the material (1) the origin of the atomic levels and (2) the surface dipole due to the surface termination [28]. The atomic levels are more or less intrinsic to a material and cannot be changed (but alloys may provide the degree of variation). This is not the case for the surface dipole. The surface dipole can be substantially affected by surface reconstructions and surface adsorbates. Because of the large effect of the surface dipole, it is essentially impossible to determine if a material is "intrinsically" NEA. Thus the surface termination is critical in describing the electron affinity (or NEA) properties of material.

One method to measure the electron affinity of a semiconductor is UV-photoemission spectroscopy [2, 5, 29, 30]. The electrons from the valence band are excited into the conduction band. Electrons scatter passing toward the surface and a large number of secondary electrons accumulate at the conduction band minimum. For materials with a positive electron affinity (PEA) these electrons cannot escape, while for NEA the electrons can be emitted directly and will be observed with a low kinetic energy. Thus the two effects, which signify NEA, are the extension of the spectral range to lower energies and the appearance of a sharp peak at low kinetic energy. This feature will appear at the largest (negative) binding energy in typical presentations of ultraviolet photoemission spectroscopy (UPS) spectra. In addition to the sharp feature that is often evident in the spectra of a NEA semiconductor, the width of the photoemission spectrum (W) can be related to the electron affinity (χ). For an NEA surface, these electrons can emit into vacuum and are detected as a sharp feature at the low energy end of a photoemission spectrum. However, for a surface with PEA, emission from the conduction band minimum will not occur, and the value of the electron affinity can thus be deduced. The electron affinity χ or the presence of NEA can be deduced from the width of the spectrum W as follows [30]

$$\chi = h\nu - E_g - W \quad \text{for a positive electron affinity,}$$

$$0 = h\nu - E_g - W \quad \text{for a negative electron affinity,}$$

where $h\nu$ is the photon energy and E_g is the bandgap.

The spectral width is obtained from a linear extrapolation of the emission onset edge to zero intensity at both the low kinetic energy cutoff and the high kinetic energy end (reflecting the valence band maximum).

Another method to study surface emission is the secondary electron emission (SEE) [31]. When a semiconducting material is exposed to high-energy (1–5 keV) electrons, electron–hole pairs are created in the conduction and valence bands. It is again obtained when one incident (high-energy) electron generates several electron–hole pairs. The electrons in the conduction band can move to the surface and emit into vacuum, the process of which can be enhanced by the presence of NEA surface. However, the electrons generated in SEE reside deeper inside the sample than those from UPS, which makes SEE less sensitive to the surface. Although the energy spectrum of emitted SEE electrons can be similarly obtained, it is more difficult to draw conclusions from spectral width measurements, because the energy of the incident electrons depends on the work function difference between the electron gun and the sample. Additionally, band bending at the surface can influence the gain in SEE, because electron–hole pairs are created deep inside the sample and will be either facilitated by downward band bending or impeded by upward band bending when they try to reach and escape the surface. Therefore, UPS is generally a more suitable technique than SEE to perform electron affinity measurements.

The NEA properties of diamond and other wide band gap semiconductors may lead to the development of cold cathode sources. The electron affinity of a semiconductor relates to the electron bands position to the vacuum ground state near the surface. The conduction band minimum of diamond and other wide band gap semiconductors drop near the vacuum level, so NEA can be achieved. The presence of NEA is strongly dependent on the surface termination of semiconductor. These effects can be understood in terms of surface dipole that is formed due to the surface adsorbates or atomic reconstruction. The experimental data show that H-termination results in NEA for all diamond surfaces studied [2, 24–26]. In contrast, oxygen termination and clean surfaces result in a small, but PEA. Thin metallic layers can also induce NEA on both clean and O-terminated diamond. The properties of other hard wide band gap semiconductors indicate that these materials will also exhibit NEA, and recent reports have established the presence of NEA in AlN and BN. Field emission from semiconductors is a complicated process that involves electron supply, transport, and emission. The results indicate that the electron supply is a critical obstacle to field emission from these wide band gap materials. The development of NEA thin films offers the possibility of construction of cold cathode structures and devices. Major issues that may limit the development of these applications including *n*-type doping are to supply electrons to the surface and obtain uniform emission over the surface.

A diamond surface must be terminated by hydrogen to have the NEA [32]. The surface C-H bonds create a dipole layer, which raises the energy of a conduction band relative to the vacuum level [33]. In reference [34], the influence of diamond surface treatments with O_2 and Cs on EFE has been studied. EFE from B-, Li-, P-, and N-doped diamond has been characterized depending on the surface treatment. The surface treated with O_2 plasma, coated with Cs, heated, and exposed to O_2 has exhibited increased emission for all the samples except for the B-doped diamond. Both Cs and Cs/O deposits are ionic species that form affinity-lowering surface dipoles. These dipoles lead to the formation of an effective NEA surface as shown in Figure 8.1b. The best emission was obtained from the Li-doped diamond and the polycrystalline P-doped diamond. Subbands formed by Li and N impurities in diamond are believed to be responsible for this enhanced emission. The surface treatment of N-doped diamond results in emission at the electric fields as low as 0.2 V/μm.

Oxygen improved the electron emission, but hydrogen had no effect. This fact was in conflict with a long-established textbook knowledge in surface science [35, 36]. If oxygen was adsorbed on diamond, it might be expected to increase the work function and electron affinity (EA), and, hence, to reduce the electron emission; hydrogen might be expected to increase the emission. According to this the most obvious interpretation of the original experiments [27, 34] was that EA was not directly affecting the observed EFE, but that something else was happening [37]. The experimental results [27, 34] provide no valid experimental scientific evidence about the benefits of NEA.

Theoretical analysis and experimental results show that the NEA does not determine the EFE from diamond and diamond films but back contact electron injection or transport through the diamond bulk can restrict the EFE.

8.2.2 *Emission from Diamond and Diamond Films*

Field emission requires the electrons to travel round a complete circuit. Diamond has a wide band gap ($E_g = 5.5\,\text{eV}$), so its resistivity is very high [38]. Therefore, the electrons have to be injected over a large barrier at the back contact of the order of 4 eV [3]. A good low-macroscopic field (LMF) emitter requires that electrons easily cross the back barrier: a poor back contact will degrade a performance. Experiments on diamond films confirm this. Field emission spectroscopy (FES) analyses [39, 40] show that annealing and plasma processing of a diamond film both lead to better behavior; including a reduced FES peak shift (i.e., a reduced potential drop, relative to the substrate Fermi level). In reference [40] it was attributed to a better electrical contact at the substrate/film interface, perhaps due to the formation of molybdenum carbide (Mo_2C). Other work [41] supports this. They showed an improved uniformity and enhanced emission when an intermediate titanium layer was used between the silicon substrate and the diamond film, and found titanium carbide (TiC) at the interface.

Further evidence is the experiment performed in [42] on a composite material similar to bulk diamond. Roughening the back surface enhanced the current, and observation of "glowing channels" demonstrated electron injection into the conduction band. It was assumed that nitrogen dopant formed a depletion region at the back metal-diamond interface which allowed electrons to tunnel into diamond. However, to extract significant current, roughening the back surface (presumably producing local field enhancement) was essential. In this experiment, the emission current was controlled by electron injection across the back junction into the conduction band, the so-called back-injection-control mechanism [3, 43].

Improving poor back contacts improves the performance. But, when the contact has been improved, the control over the emission can no longer be done at the back surface. In the experiments on EFE from bulk diamond there was a large voltage drop (thousands of volts) across the diamond, and electrons could easily escape across the front surface: so it might be expected that control remain at the back. But this is not necessarily expected for carbon materials that have significant front-surface barriers and small voltage drop across the film.

For a semiconducting field emitter such as diamond, the emitting electrons can originate either from the conduction band, the valence band, and/or surface states. Diamond has a wide bandgap (~5.5 eV), and undoped diamond is thus generally thought to be unable to produce sustained electron emission because of its insulating nature. Although electrons emitting from surface states in diamond can occur [44], there are no obvious mechanisms by which electrons can be transported through the undoped bulk to the surface states. Either the bulk or the surface of diamond must be made conductive in order to sustain the emission process.

Vacuum field emission measurements on CVD diamond samples with varying defect densities indicate that the electric field required for electron emission can be significantly reduced when CVD diamond is grown with substantial quantity of structural defects [45]. It is likely that the defects present in diamond structure create additional energy bands within the band gap of diamond and thus contribute to the electron emission at low electric fields. It has been found that the conductivity of CVD diamond films decreases by several orders of magnitude by thermal annealing and that the subsequent treatment in hydrogen plasma causes an increase in the conductivity up to the initial value [38, 45]. In reference [46], an electrical conductivity along the conducting layers between grains in CVD polycrystalline diamond film has been studied. In as-grown films, the dominating ohmic conductivity through the disordered graphite-like (disordered sp^2 bonded carbon) regions between grains with small activation energy has been observed. The annealing and hydrogenation of films strongly influence on the electrical conductivity.

Furthermore, to fully take advantage of diamond's NEA to realize electron emission at low applied fields, the Fermi level must be close to the conduction band. This would require the diamond to be n-type doped. Although p-type diamond can be readily made available by boron doping, very high electric fields are needed for emission to occur, because the emitting electrons reside deep (>5 eV) below the vacuum level in p-type diamond. An ideal emitting structure may, therefore, consist of an n-type doped semiconducting diamond with true NEA surface. However, making n-type diamond effectively and reliably and thus supplying electrons to the conduction band remains a significant challenge. As a rule, n-type diamond is obtained by ion implanting of diamond [47, 48]. Dopants such as nitrogen, phosphorus, and lithium have also been explored to produce n-type diamond, but with limited success because of either their donor levels being too deep or their incorporation in the diamond structure being unstable [49]. Reported improvements in emission from this "doped" diamond [4, 34, 42, 50–53] are likely due to the creation of in-gap defects by the ion implantation or the heavy doping, giving rise to defect enhanced electron emission.

During the investigations of EFE from CVD films of polycrystalline diamond it was noted that the emission came not from the sharply pointed tips of diamond crystallites (as might be expected) but from the areas between the crystallites [54]. It was supported that the mechanism based on graphite inclusions might operate for CVD diamond.

Nowadays, the best emitter from diamond is nano-crystalline diamond [53, 55–57]. In this case, the emission varies inversely with the grain size [53, 55]. This is because the grain boundaries provide a conduction path to front surface. Emission from grain boundaries can be explained by large enhancement of the local electric field at the emission site. The grain boundaries are sp^2-bonded carbon. They act like the internal tips.

The decreased size of grains from submicron to nanosized (~50 nm) in diamond film has allowed to decrease the threshold voltage from $E_{th} = 2.35$ to 1.88 V/μm. The effective work function ($\Phi_{ef} = \Phi_0/\beta^{2/3}$, where Φ_0 is the work function and β is the local field enhancement coefficient) has been decreased from $\Phi_{ef} = 0.028$ to 0.017 eV [58].

It has been found that for the diamond cathode (emitter), NEA does not play the determining role in EFE. The electron emission from CVD diamond film is strongly affected by several other factors. Diamond with nanosized grains has shown good EFE properties [55, 59–65]. Besides, incorporated non-diamond phases with sp^2 hybrid bonded carbon can greatly enhance the emission [66–68]. High nitrogen incorporation in polycrystalline diamond films has been also reported to reduce significantly the emission turn-on voltage [4]. Field emission results show that the threshold electric field increases, whereas the emission current density and the number of emission sites decrease with the increase of the size of diamond grains [69]. The film with a small grain size of 10 nm has a threshold electric field of 1.5 V/μm, and the emission current reaches 780 μA/cm^2 at electric field of 3.5 V/μm, demonstrating that small size nanodiamond films (NDFs) are the promising material for low field electron emitters.

Diamond materials with small grain sizes and high defect densities generally emit better than those with large crystallite sizes and low defect contents. Heavy doping, whether it is n-type or p-type in nature, enhances emission. Outstanding emission properties are discovered in both ultrafine diamond powders containing 1–20 nm crystallites produced by explosive synthesis and nanocrystalline or ultrananocrystalline diamond (UNCD) films (composed of ~1–100 nm crystallites) [60, 61, 70–73]. Emission has been found to originate from sites that are associated with defect structures in diamond rather than sharp features on the surface [54, 56, 66, 73–76]. Compared with conventional Si or metal microtip emitters, diamond emitters show lower threshold fields, improved emission stability, and robustness in low/medium vacuum environments. Attempts have also been made to apply diamond coatings on tips of silicon or metal-emitter arrays, as shown in Figure 8.2, to further enhance the emission characteristics [70, 77–82]. High emission currents of 60–100 μA per tip have been measured for Si tips conformally coated with nanocrystalline diamond films [83].

UNCD is a special form of diamond with nanometer size grain (2–5 nm) and high EFE performance as compared to micron-size diamond [60]. The UNCD films possess many excellent properties and several of them actually exceed those of diamond [84]. A very high EFE characteristic has been reported for UNCD films [70, 85–87].

To improve EFE parameters the UNCD films have been proposed as the tip coating [88–90]. The UNCD films have more possibility to integrate with various materials. Among the nanowire templates which are suitable for coating UNCD films to fabricate the nano-emitters, Si nanowires (Si-NWs) are advantageous over other nanowires for their compatibility with Si-materials and have greater potential for integration with Si devices to

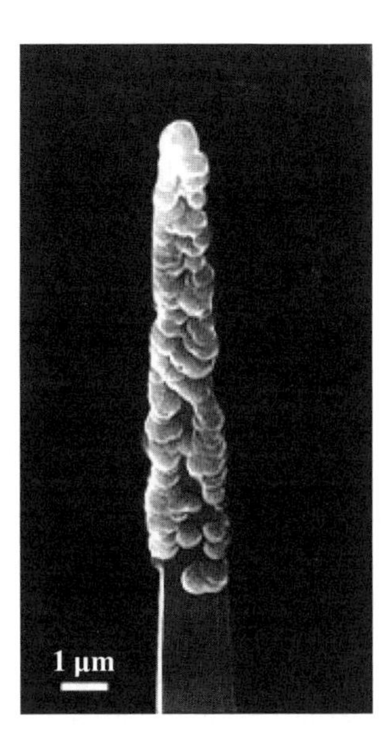

Figure 8.2 Scanning electron micrograph of a silicon tip coated with conglomerated diamond particles. Reprinted from Ref. [81]. Copyright (1995), with permission from Elsevier

form active electron field emitters [91, 92]. The Si-NWs integrated with ultrananograin size diamond can possibly be a better electron field emitter due to the NEA of diamond [4]. The direct deposition of UNCD films on Si-NW templates and the related EFE properties has been studied in [93]. However in this method, the UNCD particulates are scarcely distributed on the Si-NWs and the related EFE properties are not satisfactory. Pretreatment techniques such as ultrasonication have also been reported prior to the growth of UNCD films on Si-NWs [93, 94]. However, the tip of Si-NWs break and the length of the Si-NWs reduced due to the rigid ultrasonication process and hence the related EFE properties are poor due to the unavailability of sharp UNCD coated Si-NW tips. This phenomenon implies that the Si-NWs are not strong enough to survive the severe abrasive ultrasonication process. Hence, more gently nucleation method that is efficient enough to form active nucleation sites for further deposition of UNCD films has been required. In this context, an easy electrophoresis deposition (EPD) method, for the deposition of nanodiamond particulates on Si-NW surface, which will act as active sites for the growth of UNCD films has been proposed [88]. Systematic study of the effect of pre-treatment techniques such as electrophoresis and ultrasonication on the EFE performances of the UNCD/Si-NW nanoemitters was performed and the optimized parameters were determined. EPD of nano-diamond was found to be a very efficient process to form nucleation sites for the growth of UNCD films. Transmission electron microscopic (TEM) investigation showed that the EPD derived UNCD films contained uniform granular structure with sharp, smooth, and conductive interface layer, which improved the EFE properties of the UNCD films. Moreover, the EPD process facilitated the formation of UNCD coating on Si-NWs arrays that enhanced their EFE properties. The EPD derived UNCD/Si-NW emitters showed superior EFE performances with the turn on field of 7.18 V/μm and large EFE current density of 2.10 mA/cm^2 at 15.0 V/μm. The EFE properties of Si-nanotips coated with diamond films were significantly improved. The microcrystalline diamond (MCD) films grown on UNCD nucleation layer (MCD/UNCD) possessed low turn-on field for inducing the EFE process (E_{on} = 4.67 V/μm) with large EFE current density (J_e = 0.31 mA/cm^2) [89]. TEM studies revealed that the MCD/UNCD films contained large diamond aggregates evenly distributed among the ultra-small grain matrix, with the induction of a-few-layer graphite, surrounding the large aggregates. The presence of graphene-like phase was presumed to be the prime factor resulting in superior EFE properties for the MCD films.

The edges and corners of nanocrystalline diamond are the natural field-emitters. Much attention has been paid to the field emission application of NDF by CVD process. The CVD process and other synthesis of the large area NDF need long time and high cost, and it is difficult to adulterate impurity [74, 95–98]. It is found that through adulterated impurity electron translocated rate and electron emission efficiency [99–101] can be enhanced. A low cost nanodiamond paste was developed by mixing nano-diamond and nano-graphite with other inorganic and organic vehicles [102]. Then a high mesh to filtrate the nanodiamond paste to remove big grains was used. NDF was prepared by sol-gel method at 3000/minute.

Superb material properties of nanocrystalline diamond (nanodiamond) materials coupled with practical CVD processing of deposited nitrogen-incorporated nanodiamond on variety of substrates, have promoted further interest in the use of these diamond derived materials as electron field emitters. Experimentally, nanodiamond emitters have been observed to emit electrons at relatively low electric fields and generate useful current densities. The development in nanodiamond vacuum field emission integrated electronic devices: nanodiamond triodes, transistors, and integrated differential amplifiers have been examined in [103]. Nanodiamond vacuum field emission integrated devices, specifically nano-gap lateral field emission diode arrays, triode arrays, and integrated differential amplifiers are new configurations for robust vacuum nanoelectronic devices. These novel micro/nanostructures provide alternative and efficient means of accomplishing electronics that are impervious to temperature and radiation. Some examples of the nanodiamond vertically configured field emission integrated devices, including, ultra-sharp cathode arrays, transistors, and integrated differential amplifiers are shown in Figure 8.3.

The effective EFE from NDFs promotes the further intensive study of their growth technology and emission properties.

8.2.3 Models of EFE from Diamond

At consideration of EFE from diamond three processes have to be analyzed: (1) supplying electrons to diamond on back contact, (2) transporting them through the bulk to the surface, and (3) emitting them into vacuum. Each of them can be critical or limiting the emission process in various types of diamond material (single crystal diamond, micro- or nanocrystalls, and diamond film). According to this three main models for explanation of EFE from diamond have been proposed.

The first model considers the injection of electrons over a Schottky barrier at the back-contact interface between the metallic substrate and the diamond as the controlling mechanism [3, 42, 54, 104]. Emission current here is limited by tunneling of electrons through the metal–diamond Schottky diode into the conduction band of diamond, not by electron emission from the diamond surface into vacuum. For very thin diamond films such as those coated on sharp Si or Mo tips [82], the injected electrons could directly reach the diamond surface and

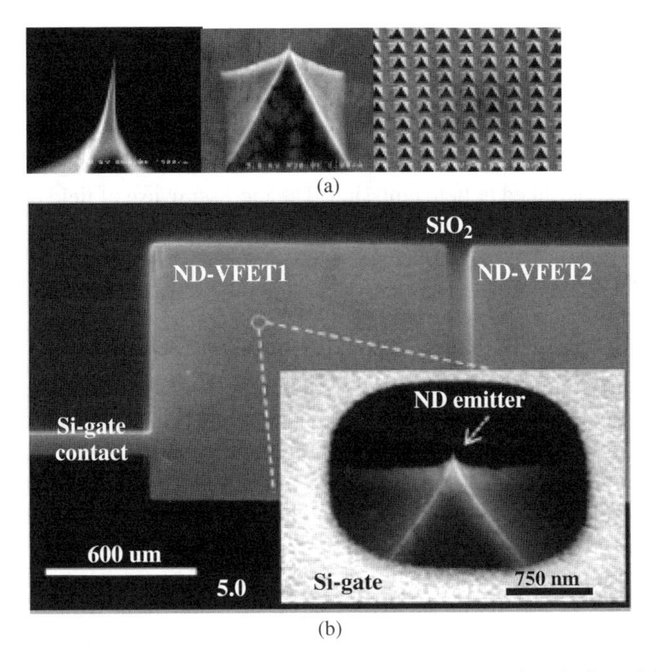

Figure 8.3 Examples of nanodiamond vertically configured field emission devices: (a) SEM images of cathode arrays with ultra-sharp tip showing high to low magnifications and (b) SEM images of transistor and differential amplifier arrays. Copyright (2012) IEEE. Reprinted with permission from Ref. [103]

emit into vacuum. This electron injection role is supported by the observation that a roughened interface between the nickel contact and nitrogen-doped diamond improved electron emission due to local field enhancement effects at the interface that facilitate tunneling through the Schottky barrier [3, 42]. Electron injection facilitated by band bending at the interface related to space charge build-up has also been proposed [105–108]. It is worth noting that the tunneling current density at the interface as a function of applied field follows a numerical formula similar to the F-N equation, and because of this, it is difficult to differentiate between this interface barrier mechanism and the surface barrier mechanism from *I-V* data alone. Additionally, it is not clear what kind of Schottky barrier structure is formed at the interface of metal and diamond that is highly defective.

The defect/impurity theory suggests that structural defects and impurities can form energy states within the band gap of diamond [34, 53, 109, 110]. When the defect density is sufficiently high, the electronic states of various defects can interact and form energy bands. If these bands are wide enough or closely spaced, the electron hopping mechanism within the bands, similar to the Poole–Frenkel conduction mechanism or the Hill type conduction [46, 111], could easily provide a steady flow of electrons to the surface and sustain a stable electron emission. The electrons can either be excited into the conduction band or unoccupied surface states from these defect/impurity bands and emit, or tunnel directly from the defect/impurity bands and emit. For example, photoelectron yield spectroscopy has detected sub-bandgap emission associated with the presence of graphite in diamond [37]. The defects/impurities essentially raise the Fermi level by acting as donors of electrons and thus reduce the tunneling barrier. This theory is supported by overwhelming experimental data indicating that defective or lower quality diamonds have better emission properties. It also appears to explain why emission characteristics are enhanced in many "doped" diamonds, not necessarily because of the electrical doping effect, but rather by the creation of defects in the doping process. However, the exact nature of the responsible defects has not been identified, and the very existence of many defect/impurity energy bands and their locations within diamond's band gap have not been positively confirmed or linked to field emission. The increasing use of field emitted energy distribution (FEED), first employed in [112], is helpful in determining the origin of the emitting electrons [74, 75, 104, 113–116].

The third model takes into account the role of graphite in the electron emission from diamond [117, 118]. There has been suggested a field-induced hot electron emission process from isolated graphitic inclusions in diamond, citing an antenna effect that leads to field concentrations on a "floating" conductive particle (i.e., graphite) embedded in an insulator matrix (i.e., diamond). This model is based on the observation that active emission sites correspond to discrete locations of defects or graphite inclusions on the diamond surface [54, 66, 74, 75, 119]. To sustain the continuous emission current, electrons are assumed to be supplied to the emitting surface through conduction channels formed in diamond via an electroforming process at high electric fields [105, 120], similar to the emission process described for composite resin–carbon materials [121]. Grain boundaries in diamond films [122, 123] and hydrogenated diamond surfaces [119, 124] have also been suggested to function as conduction channels. Similarly to this mechanism, it has been concluded from the photoelectron subband gap emission study that the diamond phase provides a thermally and mechanically stable matrix with a comparatively low work function, and graphitic phases provide the transport path for electrons to reach the surface and emit [37]. This model is the most suitable for description of EFE from NDFs.

Some other models that consider some particular parts of a complex emission process from diamond and diamond films have been proposed [31].

8.3 Diamond-Like Carbon Film Emitters

8.3.1 Electrically Nanostructured Heterogeneous Emitters

The phenomenon of LMF electron emission able to generate electrons at low macroscopic electric field was analyzed and considered in detail in [125]. LMF emitters produce the current at macroscopic fields typically of the order of 1–50 $V\,\mu m^{-1}$, much smaller than the local fields at which cold EFE normally occurs (about $5\,V \cdot nm^{-1}$). The LMF emission occurs because the dielectric or semiconductor films are, or become, the ENH materials, with quasi-filamentary conducting channels between their surfaces. These channels connect to emitting features near or on the film-vacuum surface, or act as electron emitters themselves. The film may contain conducting or semiconducting particles that assist the conductivity and/or act as emitting features.

There are many possible structures for actively field emitting ENH films. Among them there are: (1) carbon-based films [125, 126]; (2) oxide films and semiconducting inclusions on/at the surfaces of electrodes of high-voltage equipment [76, 127]; (3) manufactured composite materials with small conducting or semiconducting inclusions deliberately dispersed in resin, glass, or other dielectric [128]; (4) Si nanoclusters in SiO_2 dielectric matrix [129, 130]; (5) small clusters of gold atoms embedded in an a-C matrix [131]; (6) carbon particles in an embedded "calcium-copper-betaine" matrix [132]; and (7) surface island metal film [133, 134], and so on.

The most general type of heterogeneous emitters has conducting or semiconducting particles within a less conducting matrix that can act as a leaky dielectric. Filamentary conducting channels exist (or are formed by activation or conditioning) between the substrate and the particles, between the particles, and between the near-front-surface conducting particle and the vacuum. Films with "thick-pin-like" conducting particles, or where the conducting channel stretches from the substrate to the vacuum, or where conducting particles protrude directly into the vacuum, are also included in ENH emitters.

Field induced electron emission (FIE) from ENH materials occurs because there is a *conducting nanostructure* inside the film that causes geometrical field enhancement to take place, either at the film/vacuum interface, or near this interface but inside the film. As a result of this, the *local* field at the tip of the nanostructure becomes sufficiently high for more or less ordinary EFE to take place. In some cases the emission can be slightly hot but occur through the barrier; in some cases another mechanism may assist in enhancing or facilitating emission [125].

Various types of models for the description of EFE from cathodes with ENH films were summarized in [125]. It is possible to distinguish five principal models which are suitable for Latham-type nanocomposite emitters [76, 127] and nanostructured carbon films. They are C-D-V [3, 86, 102, 135], C-D-C_{in}-V [95, 96, 132, 136–138], C-D-C_{in}-D-V [104, 128, 139, 140], C-D-C_{on}-V [125, 141], and C-C-C_{on}-V [76] (where C means the conductor, D is the dielectric, C_{in}, C_{on} are the conducting particles within and on dielectric matrix, respectively).

It was shown in [76] that pre-break-down emission sites were associated with either foreign dielectric surface inclusions on the electrode, or anomalously thick oxide aggregations of thickness 100 nm to 1 μm. Electron emission was shown [142] to come from one or more independent sub-sites on the inclusion, each assumed to be associated with a conducting channel in the dielectric [127, 135, 143]. An analogy with the electroformed filamentary conducting channels of electrical phenomena in amorphous oxide films, discussed in review [144], had been drawn in [135] and sufficient geometrical field enhancement at the channel tip for EFE had been suggested. This idea of the conducting channel became the basis for the C-D-C (M-I-V) model of FIE [127].

As described in [76, 127], "switching on" an emitting channel is a complex process involving several physical steps that finally result in a stable "on" state. Experimentally, the corresponding electron emission often gives a two-segment F-N plot, voltage-dependent shifts, and broadening effects.

It has been proposed [76, 127] that the emission is due to the electron heating in the channel, caused by a potential drop associated with the field penetration; the resulting hot electrons have been said to get over the potential barrier at the insulator/vacuum interface, in a manner similar to the Schottky emission. In [125], however, some reservations have been added to this model.

The properties of emitting metal tips (normally tungsten tips) coated with inclusion-free dielectric films were investigated in [145–147]. A conventional field electron microscopy (FEM) and field electron spectroscopy (FES) in the tip/screen configuration were better after coating. If tips are all similar and well characterized, the attention can be focused on the effects of the coating. Dielectrics studied have been alumina, hydrocarbons, lacomit, plastic, resin, and zinc oxide. The phenomena observed have been similar in all the cases and similar to those under the vacuum breakdown.

The FEM images often show a small number of individual small image spots (arguably associated with individual emitting channels). There is a tendency for the number of spots to increase with field. It is not uncommon for individual spots to "switch on" and "switch off" randomly but this can be normal behavior for the field emitters that have "dirty" surfaces or operate under insufficiently stringent vacuum conditions. These experiments, especially the FEM images, show definitively that the conducting channels of limited lateral extent do open up across thin dielectric films under the influence of a high electric field.

The next step of the investigation of emission from ENH materials was the deliberate insertion of various inclusions into a dielectric film spun onto a broad-area copper cathode, first graphite [104], and then a systematic study involving such materials as C, MoS_2, Au, Si, SiC, and S, in various different residual gas vacuum environments (He, N_2, H_2, O_2, and CO) [128]. The main findings were as follows. First, the switch-on *macroscopic* field was very low (typically 1–2 V μm^{-1}). Second, the emitters were far more resistant to industrial vacuum conditions (in particular to ion bombardment by residual gas atoms) than uncoated tungsten emitters.

Emitter behavior has been explained by the C-D-C-D-V model [128], in which the emission is a result of a three-stage process. (1) When an electric field is applied across the dielectric, the charge on an inclusion polarizes due to the electrostatic induction, giving the positive charge at the back surface and the negative one at the front. (2) The field between the inclusion back surface and the substrate is greatest where the separation is smallest. It can be enhanced by sharp features of the substrate surface; at favorable sites the conduction opens in the dielectric. Electrons (and, possibly, holes) travel along this and raise the potential of the inclusion to the substrate level, causing an additional field enhancement at its front surface. (3) An emitting channel opens through the dielectric between the inclusion front surface and the vacuum, and the field induced emission (FIE) takes place via the two open channels and the inclusion [125]. This model can be adapted if the conducting path lies via two or more inclusions, and under appropriate circumstances could be applicable to any conducting inclusion in any dielectric (including semiconducting inclusions).

8.3.2 Nanostructured Diamond-Like Carbon Films

The DLC or amorphous carbon films are very promising for various applications, including EFE. DLC films have unique properties such as extremely high hardness, atomic-level smoothness, very low friction coefficient, excellent abrasion resistance, LEA, chemical inertness, and optical transparency [148–150]. Various types of carbon: DLC films, diamond films, and nanotubes show field emission at low applied fields of the order of 1×10^5 V cm^{-1}. This opens a way for the vacuum micro- and nanoelectronic device applications.

DLC films are nanocomposite films with graphite-like clusters embedded in amorphous diamond-like matrix. A cluster model of amorphous carbon (a-C) and hydrogenated amorphous carbon (a-C:H) was proposed in [151, 152] and further developed in [153–158]. According to this model, amorphous carbon films contain both sp^3 and sp^2 configurations of bonds. Sp^3 centers create four σ-bonds, and sp^2 centers form three σ-bonds and one weaker π-bond. The effect of π-bonding is that first π-states are stabilized by the formation of parallel oriented pairs. They are further stabilized by the formation of planar sixfold aromatic graphite-like rings and yet further if the ring fuse together into graphitic clusters. These general results have led to the cluster model of a-C and a C:H [126, 152]. The matrix of sp^3 centers controls the mechanical properties of DLC films. On the other hand, the π-states of sp^2 clusters determine the electronic properties and optical band gap of material. The band gap of created σ-states is significantly wider than that of π-states. The sp^3 matrix forms tunneling barriers between the sp^2 clusters. The existence of sp^2 aromatic clusters is confirmed by three types of experimental results: the value of the optical band gap, the Raman spectra, and the photoluminescence [154, 159, 160].

High-temperature annealing ($T = 600\,°C$) alters the carbon-carbon bonding and promotes the formation of carbon nanoclusters, as it has been observed directly using high-resolution transmission electron microscopy (HRTEM) for the case of ta-C [161, 162]. The use of silicon containing precursors during the DLC film deposition in inductively coupled plasma gives a possibility to obtain nanocomposite DLC films with crystalline SiC, Si_3N_4, or amorphous SiO_x nanoparticles, which have been observed by HRTEM [163]. The size of the Si_3N_4 clusters has been in range of 5–20 nm, SiC and SiO_x clusters have been smaller.

The optical band gap is determined by the π-states of sp^2 configuration of bonds. According to the cluster model with planar clusters, optical band gap is given by [126, 152]:

$$E_g^{opt} \approx 2\gamma/M^{1/2}, \tag{8.1}$$

where γ is the Huckel parameter, which represents the atom interaction with the nearest neighbors and M is the number of sixfold rings in the clusters.

The expression (8.1) gives higher value of E_g^{opt}. The optical band gap is determined not only by the cluster sizes, but also by the deformation of clusters. As a result, it is impossible to give simple and precise expression to determine E_g^{opt}. The theoretical calculations in the frame of continuous random network model show that the optical band gap increases with the decrease of sp^2 phase [164, 165]. Hydrogen content in the films influences weakly on E_g^{opt} because C-H bond states lie outside the band gap. Two the most popular methods are used for the determination of E_g^{opt}, namely (1) measurement of $E_{04} = E_g^{opt}$, with E_{04} being the energy at which the optical adsorption coefficient α is equal to $10^4\ cm^{-1}$ and (2) using the intersection of the energy dependence of adsorption coefficient with abscissa axis.

$$\alpha E = B(E - E_g)^2. \tag{8.2}$$

Experimentally, the optical band gap is determined by averaging the value of E_g of the sp^2 clusters in the film. The decrease of E_g^{opt} with the growth of sp^2 phase content for a-C:H, ta-C:H, and ta-C films is observed [166]. This agrees with theoretical calculations. Although many details of the cluster model are not clear, it is still correct to regard the amorphous carbon as an inhomogeneous mixture of sp^2 and sp^3 sites. In this case the band gap of film varies with position [126, 167]. Sp^3 sites determine the wide band gap between bonding σ and untibonding σ^* states, while sp^2 sites give variable E_g that depends on the size and configuration of sp^2 clusters. The sp^3 matrix is a barrier between the sp^2 clusters for the electron transport through the film. The existence of clusters of different sizes causes the appearance of adsorption "tails" but no sharp gap at optical adsorption. Urbah's energy, which is the measure of disorder, shows that average size of clusters decreases with the growth of band gap, but the distribution of the sizes of clusters is still large in wide gap a-C:H films [168].

This analysis is acceptable also in the case of amorphous carbon films with dominating sp^2 phase ($sp^2/sp^3 > 1$). It is possible to consider the film as the sp^2 matrix with sp^3 clusters embedded (Figure 8.4). This situation was analyzed in [158] and the expression for the calculation of average cluster diameter was obtained.

The ta-C film, among others amorphous carbon films, is perspective for many mechanical applications [169]. A distinguished feature of ta-C film is high content of sp^3 phase. In such

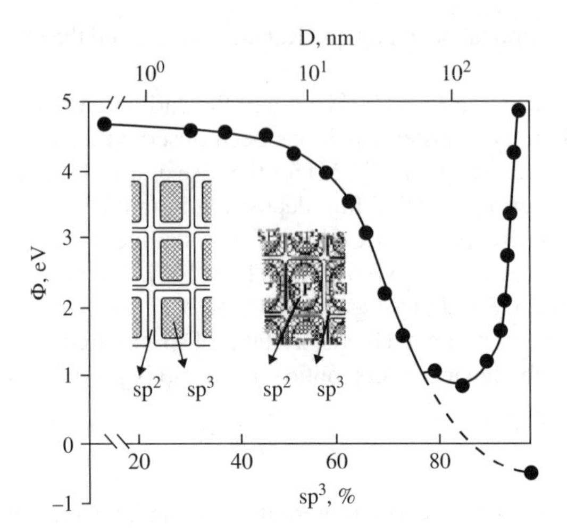

Figure 8.4 Work function on sp^3 content dependence and schematic image of DLC film structure (negative electron affinity for diamond at sp^3 = 100% is shown too)

films hydrogen is absent and they have high density and hardness like diamond [170]. The cluster model can be suitable for ta-C films too. But in this case the sp^2 site clusters are small enough.

Similar model can be used also for the analysis of diamond films. In diamond films the micro- or nanocrystals (grains) are coated with graphite-like sp^2 phase [46, 139, 171, 172]. In this case it is possible to consider the sp^3 clusters embedded into the sp^2 matrix [158].

Schematic images of energy band diagrams of different DLC films in the approximation of equal size of the clusters are shown in Figure 8.5. Electrons need to overcome the barrier Φ_b at the interface of Si-DLC film and the barrier Φ_c at the interfaces of sp^2-sp^3 phases in DLC film during the electron transport from the cathode (silicon) through the DLC film to the anode (metal) in metal-insulator-semiconductor (MIS) structure.

8.3.3 Electron Field Emission from DLC Films

A theory for the emission mechanism from DLC films supported by an energy band diagram and emission structure is given in [173]. DLC films have an energy bandgap which changes with the sp^3 phase content. The resulting energy band diagram indicates that DLC films (a-C:H type) have a PEA [173]. It has also been assumed that emission from DLC films is caused by hot electrons created in the region of strong band bending on the rear contact [174]. However, the photoemission measurements of valence band energy on thick ta-C type DLC films has not shown the presence of substantial band bending on the rear contact [175]. Measurements of the electron energy distribution also have not shown hot electrons with high-energy “tails” [176]. There are several results showing that the emission is controlled by a relatively high barrier on the surface of the DLC film [126]. DLC films like some other flat dielectric or semiconductor films can be LMF electron emitters, able to emit electrons at macroscopic electric fields in the range of $(1\text{–}50) \times 10^4\ \mathrm{V \cdot cm^{-1}}$. The phenomenon occurs because the dielectric (semiconductor) film is an ENH material [125].

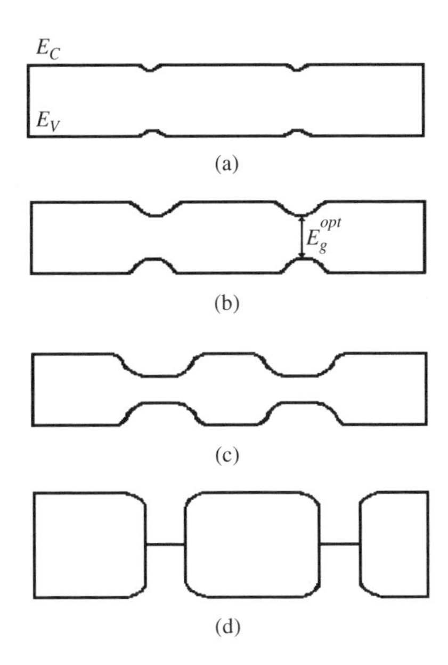

Figure 8.5 Schematic images of energy band diagrams of different DLC films: (a) ta-C; (b) a-C, a-C:H; (c) a-C, a-C:H with higher content of sp^2 phase; and (d) diamond film

In some cases, the peculiarities of EFE from DLC-coated metals or semiconductors can be explained assuming that there is the structure on the film surface (and, hence, external field enhancement), or a structure inside the film (internal field enhancement), or a combination of both (hybrid field enhancement). Both hybrid and internal field enhancements may operate at different locations on the same film. In reference [125], such DLC films have been determined as partially graphite-like (PGL) films.

EFE that can be explained by above described emission model based on the cluster model of DLC film has been observed in [177, 178]. But in the EFE investigations, the effect of sharp growth of emission current and decrease of threshold voltage under repeating measurements at high electric fields has been revealed after DLC film conditioning. The detailed analysis of EFE peculiarities from ENH materials is given in [125].

In reference [96], the results of experiments with DLC films deposited by laser ablation using a conductive atomic force microscopy (AFM) probe have been reported. It has been shown that field-induced electron emission (FIE) sites are associated with the small morphological features (inclusions) on the film front surface. These sites exhibit high conductivity between the probe and the underlying substrate. In films with high sp^2 content these features are found without any need to activate the film. For a film with low sp^2 content initially insulating and flat inclusions could be found only after the film activation by a vacuum-arc discharge between the film and the counter-electrode, as described in [177, 178]. It has been assumed that the observed increase in conductivity (by at least 2 orders of magnitude) can only be explained by the formation of conductive channels through the whole thickness of the film down to the silicon substrate. They have assumed that the surface inclusions (and the channels) have been created by the transformation of sp^3 bonding to conducting sp^2-bonded carbon. For the films with high sp^2 content they have assumed that channels already exist, or can be formed by applying a field without the need for arc-based activation process. Like a free-standing conductive tip in the vacuum, a conductive channel in the insulating matrix leads to a field enhancement and enhanced electron emission.

In reference [113], the emission from 300 nm thick nitrogen-containing DLC films (ta-C:N) deposited by filtered arc deposition in nitrogen atmosphere onto *p*-doped silicon was studied. The nitrogen content was 12% before and 17% after the plasma cleaning. The aim was to compare their results with those of [106]. It was shown that the surface had high density of conductive clusters embedded in otherwise smooth film. Two types were observed: apparently spherical clusters 20–40 nm in size with the density about $3 \times 10^{12}\ m^{-2}$, and larger irregular clusters with the size of 100–400 nm and the density of about $10^{11}\ m^{-2}$.

Field emission spectroscopy (FES) coupled with UPS clearly showed that the emission was of tunneling type and occurred from states nearly 5 eV below the vacuum level and indicated a local field of about $6.5\ V \cdot nm^{-1}$. It was noted that these results were incompatible with the emission mechanism proposed in [106], but also noted that these ta-C:N films were different from the a-C:H:N films [106]. The macroscopic applied fields were about $20\ V \cdot \mu m^{-1}$. A conclusion was drawn that EFE observed from this particular sample was due to the field-enhancing structures on the sample surface with field enhancement factors in the range of 150–300.

The results of high-resolution scanning electron microscopy (HRSEM) and micro-Raman spectroscopy of emitting clusters on a DLC film were reported in [179]. The HRSEM showed small, sharp, and irregularly shaped structures typically 20–50 nm in size, clearly able to generate some geometrical field enhancement; micro-Raman investigations showed that these particles were graphitic. The structures were presumably formed during the film deposition process. It was found that good emission sites correlated with the presence of particles. In general, nonactivated EFE from DLC was observed only when field-enhancing structures were present on the surface.

With the last structures, there seem to be two possibilities as regards to the emission process. Either the DLC film is so highly conductive that it becomes quasi-metallic, and it is possible to regard these structures as equivalent to metal micro-protrusions. Or, possibly, these structures have made the film self-activating by a process step similar to the first stage of C-D-C-D-V emission process. If so, the emission process is analogous to the C-D-C-V process described in [76], and the channel formation in the film could be significant because it would contribute to the field enhancement and make this a hybrid situation.

8.3.4 Model of EFE from Si Tips Coated with DLC Film

During EFE from Si tips coated with DLC films, many various mechanisms are involved in a current transport of electrons from the negatively biased substrate through different interface boundaries and the film bulk to its surface with following emission into vacuum gap, and finally, to the anode. The situations are possible, when the contact of a semiconductor (metal) and a film or a film volume significantly influences the emission characteristics. Current flow through the bulk of the film can be limited by a current transport mechanism in the case of low conductivity and relatively thick (>0.1 μm) films. The analysis of the experimental results on EFE performed in [180–182] has showed that along with generally used Fowler-Nordheim tunneling they can be satisfactorily described also by other transport mechanisms, such as Schottky emission, space charge limited currents (SCLCs), and Frenkel-Pool mechanism modified with SCLC. To clarify which mechanism limits the current transport it is necessary to analyze *I-V* dependences of EFE in detail.

The model of EFE from Si tips coated with DLC films is developed based on the cluster model of DLC films [152, 153, 159, 164] and the peculiarities of electron transport through DLC film. In the case of electron transport through DLC film in MIS structure, the electrons are injected from a semiconductor, pass through the DLC film and freely go out into the metal electrode (anode). At EFE, there is a vacuum-induced barrier for electrons to escape the DLC film. This is the main distinction of a current transport in MIS structure through the DLC film and EFE with DLC film on silicon models. The model of EFE is developed for the definite case of silicon tips, but it is possible to use it also for cathodes of other semiconductors and metals. A schematic image of the energy band diagram at EFE of Si tips coated with DLC film is shown in Figure 8.6. As it can be seen from comparison with the diagram of MIS structure there is an additional output barrier for electrons, Φ_0, at the DLC film-vacuum interface. So, at EFE the electrons pass from cathode (semiconductor, metal) to anode through the vacuum barrier. Three types of barriers are met on the way of electrons: at the Si-DLC film interface, Φ_b, at the sp^2-sp^3 phase interface of DLC film, Φ_c, and at the DLC film-vacuum interface, Φ_0. Since the current passes through the system in the region of high electric fields, EFE will be determined (limited) by the barrier with the lowest tunneling probability.

The barrier heights at interfaces of Si-DLC film and at sp^2-sp^3 of DLC film depend on the sp^2/sp^3 ratio. This ratio influences also the barrier height at DLC film–vacuum interface. Theoretical calculations of current transport in MIS structure show that both the barrier at Si-sp^2 phase of DLC film interface and the barrier at sp^2-sp^3 phase interface can control (limit) the current. In the case of EFE the barrier at DLC film-vacuum interface can restrict also the current transport. To determine which barrier controls the EFE it is necessary to know the heights of these barriers and their ratio. The work function, that is, the barrier height during tunneling can be determined measuring EFE and taking into account the electric field enhancement coefficient. But it is impossible to say definitely which barrier restricts the EFE: Si-DLC, sp^2-sp^3, or DLC-vacuum. As a result, the value of the work function obtained from electron EFE experiments can differ from that obtained by others methods. The term “effective work function,” Φ_{ef} is often used in literature for the values obtained from EFE experiments in the case when it is impossible to separate the influence of work function and electric field enhancement coefficient on EFE: $\Phi_{ef} = \Phi_0/\beta^{2/3}$ [183, 184].

The proposed model of EFE allows us to understand the physical nature of a decrease of the work function at EFE from Si tips coated with DLC films (Figure 8.6). Φ_0 on sp^2/sp^3 phase ratio in DLC films and doping dependences are also clarified. In the case of clean silicon surface there is the barrier $\Phi_0(Si)$ for electrons at Si-vacuum interface. The thermodynamic work function coincides with the height of Si-vacuum barrier, $\Phi_0(Si) = 4.15$ eV, at high doping with the donor impurity ($N_d \geq 10^{18}$ cm^{-3}) [185]. The positions of the Fermi level of Si and DLC film under thermodynamic equilibrium coincide after the DLC film deposition (both non-doped and in-situ doped). As can be seen from Figure 8.6, in general case the barrier $\Phi_0(Si)$ is divided in three barriers: at the Si-DLC film interface, Φ_0, at the sp^2-sp^3 phase interface, Φ_c, and at the DLC film-vacuum, $\Phi_0(DLC)$. It is possible to show (Figure 8.6) the fulfillment of such relations

$$\Phi_0(Si) = \Phi_b + \Phi_0(DLC) \tag{8.3a}$$

$$\Phi_0(Si) = \Phi_b + \Phi_c + \Phi_0(DLC) \tag{8.3b}$$

$$\Phi_0(Si) = \Phi_b - \Phi_c + \Phi_0(DLC) \tag{8.4a}$$

$$\Phi_0(Si) = \Phi_b + \Phi_0(DLC). \tag{8.4b}$$

If the electrons are injected from the Si into the sp^3 phase of DLC film then Equations (8.3a) and (8.4a) represent the case of electron emission into the vacuum from the sp^3 and the sp^2 phases respectively. Equations (8.3b) and (8.4b) describe the injection into the sp^2 phase of DLC film and electron emission into the vacuum from the sp^3 and the sp^2 phases respectively.

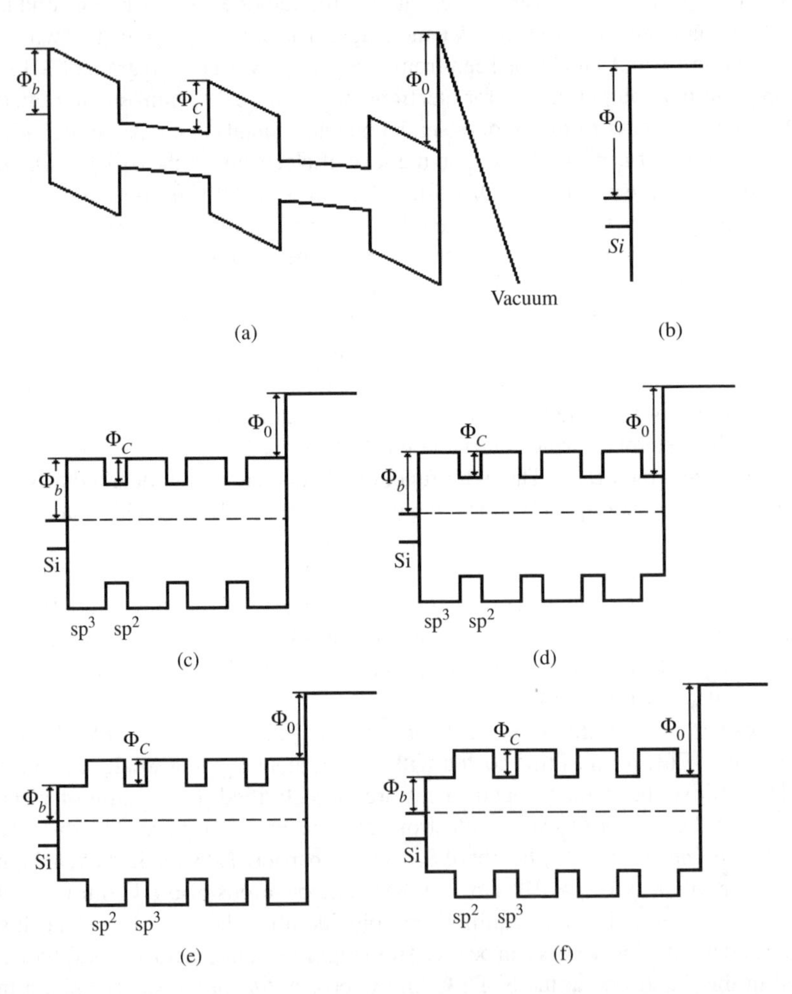

Figure 8.6 Schematic images of energy band diagrams of Si-DLC-vacuum system: (a) in electric field; (b) Si-vacuum system; (c–f) system Si-DLC-vacuum (different possible cases): (c) electron injection from Si into sp^3 phase, emission from sp^3 phase in vacuum; (d) electron injection from Si into sp^3 phase, emission from sp^2 phase in vacuum; (e) electron injection from Si into sp^2 phase, emission from sp^3 phase in vacuum; and (f) electron injection from Si into sp^2 phase, emission from sp^2 phase in vacuum

As the calculations show, the electron injection from the Si into the sp^2 phase is realized at relatively large sizes of sp^2 inclusions (≥ 5 nm). The electron injection into the sp^3 phase takes place at low sizes of sp^2 inclusions (≈ 1 nm). As a rule, sp^2 phase is present on the DLC film surface, as shown in Figure 8.6d,f. This situation corresponds to Equations (8.4a,b). This is caused by peculiarities of DLC film growth. Independently on the growth method the formation of sp^3 phase in DLC film is realized according to the subplantation mechanism (low energy subsurface implantation of C^+) [126]. It causes the formation of the sp^2 surface layer several nanometer thick in DLC film. The influence of sp^2 phase on DLC film parameters is more significant at relatively high size of inclusions (≥ 5 nm). In this case the probability of electron injection into the sp^2 phase at the Si-DLC film interface is higher than in the sp^3 phase (due to the lower barrier Φ_b). This case is shown in Figure 8.6f (Equation (8.4b)) and as a rule is realized in practice at $sp^3/sp^2 > 1$. From the model described above one may conclude that the physical nature of lowering Si work function after coating with DLC film is understood. The barrier for electron emission into vacuum is divided in two or three barriers (parts), each of them being naturally lower than the initial one. The emission is determined by the electron transport through a barrier with the lowest tunneling probability. The barrier heights and electron effective masses in the sp^2 and sp^3 phases should be known to calculate the tunneling probability through each barrier during EFE. It is difficult to determine these parameters. The energy barrier heights at the interface of Si-sp^2 phase and sp^2-sp^3 phases were obtained by the internal photoemission method [186]. The values of $\Phi_b = 0.45$ eV and $\Phi_c = 0.55$ eV were determined for the case of nitrogen doped DLC film (25% N_2 in gas mixture and 3 at% N in the film). According to Equation (8.3a), $\Phi_0(DLC) = 3.7$ eV. Taking into account the lowering of energy barrier due to the image force the value $\Phi_0(DLC) = 2.85$ eV has been obtained. The work function obtained from the slope of experimental *I-V* characteristic in Fowler-Nordheim plot is 1.80 eV. This value is noticeably lower than that obtained from the internal photoemission measurements and analysis of EFE model. The difference between the work function values obtained by two independent methods is 1.05 eV. This points out on the additional lowering of the work function. The calculation of the emission current density in the case of an experimental value of $\Phi_0(DLC) = 1.80$ eV shows that the EFE is limited by tunneling through the barrier sp^2 phase-vacuum.

The electron work function of the material has two components: the internal work function, Φ_{in} and the surface (dipole) one, D:

$$\Phi_0 = \Phi_{in} + D. \tag{8.5}$$

According to quantum mechanics theory, the internal work function is the difference between the energies of the base neutral state of the body with N electrons (E_n) and the charged to $+1\ q$ state with $N-1$ electrons (E_{n-1})

$$\Phi_{in} = E_n - E_{n-1} = -\mu, \tag{8.6}$$

where μ is the electrochemical potential [36]. The existence of dipoles on the surface changes the total work function.

The influence of the dipole layer explains the orientation dependence of work function of crystalline solid states. At positive dipoles on the surface ("-" on the surface) the work function decreases in accordance with Equation (8.5). For DLC films such a situation is realized at

C-O bonds on the surface. Carbon electronegativity is equal to 2.5, and oxygen is equal to 3.5. The negative charge shifts to surface direction (to oxygen). On the contrary, at surface negative dipole layer (dipoles directed to vacuum) the work function decreases. This is realized with C-H bonds at the surface of DLC film. The a-C:H films obtained by plasma enhanced chemical vapor deposition (PECVD) method from CH_4:H_2 or CH_4:H_2:N_2 mixtures have a hydrogen-terminated surface [126]. The hydrogen electronegativity is 2.1, which is lower than that of carbon. The electron charge shifts to solid, the dipoles are negative, and directed from the bulk to the surface. In the case of diamond films, the surface termination with hydrogen allows even to obtain the effective NEA [187, 188]. The surface passivation with different atoms can strongly influence the emission properties. The calculation of the influence of dipole layer on the work function is rather difficult. The estimation of the dipole component of the work function in the case of hydrogen-terminated DLC film gives a value $D = -0.67$ eV. As can be seen, the dipole layer has strong influence. The account of the influence of dipole layer decreases the discrepancy between the results of theoretical calculations and experimental findings down to 0.38 eV. This difference can be caused by approximation about the equality of sp^2 cluster sizes and correspondingly of energy barriers Φ_b, Φ_c, and Φ_0 in the film. In fact, there is some distribution in sp^2 nanocluster sizes and barrier heights [126].

8.3.5 Electron Field Emission from Tips Coated with Ultrathin DLC Films

Thin undoped and in situ nitrogen-doped 8–80 nm thick DLC films were deposited on emission cathodes by the method of PECVD from CH_4:H_2 or CH_4:H_2:N_2 mixtures, respectively [7, 189, 190]. Gas pressure in the chamber was in the range of 0.2–0.8 Torr. The film deposition was carried out at room temperature and in some cases at 60 °C. The substrates for DLC film deposition were put directly on the 200 mm-diameter cathode cooled by water and capacitive connected to a 13.56 MHz generator. During the plasma decomposition the radio frequence (RF) bias voltage was 1900 V. The DLC coatings were smooth enough and had reproducible properties from sample to sample under the same deposition conditions.

The results of experimental investigations of EFE from DLC films coated Si tip arrays are presented in Figure 8.7a [190–192]. As it can be seen, the resonant peak in emission current is observed. The resonant enhancement of current density is by factor of 2 for tips and 1.7 for a flat Si-DLC film cathode. Obtained *I-V* characteristics also point out on the enhancement of emission current with the increase of DLC film thickness. This is connected with a pronounced lowering of the output barrier for thicker DLC films. In this case the resonant peak position shifts to smaller electric fields and disappears as the thickness increases further.

The data on EFE from GaAs tips coated by ultrathin (10 nm) DLC film are shown in Figure 8.7b. Two peaks are observed on *I-V* characteristics of field emitters with DLC films. Analysis of the position of resonance peaks on *I-V* characteristic taking into account the electric field distribution and electric field enhancement coefficient allowed estimating the distance between the local energy levels in a quantum well. It is equal to 360 meV. The theoretical calculation of the position of energy level in a triangular potential band [193] gives the value $\Delta E = 650$ meV. The discrepancy between the experimental and theoretical results is connected with electric field dependence of energy level in triangular quantum well position $E_n \sim F_s^{2/3}$. The average field value was accepted during calculation.

The resonance peaks at EFE from Si tip coated with ultrathin (d ≈ 2 nm) DLC films were also observed in [194] at macro-fields of (1.1–1.2) × 10^5 V/cm.

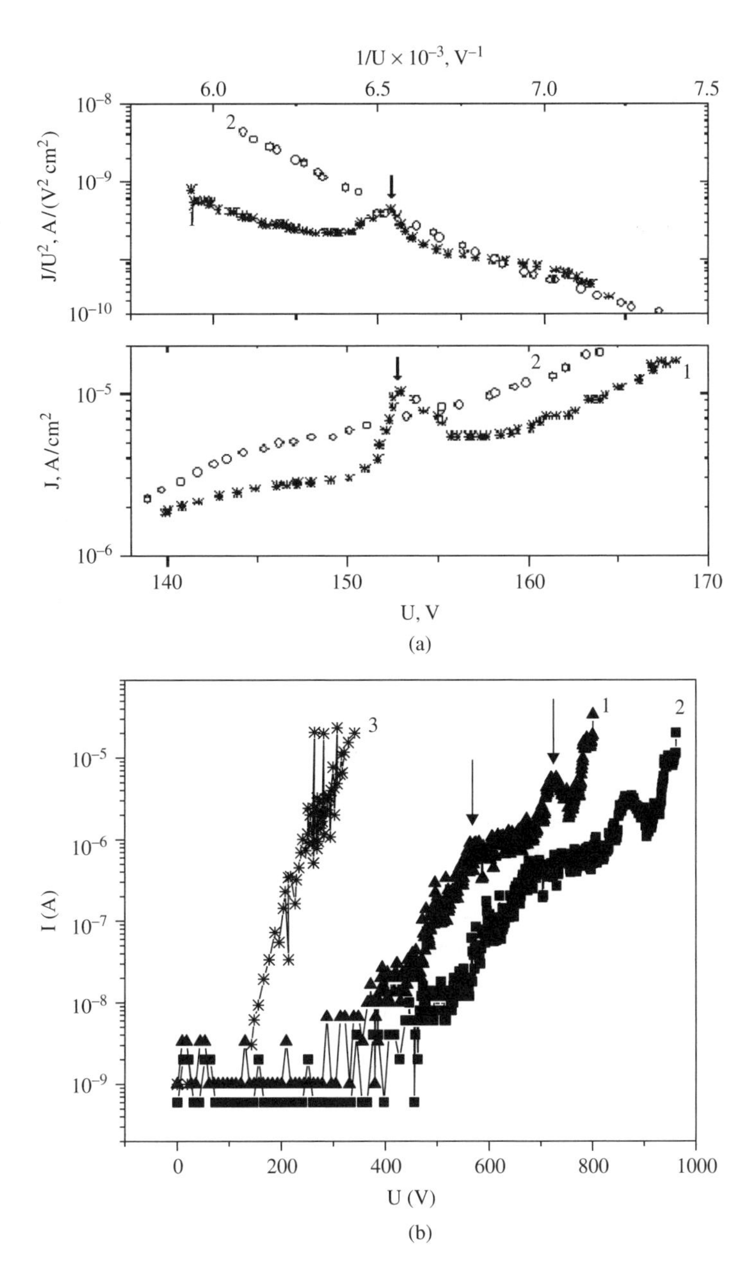

Figure 8.7 *I-V* characteristics of electron field emission current from (a) Si and (b) GaAs tips: 1,2: coated with DLC film; 3: without DLC film (a) 1: $d_{DLC} = 8$ nm, 2: $d_{DLC} = 30$ nm; (b) 1, 2: $d_{DLC} = 10$ nm, 2: repeated measurement). Reprinted from Ref. [191]. Copyright (2003), with permission from Elsevier

8.3.6 Formation of Conductive Nanochannels in DLC Film

The formation of conducting nanochannels in DLC films has been experimentally observed and the method for nanochannel formation has been proposed [195].

The arrays of initial silicon emitter tips were fabricated by wet chemical etching. The cathodes were formed on (100) n-type Si wafers ($N_d = 10^{15}\ cm^{-3}$) by patterning with Si_3N_4 as the masking material. The sharpening of tips was performed by oxidation of the as-etched tips at 900 °C in wet oxygen and etching in $HF{:}H_2O$ solution. This sharpening technique allowed the production of tips with a curvature radius of 10–20 nm. The height of the silicon tips was 4 μm. The density of tips was 2.5×10^5 tips cm^{-2}. The radii of the tips and their height were estimated by scanning electron microscopy (SEM). At formation of GaAs vertical wedge emitters the starting material was 10 μm thick n-GaAs layer ($N_d = 2 \times 10^{15}\ cm^{-3}$) on n^+-GaAs substrate covered by 300 nm thick PECVD SiO_2 passivation layer. First, the photoresist mask for the vertical wedge emitter was aligned along the (110) direction. After etching the passivation around the photoresist with reactive ion etching (RIE), GaAs spacer structures and wedge emitters were anisotropically etched using citric acid solution with H_2O_2. This process continued until the passivation over the emitter was removed. An emitter radius and height of about 10 nm and 9 μm could be reached, respectively.

Thin undoped and in situ nitrogen-doped DLC films of 8–80 nm thickness were deposited on the emission cathodes by PECVD from $CH_4{:}H_2{:}N_2$ mixtures. The DLC film used in these experiments was in situ nitrogen-doped amorphous semiconductor a-C:H. The film refractive index was $n = 1.83$ ($\lambda = 6328$ nm). Its electrical conductivity measured at electrical field $F = 1 \times 10^6\ V \cdot cm^{-1}$ was $\sigma = 1 \times 10^{-11}\ S \cdot cm^{-1}$ and was in the range typical for hydrogenated amorphous semiconductors. The DLC films were deposited on Si tips, GaAs wedges, and on flat silicon wafers.

The measurements of the emission current from the samples were performed in the vacuum system that could be pumped out to the stable pressure of 10^{-6}–10^{-7} Torr. The emission current was measured at ungated cathode–anode diode structures. These were realized by inserting a spacer of defined thickness between the emitting cathode and the anode structure. The emitter–anode spacing L was constant and chosen in the range of 7.5–25 μm. Teflon film was used as a spacer material and heavily doped silicon wafer or molybdenum wire was used as anode. The emission current–voltage characteristics were obtained with the current sensitivity of 1 nA over the voltage range up to 1500 V. A 0.56–1 MΩ resistor was placed in series with the cathode to provide short-circuit protection.

Measurements of EFE from semiconductor tips and wedges coated with DLC film showed a sharp increase of the emission current at high electric fields and a considerable reduction of the threshold voltage after the pre-breakdown conditioning of the DLC film (Figure 8.8). At pre-breakdown conditioning during EFE measurements the applied voltage was continuously increased (at a rate of $0.5\ V \cdot s^{-1}$) till reaching the current level of 10^{-6} A (current density $J = 10^{-4} A \cdot cm^{-2}$). This effect was observed for DLC coated silicon tips and GaAs wedges. The corresponding Fowler–Nordheim plots for Si tips are shown in Figure 8.9. As it can be seen, after pre-breakdown conditioning the slopes of the lines are significantly reduced. This shows a decrease of the effective work function.

During EFE at high electric fields the barriers determined by sp^3 phase between sp^2 inclusions can be broken, resulting in conducting channels between the substrate (Si or GaAs) and the surface of the DLC film. At pre-breakdown with large current densities a local heating occurs which may transform the sp^3 phase of the film into sp^2 one. As a result of the

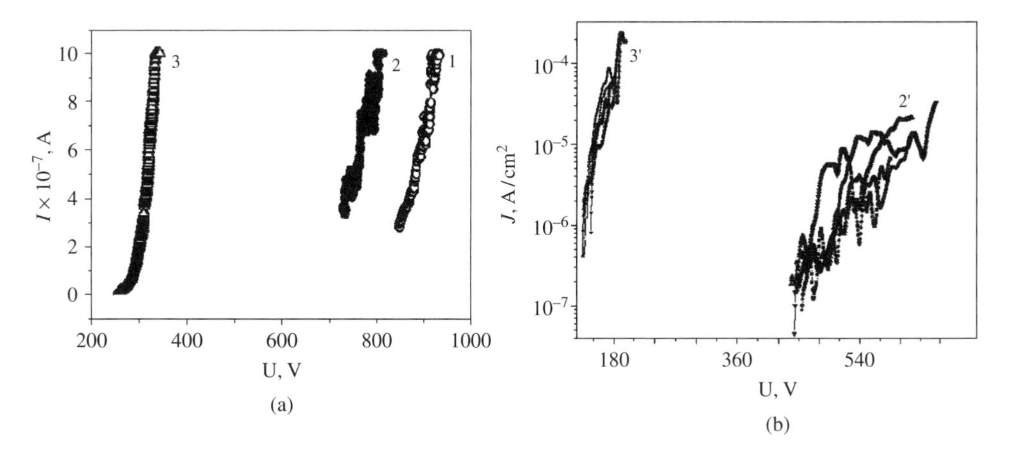

Figure 8.8 Characteristics of emission currents from (a) GaAs wedges and (b) Si tips: 1: emission of GaAs wedges, 2: emission of GaAs wedges covered with thin DLC film (27 nm), 3: emission of DLC-covered GaAs wedges after pre-breakdown conditioning; 2′: emission of DLC-coated (60 nm) Si tips, 3′: emission of DLC-coated Si tips after pre-breakdown conditioning. Curves of subsequent measurements are included in 2′ and 3′ to show the reproducibility of characteristics [195]. Copyright (2006) IOP Publishing. Reproduced with permission. All rights reserved

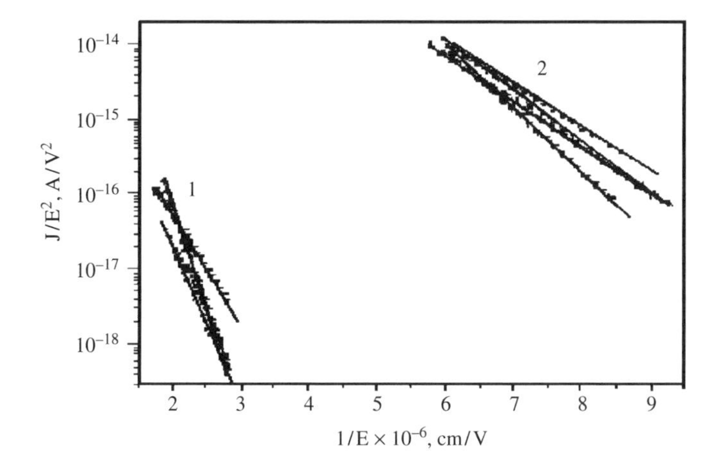

Figure 8.9 Fowler–Nordheim emission characteristics from DLC-coated Si tips before (1) and after (2) pre-breakdown conditioning. The set of subsequent measurements, as shown in Figure 8.8, is included [195]. Copyright (2006) IOP Publishing. Reproduced with permission. All rights reserved

pre-breakdown, an electrical conducting nanochannel is formed in the DLC film. The decrease of the threshold voltage (Figure 8.8) can be used for the estimation of the size of the conducting nanochannel according to [196]

$$\beta^* \approx \frac{h}{r}, \tag{8.7}$$

where β^* is the local electric field enhancement coefficient, r is the curvature radius of the emitter tip (radius of the conducting channel), and h is the tip height (DLC film thickness).

DLC films are heterogeneous materials which contain sp^3-bonded (diamond-like) and sp^2-bonded (graphite-like) phases. According to the work of [158] the DLC film is considered as a diamond-like matrix with graphite-like inclusions. The diameter of the sp^2 phase inclusion can reach $\sim$10 nm. In the case of a low value of the sp^3/sp^2 phase ratio the DLC film consists of a sp^2 matrix with sp^3 phase inclusions [158]. The films under investigation have high sp^3/sp^2 ratio and contain nanometer-size graphite-like inclusions. At electrical conditioning conducting channels through DLC film between the substrate and the DLC surface are formed. They act as electron emitters themselves.

At electrical conditioning the flow of high currents induces a local heating process which promotes the transformation of the sp^3 phase into more conductive sp^2 phase and, in such a way, gives rise to the formation of conductive nanochannels. Besides, a lot of structure defects appear in the channel under pre-breakdown conditioning. These defects are mainly connected with slightly bonded atoms and different structures of carbon (deformed ring, etc.). The defects create local energy states in the bandgap of the sp^3 phase and increase in such a way its conductivity. Referring to the mentioned processes of sp^3 phase graphitization and defect creation, two types of conducting channels between the substrate and the DLC film surface can be considered (Figure 8.10). The first one (type I) is a conductive graphite (graphite-like) channel formed as a result of the transformation of the sp^3 phase into the sp^2 phase (Figure 8.10b). The second type (type II) of conduction channel is based on the conductivity increase of the sp^3 phase due to structure defects' generation between sp^2 phase inclusions in the DLC film (Figure 8.10c). For this channel type the surface of the film consists of a diamond-like sp^3 phase which has a significantly lower work function in comparison with graphite. Values of less than 1 eV have been reported in [8]. From the Fowler–Nordheim plots the effective work function

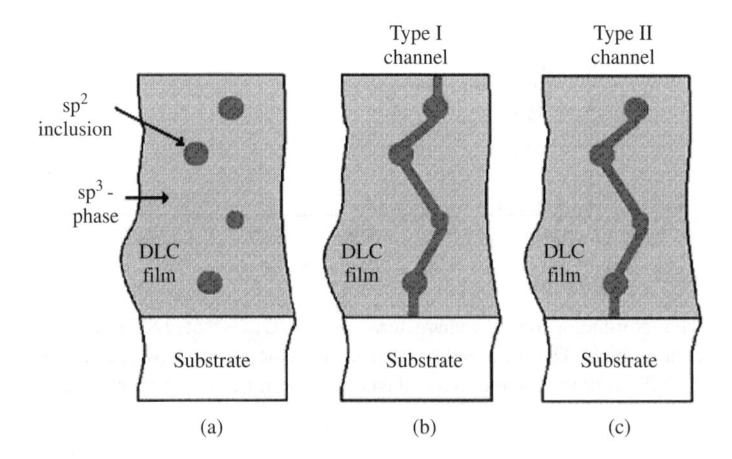

Figure 8.10 Model of DLC film (a) and formed nanochannels: the type I (b) is a result of the transformation of sp^3 phase into sp^2 phase and the type II (c) is due to structure defects' generation between sp^2 phase inclusions in the DLC film [195].

can be determined $\Phi^* = \Phi/\beta^{*\,2/3}$ (Φ is the work function of DLC, β^* is the field enhancement coefficient). The obtained effective work-function values before and after conditioning are $\Phi_1^* = 1.1 \times 10^{-1}$ eV and $\Phi_1^* = 4.1 \times 10^{-2}$ eV, respectively. They are in agreement with values reported in [9]. For the used silicon tips ($r = 10$ nm, $h = 4$ μm) and DLC coating thickness ($d = 60$ nm) the work function of 1.63 eV before pre-breakdown conditioning has been calculated. In the case of DLC film Φ does not denote the real work function but the electron affinity. The field enhancement coefficient was estimated from relationship similar to Equation (8.7) that included the geometrical parameters (r, h, d)

$$\beta^* \approx \frac{h+d}{r+d}. \tag{8.8}$$

The obtained value of β was 58.

The conducting channel diameter was calculated for type I and type II channels. In the first case the graphite work function $\Phi \approx 5$ eV was considered. Values of conducting channel diameter in the range of 5–9 nm were obtained. For type II the lower work function of the DLC film was used ($\Phi = 1.63$ eV). The channel diameter of about 25 nm was obtained. These channel diameter values for type I and type II are in good agreement with the size of the sp^2 phase inclusion in the sp^3 phase matrix [5, 125, 126] and with the model proposed in [95, 125]. It cannot be excluded that the work function of the DLC film on the surface is also changed. For this case the channel diameter will be taken between 5 and 25 nm.

Based on the experimental results above the method for the formation of conducting channel arrays in DLC films for significantly enhanced EFE properties has been proposed [195]. It will allow obtaining the effective EFE from DLC films deposited on flat surfaces (without tips or wedges). The formation method of graphite-like conducting channels is schematically shown in Figure 8.11 for a DLC film on a flat Si substrate. As a catalyst, the Si tip array formed by wet etching and with Al backside metallization was used. The measured I–V characteristics of the

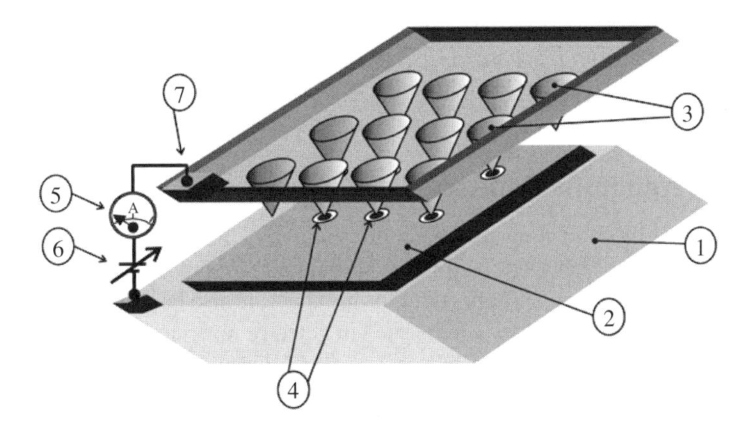

Figure 8.11 Schematic of a setup for the formation of conducting channels in DLC films. 1: semiconductor substrate; 2: deposited DLC film; 3: silicon tip array on a Si substrate; 4: sites of conducting nanochannels; 5: amperemeter; 6: voltage source; and 7: contact to the silicon substrate [195]. Copyright (2006) IOP Publishing. Reproduced with permission. All rights reserved

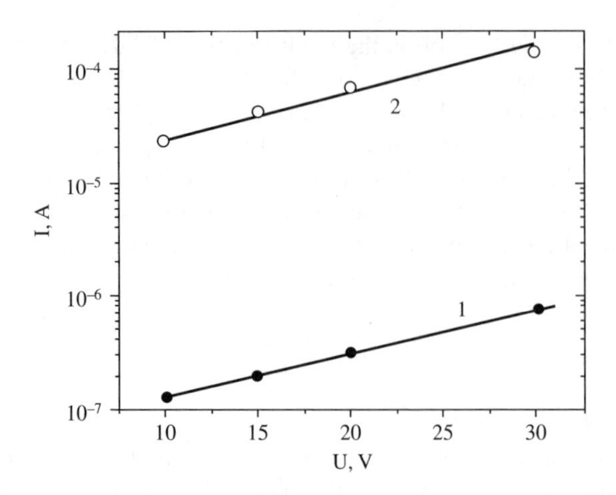

Figure 8.12 Current-voltage characteristics obtained with the setup of Figure 8.11 on DLC film before (1) and after (2) pre-breakdown conditioning. The current increases to more than 2 orders of magnitude value [195].

structure in Figure 8.11 before and after electrical conditioning are shown in Figure 8.12. As can be seen, the conductivity of the DLC film increased significantly. After electrical conditioning the current value increases to more than 2 orders of magnitude. The conducting nanochannels were formed at the sites of Si tip contacts with the DLC film due to electric field enhancement on the tips. In this case the conditioning was also performed by the continuous increase of voltage till reaching the current of $I \approx 10^{-6}$ A through the film.

The definitive proof of the conducting channel formation will be provided by their direct observation. It is necessary to point out that direct observation of the channel in the DLC film is very difficult even by TEM. On the one hand, this is due to the fact that graphite-like and diamond-like phases coexist in the same carbon material. On the other hand, the channels have a diameter in the nanometer range. Usually a surface modification of carbon films is observed after breakdown or vacuum discharge [197, 198]. Unfortunately, no surface nanometer-size peculiarities before and after DLC conditioning could be observed by SEM.

8.4 Carbon Nanotube Emitters

There are several methods of preparing nanotube emitters. CNTs can be fabricated with random orientation; randomly, but oriented in the vertical direction with respect to the substrate and vertically oriented with a well-defined footprint. They can also be mixed in media for easier application to the substrate and they can be used in gated or nongated configurations [199].

CNTs can be grown by dc or rf PECVD or thermal CVD from hydrocarbon source gases at 600–800 °C in the presence of a catalyst [200]. The catalyst is a transition metal like Co, Ni, or Fe. The growth location is controlled by the location of the catalyst on a surface. An electric field in the PECVD causes the nanotubes to be aligned perpendicular to the surface [201]. The PECVD process allows one to make nanotube thin film emitters. The concept consists of the deposition of phase pure nanotube films for emission, without the need of any post-deposition treatment. Furthermore, PECVD deposition techniques have proven to be scalable for the deposition on large surfaces [202].

8.4.1 The Peculiarities of Electron Field Emission from CNTs

Nanotubes make good field emitters because of their extreme shape. They are conductive and can carry very large current densities if needed [11]. Individual nanotubes can stably emit up to 10 μA [10]. It is sufficiently to yield an apparent current density of 0.1 A/cm^2. The current density from randomly oriented films routinely reaches 1 A/cm^2 and can be as high as 4 A/cm^2 [11]. These are the highest emission current densities ever reported for any carbon-based emitters. Upon operating at such high current densities, the emission remains robust with no apparent structural degradation or surface damages occurring to the nanotube emitters.

Single nanotubes or mats of nanotubes can be used (Figure 8.13). Although the emission properties reported for nanotubes vary depending on the nanotube content and size distribution in the specific samples measured, CNTs, in general, exhibit excellent emission properties, regardless of their structures (single-wall vs. multi-wall), orientations (randomly distributed vs. highly aligned), or production techniques (arc discharge, laser ablation, or CVD). Emission

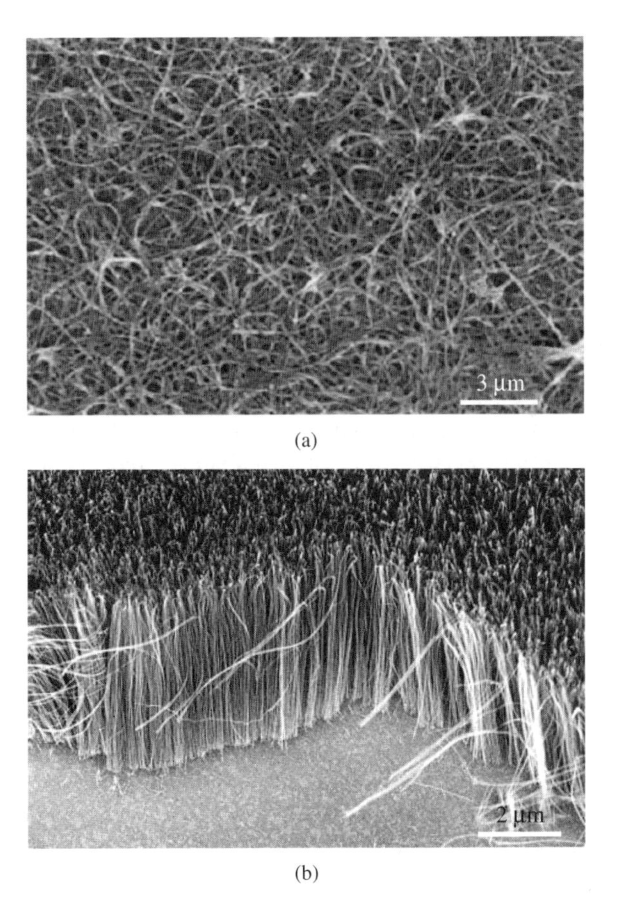

Figure 8.13 Scanning electron micrographs of (a) spray-coated, randomly oriented SWNT ropes and (b) highly oriented MWNTs grown using CVD techniques. Reproduced with permission from Ref. [201]. Copyright (2000), AIP Publishing LLC

is identified to originate from the tips of nanotubes. Loose ends of randomly oriented nanotubes can align themselves in applied electric fields, emitting electrons in a current level similar to that of oriented nanotubes.

The emission occurs from the nanotube tip from the Fermi level. Their work function is similar to that of graphite, 4.7–5 eV [14, 203]. The emission distribution from mats of nanotubes shows that the ESD is surprisingly much smaller than the total density of nanotubes [14, 16]. Most nanotubes do not participate in the emission. This is because the adjacent tubes screen the field enhancement at a tip [16, 204]. The nanotube density must be lowered, so that the mean separation between the tubes is about two times the tube length.

The ESD N from nanotube mats is also found to vary exponentially with applied electric field F, as in ta-C or nanostructured carbon [16, 205, 206].

$$N(F) = N_0 \exp\left(-\frac{b}{F}\right). \tag{8.9}$$

This means that the distribution of Equation (8.9) is equivalent to an exponential distribution of the field enhancement factors

$$N(\beta) = \exp\left(-\frac{\beta}{\beta_0}\right). \tag{8.10}$$

This distribution seems to describe the ESD in all forms of carbon film. The width of the distribution is the weakness of field emission when considering applications. The broad β distribution in nanotubes and nanostructures arises from the field screening by adjacent tips. This means that most nanotubes or sharp features in nanostructured carbon do not emit. This can be improved only by controlling the tip spacing. This can be achieved, with effort, for aligned nanotube mats, but is the weakness for nanostructured carbon where the means to control the various structural features is not so apparent. Ballast resistors can be used to distribute current between the emitting sites, but at the expense of maximum current.

Plasma etching is found to enhance field emission efficiency by decreasing the density of CNTs and sharpening CNT-tips. Generally, plasma treatment reduces amorphous carbon and graphitic particles, opening the tips, and reducing the number of sidewalls of CNTs. Finally, plasma was chemically terminating the entangled region of CNTs on the surface and modifying CNTs surface. The effects of plasma treatment are usually as follows. Plasma treatment, which is likely to make external surface of CNT more active to emit electrons has been changed. In addition, the sharpening CNT-tips, after removal of the catalyst particles, may increase the local electronic field more effectively [207–210].

Field-emission behavior of CNT strongly depends upon its morphology, diameter, spatial distribution, alignment, and contact between CNT and substrate, as well as the condition of the CNT tip (open or capped end, defects, adsorption, Cs doping, and so on). Generally, Fowler–Nordheim (F-N) theory is used to describe field emission behavior of metals. The emission behavior of CNTs fabricated on a metal tip is in excellent agreement with Fowler–Nordheim theory and no current saturation has been found even with an emission current reaching 1 A/cm^2 [211].

But as a rule I-V characteristics of the nanotubes do not follow the Fowler-Nordheim behavior over whole current range. This is apparent for both films [212–214] and single emitters. The electron emission of each multi-wall nanotube (MWNT) followed the conventional

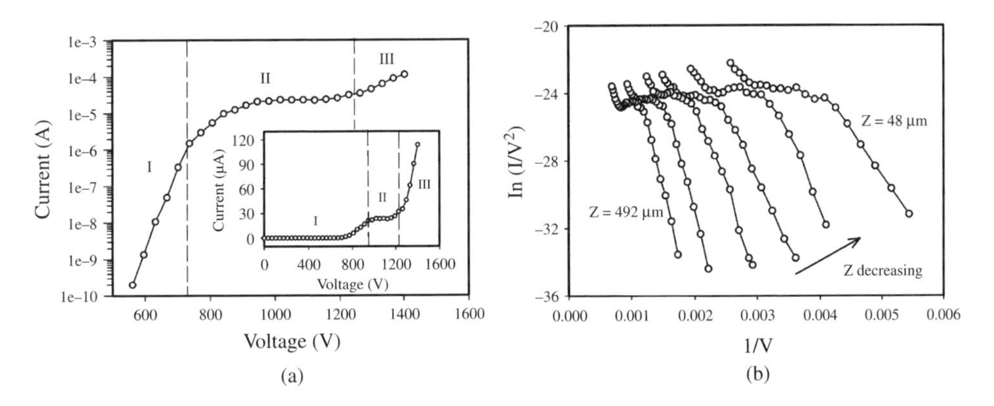

Figure 8.14 (a) Single emission *I–V* curve from SWNT sample shows three distinct regions, adsorbate enhanced emission (region I), current saturation associated with the removal of the adsorbates (region II), and emission from the clean SWNTs (region III). (b) A series of *I–V* curves taken at different anode–cathode spacing (*Z*) from a SWNT cathode. The data are plotted as $\ln(I/V^2)$ vs. $1/V$ and should fall on a straight line if the emission obeys the F-N equation [31]. Copyright (2001) by John Wiley & Sons, Inc. This material is reproduced with permission of John Wiley & Sons, Inc

Fowler–Nordheim field emission mechanism after their apexes were freed from the erratic adsorption species using a conditioning process at room temperature. The conditioning process led to stable emission currents and reduced their variations $\Delta I/I$ to less than 30% between different MWNTs of the array [215].

There are some interesting and unique *I–V* characteristics of nanotube emitters that have not been observed in other carbon-based emitters. As it is shown in Figure 8.14, the *I–V* behavior typically follows the F-N tunneling mechanism at low voltages and currents (region I). However, at high voltages and increased current (>100 nA for individual nanotube emitter or >100 mA/cm^2 for groups of emitters), the *I–V* data exhibit a current saturation region that sharply deviates from the tunneling mechanism (region II). At further higher voltages and currents, *I–V* characteristics return to the electron tunneling behavior governed by the F-N theory (region III). These distinct *I–V* regions have been observed in emissions from both SWNTs and MWNTs in the form of either an individual emitter or large groups of nanotubes (such as in a film) [200, 214, 216–218]. Saturation in emission current is of great interest for device applications, because devices operating in the current saturation region are more stable and less influenced by external factors. The current saturation characteristics of nanotubes, if proven and understood, therefore, make them better suited for building stable devices.

Possible explanations for the saturation phenomenon in nanotube emitters range from space charge effects [219, 220] to interactions between neighboring nanotubes [214] and existence of nonmetallic, localized states at the nanotube tips [217]. However, it has been indicated [13] that the current saturation is actually a surface adsorbate effect that is not observed with clean nanotubes. As has been reported [13, 220–222], nanotubes emit electrons through adsorbate states at room temperature. These adsorbate states, mostly water-related, enhance the field emission current of nanotubes by 2–4 orders of magnitude. As the field and emission current increase, the adsorbate states are perturbed, resulting in the reduction of tunneling enhancement at the adsorbates and the accompanying current plateau. The current saturation was found to be concurrent with rapid fluctuations in emission current (100-fold increase in fluctuation, compared to a typical 5–10% [217) and distinctive changes in field emission patterns from lobed configurations to circular ones, consistent with physical changes occurring at the adsorbate sites. At even higher fields, the adsorbate states are completely removed, and *I–V* behavior now represents that of clean nanotubes, which shows no evidence of current saturation for emission current reaching 1 μA per tip. Under nonideal vacuum conditions where the adsorbates return to the emitter surface when the applied field is reduced, the *I–V* characteristics, including the saturation region, are completely reversible. However, this adsorbate effect, while it is likely present, does not fully explain why the current saturation data consistently follow a certain slope. It has been found that the I–V data in the saturation region fit nicely with the Child–Langmuir law [223, 224] that governs the vacuum space charge phenomenon. It is, therefore, tempting to attribute the current saturation to the local space charges established by the adsorbates. Considering the molecular nature of these adsorbates that can induce very high-field concentrations and generate very high current densities locally, such a space charge effect appears plausible.

The field emission properties of nanotube thin films deposited by the PECVD process from 2% CH_4 in H_2 atmosphere were investigated [14]. Depending on the deposition of the metallic catalyst $Fe(NO_3)_3$ (in an ethanol solution or sputtered) the nanotube films showed a nested or continuous dense distribution of tubes. The films consisted of MWNTs with diameters ranging from 40 down to 5 nm, with a large fraction of the tubes having open ends. The nanotube thin film emitters showed a turn-on field of less than 2 V·μm^{-1} for the emission current of 1 nA. The ESD of 10 000 emitters per cm^{-2} is achieved at fields around 4 V·μm^{-1}. The emission spots, observed on a phosphorous screen, show various irregular structures, which have been attributed to open-ended tubes. A combined measurement of the field emitted electron energy distribution (FEED) and current-voltage characteristic allowed the authors to determine the work function at the field emission site. In the case of the MWNT thin films and arc discharge grown MWNTs the work function values around 5 eV was found, which agrees well with the global work function of 4.85 eV determined by photoelectron spectroscopy. From the shape of the FEED peaks it is possible to conclude that the field emission has originated from the continuum at the Fermi energy, indicating the metallic character of the emission site. In the case of SWNTs the significantly lower work function values of around 3.7 eV compared to those of MWNTs have been founded. This has been attributed to a size dependent electrostatic effect of the image potential, which lowers the work function for small (<5 nm) structures.

Field emission capability of the CNTs has been improved by hydrogen plasma treatment [208]. Hydrogen concentration in the samples increases with increasing plasma treatment duration. $C^{\delta-}$–$H^{\delta+}$ dipole layer may be formed on CNTs' surface and a high density of defects results from the plasma treatment, which is likely to make the external surface of CNTs more active to emit electrons after treatment. In addition, the sharp edge of CNTs' top, after removal of the catalyst particles, may increase the local electric field more effectively. This study suggests that hydrogen plasma treatment is a useful method for improving the field electron emission property of CNTs. The large increases in field emission current when operating CNTs in substantial pressures of hydrogen, especially when the nanotubes are contaminated have been observed [225]. For CNT field emission array (FEA) intentionally degraded by oxygen, the operation in hydrogen has resulted in a 340-fold increase in emission current at constant gate voltage. The results suggested a dependence on atomic hydrogen produced from the interaction between emission electrons and molecular hydrogen. The observed emission enhancement could be due to a surface dipole formation, hydrogen doping, or removal of oxygen-containing surface species.

The influence of oxygen has been investigated at treatment of vertically aligned carbon nanotube (VACNT) arrays by chemical reagents, such as oxygen (O_2) and ozone (O_3) [226]. When treated by O_2 and O_3, the emission current of the CNT array was increased ~800% along with the decrease of the onset field emission voltage from 0.8 to 0.6 V/μm. TEM showed that the end tips of many CNTs were opened or partially opened after the O_2 oxidation process. Evidently, the increase in the field emission current and the decrease in the onset field emission voltage are due to opening of the CNT end tip by the O_2 oxidation process. Open-ended CNTs have sharper tips, and thus lower onset field emission voltage as well as higher emission current [207, 227, 228]. The emission current is not only determined by the sharpness of the emitter tips, it is also affected by the conductivity of the emitters. In the array format, all highly strained CNT tips are exposed to air and readily oxidized by O_2 first before oxidation of the less strained CNT walls, which are partially protected from being oxidized by the array configuration. However, prolonged O_2 oxidation will lead to damage along the CNT walls, and thus a gradual decrease in emission current. Selective and preferential oxidation at the CNT tips is the feature of the current process. Open-ended MWNTs have much poorer field emission properties than close-ended MWNTs, when the open-ended MWNTs have been prepared by O_2 oxidation of carbon soot at high temperatures. Under the carbon soot-O_2 oxidation condition, most of the CNT tips were buried within the powder, and a lot of CNT walls were exposed to O_2 oxidation [229]. Only a small percent of open-ended MWNTs was left after the oxidation process. Serious oxidative damage along the tube walls of the surviving MWNTs is most likely responsible for the reported poor field emission performance. The F-N plots of O_2 treated samples show a linear slope at the low applied voltage region and a smaller slope at the high voltage region. This type of F-N pattern for MWNT field emitters has been observed before, and the smaller F-N slope at high voltage region is attributed to the saturation effect of the emitters [229].

Thermal-assisted field emission has been investigated from screen printed CNT cathode [230, 231]. It was found that the field emission properties of the CNTs cathode, such as turn-on field and the field emission current density, were largely related to the temperature of the cathode. The field emission properties of the CNTs were enhanced as by increasing the substrate temperature as well as by irradiating pulsed laser. The decrement of work function was the main reason for the variation in the field emission properties.

Field emission properties of MWNTs composite films have been investigated under mechanical stress [232]. The emission threshold fields initially decrease from 2.3 to 0.6 V/μm before rising back to 3.1 V/μm with increasing mechanical stress applied externally to the film. This behavior from nanotube composites is believed to be associated with modification to the work function of the nanotubes.

Interesting observation was the field-emission-induced luminescence from CNT emitters. The light emission was detected from MWNT nanotube emitters at a current density of 2 mA/cm^2 or higher [233]. By analyzing the spectra of the emitted light, it was concluded that the luminescence did not come from blackbody radiation or current-induced heating rather it resulted from electronic transitions between different electronic states participating in the field emission at the tips of nanotubes. The electronic structure at the nanotube tips is known to be different from that of the bulk [234, 235], and electron energy distribution measurements have confirmed the existence of nonmetallic, localized electronic states above the Fermi level at the tips of SWNTs [236].

8.4.2 Stability of Electron Field Emission from CNTs

The stability of field emitters in various vacuum environments is of critical importance to device applications [31]. It is mostly associated with the resistance of emitter to sputtering and oxidation. The emission from CNTs is stable, but it depends on the presence or absence of molecules, which can be absorbed on the nanotube tip [12, 13].

Field emission sources possess both intrinsic and extrinsic environmental stability. The intrinsic environmental stability is a material property and it includes the resistance of the field emitter to oxidation and sputtering. Sputtering of the typical refractory metal emitters occurs in background gas pressures above 10^{-11} Torr [237]. It was reported that high mass gases sputter-etch typical wire emitters (radius 100 nm), leading to a sharper morphology and higher current [238, 239]. High mass gases can also dull extremely sharp tips and microprotrusions. Light gases like He and H_2 cause sputter-assisted atomic diffusion. When unballasted metal emitters are operated in poorly outgassed systems, the formation of field emission patterns typical of adsorbates on the metal surface is frequently the high applied field, this diffusion grows nanoscale protrusions on the emitter surfaces [237, 240]. These protrusions lead to increased emission current and more rapid growth of the protrusions. This positive feedback leads to runaway emission current resulting in the destruction of the emitter through a vacuum arc [237, 240]. In metal emitters, the easy formation of microprotrusions on the emitter surface under the bombardment of residual gas species is often realized. To reduce or eliminate device failures due to arcing, many vacuum microelectronic devices employ a series resistance in the emitter circuit [241]. This ballasting resistor adds negative feedback and creates extrinsic stability of the system's property.

The degradation of emission is usually due to several phenomena and can be either reversible or permanent. Irreversible damage can occur through resistive heating, bombardment from gas molecules ionized by the emitted electrons, or arcing. Electrostatic deflection or mechanical stresses can cause alterations in the shape and/or surrounding of the emitter and lead to a decrease of the local field amplification. Other degradation phenomena are of chemical origin (adsorption or desorption of molecules on the emitter surface) and modify the work function.

CNT emitters do not show this type of emitter destruction, even though the emission is from adsorbate states at room temperature [12]. They are also generally more durable and stable than other carbon emitters such as diamond and carbon fibers, possibly because nanotubes have a relatively defect-free graphite structure with small sputtering yield and low carbon atom mobility. They do degrade, however, depending on the emission current level and the environments. The field emission changes resulting gas-field emitter interaction are directly proportional to the emission current [240, 242].

The intrinsic properties of nanotubes also have an importance. A comparison between films of SWNTs and MWNTs at comparable chamber pressure (10^{-7} mbar) and emitted current density (0.2 mA/cm^2) showed that the degradation was a factor 10 faster for SWNTs [217, 243]. The faster degradation of SWNTs has been attributed to the fact that their single shell makes them more sensitive to ion bombardment and irradiation, while the multiple shells of MWNTs tend to stabilize their structure. SWNT emitters to be stable and robust at 20 mA/cm^2 [11], but the degradation over time at current densities in excess of 500 mA/cm^2 has been observed. The emitter failure has been attributed primarily to the "uprooting" of nanotubes from substrate surfaces due to the poor adhesion strength, rather than to the intrinsic structural degradation and failure as found in diamond emitters. The environmental stability of SWNT emitters in a number of different gases has been studied in [12]. An exposure to 10^{-6} Torr of hydrogen showed no significant effects on the emission (Figure 8.15). While an initial drop in current of ~8% occurred in the first 10 hours of operation, no significant degradation in current was observed during the next 36 hours of emission. This initial drop in current is commonly observed on other nanotube samples exposed to 10^{-6} Torr of hydrogen. Flash cleaning the emitter does not restore the current, but introduction of water does.

When the water (10^{-7} Torr) was introduced, the emission current experienced a rapid increase due to the establishment of the adsorbate tunneling states. During extended exposure (around 45 hours), the current started to decay due to ion bombardment that removed this state. But nanotubes field emitters could operate in high water environments without degradation under less extreme conditions [12].

Operation of nanotubes in 10^{-7} Torr of argon showed little effect aside from small initial decrease in current, consistent with the sputter cleaning of some adsorbates by argon. In 10^{-7} Torr of oxygen, the current steadily decreased over the duration of the exposure with a drop of 75% over 48 hours. This decrease in current was related to chemical interactions at the surfaces such as the formation of C-O dipoles. Oxygen dipoles are known to reduce emission currents from metal and diamond emitters [183, 244]. After the rapid initial decrease, long exposures (40 hours) lead to substantial irreversible damage. Neither operation in high vacuum,

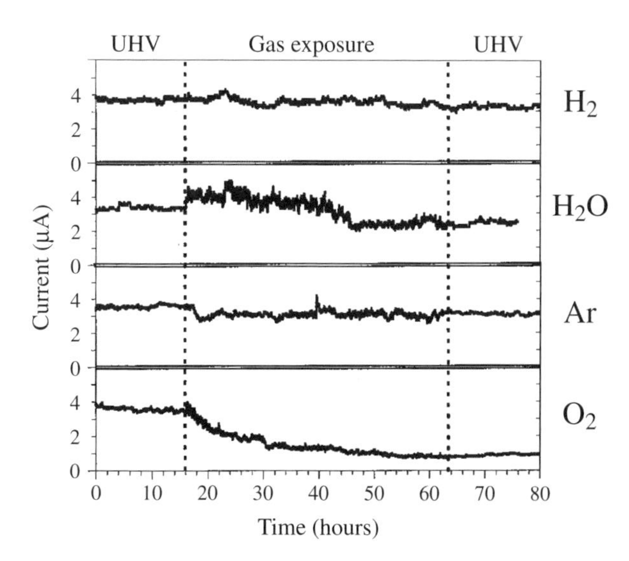

Figure 8.15 Operation of a single nanotube emitter sample under several gas ambient. Total exposures were 0.15 Torrs for H_2 and 0.015 Torrs each for H_2O, Ar, and O_2. To maintain an initial current above 3 μA for each gas exposure, it was necessary to increase the voltage by a few percent between some exposures. Reproduced with permission from Ref. [12]. Copyright (1999), AIP Publishing LLC

reintroduction of water at 10^{-7} Torr, nor high temperature flash heating above 1000 K induce complete recovery of the emission current. Since the equivalent exposure to Ar or H_2 produced little degradation, the irreversible decrease involved the chemically enhanced interaction.

Studies on the sputter etching of graphite show much faster etching in O_2 than in Ar, demonstrating a reactive ion etching effect [245]. Presumably, O_2 forms C-O bonds on the surface which are easily liberated by further bombardment. The decrease in nanotube emission current is probably due to etching which leads to a decrease in height of the nanotube, opening of the nanotube cap, disruption of electronic states in the cap, or all of the above. In addition, carbon etching studies have also found that water chemically enhances the sputter etching of carbon [245]. Thus, reactive sputter etching explains the irreversible current decrease observed with both O_2 and H_2O, although O_2 exposure is considerably more severe.

The existence of a field emission-assisted surface reaction process (EASRP) which produces different behavior in low and high emission regimes has been observed in work [246] under investigation of EFE from multi-walled carbon nanotubes (MWCNTs). The emission Fowler-Nordheim curves for four different test cycles are shown in Figure 8.16. Emission currents were measured by cycling the voltage. After the first emission cycle the remaining three cycles showed repeatable data in the high emission range but a large deviation in low emission region. A straight line was drawn from the high emission data points. Emission data could not be fit into a straight F-N line. It was suggested that the space-charge effect resulted in emission saturation in the high current range [212] and that the catalyst played an important role [247]. In work [13] it has been reported that the surface adsorbate tunneling state causes enhanced low current emission for SWNT samples. The second cycle which was conducted downward by decreasing the voltage after the first cycle, showed F-N data closer to the straight line in low current regions. This improvement occurred primarily because the emission was tested after the cleaning of the nanotube surface by the high emission Joule heating from the first cycle. Baking of the chamber and sample under 150 °C did not effectively clean surface contaminations. The

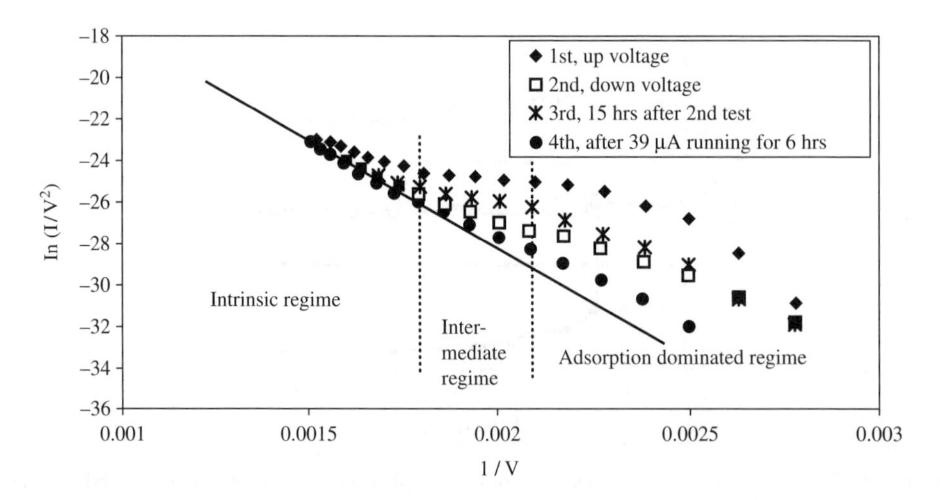

Figure 8.16 Emission F–N curves tested under a 10^{-9} Torr vacuum. Straight line was acquired from data of the fourth curve in the intrinsic regime by the least-squares fitting. Reproduced with permission from Ref. [246]. Copyright (2003), AIP Publishing LLC

results showed that the change of the F-N slope was mainly caused by the variation of surface chemical and adsorption states during emission. It is suggested that the emission in the low current regime is dominated by the surface adsorption state. In the intermediate range, gas desorption due to the Joule heating, the gas-nanotube reactions dominate. In the high emission current range, electrons emit from the clean MWNT surface. F-N curves from different measurement cycles have merged in this range. These three emission regimes have been labeled as adsorption dominated, intermediate, and intrinsic emission, respectively. The low emission data of the third curve, which was tested one day after the second cycle, deviated from the intrinsic line further compared with the second curve. This means that the reoccupation of the emission sites by the physical adsorption of gases has caused low emission current increase. The fourth cycle, which was tested after the surface cleaning by 6 hours of 38 μA emission, presented a minimum low emission deviation from the intrinsic line. The intrinsic emission regime may not appear if heavy surface chemical and/or physical reactions exist.

Although the surface emission sites have been cleaned by high emission Joule heating, such as in second and fourth cycles, still there exist deviations in low emission current from intrinsic values. Besides the effects of gas reoccupation on the emission sites (physical adsorption) under low emission and nonemission conditions, there exists an EASRP.

To identify the main gas components which caused the current increase in the low emission regime, the emission performance was studied by introducing different gases into the system [246]. Experiments were carried out on the same sample under 420 V with exposing N_2, O_2, and H_2 under the same partial pressure (6.0×10^{-9} Torr). The sample was cleaned by high emission Joule heating before the gas introduction. Emission increase rates in N_2 and O_2 were close to that under a 10^{-10} Torr vacuum, revealing that N_2 and O_2 did not play role. The current increase was much faster in H_2, especially in the initial period. For graphite structures, reactions with gases or vapors occur preferentially at "active sites," that is, the end of the basal planes and the defects due to high surface energy [248]. Electrons emit mainly from the graphite structure tube ends and defects due to high local fields and high local density of states (DOSs) [229, 235, 249]. Graphite structure does not react with hydrogen at ordinary temperatures, but reacts in the temperature of 1000–1500 °C. With nickel catalyst, the hydrogen reaction begins at approximately 500 °C. Thus, carbon-hydrogen reaction may occur in emission sites during emission because of surface temperature increase by Joule heating. The complexity of carbon-hydrogen reaction suggests the diverse reactions under field emission conditions in H_2, because (i) the content of nickel catalyst in nanotubes could be different between samples and (ii) the emission site temperature would vary for different MWNTs and under different currents and fields. Stable emissions have been reported in H_2 in works [12, 250]. In the work [246], when operating MWNT emitters under high emission currents, emissions were stable in either a 10^{-10} Torr vacuum or in 10^{-9} Torr with exposure to H_2, N_2, O_2, and H_2O, because the desorption EASRP dominated the gas-MWNT reaction. When the emission stability was tested in a water dominated vacuum (5×10^{-9} Torr of H_2O, partial pressures of H_2, N_2, and $O_2 < 1 \times 10^{-9}$ Torr), currents reached a relatively stable level after a rapid initial increase, which indicated the establishment of the adsorbate tunneling state in water [221]. The water-nanotube reaction under emission conditions plays a key role for the current rise in a low emission range. Under the presence of a large electric field at the tip, the adsorbate is stabilized due to the increase of binding energy and the ionization potential is lowered, resulting in low emission current increase [251]. The high current emission may break up the adsorbent-nanotube bonds to degas the emission sites.

8.4.3 Models of Field Emission from CNTs

The very low turn-on fields measured for nearly all CNTs emitters originate certainly from the small diameter and elongated shape of the tubes that lead to a high geometrical field enhancement. In fact, the local electric field just above the emitter surface needed for field emission is around 2–3 V/nm as for metallic emitters, as can be estimated from the applied field and the field enhancement factor [212, 252]. On the other hand, nanotubes do not behave like very sharp metallic tips.

In some investigations *I-V* characteristics follow the Fowler-Nordheim law (at least over a certain current range), from which it was concluded that CNTs behave as metallic emitters [207, 253, 254].

As has been proposed in [255–257], graphene layers have sharp bending at the open end of the nanotube, where the carbon atoms show sp^3-like atomic bonds instead of sp^2 configuration typical for graphene. This change in coordination would decrease the height of the potential barrier and could explain the very low work function that the authors estimated from the slope of Fowler-Nordheim plots. Again, this model would be valid only for open MWNT.

From other observations [243] it was concluded that the electron were emitted from sharp energy levels due to localized states at the tube cap. Luminescence induced by the electron emission could arise from irradiative transitions between two levels participating to the field emission [233]. Actually, theoretical calculations predict the presence of localized states at the tube cap, with the DOSs that differs markedly to that of the tube body [234, 235, 258]. Two points have to be mentioned concerning this hypothesis. If several energy levels participate in the emission, the occupied level nearest to the Fermi energy will supply nearly all the emitted electrons. Since the position of this level would depend strongly on the local atomic configuration (tube diameter, chirality, presence of pentagons, and other defects) significant differences of the emitted currents can be expected from one tube to another. Second, these localized states often show far higher carrier densities as compared to the tube body at the Fermi level [235]. As the field emission current depends directly on this carrier density, it has been speculated that the emitted current will be far lower for a nanotube without such states.

A complete study in [13, 221] suggests complementary mechanisms, and shows clearly that the emission behavior of nanotubes is far more complex than the one expected from a very sharp metallic tip with the workfunction of ~5 eV. Different emission regimes on single SWNTs were identified and depended on applied field and temperature. A first regime corresponding to resonant tunneling through an adsorbate was found under "usual" experimental conditions at low temperatures and applied fields. The involved molecule was identified as water and it appeared that this adsorbate-assisted tunneling was the stable field emission mode at room temperature (Figure 8.17). These molecules desorb either at high fields and emitted current or at temperatures higher than 400 °C. The other regimes correspond to the intrinsic emission from the cleaned tube and show far lower emitted current for comparable voltages with strongly reduced current fluctuations. The origin of these (at least two) intrinsic regimes is not clear yet but the emission mechanism involves probably non-metallic electronic states, such as enhanced field emission states above the Fermi level or nonmetallic DOS.

Repeatable staircase like current-field curves were observed in the field emission from acid functionalized MWNTs [259]. Acid oxidization of 10 nm diameter MWNTs was performed according to procedure described in [260]. The model for explanation of observed emission *I-V* characteristics was proposed. These atypical curves were attributed to resonant tunneling (RT) through localized surface states in a quantum well structure, which arose due to the

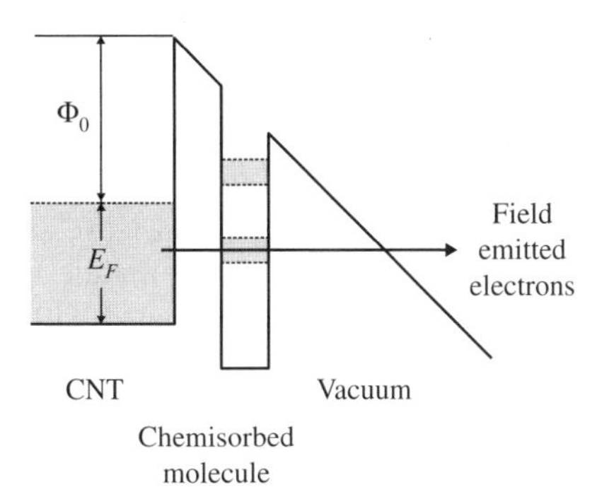

Figure 8.17 Idealized representation of resonance tunneling FE from the CNTs with a narrow band adsorbate

presence of the surface carboxylic functional group. A probable source of this nonlinearity is RT through quantized subbands at the MWNT surface. Nonlinearity can arise in the field emission (FE) current due to resonance when the electronic sublevels in quantum wells are aligned and negative differential resistance when misaligned [261, 262].

It is suggested that the oxidized MWNT carboxylic moiety gives rise to a quantum well structure, providing the localized surface states necessary for RT to occur. Adsorbed water molecules may also contribute to this effect. An idealized potential diagram in which the potential of an adsorbate is taken as a square well is shown in Figure 8.17. In this case, electrons are more likely to tunnel from CNT to vacuum through the energy levels of the adsorbate due to a reduction in the effective thickness of the triangular barrier.

RT through localized surface states in chemisorbed molecules has been used to explain the lobed patterns observed in FE microscopy from CNT caps [222]. As the temperature of the emitter increases, the lobes disappear and the emission current is reduced due to thermal removal of the adsorbates. RT through surface states is also widely observed and utilized in scanning tunneling microscopy [263–265]. These RT effects have traditionally been observed via fine structure in the electron energy distribution of emitted electrons, whereby energy peaks visible in the energy spectra represent resonances. Otherwise, lobed patterns observed in FE microscopy have been explained by RT and scattering of electrons through adsorbates [266].

Given the theoretical predictions of RT in FE and the experimental observations of RT effects from surface adsorbates, it is likely that the experimental results presented in [259] indeed represent RT through the localized electronic energy levels of a chemisorbed molecule (i.e., the carboxyl moiety and/or adsorbed water molecules). The steps observed in the current-field data represent multiple electronic energy levels confined in the carboxylic potential well. As each energy level aligns with the Fermi level of the MWNT, an increase in the FE current is expected, as observed experimentally. The disparity between the up and down cycles can be explained by slow discharging due to trapped states in the quantum wells. Chemisorbed and adsorbed atoms or molecules on the surfaces enhance the tunneling efficiency of electrons into vacuum, thereby improving FE efficiency.

The analysis shows that the emission involves a non-metallic DOS and/or adsorbate resonant tunneling.

The simulation of EFE from semiconducting CNTs was performed in [267]. Three-dimensional simulations of field emission from an ideal open (10,0) CNT without adsorption have been performed using a transfer-matrix methodology. Hind-structure effects are manifested in the distribution of energies by introducing pseudopotentials for the representation of carbon atoms and by repeating periodically a basic unit of the nanotube. The total-energy distributions of both the incident and field-emitted electrons present features, which are related to the gap of semiconducting (10,0) nanotube, to a van Hove singularity, and to stationary waves in the structure. The transmission through the middle of the gap is exponentially decreasing with each basic unit of the nanotube associated with a reduction of the current density at this particular energy value by a factor of 5.4. Except for the contribution associated with the van Hove singularity, all peaks are displaced to lower energies when the extraction field is increased.

The modeling of EFE from CNTs without taking into account adsorbed molecules was performed using the tunneling theory in [268, 269]. Using a tunneling approach for the field emission from a single CNT, in a simple triangular model for the potential energy barrier at the tube end the expressions for the emission current as a function of the anode voltage and of the emitted electron energy spectrum were obtained. The low dimensionality of the electronic system of a CNT was taken into account. Starting with a two-dimensional graphene sheet, the model treats the conduction electrons system as a one-dimensional continuum. The extraction field on the nanotube's tip was evaluated using numerical computations and its dependence on the CNT diameter's size was established. For nanotubes of practical interest, having large enough diameters, it was demonstrated that the influence of the detailed form of the electron energy dispersion relations was not of major importance. This influence could be generally embedded in a numerical factor entering the expression of the emission current. The influence of the various tube parameters on the characteristics was also identified and analyzed. An approximate formula for use in practical analysis in field emission was deduced and its validity for different nanotube sizes was verified.

In the case of an aligned CNT film a statistical model taking into account the scattered values of the geometrical and electrical parameters of the individual tubes have been proposed [269, 270]. The screening effect, statistical distribution of electric field enhancement coefficient and nanotubes radius was included as the parameters into the expressions for the emission current as a function of anode voltage and emitted electron energy spectrum. The constructed statistical model was applied to some practical CNT films of excellent alignment obtained by PECVD. This method provides thus a simple mean to get the average electrical parameters of the film by using a limited amount of experimental data on the film.

8.5 Electron Emission from Graphene and Nanocarbon

8.5.1 Electron Emission from Graphene

The different ways to obtain the graphene sheets (GSs), such as CVD [271] and chemical oxidation [272], had been intensively studied. PECVD technique is a widely used approach to grow nanocarbon materials, which includes microwave plasma CVD, RF-plasma CVD, DC plasma CVD, and so on.

Recent experiments on field emission from graphene fabricated by different methods show that graphene is a good field emission material [273–282].

During the investigation of the EFE from graphene in work [283] the multilayer graphene emitter was attached on the tip of tungsten (W) needle by peeling off a piece of graphene layers (several to 20 layers) from a highly oriented pyrolytic graphite (HOPG) crystal inside a scanning electron microscope. TEM revealed that edges of multilayer graphene were open before EFE measurements changed to closed edges after EFE under high emission current over a few tens of microamperes. In previous experimental study there was observed the change of edge structure of graphene from open to closed edges by heat treatment of above 1500 °C [284]. The structural change of the multi-layered graphene is considered to be caused by the current-induced Joule heating. The current density of multilayer graphene at which the tip structure changed is 7×10^7 A/cm^2. This current density is larger than the maximum current density of single-wall and double-wall CNTs, and is comparable to that of multiwall CNTs.

The field emission current and electron beam focus performance can be improved significantly by using single layer or double layer graphene sheets [285]. Graphene sheet has good conductivity, so it can be used to modify the distribution of electric field. Because the graphene sheet may have one atomic layer, the field emission electrons can pass through the graphene sheet. Single- and double-layer graphene sheet was grown on Si substrate by CVD method. The press lamination and the chemical etching process were used for transferring graphene grown on Si substrate to the metal electrode of emitting gate triode structure. Both the simulation and experimental results have shown that the focus performance of electron source has been improved largely [285].

The graphene synthesized by using CVD is in high quality and easy for large area fabrication. However, the flat type of film is considered to be not good for the field emission [286]. The EFE from graphene film on nickel substrate was investigated in Ref. [287]. The graphene film on nickel substrate in 1×1 cm^2 area was synthesized by using thermal CVD method. The polished nickel substrate was placed in a quartz tube and heated up to the temperature of 900 °C in the argon gas atmosphere. Then hydrogen gas was inlet to activate the nickel surface. After that methane and argon with the flow rate 5 : 100 (sccm) was introduce to synthesize the graphene. Finally the sample was fast cool down to room temperature in argon atmosphere. The Raman spectra showed that the graphene film had the 2D peak in around 2700 cm^{-1} with sharp full width at half maximum (FWHM) of 60 cm^{-1}, the intensity ration of 2D:G was larger than 4, the D peak almost disappeared. It confirms that the grown graphene film is single layer graphene.

The field emission character of graphene film was measured using the transparent diode method. The gap between anode and cathode was 250 μm. The EFE characteristics of the CVD grown graphene is shown in Figure 8.18. The turn-on field is 6.4 V/μm and emission current reaches 1.5 mA. The number of emission sites is about 200 which are counted in the emission site image in the inset of Figure 8.18. So, the average current emitted from each site is as large as 7.5 μA. The F-N plot showed the traditional field emission behavior. The current stability of the sample was very good with a current fluctuation less than 2.4% in 4 hours continuous test. This result indicated that the graphene film could carry large current in high emission stability which was essential for the application in high power electron sources.

The origin of emission sites is related to the micromorphology of the graphene film. Observed from the SEM image some protrusion and crumple pieces of graphene appear on the surface of graphene film which is induced by the intrinsic stress and thermal expansion. They provide plenty of tips for field emission. Under external field attraction, electrons emit from these kinds of tips. Due to its short height, the turn-on field is high. Due to its large

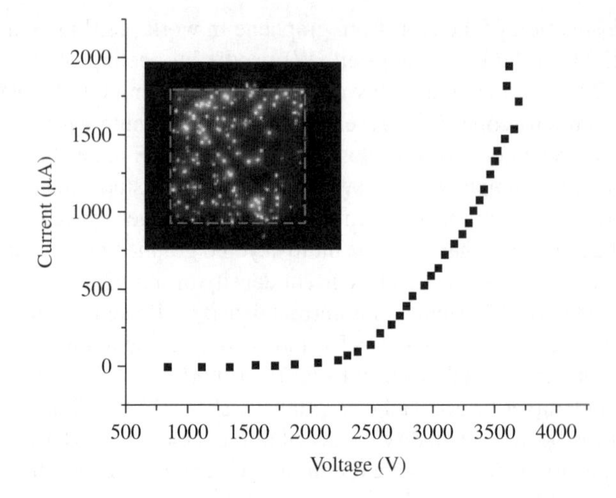

Figure 8.18 Field emission current vs. voltage characteristics of graphene film. Copyright (2012) IEEE. Reprinted with permission from Ref. [287]

thermal conductivity and two-dimensional continuous structure the heat dissipation is fast and the probability of vacuum breakdown is reduced, so the graphene film obtains an excellent current stability.

One of the critical issues for the application of flexible field emitters using graphene nanosheets is to realize graphene nanosheets to be vertically aligned to the polymeric substrate at relatively low temperatures. The simple methods for fabrication of graphene field emitters are the use of thermal welding and filtration technique. The structural properties of graphene emitters such as vertical alignment, density, and length were systematically controlled using various experimental conditions [22]. Two different techniques for fabrication of the graphene emitters have been developed in work [22]. One is a thermal welding and another is a filtering. For the thermal welding method the graphene oxide synthesized by the modified Brodies method has been used [288]. Synthesized graphene oxide showed a relatively large size with very high crystallinity. After reduction using a hydrazine monohydrate, the reduced graphene oxide (RGO) nanosheets were filtered on a polytetrafluoroethylene (PTFE) membrane. Then, the RGO film was thermally welded on the organic film which is prepared using a simple spray method [289]. After cooling to room temperature, the PTFE membrane was slowly peeled away from the welded sample. During this process, cohesive failure occurred at the middle of the RGO film, yielding graphene emitter with vertical alignment [22]. For a filtered method, the RGO nanosheets were dispersed in dimethylformamide (DMF) using a simple sonication. Then the RGO emitters were formed on a polycarbonate (PC) membrane with a pore size of 3 μm via filtration of the dispersed RGO solutions. After drying, the PC membrane was selectively etched in chloroform. The floating RGO arrays were transferred by direct contact with polyethylene terephthalate (PET) substrates, following by rinsing and drying [290]. The SEM images of vertically aligned graphene emitters are shown in Figure 8.19. The turn-on field at EFE was of ~2.0 V/μm for undoped and ~1.6 V/μm for Au-doped RGO. The fabricated graphene emitters showed a stable field emission even at high bending angles.

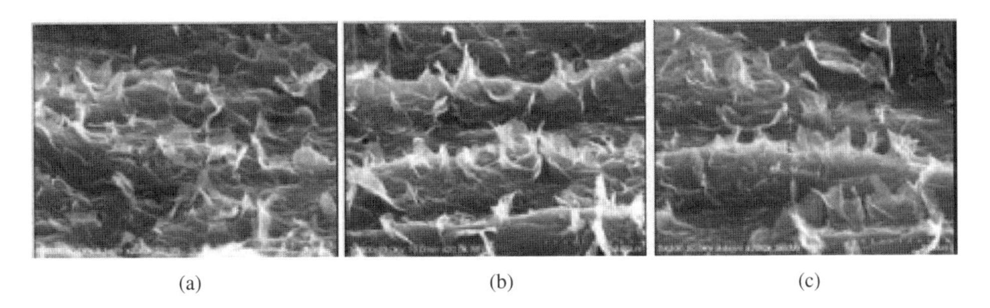

Figure 8.19 SEM images of vertically aligned (a) RGO emitters, (b) Au-doped RGO emitters, and (c) Al-doped RGO emitters fabricated by thermal welding method (RGO – reduced graphene oxide) [22]. Copyright (2012) WILEY-VCH Verlag GmbH & Co. KGaA, Weinheim

8.5.2 Electron Emission from CNT-Graphene Composites

A novel brush-type carbon nano materials graphene-CNT (brush type carbon nanotube (BCNT)) has been proposed in [291]. The VACNT arrays are covered by a layer of graphene sheet like a flat lid. The BCNT provides self-taken cathode graphene, increases the effective emission area significantly, and the emission emitters are very even. The VACNT arrays covered by a few layer graphene (FLG) sheets was grown on SiO_2/Si substrates by high-temperature pyrolysis CVD, in which iron phthalocyanine was used as precursor (catalyst) and carbon source at 800–900 °C, and hydrogen was used as reductant. When the hydrogen flow was 25 sccm and protect gas argon flow was 60 sccm, it was possible to get VACNT arrays [292–294]. Fortunately, when the hydrogen flow was decreased to 5 sccm or lower enough and increase the argon flow was increased to 80 sccm, there was found a new nanostructure like BCNT that VACNT arrays covered by FLG (Figure 8.20). Then the VACNT and as-grown BCNT samples were put into hydrogen fluoride solution, respectively. The HF etched the SiO_2 in few second, so the BCNT and CNT films were floating on the surface of solution. The floating film was transferred on or upside down glass substrate with transparent electrode indium tin oxide (ITO) to complete a field emission cathode. The schematic diagram of BCNT transfer is shown in Figure 8.20. Four kinds of cathode emitters array that are made of as-grown VACNT tip (named as CNT tip), transfer VACNT root (named as CNT root), as-grown BCNT tip (named as BCNT tip), and transfer BCNT root (named as BCNT root), respectively have been formed. As-grown VACNT tips are too dense to cause screen

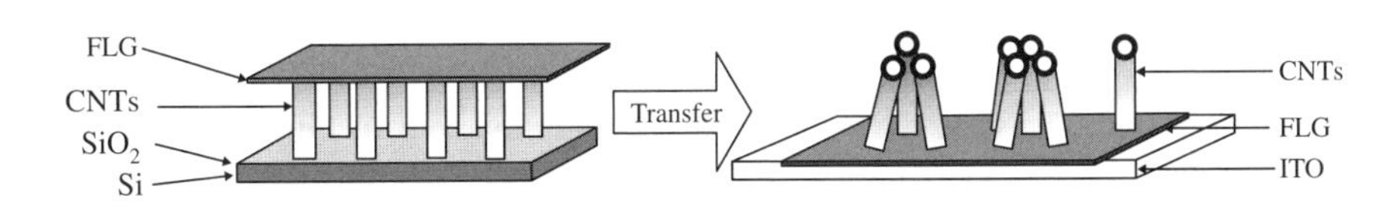

Figure 8.20 Schematic diagram of as-grown BCNT and transfer BCNT. Copyright (2012) IEEE. Reprinted with permission from Ref. [291]

effect and not even in height to lead to emission current distribution heavy nonuniform. CNT root array has relatively lower tip density and the same length because of the transfer process.

In BCNT array the wave-style graphene is supposed by VACNT like a thin blanket with the domain area over 5 μm square, and the FLG thickness of about 10 nm. FLG sheet is a perfect transparent conductance material. However, the FLG covering on the BCNT tips would reduce the probability of electron overflows from the surface.

After transferring the BCNT upside down on the ITO glass, the FLG sheets serve as conductance cathode and BCNT roots are made to be the field emission emitters. There are both good electrical and mechanical connections between VACNT arrays and graphene sheet after transfer. The distance between CNT roots is far enough to avoid screening effect because of BCNT roots flock together. And also BCNT roots have the same length just like that of CNT roots. The emission current density (J) versus electric field (E) characteristics and Fowler-Nordheim curve for BCNT root and CNT root revealed that the turn on electric field of the BCNT root is lower than that of CNT root, and the current density is larger enough. This is because there is enough far distance between emission emitters to decrease the screening effect and the local field of BCNT root is high enough for electrons overflow from BCNT root surface. On the other hand, the FLG has good conductance to provide better contact for BCNT roots and cathode substrate.

The field emission characteristic of BCNT tips, in which graphene cracks or wrinkles are used as emission emitters, has been also measured. The current density of the BCNT tip is lower than that of BCNT root, because the emitters on as-grown BCNT tips are less than that of transfer BCNT roots.

The field emission properties of various carbon nanostructures, including CNT film, graphene, amorphous graphene (a-G), and CNT/graphene composite films have been investigated and compared in [295]. CNT were prepared using a floating CVD method. Freestanding purified CNT films were obtained by a post-treatment involved air oxidation and acid purification [296]. Graphene films were prepared by an atmospheric pressure CVD on copper foils using methane as carbon source [297, 298]. The graphene-on-copper sample can be directly used for field emission test. Otherwise, after etching away the copper foil with a solution of 0.5 mol/l $FeCl_3$ and 0.5 mol/l HCl, the graphene film was transferred onto a target substrate. To investigate the structural effect of graphene on its field emission behavior, CVD growth parameters (temperature, gas flow, and cooling rate) were intentionally varied to yield graphene films of different qualities, including high-quality graphene (G) and a-G. Composite films of CNTs and graphene (CNTs/G) were assembled by a solid-phase layer stacking approach [299], possessing the structural integrity and continuity of both components and preserving their intrinsic electrical and mechanical properties. Graphene-based woven fabrics (GWFs) were prepared by CVD using a woven metal mesh as a template [300].

For field emission test, above-mentioned nano-carbon films were transferred to a highly polished copper electrode (cathode holder) to make a conformal and uniform coating. The emission experiment was carried out in a conventional field emission system at the pressure of 5×10^{-6} Torr. The field emission results are summarized in Figure 8.21 which can be fitted well with the Fowler-Nordheim theory [301]. CNT materials have relatively better geometrical similitude formed by the direct deposition method, showing the best field emission properties with turn-on electric field (defined as the electric field required to generate the emission current density of 10 μA/cm^2) of about 0.15 V/μm. In contrast, graphene-based

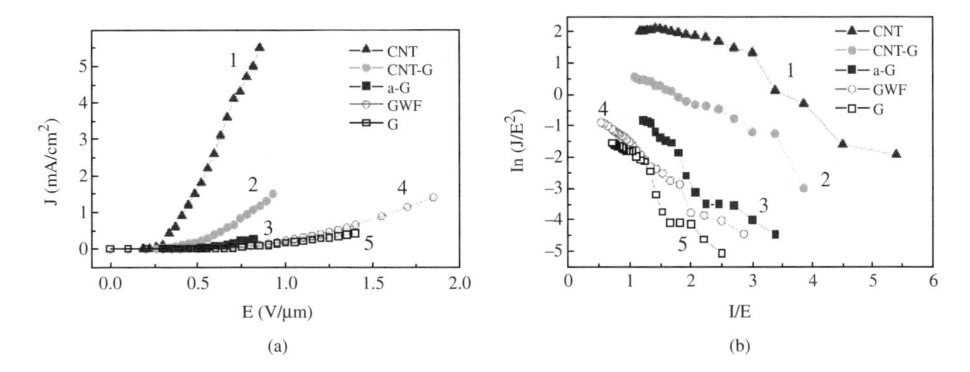

Figure 8.21 Field emission properties of graphene and CNTs. (a) *J-E* curves and (b) F-N plots (1: CNT, 2: CNT-G, 3: a-G, 4: GWF, and 5: G). Copyright (2012) IEEE. Reprinted with permission from Ref. [295]

materials show relatively weak field emission due to the intrinsic flatness of the 2D structure. Comparatively, the a-G sample behaves better thanks to the presence of high density surface wrinkles.

The graphene on CNT surface was synthesized in [302] by microwave CVD. At the beginning CNTs were grown in a tubular furnace with a horizontal quartz tube at atmospheric pressure by floating catalyst method. Then the graphene was synthesized on CNT surface. Before graphene was grown, MWCNTs were treated by H^+ plasma. Graphene growth time was 3 and 6 hours. The growth conditions were the following: the microwave power was 800 W, the chamber pressure was 13.26 kPa, $H_2 : Ar : CH_4 = 2 : 40 : 2$, and the growth temperature was 925 K. The diameter of MWCNTs was about 50 nm. Then the diameter of CNT became about 300 nm after H^+ treatment and 3 hours growth. In the sample grown 3 hours there was observed only a few graphene on CNT surface. In the sample grown 6 hours there was found the surface of CNT covered by graphene sheets. It was found that graphene on CNT's surface could improve efficiently the field emission performance of CNTs.

Chemical exfoliation was proved to be mass producible and showed the great expectation for application. However, there are still challenges for chemical oxide GS to be used in electronic devices, because the chemical exfoliation leads to obtain the separated GS. Moreover, the conductivity of the graphene reduced from the graphene oxide is not as good as expected.

CVD has been widely used to grow carbon nanotubed, graphene sheets, and relative carbon nanostructures. Due to high temperature growth, these carbon based nanostructures had good conductivity in nature. The vertically aligned graphene sheets on CNTs were synthesized on one step in work [303]. Low resistance silicon was used as substrate to grow the CNTs and graphene. Prior the experiment, the substrates were washed and ultrasonicated by deionized water. Microwave CVD (with power at 5 kW and frequency at 2.45 GHz) was used to prepare the sample. Methane and hydrogen were used as feedstock and reducing gas respectively. Substrate was put on a platform at the center of reaction cavity, where the electric field was mostly assembled. Ferrocene was put around the substrate as floating catalyst. Firstly the substrate was heated up to 850 °C under the protection of hydrogen at 1000 Pa. Then the same flow rate

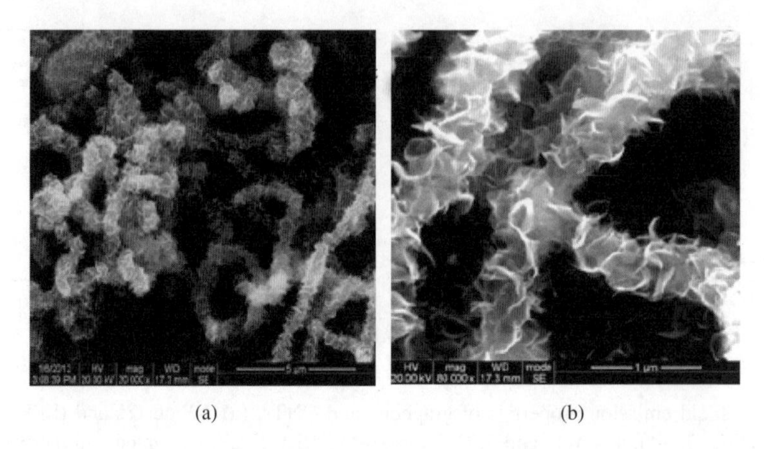

Figure 8.22 Carbon nanotubes and graphene sheets (a) low magnification image and (b) vertically aligned graphene sheets on carbon nanotubes. Copyright (2012) IEEE. Reprinted with permission from Ref. [303]

of methane was induced for 10 minutes to grow the CNTs and graphenes. The SEM images of the morphology of CNTs and graphene sheets grown on the CNTs are shown in Figure 8.22.

The graphene sheets on the CNT because of the multistage enhancement [304–306] have good emission properties. The turn-on field was about 1.4 V/μm, and the threshold voltage was 2.9 V/μm. This was attributed to the many edges, which stuck to the CNTs and resulted in multistage geometry enhancement. The enhancement coefficient was estimated from slope of F-N curve at assumption that work function of graphene was 5 eV. Such high value as 8570 was obtained. Both multistage field enhancement and low resistance made the graphene on CNT as a low voltage field emitter.

8.5.3 *Electron Emission from Nanocarbon*

Most CVD systems for growth of nanocarbon are generally equipped with the flat plate type sample holders, and their plasma geometry is mainly favorable in depositing thin films on the plate substrates. In work [307] the novel CVD system can generate 1000 mm long uniform column type plasma which enables the growth of thin films such as carbon nanowalls (CNWs) films on the entire surface (360°) of the long wire substrate. Recently, two-dimensional CNWs have attracted much attention. Their large surface area and thin edges provide opportunities for numerous applications, especially, as the fuel cell catalyst supporter and the efficient electron field emitter [308]. The CNW films were grown on long steel wire substrates (1 mm in diameter) by means of confined plasma CVD. CH_4 mixed with H_2 was introduced into the growth chamber as reaction source, and CH_4 concentration in mixing as was 8%. The gas pressure was about 4 kPa, and the growth time was kept at 30 minutes. SEM images clearly showed that as-deposited carbon film presented a wall-like surface morphology. Most primary carbon walls are aligned nearly perpendicularly to the substrate surface and the number of secondary walls nucleate and grow on the side surface of the primary walls, which greatly increase the film total surface area. The thickness of each wall is of the order of nanometers. It is a typical CNW morphology. One of the major advantages of the wire type electron emitters is to be able to effectively mitigate the so-called "edge-effect" that is a critical issue causing nonuniform electron emission for the plate type emitters because the electrical field tends to overconcentrate on the sample sharp edges rather than flat surface. Moreover, wire type emitters are supposed to have the ideal geometry facilitating to develop tube type field emission lamp and field emission electron sources. The field emission properties of as-deposited CNW films on the long wire substrates have been examined.

The unique geometrical shape is considered to be one of the key factors accounting for the good field emission properties of CNWs. The thin thickness of each CNW can lead to the strong field enhancement effect, making electron escape easily into vacuum. In addition, the high degree graphitization as evidence by Raman spectrum is also a very important factor for carbon emitters since the higher electric conductivity is necessary for generation of the higher electron emission current [309].

A unique three-dimensional nano-carbon electron emitter structure that combines the carbon nanoneedles (CNNs) and CNWs together has been developed by direct current plasma CVD system [310]. During growth, CH_4 and H_2 were used as reaction gases, whose typical flow rates were 40 and 500 sccm, respectively. The substrates were firstly treated at 873 K for 10 minutes in H_2 plasma with a process pressure of 30 Torr. Then, the process pressure and the temperature were increased to grow nano-carbon films in the CH_4 and H_2 condition. The process pressure was maintained at 75 Torr, and the growth time was 2 hours. In order to investigate the effect of growth temperature on the nano-carbon film growth mechanism, three samples were grown at the temperature of 1223, 1323, and 1373 K. They were correspondingly denoted as sample A, sample B, and sample C. The typical SEM images of obtained samples are shown in Figure 8.23. When the growth temperature was set at 1223 K, the carbon film grew all over the substrate surface of sample (Figure 8.23a). The insert indicates that this carbon film is CNWs. When the temperature is increased to 1323 K, several cone like nano-carbon structures can be observed on the surface of sample B (Figure 8.23b). When the temperature is further increased to 1373 K, many cone shape nano-carbon structures appear on the surface of sample C (Figure 8.23c). Based on the large amount of SEM observation, the density of cone like nano-carbon structures has been established to be $3 \times 10^6\ cm^{-2}$, which is very important for field emission devices. Figure 8.23d is a magnified SEM image of the cone nano-carbon emitter in sample C film. A straight carbon needle stands upward, around which there are many pieces of CNWs. The CNWs at the bottom of the needle are larger, and those at the upside are smaller. The total height of the needle is about 20 μm. As has been determined from emission measurement results of CNN-CNW film (sample C), the threshold field is about 1.0 V/μm (at $J = 1\ mA/cm^2$). Also the emission current density of CNN-CNW emitter could get to 143 mA/cm^2 at the electric field of 2.2 V/μm, indicating excellent emission efficiency. The field emission current density–voltage relation follows the classical F-N behavior.

The graphene structures with carbon sheets nanoclusters obtained by detonation synthesis were investigated in works [311, 312]. The graphene flakes were flat enough, had the size 0.2–0.3 mm, with extremely low roughness coefficients. As was shown by field emission microscopy, the field emission current was homogeneously distributed on the flat surface of the flakes. The effective work function determined from the slope of *I-V* characteristics in F-N coordinates was extremely low ($\Phi = 5 \times 10^{-2} - 3 \times 10^{-1}$ eV). Such low value of work function cannot be explained by high field enhancement coefficient because the surface of carbon flake obtained by detonation synthesis is smooth enough. The explanation of threshold voltage lowering based on the size quantization in the under-surface space charge layer has been proposed [311, 312]. The resonance levels of the size quantization serve as bridges helping the electrons

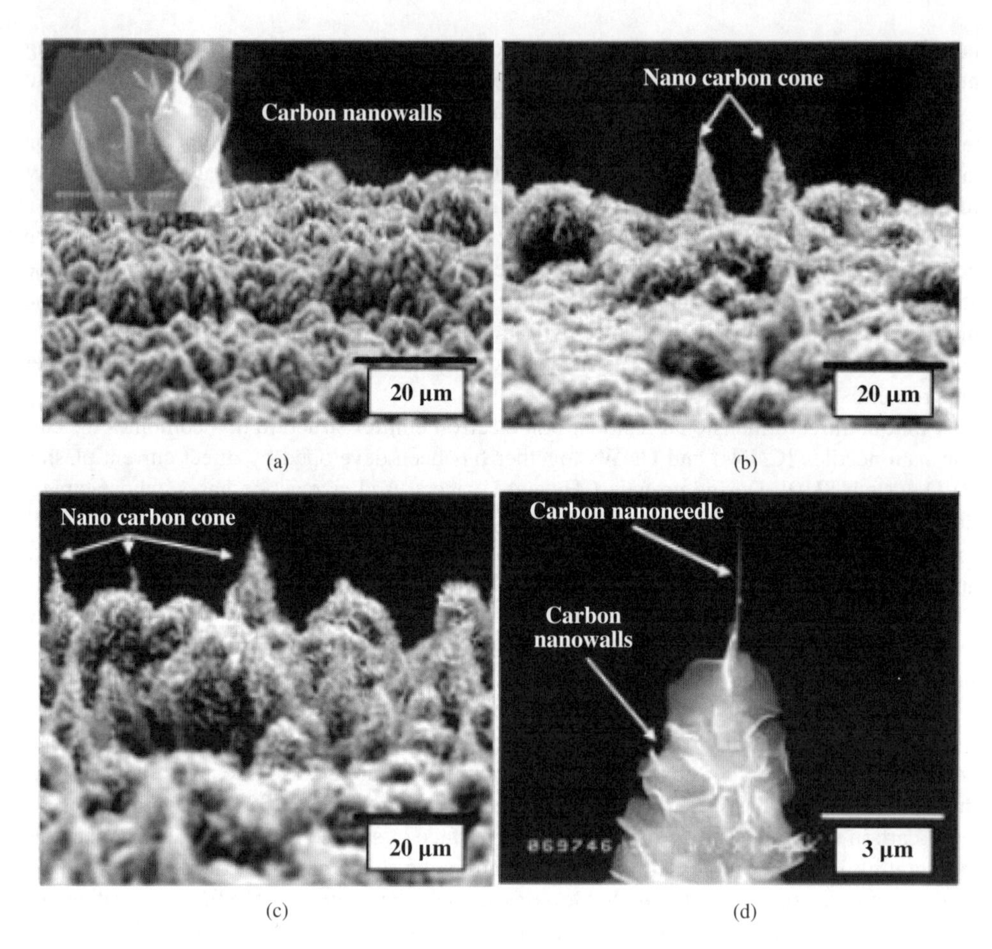

Figure 8.23 Typical SEM images of nano-carbon film grown at temperature of (a) 1223 K, the inset is the enlarged image, (b) 1323 K, (c) 1373 K, and (d) magnified SEM image of one cone nano-carbon emitter. Copyright (2012) IEEE. Reprinted with permission from Ref. [310]

with Fermi level to exit from the carbon nanostructure into vacuum. Due to presence of electric field the space charge is generated on the interface of the carbon nanostructures and vacuum. The horizontal periodicity of the structure remains unperturbed by the field, but the vertical periodicity (in direction of emission) is affected by the field within the space charge region, resulting in the corresponding transformation of the bandgap structure with arising flat thin spectral bands of size quantization.

Besides graphene sheets, the one more carbon nanoelectronic material, namely, graphite nanoplatelets (GNPs) that consist of platelets of graphite sheet with thickness of nano-scaled has been investigated [313]. GNPs have recently attracted considerable attention as a viable and inexpensive carbon nanoelectronic material. It has the predicted excellent in-plane mechanical, structural, thermal, and electrical properties of graphite [314–316]. GNPs are fundamentally different from other carbon nanoelectronic materials such as fullerenes and CNTs because of open surfaces and sharp edges, and also the high surface-to-volume ratio and unique electrical properties could serve as an ideal material for EFE. Furthermore, GNPs are able to achieve complete dispersion in organic solvent not like the poor dispersion and easy bundling properties of CNTs in solvent [317, 318].

The GNP field emitters were fabricated by screen printing method on ITO glass. The screen printed GNPs performed good field emission properties with long-term stability and high emission current density because the high surface-to-volume ratio avoided field screen effect and most electrons emitted from the sharp edge with few degradation. GNP was deposited on silicon substrate by the DC flow discharge of methane-containing gas mixture at about 900 °C. GNP paste was fabricated with GNP, ethyl cellulose, and SiO_2 nanopowder. Ethyl cellulose and SiO_2 nanopowder were used for organic binder and filler, respectively. Firstly, GNPs were dispersed in ethanol by tip-sonication and mixed with filler nanoparticles by magnetic stirrer. Then, the GNP solution was filtrated on filter membrane, and the film was transfer into the binder. At last, the composite was mixed and deformed by mechanical mixer. GNP field emitter was fabricated by printing the paste materials onto ITO glass through screen mask to form a thin field emitter film. In order to remove the organic materials, thermal annealing was induced in air at 450 °C for 30 minutes. Then, the screen printed GNP field emitter was peeled off with a tape to remove the impurities on the emitter surface and confirm good mechanical adhesion between emitter and substrate. SEM images of emitters showed sharp edges of GNPs and most of them were vertical aligned without bundling. The distance from edge to edge was suitable to avoid screen effect. The surface was cleaned with few binder materials, because the large surface area was good for evaporating the organic binder during thermal annealing process. From *I-V* characteristics of emission current the turn-on electric field (at 0.1 μA/cm^2) $F_{to} = 2.43$ V/μm and threshold field (at 1 mA/cm^2) $F_{th} = 5$ V/μm were determined. The good field emission properties were attributed to the vertical aligned GNPs with high aspect ratio and sharp edge. Moreover, the most of the GNP were well aligned with edge part exposed in the surface. As to emission stability the current density was held 1 mA/cm^2 for 20 hours, and the emission current degradation was less than 30%. That is because few organic materials attached on the surface of GNP after thermal annealing and also the taping treatment is necessary for the stable emission of GNP field emitter. The maximum current density can reach of 39 mA/cm^2 before stability test and 33 mA/cm^2 after stability test, respectively. The turn-on electric field increased after stability due to destruction of few active GNP emitters. GNP emitters are not easy destroyed because of their sharp and long edges. This planar type GNP field emitter is a good candidate for field emission applications such as flat lamp, field emission display, and X-ray source.

New type of nano-carbon material, namely inflamed graphite at high temperature (IGHT) has been proposed and investigated for field emission application in works [319, 320]. During obtaining the IGHT a graphite rod was set in a lathe and a tip of the graphite rod was inflamed under hydrogen and oxygen mixture gas. The containing ratio of hydrogen and oxygen was 2 : 1. The tip of the graphite rod was physically and chemically etched in the mixture flame with the temperature of 2000 °C and nanostructure on the graphite tip was formed. As SEM images showed, the surface of carbon rod before inflammation process was flat. On the other hand, numerous juts and edges were observed on the inflamed surfaces. The apical sizes of the juts on IGHT were estimated to be from 10 to 1000 nm. These juts and edges were fabricated during the inflammation process. The large electron emission current exceeding 2 mA was obtained from an IGHT miniaturizing of the curvature radius. The calculated current density was 6.4 A/cm^2.

For explanation of the high emission efficiency from graphene and nano-carbon materials, as a rule, the high electric field enhancement on edges and sharp tips are used. In works [311, 312] new mechanism of emission from flat graphene flakes based on the size quantization in the under-surface space charge layer has been proposed. The work function is the most important feature of the edge (surface) potential. It is known as a critical quantity in understanding the field emission properties. The vacuum potential barriers of various graphene edges have been investigated theoretically [321]. The simulation was done via the Vienna Ab initio Simulation Package (VASP) [322–325]. Clean, H terminated, OH, O, and NH terminated edges were investigated. The usual definition of the work function [326] is not applicable to graphene's edge, as the crystal face scale is the same that the atomic scale at the edge. Therefore, the work function has been determined as $\Phi_m = E_F - \phi_{max}$, were E_F is the Fermi level and ϕ_{max} is the maximum of vacuum potential energy in the path of electron moving out of graphene to vacuum (i.e., the barrier height in field emission). It was found that the exchange-correlation potential correction was significant for the edge structures with electronegativity higher than carbon. The correction leads to the local work function decreased by more than 1 eV for the O terminated edge. An extraordinary low vacuum barrier height of 2.0 eV has been found on the zigzag-edge of graphene terminated with the secondary amine. This edge structure has a flat band of edge states attached to the gamma point where the transversal kinetic energy is vanishing. It has been shown that the field electron emission is dominated by the flat band. The edge states pin the Fermi level to a constant, leading to an extremely narrow emission energy width. The graphene with such edge is a promising line field electron emitter that can produce highly coherent emission current.

8.6 Conclusion

The peculiarities of EFE from various carbon nanostructures have been reviewed. The models for explanation of the efficient EFE from diamond, DLC films, CNTs, graphene, and nanocarbon have been considered and analyzed. The carbon based cathode materials potentially offer the important advantage of ease in fabrication and low-cost manufacturing. They have high emission parameters. Nowadays they are considered as the most promising for obtaining high current density, effective, and stable emission. But there are still problems connected with emission uniformity, reliability, and reproducibility of carbon based cathodes for their wide implementation in working devices. Some processing methods for further improvement of the field emission properties of carbon cathodes are under investigation. In the case of CNTs these methods include tapering, peeling, truncation, IR exposure, plasma etching, protective coating, burning by electron breakdown, electrical and magnetic field alignment, and so on. For example, plasma treatment on SWNT field emitters significantly improves the stability of the emission current at a constant electric field. Current applications of nanotube-based field emitters already include light sources, miniaturized X-ray tubes, field emission displays, and vacuum microelectronic devices.

References

1. J.E. Jaskie, Diamond-based field-emission displays, *MRS Bulletin* **21**, 59 (1996).
2. F.J. Himpsel, J.A. Knapp, J.A. Van Vechten, and D.E. Eastman, Quantum photoyield of diamond(111) – A stable negative-affinity emitter, *Physical Review B* **20**, 624 (1979).

3. M.W. Geiss, J.C. Twichell, and T.M. Lyszczarz, Diamond emitters fabrication and theory, *Journal of Vacuum Science and Technology B* **14**, 2060 (1996).
4. K. Okano, S. Koizumi, S. Ravi, *et al.*, Low-threshold cold cathodes made of nitrogen-doped chemical-vapour-deposited diamond, *Nature (London)* **381**, 140 (1996).
5. V.G. Litovchenko, A.A. Evtukh, Yu. M. *et al.*, The model of photo- and field emission from thin DLC films, *Materials Science and Engineering A* **353**, 47 (2003).
6. V.G. Litovchenko, A.A. Evtukh, R.I. Marchenko, *et al.*, Parameters of the tip arrays covered by low work function layers, *Journal of Vacuum Science and Technology B* **14**, 2130 (1996).
7. V.G. Litovchenko, A.A. Evtukh, R.I. Marchenko, *et al.*, Enhancement of field emission from cathodes with superthin diamond-like carbon films *Applied Surface Science* **111**, 213 (1997).
8. A.A. Evtukh, V.G. Litovchenko, N.I. Klyui, *et al.*, Property of plasma enchanced chemical vapor deposition diamond-like carbon films as field electron emitters prepared in different regimes, *Journal of Vacuum Science and Technology B* **17**, 679 (1999).
9. B.S. Satyanarayana, A. Hart, W.I. Milne and J. Robertson, Field emission from tetrahedral amorphous carbon, *Applied Physics Letters* **71**, 1430 (1997).
10. J.M. Bonard, H. Kind, T. Stockli, and L.O. Nulson, Field emission from carbon nanotubes: the first five years, *Solid State Electronics* **45**, 893 (2001).
11. W. Zhu, C. Bower, O. Zhou, *et al.*, Large current density from carbon nanotube field emitters, *Applied Physics Letters* **75**, 873 (1999).
12. K.A. Dean and B.R. Chalamala, The environmental stability of field emission from single-walled carbon nanotubes, *Applied Physics Letters* **75**, 3017 (1999).
13. K.A. Dean and B.R. Chalamala, Current saturation mechanisms in carbon nanotube field emitters, *Applied Physics Letters* **76**, 375 (2000).
14. O. Groning, O.M. Kuttel, C. Emmenegger, *et al.*, Field emission properties of carbon nanotubes, *Journal of Vacuum Science and Technology B* **18**, 665 (2000).
15. L. Nilsson, O. Groning, C. Emmenegger, *et al.*, Scanning field emission from patterned carbon nanotube films, *Applied Physics Letters* **76**, 2071 (2000).
16. L. Nilsson, O. Groning, P. Groning, *et al.*, Characterization of thin film electron emitters by scanning anode field emission microscopy, *Journal of Applied Physics* **90**, 768 (2001).
17. J.W.G. Wildoer, L.C. Venema, A.G. Rinzler, *et al.*, Electronic structure of atomically resolved carbon nanotubes, *Nature (London), Physical Science* **391**, 59 (1998).
18. A. van Oostrom, Field emission cathodes, *Journal of Applied Physics* **33**, 2917 (1962).
19. M. Bockrath, D.H. Cobden, P.L. McEuen, *et al.*, Single-electron transport in ropes of carbon nanotubes, *Science* **275**, 1922 (1997).
20. A. Bachtold, M. Henny, C. Strunk, *et al.*, Contacting carbon nanotubes selectively with low-ohmic contacts for four-probe electric measurements, *Applied Physics Letters* **73**, 274 (1998).
21. S.J. Tans, M.H. Devoret, R.J.A. Groeneveld, and C. Dekker, Electron–electron correlations in carbon nanotubes, *Nature (London), Physical Science* **394**, 761 (1998).
22. H.J. Jeong, H.D. Jeong, H.Y. Kim, *et al.*, Flexible field emission from thermally welded chemically doped graphene thin films, *Small* **8**, 272 (2012).
23. S. Watcharotone, R.S. Ruoff. and F.H. Read, Possibilities for graphene for field emission: modeling studies using the BEM, *Physics Procedia* **1**, 71 (2009).
24. B.B. Pate, B.J. Waclawacki, P.H. Stefan, *et al.*, The diamond (111) surface: A dilemma resolved, *Physics B* **117/118**, 783 (1983).
25. M.W. Geis, J.A. Gregory, and B.B. Pate, Capacitance-voltage measurements on metal-SiO_2-diamond structures fabricated with (100)- and (111)-oriented substrates, *IEEE Transactions on Electron Devices* **38**, 619 (1991).
26. J. van der Weide and R.J. Nemanich, Influence of interfacial hydrogen and oxygen on the Schottky barrier height of nickel on (111) and (100) diamond surfaces, *Physical Review B* **49**, 13629 (1994).
27. M.W. Geis, N.N. Efremow, J.D. Woodhouse, *et al.*, Diamond cold cathode, *IEEE Transactions on Electron Devices Letters* **12**, 456 (1991).
28. A. Zangwill, *Physics of Surface*, Cambridge University Press Cambridge (1988).
29. B.B. Pate, The diamond surface: atomic and electronic structure, *Surface Science* **165**, 83 (1986).
30. J. van der Weide and R.J. Nemanich, Argon and hydrogen plasma interactions on diamond (111) surfaces: Electronic states and structure, *Applied Physics Letters* **62**, 1978 (1993).
31. W. Zhu, P.K. Baumann, and C.A. Bower, Novel cold cathode material, in *Vacuum Microelectronics*, ed. W. Zhu, John Wiley &Sons, Inc., New York (2001), pp. 247–87.

32. J.B. Cui, J. Ristein, and L. Ley, Electron affinity of the bare and hydrogen covered single crystal diamond (111) surface, *Physical Review Letters* **81**, 429 (1998).
33. M.J. Rutter and J. Robertson, Ab initio calculation of electron affinities of diamond surfaces, *Physical Review B* **57**, 9241 (1998).
34. M.W. Geis, J.C. Twichell, J. Macaulay, and K. Okano, Electron field emission from diamond and other carbon materials after H2, O2, and Cs treatment, *Applied Physics Letters* **67**, 1328 (1995).
35. D.P. Woodruff and T.A. Delchar, *Modern Techniques of Surface Science*, Cambridge University Press, Cambridge (1986).
36. M. Cardona and L. Ley, *Photoemission in Solids I. General Principles,* Springer-Verlag, Berlin (1978).
37. J.B. Cui, J. Ristein, and L. Ley, Low-threshold electron emission from diamond, *Physical Review B* **60**, 16135 (1999).
38. M.I. Landstrass and K.V. Ravi, Hydrogen passivation of electrically active defects in diamond, *Applied Physics Letters* **56**, 1391 (1989).
39. R. Schlesser, M.T. McClure, B.L. McCarson, and Z. Sitar, Bias voltage dependent field-emission energy distribution analysis of wide band-gap field emitters, *Journal of Applied Physics* **82**, 5763 (1997).
40. W.B. Choi, R. Schlesser, G. Wojak, *et al.*, Electron energy distribution of diamond-coated field emitters, *Journal of Vacuum Science and Technology B* **16**, 716 (1998).
41. D.S. Mao, J. Zhao, W. Li, *et al.*, Enhanced electron field emission properties of diamond-like carbon films using a titanium intermediate layer, *Journal of Physics D: Applied Physics* **32**, 1570 (1999).
42. M.W. Geiss, J.C. Twichell, N.N. Efremow, *et al.*, Comparison of electric field emission from nitrogen-doped, type Ib diamond, and boron-doped diamond, *Applied Physics Letters* **68**, 2294 (1996).
43. P. Lerner, P.H. Cutler, and N.N. Miskovsky, Theoretical analysis of field emission from a metal diamond cold cathode emitter, *Journal of Vacuum Science and Technology B* **15**, 337 (1997).
44. B.B. Pate, P.M. Stefan, C. Binns, *et al.*, Formation of surface states on the (111) surface of diamond, *Journal of Vacuum Science and Technology* **19**, 349 (1981).
45. S. Albin and L. Watkins, Electrical properties of hydrogenated diamond, *Applied Physics Letters* **56**, 1454 (1990).
46. Y. Muto, T. Sugino, J. Shirafuji, and K. Kobashi, Electrical conduction in undoped diamond films prepared by chemical vapor deposition, *Applied Physics Letters* **58**, 843 (1991).
47. J. Prins, Ion implantation and diamond: some recent results on growth and doping, *Thin Solid Films* **212**, 11 (1992).
48. J. Prins, Ion-implanted structures and doped layers in diamond, *Materials Science Reports* **7**, 271 (1992).
49. K.K. Das, in *Diamond Films and Coatings*, in R.F. Davis (ed.), Noyes Publications: Park Ridge, NJ, p. 381, 1993.
50. K. Okano and K.K. Gleason, Electron emission from phosphorus- and boron-doped polycrystalline diamond films, *Electronics Letters* **31**, 74 (1995).
51. K. Okano, T. Yamada, H. Ishihara, *et al.*, Electron emission from nitrogen-doped pyramidal-shape diamond and its battery operation, *Applied Physics Letters* **70**, 2201 (1997).
52. A.T. Sowers, B.L. Ward, S.E. English, and R.J. Nemanich, Field emission properties of nitrogen-doped diamond films, *Journal of Applied Physics* **86**, 3973 (1999).
53. W. Zhu, G.P. Kochanski, S. Jin, and L. Seibles, Defect-enhanced electron field emission from chemical vapor deposited diamond, *Journal of Applied Physics* **78**, 2707 (1995).
54. C. Wang, A. Garcia, D.C. Ingram, *et al.*, Cold field emission from CVD diamond films observed in emission electron microscopy, *Electronics Letters* **27**, 1459 (1991).
55. F. Lacher, C. Wild, D. Behr, and P. Koidl, Electron field emission from thin fine-grained CVD diamond films, *Diamond Related Materials* **6**, 1111 (1997).
56. A.A. Talin, L.S. Pan, K.F. McCarty, *et al.*, The relationship between the spatially resolved field emission characteristics and the raman spectra of a nanocrystalline diamond cold cathode, *Applied Physics Letters* **69**, 3842 (1996).
57. O. Groning, O.M. Kuttel,P. Groning, and L. Schlapbach, Field emission spectroscopy from discharge activated chemical vapor deposition diamond, *Journal of Vacuum Science and Technology B* **17**, 1064 (1999).
58. Y.-C. Yu, J.-H. Huang, and I.-N. Lin, Electron field emission properties of nanodiamonds synthesized by the chemical vapor deposition process, *Journal of Vacuum Science and Technology B* **19**, 975 (2001).
59. V.V. Zhirnov and J.J. Hren, Electron emission from diamond films, *MRS Bulletin* **23**, 42 (1998).
60. D. Zhou, A.R. Krauss, L.C. Qin, *et al.*, Synthesis and electron field emission of nanocrystalline diamond thin films grown from N_2/CH_4 microwave plasmas, *Journal of Applied Physics* **82**, 4546 (1997).

61. W. Zhu, G. P. Kochanski, and S. Jin, Low-field electron emission from undoped nanostructured diamond, *Science* **282**, 1471 (1998).
62. J.B. Cui, M. Stammler, J. Ristein, and L. Ley, Role of hydrogen on field emission from chemical vapor deposited diamond and nanocrystalline diamond powder, *Journal of Applied Physics* **88**, 3667 (2000).
63. V.D. Frolov, A.V. Karabutov, S.M. Pimenov, and V.I. Konov, Electronic properties of the emission sites of low-field emitting diamond films, *Diamond Related Materials* **9**, 1196 (2000).
64. K. Wu, E.G. Wang, Z.X. Cao, *et al.*, Microstructure and its effect on field electron emission of grain-size-controlled nanocrystalline diamond films, *Journal of Applied Physics* **88**, 2967 (2000).
65. J.Y. Shim, H.K. Baik, and K.M. Song, Mechanism of field emission from chemical vapor deposited undoped polycrystalline diamond films, *Journal of Applied Physics* **87**, 7508 (2000).
66. N.S. Xu, Y. Tzeng, and R.V. Latham, A diagnostic study of the field emission characteristics of individual micro-emitters in CVD diamond films, *Journal of Physics D: Applied Physics* **27**, 1988 (1994).
67. Y.K. Hong, J.J. Kirn, C. Park, *et al.*, Field electron emission of diamondlike carbon films deposited by a laser ablation method, *Journal of Vacuum Science and Technology B* **16**, 729 (1998).
68. W.P. Kang, A. Wisitsora-at, J.L. Davidson, *et al.*, Effect of sp^2 content and tip treatment on the field emission of micropatterned pyramidal diamond tips, *Journal of Vacuum Science and Technology B* **16**, 684 (1998).
69. S.G. Wang, Q. Zhang, S.F Yoon, *et al.*, Growth and electron field emission characteristics of nanodiamond films deposited in $N_2/CH4/H_2$ microwave plasma-enhanced chemical vapor deposition, *Journal of Vacuum Science and Technology B* **20**, 1982 (2002).
70. D.M. Gruen, Nanocrystalline diamond films, *Annual Review of Materials Science* **29**, 211 (1999).
71. T.G. McCauley, T.D. Corrigan, A.R. Krauss, *et al.*, Electron emission properties of Si field emitter arrays coated with nanocrystalline diamond from fullerene precursors, *Materials Research Society Symposium Proceedings* **498**, 227 (1998).
72. A.R. Krauss, D.M. Gruen, D. Zhou, *et al.*, Morphology and electron emission properties of nanocrystalline CVD diamond thin films, *Materials Research Society Symposium Proceedings* **495**, 299 (1998).
73. O. Groning, O.M. Kuttel, P. Groning, and L. Schlapbach, Field emission properties of nanocrystalline chemically vapor deposited-diamond films, *Journal of Vacuum Science and Technology B* **17**, 1970 (1999).
74. N.S. Xu, Y. Tzeng, and R.V. Latham, Similarities in the "cold" electron emission characteristics of diamond coated molybdenum electrodes and polished bulk graphite surfaces, *Journal of Physics D: Applied Physics* **26**, 1776 (1993).
75. N.S. Xu, R.V. Latham, and Y. Tzeng, Field-dependence of the area-density of 'cold' electron emission sites on broad-area CVD diamond films, *Electronics Letters* **28**, 1596 (1993).
76. R.V. Latham, *High Voltage Vacuum Insulation: Basic Concepts and Technological Practice*, London: Academic Press (1995).
77. J. Liu, V.V. Zhirnov, G.J. Wojak, *et al.*, Electron emission from diamond coated silicon field emitters, *Applied Physics Letters* **65**, 2842 (1994).
78. W.B. Choi, J.J. Cuomo, V.V. Zhirnov, *et al.*, Field emission from silicon and molybdenum tips coated with diamond powder by dielectrophoresis, *Applied Physics Letters* **68**, 720 (1996).
79. W.B. Choi, J. Liu, M.T. McClure, *et al.*, Field emission from diamond coated molybdenum field emitters, *Journal of Vacuum Science and Technology B* **14**, 2050 (1996).
80. V. Raiko, R. Spitzl, B. Aschermann, *et al.*, Field emission observations from CVD diamond-coated silicon emitters, *Thin Solid Films* **290/291**, 190 (1996).
81. E.I. Givargizov, V.V. Zhirnov, A.N. Stepanova, *et al.*, Microstructure and field emission of diamond particles on silicon tips, *Applied Surface Science* **87/88**, 24 (1995).
82. V.V. Zhirnov, On the cold emission mechanism of diamond coated tips, *Journal de Physique IV* **6**, C5 (1996).
83. A.R. Krauss, O. Auciello, M.Q. Ding, *et al.*, Electron field emission for ultrananocrystalline diamond films, *Journal of Applied Physics* **89**, 2958 (2001).
84. J.A. Carlisle, and O. Auciello, Ultrananocrystalline diamond, *Electrochemical Society Interface* **12**, 2003, 28-31.
85. Y.C. Lee, S.J. Lin, C.T. Chia, *et al.*, Synthesis and electron field emission properties of nanodiamond films, *Diamond Related Materials* **13**, 2100 (2004).
86. D. Pradhan, Y.C. Lee, C.W. Pao, *et al.*, Low temperature growth of ultrananocrystalline diamond film and its field emission properties, *Diamond Related Materials* **15**, 2001, (2006).
87. Y.C. Lee, S.J. Lin, D. Pradhan, and I.N. Lin, Improvement on the growth of ultrananocrystalline diamond by using pre-nucleation technique, *Diamond Related Materials* **15**, 353 (2006).

88. Chang, T.-H., Tai, N.-H., Panda, K. *et al.* (2012) The enhancement on the characteristics of the microplasma devices fabricated by coating the diamond films on silicon nanowires templates, *Technical Digest of the 25th International Vacuum Nanoelectronics Conference, 2012*, pp. 34–42.
89. Lou, S.-C., Chen, C., Teng, K.-Y. *et al.* (2012) The synthesis of diamond nano-tips for enhancing the plasma illumination characteristics of the capacitive type plasma devices, *Technical Digest of the 25th International Vacuum Nanoelectronics Conference, 2012*, pp. 162–170.
90. Chang, T.-H., Lou, S.-C., and Tai, N.-H. (2012) Fabrication of nitrogen-doped ultrananocrystalline diamond nanowire arrays with enhanced field emission and plasma illumination performance, *Technical Digest of the 25th International Vacuum Nanoelectronics Conference, 2012*, pp. 332–339.
91. J.C. She, S.Z. Deng, N.S. Xu, *et al.*, Fabrication of vertically aligned Si nanowires and their application in a gated field emission device, *Applied Physics Letters* **88**, 13112 (2006).
92. Y.F. Tseng, Y.C. Lee, C.Y. Lee, *et al.*, On the enhancement of field emission performance of ultrananocrystalline diamond coated nanoemitters, *Applied Physics Letters* **91**, 63117 (2007).
93. Y.F. Tzeng, Y.C. Lee, C.Y. Lee, *et al.*, Electron field emission properties on UNCD coated Si-nanowires, *Diamond Related Materials* **17**, 753 (2008).
94. P.T. Joseph, N.H. Tai, Y.F. Cheng, *et al.*, Growth and electron field emission properties of ultrananocrystalline diamond on silicon nanostructures, *Diamond Related Materials* **18**, 169 (2009).
95. A.V. Okotrub, L.G. Bulusheva, A.V. Gusel'nikov, *et al.*, Field emission from products of nanodiamond annealing, *Carbon* **42**, 1099 (2004).
96. Y. Show, F. Matsuoka, M. Hayashi, *et al.*, Influence of defects on electron emission from diamond films, *Journal of Applied Physics* **84**, 6351 (1998).
97. M. Yoshimoto, K. Yoshida, H. Maruta, *et al.*, Epitaxial diamond growth on sapphire in an oxidizing environment, *Nature* **399**, 340 (1999).
98. Li, J.J., Zheng, W.T., Gu, C.Z., *et al.* (2004) Field emission enhancement of carbon nitride films by annealing with different durations, *Materials Science and Processing* **28**, 279.
99. H.-F. Cheng, T. Tong, C.-H. Tsai, and I.-N. Lin, Nanocarbonaceous materials synthesized by microwave CVD and their characteristics of electron field emission, *Microelectronic Engineering* **66**, 2 (2003).
100. S. Gupta, M. Pal Chowdhury, and A.K. Pal, Synthesis of DLC films by electrodeposition technique using formic acid as electrolyte, *Diamond Related Materials* **13**, 1680 (2004).
101. W.P. Kang, J.L. Davidson, A. Wisitsora-at, *et al.*, Diamond vacuum field emission devices, *Diamond Related Materials* **13**, 1944 (2004).
102. Zhang, X., Wei, S., Yang, X. *et al.* (2012) Luminescent of nano-diamond field emission, *Technical Digest of the 25th International Vacuum Nanoelectronics Conference, 2012*, pp. 172–173.
103. Kang, W.P., Hsu, S.H., Ghosh, N. *et al.* (2012) Nanodiamond vacuum field emission integrated devices, *Technical Digest of the 25th International Vacuum Nanoelectronics Conference, 2012*, pp. 88–89.
104. R. Schlesser, M.T. McClure, W.B. Choi, *et al.*, Energy distribution of field emitted electrons from diamond coated molybdenum tips, *Applied Physics Letters* **70**, 1596 (1997).
105. K.H. Bayliss and R.V. Latham, An analysis of field-induced hot-electron emission from metal-insulator microstructures on broad-area high-voltage electrodes, *Proceedings of the Royal Society of London, Series A* **403**, 285 (1986).
106. G.A.J. Amaratunga and S.R.P. Silva, Nitrogen containing hydrogenated amorphous carbon for thin-film field emission cathodes, *Applied Physics Letters* **68**, 2529 (1996).
107. V.S. Veersamy, J. Yuan, G.A.J. Amaratunga, *et al.*, Nitrogen doping of highly tetrahedral amorphous carbon, *Physical Review B* **48**, 17954 (1993).
108. S.R.P. Silva, G.A. Amaratunga, and K. Okano, Modeling of the electron field emission process in polycrystalline diamond and diamond-like carbon thin films, *Journal of Vacuum Science and Technology B* **17**, 557 (1999).
109. W. Zhu, G.P. Kochanski, S. Jin, *et al.*, Electron field emission from ion-implanted diamond, *Applied Physics Letters* **67**, 1157 (1995).
110. Z.H. Huang, P.H. Cutler, N.M. Miskovsky, and T.E. Sullivan, Theoretical study of field emission from diamond, *Applied Physics Letters* **65**, 2562 (1994).
111. R.M. Hill, Poole-Frenkel conduction in amorphous solids, *Philosophical Magazine* **23**, 59 (1971).
112. J.E. Henderson and R.E. Badgley, The work required to remove a field electron, *Physical Review* **38**, 5401 (1931).

113. O. Groning, O.M. Kuttel, P. Groning, and L. Schlapbach, Field emitted electron energy distribution from nitrogen-containing diamondlike carbon, *Applied Physics Letters* **71**, 2253 (1997).
114. O. Groning, O.M. Kuttel, P. Groning, and L. Schlapbach, Field emission from DLC films, *Applied Surface Science* **111**, 135 (1997).
115. C. Bandis and B.B. Pate, Simultaneous field emission and photoemission from diamond, *Applied Physics Letters* **69**, 366 (1996).
116. M.L. Yu, H.S. Kim, B.W. Hussey, *et al.*, Energy distributions of field emitted electrons from carbide tips and tungsten tips with diamondlike carbon coatings, *Journal of Vacuum Science and Technology B* **14**, 3797 (1996).
117. C.S. Athwal, K.H. Bayliss, R. Calder, and R.V. Latham, Field-induced electron emission from artificially produced carbon sites on broad-area copper and niobium electrodes, *IEEE Transactions on Plasma Science* **13**, 225 (1985).
118. N.S. Xu and R.V. Latham, Coherently scattered hot electrons emitted from MIM graphite microstructures deposited on broad-area vacuum-insulated high-voltage electrodes, *Journal of Physics D: Applied Physics* **19**, 477 (1986).
119. O.M. Kuttel, O. Groening, and L. Schlapbach, Surface conductivity induced electron field emission from an indium cluster sitting on a diamond surface, *Journal of Vacuum Science and Technology A* **16**, 3464 (1998).
120. J.D. Shovlin and M.E. Kordesch, Electron emission from chemical vapor deposited diamond and dielectric breakdown, *Applied Physics Letters* **65**, 863 (1994).
121. S. Bajic and R.V. Latham, Enhanced cold-cathode emission using composite resin-carbon coatings, *Journal of Physics D: Applied Physics* **21**, 200 (1998).
122. Y.D. Kim, W. Choi, H. Wakimoto, *et al.*, Direct observation of electron emission site on boron-doped polycrystalline diamond thin films using an ultra-high-vacuum scanning tunneling microscope, *Applied Physics Letters* **75**, 3219 (1999).
123. B. Fiegl, R. Kuhnert, M. Ben-Chorin, and F. Koch, Evidence for grain boundary hopping transport in polycrystalline diamond films, *Applied Physics Letters* **65**, 371 (1994).
124. I. Andrienko, A. Cimmino, D. Hoxley, *et al.*, Field emission from boron-doped polycrystalline diamond film at the nanometer level within grains, *Applied Physics Letters* **77**, 1221 (2000).
125. R.G. Forbes, Low-macroscopic-field electron emission from carbon films and other electrically nanostructured heterogeneous materials: hypotheses about emission mechanism, *Solid State Electronics* **45**, 779 (2001).
126. J. Robertson, Diamond-like amorphous carbon, *Materials Science and Engineering R* **37**, 129 (2002).
127. R.V. Latham, *High Voltage Vacuum Insulation: The Physical Basis*, Academic Press, New York (1981).
128. S. Bajic, M.S. Mousa, and R.V. Latham, Factors influencing the stability of cold-cathodes formed by coating a planar electrode with a metal-insulator composite, *Collogue de Physique* **50**(C8), 79, (1989).
129. A.A. Evtukh, E.B. Kaganovich, V.G. Litovchenko, *et al.*, Silicon tip arrays with nanocomposite films for electron field emission applications, *Materials Science and Engineering C.* **19**, 401 (2002).
130. A.A. Evtukh, I.Z. Indutnyy, I.P. Lisovskyy, *et al.*, Electron field emission from SiO_x films, *Semiconductor Physics, Quantum Electronics & Optoelectronics* **6**, 32 (2003).
131. C. Schossler, A.J. Kaya, A.J. Kretz, *et al.*, Electrical and field emission properties of nanocrystalline materials fabricated by electron-beam induced deposition, *Microelectronic Engineering* **30**, 471 (1996).
132. J.C. She, J Chen, S.Z. Deng, *et al.*, An experimental study of the field emission characteristics of thin films of a new organic compound Ca–Cu–B, *Ultramicroscopy* **79**, 149 (1999).
133. R.D. Fedorovich, A.G. Naumovets, and P.M. Tomchuk, Electronic properties of island thin films caused by surface scattering of electrons, *Progress in Surface Science* **42**, 189 (1993).
134. R.D. Fedorovich, A.G. Naumovets, and P.M. Tomchuk, Electron and light emission from island metal films and generation of hot electrons in nanoparticles, *Physics Reports* **328**, 73 (2000).
135. R.E. Hurley, Electrical phenomena at the surface of electrically stressed metal cathodes. I. Electroluminescence and breakdown phenomena with medium gap spacings (2-8 mm), *Journal of Physics D: Applied Physics* **12**, 2229 (1979).
136. Hie, A., Ferrari, A.C., Yagi, T., and Robertson, J. (1999) Effect of sp^2-phase nanostructure on field emission from amorphous carbons, *Proceedings of the Diamond '99, Prague, Czech Republic, September 1999.*
137. A. Ilie, A.C. Ferrari, T. Yagi, and J. Robertson, Effect of sp^2-phase nanostructure on field emission from amorphous carbons, *Applied Physics Letters* **76**, 2627 (2000).
138. Forbes, R.G. (1999) A look at the basic physics of cold field electron emission, *Technical Digest of 18th International Vacuum Nanoelectronic Conference (Oxford, UK, July 10–14, 2005)*, 2005, pp. 1–2.

139. R.A. Tuck, W. Taylor, P.G.A. Jones, and R.V. Latham (1997) The pFED – a visible route to large field emission displays, *Technical Digest of 18th International Vacuum Nanoelectronic Conference (Oxford, UK, July 10–14, 2005)*, 2005, pp. 80–81.
140. G.A.J. Amaratunga, M. Baxendale, N. Rupesinghe, *et al.*, Performance of thin film carbon materials and carbon nanotubes as cold cathodes, *Technical Digest of 11th International Vacuum Nanoelectronic Conference (Asheville, NC, USA, July 19–24, 1998)*, 1998, pp. 184–185.
141. N.S. Xu and R.V. Latham, The application of an energy-selective imaging technique to a study of field-induced hot electrons from broad-area high-voltage electrodes, *Surface Science* **274**, 147 (1992).
142. K.H. Bayliss and R.V. Latham, An analysis of field-induced hot-electron emission from metal-insulator microstructures on broad-area high-voltage electrodes, *Vacuum* **35**, 211 (1985).
143. R.V. Latham, The origin of prebreakdown electron emission from vacuum-insulated high voltage electrodes, *Vacuum* **32**,137 (1982).
144. G. Dearnaley, A.M. Stoneham, and D.V. Morgan, Electrical phenomena in amorphous oxide films, *Reports on Progress in Physics* **33**, 1129 (1970).
145. R.V. Latham and M.S. Mousa, Hot electron emission from composite metal-insulator micropoint cathodes, *Journal of Physics D: Applied Physics* **19**, 699 (1986).
146. M.S. Mousa, A new perspective on the hot-electron emission from metal-insulator microstructures, *Surface Science* **231**, 149 (1990).
147. M.S. Mousa, Field electron emission studies on zinc oxide coated tungsten microemitters, *Surface Science* **266**, 110 (1992).
148. D. Ruter and W. Bauhofer, Light generation and transportation in luminescent a-SiC:H thin film waveguides, *Journal of Non-Crystalline Solids* **198-200**, 619 (1996).
149. H. Lee, I.-Y. Kim, S.-S. Han, *et al.*, Spectroscopic ellipsometry and Raman study of fluorinated nanocrystalline carbon thin films, *Journal of Applied Physics* **90**, 813 (2001).
150. E. Martinez, J.L. Andujar, M.C. Polo, *et al.*, Study of the mechanical properties of tetrahedral amorphous carbon films by nanoindentation and nanowear measurements, *Diamond Related Materials* **10**, 145 (2001).
151. J. Robertson, Amorphous carbon, *Advances in Physics* **35**, 317 (1986).
152. J. Robertson and E.P. O'Reilly, Electronic and atomic structure of amorphous carbon, *Physical Review* **B35**, 2946 (1987).
153. J.C. Angus, Diamond and diamond-like films, *Thin Solid Films* **216**, 126 (1992).
154. J. Robertson, Mechanical properties and coordinations of amorphous carbons, *Physical Review Letters* **68**, 220 (1992).
155. J. Robertson, Structural models of a-C and a-C:H, *Diamond Related Materials* **4**, 297 (1995).
156. V.G. Litovchenko, The enhanced field emission from microtips covered by ultrathin layers, *Ukrainian Journal of Physics* **42**, 228(1997).
157. A.A. Evtukh, V.G. Litovchenko, N.I. Klyui, *et al.*, Property of plasma enchanced chemical vapor deposition diamond-like carbon films as field electron emitters prepared in different regimes, *Journal of Vacuum Science and Technology* **B** 17, 679 (1999).
158. V.G. Litovchenko and A.A. Evtukh, Effects of electron field emission enhancement in structures with quantum well, *Physics of Low-Dimensional Structures* **3/4**, 227 (1999).
159. J.L. Bredas and G.B. Street, Electronic properties of amorphous carbon films, *Journal of Physics C* **18**, L651 (1985).
160. M.A. Tamor and C.H. Wu, Graphitic network models of "diamondlike" carbon, *Journal of Applied Physics* **67**, 1007 (1990).
161. M.P. Siegal, J.C. Barbour, P.N. Provencio, *et al.*, Amorphous-tetrahedral diamondlike carbon layered structures resulting from film growth energetic, *Applied Physics Letters* **73**, 759 (1998).
162. M.P. Siegal, D.R. Tallant, P.N. Provencio, *et al.*, Ultrahard carbon nanocomposite films, *Applied Physics Letters* **76**, 3052 (2000).
163. L.-Y. Chen and F. C.-N. Hong, Diamond-like carbon nanocomposite films, *Applied Physics Letters* **82**, 3526 (2003).
164. D.G. McCulloch, D.R. McKenzie, and C.M. Goringe, Ab initio simulations of the structure of amorphous carbon, *Physical Review B* **61**, 2349 (2000).
165. G. Jungnickel, Th. Frauenheim, D. Porezag, *et al.*, Structural properties of amorphous hydrogenated carbon. IV. A molecular-dynamics investigation and comparison to experiments, *Physical Review B* **50**, 6709 (1994).

166. J. Robertson, Recombination and photoluminescence mechanism in hydrogenated amorphous carbon, *Physical Review B* **53**, 16302 (1996).
167. J. Robertson, π-bonded clusters in amorphous carbon materials, *Philosophical Magazine B* **66**, 199 (1992).
168. H. Pan, M. Pruski, B.C. Gertstein, *et al.*, Local coordination of carbon atoms in amorphous carbon, *Physical Review B* **44**, 6741 (1991).
169. K.C. Park, J.H. Moon, S.J. Chung, *et al.*, Field emission properties of ta-C films with nitrogen doping, *Journal of Vacuum Science and Technology B* **15**, 431 (1997).
170. D.R. McKenzie, D.A. Muller, E. Kravtchinskaia, *et al.*, Synthesis, structure and applications of amorphous diamond, *Thin Solid Films* **206**, 198 (1991).
171. J. Birrell, J.A. Carlisle, O. Auciello, *et al.*, Morphology and electronic structure in nitrogen-doped ultrananocrystalline diamond, *Applied Physics Letters* **81**, 2235 (2002).
172. H.-F. Cheng, Y.-C. Chen, Y.-L. Wang, *et al.*, Comparison of structure and electron-field-emission behavior of chemical-vapor-deposited diamond and pulsed-laser-deposited diamond-like carbon films, *Japanese Journal of Applied Physics* **39**, 1866 (2000).
173. J. Robertson, Mechanisms of electron field emission from diamond, diamond-like carbon, and nanostructured carbon, *Journal of Vacuum Science and Technology B* **17**, 659 (1999).
174. J.P. Zhao, Z.Y. Chen, X. Wang, *et al.*, Thickness-independent electron field emission from tetrahedral amorphous carbon films, *Applied Physics Letters* **76**, 191 (2000).
175. N.L. Rupersinghe, M. Chhowalla, G.A.J. Amaratunga, *et al.*, Influence of the heterojunction on the field emission from tetrahedral amorphous carbon on Si, *Applied Physics Letters* **77**, 1908 (2000).
176. K.C. Park, J.H. Moon, S.J. Chung, *et al.*, Deposition of n-type diamondlike carbon by using the layer-by-layer technique and its electron emission properties, *Applied Physics Letters* **70**, 1381 (1997).
177. Evtukh, A.A. (2004) The tunnel injection and emission of electrons in layered semiconductor structures on silicon. DrS. thesis. NAS of Ukraine, Kiev.
178. V. Litovchenko and A. Evtukh, Vacuum nanoelectronics, In *Handbook of Semiconductor Nanostructures and Nanodevices*, Vol. 3. Spintronics and Nanoelectronics, ed. A.A. Balandin and K.L. Wang, American Scientific Publishers, Los Angeles, CA. 2006, pp. 153–234.
179. Groning, O. (1999) Field emission properties of carbon thin films and carbon nanotubes. PhD thesis. University of Fribourg.
180. W.P. May, S. Hohn, W.N. Wang, and N.A. Fox, Field emission conduction mechanisms in chemical vapor deposited diamond and diamondlike carbon films, *Applied Physics Letters* **72**, 2182 (1998).
181. W.P. May, S. Hohn, M.N.R. Ashfold, *et al.*, Field emission from chemical vapor deposited diamond and diamond-like carbon films: Investigations of surface damage and conduction mechanisms, *Journal of Applied Physics* **84**, 1618 (1998).
182. W.P. May, M.-T. Kuo, and M.N.R. Ashfold, Field emission conduction mechanisms in chemical-vapour-deposited diamond and diamond-like carbon films, *Diamond Related Materials* **8**, 1490 (1999).
183. I. Brodie and C.A. Spindt, Vacuum microelectronics, in *Advances in Electronics and Electron Physics*, Vol. 83 (P.W. Hawkes, ed.), Academic Press: New York, p. 1, 1992.
184. D.W. Branston and D. Stephani, Field emission from metal-coated silicon tips, *IEEE Transactions on Electron Devices* **38**, 2329 (1991).
185. S.M. Sze, *VLSI Technology*, McGraw-Hill Book Company, New York 1983.
186. A.A. Evtukh , N.L. Dmitruk, Kizjak, *et al.* (2002) To energy band diagram of silicon-DLC-vacuum system: determination of barrier height at Si-DLC interface, *Proceedings of the 15th International Vacuum. Microelectronics Conference, Lyon, France, 2002*, p. PT.02.
187. J. Weide, Z. Zhang, P.K. Baumann, *et al.*, Negative-electron-affinity effects on the diamond (100) surface, *Physical Review B* **50**, 5803 (1994).
188. F. Demichelis, X.F. Rong, S. Schreiter, *et al.*, Deposition and characterization of amorphous carbon nitride thin films, *Diamond Related Materials* **4**, 361 (1995).
189. A.A. Evtukh, V.G. Litovchenko, R.I. Marchenko *et al.* (1995) Field emitter tip arrays with high efficiency for flat panel display application, *Abstract of MRS Spring Meeting, San-Francisco, 1995*, p. 396.
190. A.A. Evtukh, V.G. Litovchenko, Yu.M. Litvin, *et al.*, Electron field emission from silicon emitters coated with a thin DLC films, *Physics of Low-Dimensional Structures* **5/6**, 117 (2001).
191. A. Evtukh, H. Hartnagel, V. Litovchenko, O. Yilmazoglu, Two mechanisms of negative dynamic conductivity and generation of oscillations in field-emissions structures, *Materials Science and Engineering* **A353**, 27 (2003).

192. V.G. Litovchenko, A.A. Evtukh, Yu.M. Litvin, *et al.*, Peculiarities of the electron field emission from quantum-size structures, *Applied Surface Science* **215**, 160 (2003).
193. A.F. Kravchenko, V.N. Ovsyuk, *Electron Processes in Solid State Low-Dimensional Systems*, Novosibirsk University, Novosibirsk (2000) (in Russia)
194. S.Z. Deng, N.S. Xu, J.B. Liu, *et al.*, Field electron emission properties from aligned carbon nanotube bundles of different density, *Surface and Interface Analysis* **36**, 501 (2004).
195. A. Evtukh, V. Litovchenko, M. Semenenko, *et al.*, Formation of conducting nanochannels in diamond-like carbon films, *Semiconductor Science and Technology* **21**, 1326 (2006).
196. T. Utsumi, Vacuum microelectronics: What's new and exciting, *IEEE Transactions on Electron Devices* **38**, 2276 (1991).
197. O. Groning, O.M. Kuttel, P. Groning, and L. Schlapbach, Vacuum arc discharges preceding high electron field emission from carbon films, *Applied Physics Letters* **69**, 476 (1996).
198. N. Missert, T.A. Friedmann, J.P. Sullivan, and R.G. Copeland, Characterization of electron emission from planar amorphous carbon thin films using in situ scanning electron microscopy, *Applied Physics Letters* **70**, 1995 (1997).
199. H. Busta, Z. Tolt, J. Montgomery, and A. Feinerman, Field emission from teepee-shaped carbon nanotube bundles, *Journal of Vacuum Science and Technology B* **23**, 676 (2005).
200. Z.F. Ren, Z.P. Huang, J.W. Xu, *et al.*, Synthesis of large arrays of well-aligned carbon nanotubes on glass, *Science* **282**, 1105 (1998).
201. C. Bower, W. Zhu, S. Jin, and O. Zhou, Plasma-induced alignment of carbon nanotubes, *Applied Physics Letters* **77**, 830 (2000).
202. J.V. Busch and J.P. Dismukes, Trends and market perspectives for CVD diamond, *Diamond Related Materials* **3**, 295 (1994).
203. J.S. Suzuki, C. Bower, Y. Watanabe, and O. Zhou, Work functions and valence band states of pristine and Cs-intercalated single-walled carbon nanotube bundles, *Applied Physics Letters* **76**, 4007 (2000).
204. K.B.K. Teo, M. Chhowalla, G.A.J. Amaratunga, *et al.*, Field emission from dense, sparse, and patterned arrays of carbon nanofibers, *Applied Physics Letters* **80,** 2011 (2002).
205. J.B. Cui, J. Robertson, and W.I. Milne, Field emission site densities of nanostructured carbon films deposited by a cathodic arc, *Journal of Applied Physics* **89**, 5707 (2001).
206. B.S. Satyanarayana, J. Robertson, and W.I. Milne, Low threshold field emission from nanoclustered carbon grown by cathodic arc, *Journal of Applied Physics* **87**, 3126 (2000).
207. Q.H. Wang, T.D. Corrigan, J.Y. Dai, *et al.*, Field emission from nanotube bundle emitters at low fields, *Applied Physics Letters* **70**, 3308 (1997).
208. C.Y. Zhi, X.D. Bai, and E.G. Wang, Enhanced field emission from carbon nanotubes by hydrogen plasma treatment, *Applied Physics Letters* **81**, 1690 (2002).
209. H. Burbert, S. Haiber, W. Brandl, *et al.*, Characterization of the uppermost layer of plasma-treated carbon nanotubes, *Diamond Related Materials* **12**, 811 (2003).
210. Yu, S.G., Kim, J.-Y., and So-Yeon (2012) Structure change and field emission of carbon nanotubes treated by plasma, *Technical Digest of IVNC, 2012*, pp. 190–191.
211. D.Y. Zhong, G.Y. Zhang, S. Liu, *et al.*, Universal field-emission model for carbon nanotubes on a metal tip, *Applied Physics Letters* **80,** 506 (2002).
212. J.-M. Bonard, F. Maier, T. Stockli, *et al.*, Field emission properties of multiwalled carbon nanotubes, *Ultramicroscopy* **73**, 9 (1998).
213. S. Dimitrijevic, J.C. Withers, V.P. Mammana, *et al.*, Electron emission from films of carbon nanotubes and ta-C coated nanotubes, *Applied Physics Letters* **75**, 2680 (1999).
214. P.G. Collins and A. Zettl, Unique characteristics of cold cathode carbon-nanotube-matrix field emitters, *Physical Review* **55**, 9391 (1997).
215. V. Semet, V.T. Binh, P. Vincent, *et al.*, Field electron emission from individual carbon nanotubes of a vertically aligned array, *Applied Physics Letters* **81,** 343 (2002).
216. P.G. Collins and A. Zettl, A simple and robust electron beam source from carbon nanotubes, *Applied Physics Letters* **69**, 1969 (1996).
217. J. Bonard, J. Salvetat, T. Stockli, *et al.*, Field emission from single-wall carbon nanotube films, *Applied Physics Letters* **73**, 918 (1998).
218. Y. Saito, K. Hamaguchi, S. Uemura, *et al.*, Field emission from multi-walled carbon nanotubes and its application to electron tubes, *Applied Physics A* **67**, 95 (1998).

219. J.P. Barbour, W.W. Dolan, J.K. Trolan, *et al.*, Space-charge effects in field emission, *Physical Review* **92**, 45 (1953).
220. W.A. Anderson, Role of space charge in field emission cathodes, *Journal of Vacuum Science and Technology B* **11**, 383 (1993).
221. K.A. Dean, P. von Allmen, and B.R. Chalamala, Three behavioral states observed in field emission from single-walled carbon nanotubes, *Journal of Vacuum Science and Technology B* **17**, 1959 (1999).
222. K.A. Dean and B.R. Chalamala, Field emission microscopy of carbon nanotube caps, *Journal of Applied Physics* **85**, 3832 (1999).
223. C.D. Child, Discharge from hot CaO, *Physical Review* **32**, 492 (1911).
224. I. Langmuir, The effect of space charge and residual gases on thermionic currents in high vacuum, *Physical Review* **2**, 450 (1913).
225. D.S.Y. Hsu and J.L. Shaw, Regeneration of gated carbon nanotube field emission, *Journal of Vacuum Science and Technology B* **23**, 694 (2005).
226. S.-C. Kung, K. Chu Hwang, and I. Nan Lin, Oxygen and ozone oxidation-enhanced field emission of carbon nanotubes, *Applied Physics Letters* **80**, 4819 (2002).
227. A.G. Rinzler, J.H. Hafner, P. Nikolaev, *et al.*, Unraveling nanotubes: Field emission from an atomic wire, *Science* **269**, 1550 (1995).
228. Z.W. Pan, F.C.K. Au, H.L. Lai, *et al.*, Very low-field emission from aligned and opened carbon nanotube arrays, *Journal of Physical Chemistry B* **105**, 1519 (2001).
229. J.M. Bonard, J.P. Salvetat, T. Stockli, *et al.*, Field emission from carbon nanotubes: perspectives for applications and clues to the emission mechanism, *Applied Physics A: Materials Science & Processing* **A69**, 245 (1999).
230. Y.-C. Chen, H.-F. Cheng, Y.-S. Hsieh, and Y.-M. Tsau, Electron field emission properties of carbon nanotubes during thermal heating and laser irradiation, *Journal of Applied Physics* **94**, 7739 (2003).
231. Cui, Y., Zhang, X., Lei, W. *et al.* (2012) Thermal-assisted field emission from carbon nanotube cathode, *Technical Digest of IVNC, 2012*, pp. 192–193.
232. C.H.P. Poa, R.C. Smith, S.R.P. Silva, and C.Q. Sun, Influence of mechanical stress on electron field emission of multiwalled carbon nanotube–polymer composites, *Journal of Vacuum Science and Technology B* **23**, 698 (2005).
233. J. Bonard, T. Stockli, F. Maier, *et al.*, Field-emission-induced luminescence from carbon nanotubes, *Physical Review Letters* **81**, 1441 (1998).
234. R. Tamura and M. Tsukada, Electronic states of the cap structure in the carbon nanotube, *Physical Review B* **52**, 6015 (1995).
235. D.L. Carroll, P. Redlich, P.M. Ajayan, *et al.*, Electronic structure and localized states at carbon nanotube tips, *Physical Review Letters* **78**, 2811 (1997).
236. K.A. Dean, O. Groening, O.M. Kuttel, and L. Schlapbach, Nanotube electronic states observed with thermal field emission electron spectroscopy, *Applied Physics Letters* **75**, 2773 (1999).
237. W.P. Dyke and W.W. Dolan, Field emission, *Advances in Electronics and Electron Physics* **8**, 89 (1956).
238. A.P. Jansen and J.P. Jones, The sharpening of field emitter tips by ion sputtering, *Journal of Physics D: Applied Physics* **4**, 118 (1971).
239. A. Zeitoun-Fakiris and B. Junner, On the dose of bombarding residual gas ions for influencing pre-breakdown field emission in a vacuum, *Journal of Physics D: Applied Physics* **24**, 750 (1991).
240. J.Y. Cavaille and M. Dechsler, Surface self-diffusion by ion impact, *Surface Science* **75**, 342 (1978).
241. A. Chis, R. Meyer, P. Rambaud, *et al.*, Sealed vacuum devices: fluorescent microtip displays, *IEEE Transactions on Electron Devices* **38**, 2320 (1991).
242. H. Adachi, K. Fudjii, S. Zaima, *et al.*, Stable carbide field emitter, *Applied Physics Letters* **43**, 702 (1983).
243. P.J. de Pablo, S. Howell, S. Crittenden, *et al.*, Correlating the location of structural defects with the electrical failure of multiwalled carbon nanotubes, *Applied Physics Letters* **75**, 3941 (1999).
244. P.K. Baumann and R.J. Nemanich, Surface cleaning, electronic states and electron affinity of diamond (100), (111) and (110) surfaces, *Surface Science* **409**, 320 (1998).
245. L. Holland and S.M. Olha, rf sputtering of graphite in argon-oxygen mixtures, *Vacuum* **26**, 233 (1976).
246. C. Dong and M.O. Gupta, Influences of the surface reactions on the field emission from multiwall carbon nanotubes, *Applied Physics Letters* **83**, 159 (2003).
247. H. Murakami, M. Hirakawa, C. Tanaka, and H. Yamakawa, Field emission from well-aligned, patterned, carbon nanotube emitters, *Applied Physics Letters* **76**, 1776 (2000).
248. H.O. Pierson, *Handbook of Carbon, Graphite, Diamond, and Fullerenes* (Noyes, Park Ridge, NJ, 1993).

249. Y. Chen, D.T. Shaw, and L. Gou, Field emission of different oriented carbon nanotubes, *Applied Physics Letters* **76**, 2469 (2000).
250. S.C. Lim, H.J. Jeong, Y.M. Shin, *et al.*, Saturation of emission current from carbon nanotube field emission array, *AIP Conference Proceedings* **590**, 221 (2001).
251. A. Maiti, J. Andzelm, N. Tanpipat, and P. von Allmen, Saturation of emission current from carbon nanotube field emission array, *Physical Review Letters* **87**, 155502 (2001).
252. O.M. Kuttel, O. Groning, C. Emmenegger, and L. Schlapbach, Electron field emission from phase pure nanotube films grown in a methane/hydrogen plasma, *Applied Physics Letters* **73**, 2113 (1998).
253. Y. Saito, K. Hamaguchi, K. Hata, *et al.*, Field emission from carbon nanotubes; purified single-walled and multi-walled tubes, *Ultramicroscopy* **73**, 1 (1998).
254. O.M. Kuttel, O. Groning, C. Emmenegger, *et al.*, Field emission from diamond, diamond-like and nanostructured carbon films, *Carbon* **37**, 745 (1999).
255. E.D. Obraztsova, J.-M. Bonard, V.L. Kuznetsov, *et al.*, Structural measurements for single-wall carbon nanotubes by Raman scattering technique, *Nanostructured Materials* **12**, 567 (1999).
256. A.N. Obraztsov, A.S. Volkov, A.L. Chuvilin, *et al.*, Curvature role of atom layers in field emission from graphite-like nanostructured carbon, *JETP Letters* **69**, 411 (1999).
257. A.N. Obraztsov, I.Y. Pavlovskii, A.S. Volkov, *et al.*, Aligned carbon nanotube films for cold cathode applications, *Journal of Vacuum Science and Technology B* **18**, 1059 (2000).
258. A. de Vita, J.C. Charlier, X. Blase, and R. Car, Electronic structure at carbon nanotube tips, *Applied Physics A* **68**, 283 (1999).
259. S.M. Lyth and S.R.P. Silva, Resonant behavior observed in electron field emission from acid functionalized multiwall carbon nanotubes, *Applied Physics Letters* **94**, 123102 (2009).
260. S.M. Lyth, R.A. Hatton, and S.R.P. Silva, Efficient field emission from Li-salt functionalized multiwall carbon nanotubes on flexible substrates, *Applied Physics Letters* **90**, 013120 (2007).
261. Y.V. Kryuchenko and V.G. Litovchenko, Computer simulation of the field emission from multilayer cathodes, *Journal of Vacuum Science and Technology B* **14**, 1934 (1996).
262. V.G. Litovchenko, A.A. Evtukh, Y.M. Litvin, *et al.*, Observation of the resonance tunneling in field emission structures, *Journal of Vacuum Science and Technology B* **17**, 655 (1999).
263. C.J. Chen, Origin of atomic resolution on metal surfaces in scanning tunneling microscopy, *Physical Review Letters* **65**, 448 (1990).
264. I. Martin, A.V. Balatsky, and J. Zaanen, Impurity states and interlayer tunneling in high temperature superconductors, *Physical Review Letters* **88**, 097003 (2002).
265. K. Bobrov, A.J. Mayne, and G. Dujardin, Atomic-scale imaging of insulating diamond through resonant electron injection, *Nature (London)* **413**, 616 (2001).
266. J.W. Gadzuk and E.W. Plummer, Field emission energy distribution (FEED), *Reviews of Modern Physics* **45**, 487 (1973).
267. A. Mayer, N.M. Miskovsky, and P.H. Cutler, Simulations of field emission from a semiconducting (10,0) carbon nanotube, *Journal of Vacuum Science and Technology B* **20**, 100 (2002).
268. V. Filip, D. Nicolaescu, and F. Okuyama, Modeling of the electron field emission from carbon nanotubes, *Journal of Vacuum Science and Technology B* **19**, 1016 (2001).
269. V. Filip, D. Nicolaescu, M. Tanemura, and F. Okuyama, Modeling the electron field emission from carbon nanotube films, *Ultramicroscopy* **89**, 39 (2001).
270. M. Tanemura, V. Filip, K. Iwata, *et al.*, Field electron emission from carbon nanotubes grown by plasma-enhanced chemical vapor deposition, *Journal of Vacuum Science and Technology B* **20**, 122 (2002).
271. P. Hojati-Talemi and G.P. Simon, Field emission study of graphene nanowalls prepared by microwave-plasma method, *Carbon* **49**, 2875 (2011).
272. W.S. Hummers and R.E. Offeman, Preparation of graphitic oxide, *Journal of the American Chemical Society* **80**, 1339 (1958).
273. S. Santandrea, F. Guibileo, V. Grossi, *et al.*, Field emission from single and few-layer graphene flakes, *Applied Physics Letters* **98**, 163109 (2011).
274. H.M. Wang, Z. Zheng, Y.Y. Wang, *et al.*, Fabrication of graphene nanogap with crystallographically matching edges and its electron emission properties, *Applied Physics Letters* **96**, 023106 (2010).
275. Z.S. Wu, S.F. Pei, W.C. Ren, *et al.*, Field emission of single-layer graphene films prepared by electrophoretic deposition, *Advanced Materials* **21**, 1756 (2009).
276. Y. Zhang, J. Du, S. Tang, *et al.*, Optimize the field emission character of a vertical few-layer graphene sheet by manipulating the morphology, *Nanotechnology* **23**, 015202 (2012).

277. G. Eda, H.E. Unalan, N. Rupesinghe, *et al.*, Field emission from graphene based composite thin films, *Applied Physics Letters* **93**, 233502 (2008).
278. S.W. Lee, S.S. Lee, and E.H. Yang, A study on field emission characteristics of planar graphene layers obtained from a highly oriented pyrolyzed graphite block, *Nanoscale Research Letters* **4**, 1218 (2009).
279. A. Malesevic, R. Kemps, A. Vanhulsel, *et al.*, Field emission from vertically aligned few-layer grapheme, *Journal of Applied Physics* **104**, 084301 (2008).
280. M. Qian, T. Feng, H. Ding, *et al.*, Electron field emission from screen-printed graphene films, *Nanotechnology* **20**, 425702 (2009).
281. Z. Shpilman, B. Philosoph, R. Kalish, *et al.*, Enhanced electron field emission from preferentially oriented graphitic films, *Applied Physics Letters* **89**, 252114 (2006).
282. W.T. Zheng, Y.M. Ho, H.W. Tian, *et al.*, Field emission from a composite of graphene sheets and ZnO nanowires, *Journal of Physical Chemistry C* **113**, 9164 (2009).
283. Saito, Y., Nakakubo, K., Asai, T., and Asaka, K. (2012) Dynamic behaviour of multilayered graphene and metal-coated carbon nanotube field emitters, *Technical Digest of the 25th International Vacuum Nanoelectronics Conference, 2012*, pp. 24–25.
284. J.C. Delgado, Y.A. Kim, T. Hayashi, *et al.*, Thermal stability studies of CVD-grown graphene nanoribbons: Defect annealing and loop formation, *Chemical Physics Letters* **469**, 177 (2009).
285. Lei, W., Zhang, X., and Wang, B. (2012) Field emission electron source with graphene layer, *Technical Digest of the 25th International Vacuum Nanoelectronics Conference, 2012*, pp. 78–79.
286. Z. Xiao, J. She, S. Deng, *et al.*, Field electron emission characteristics and physical mechanism of individual single-layer graphene, *ACS Nano* **4**, 6332 (2010).
287. Zhang, Y., Zhou, O., Deng, S.Z. *et al.* (2012) Field emission characteristics of grapheme film on nickel substrate, *Technical Digest of the 25th International Vacuum Nanoelectronics Conference, 2012*, pp. 126–127.
288. S.Y. Jeong, S.H. Kim, J.T. Han, *et al.*, High-performance transparent conductive films using rheologically derived reduced graphene oxide, *ACS Nano* **5**, 870 (2011).
289. H.J. Jeong, H.D. Jeong, H.Y. Kim, *et al.*, All-carbon nanotube-based flexible field-emission devices: From cathode to anode, *Advanced Functional Materials* **21**, 1526 (2011).
290. H.J. Jeong, H.Y. Kim, H.D. Jeong, *et al.*, Arrays of vertically aligned tubular-structured graphene for flexible field emitters, *Journal of Materials Chemistry* **22**, 11277 (2012).
291. Li, X., Zhao, D., Pang, J. *et al.* (2012) Transparent brush-type carbon nanotube cathode, *Technical Digest of the 25th International Vacuum Nanoelectronics Conference, 2012*, pp. 114–116.
292. X. Li, F. Ding, W. Liu, *et al.*, Luminescence uniformity studies on dendrite bamboo carbon submicron-tube field-emitter arrays, *Journal of Vacuum Science and Technology B* **26**, 171 (2008).
293. S. Zuo, X. Li, W. Liu, *et al.*, Field emission properties of the dendritic carbon nanotubes film embedded with ZnO quantum dots, *Journal of Nanomaterials*, **2011**, 382068 (2011).
294. H. Liu, H. Ma, W. Zhou, *et al.*, Synthesis and gas sensing characteristic based on metal oxide modification multi wall carbon nanotube composites, *Applied Surface Science* **258**, 1991, (2012).
295. Li, X., Wang, K., Wei, J. *et al.* (2012) Field emission of graphene and carbon nanotubes, *Technical Digest of the 25th International Vacuum Nanoelectronics Conference, 2012*, pp. 226–227.
296. H.W. Zhu and B.Q. Wei, Direct fabrication of single-walled carbon nanotube macro-films on flexible substrates, *Chemical Communications* **29**, 3042 (2007).
297. L.L. Fan, Z. Li, Z.P. Xu, *et al.*, Step driven competitive epitaxial and self-limited growth of graphene on copper surface, *AIP Advances* **1**, 032145 (2011).
298. L.L. Fan, Z. Li, X. Li, *et al.*, Controllable growth of shaped graphene domains by atmospheric pressure chemical vapour deposition, *Nanoscale* **3**, 4946 (2011).
299. C.Y. Li, Z. Li, H.W. Zhu, *et al.*, Graphene nano-"patches" on a carbon nanotube network for highly transparent/conductive thin film applications, *Journal of Physical Chemistry C* **114**, 14008 (2010).
300. X. Li, P.Z. Sun, L.L. Fan, *et al.*, Multifunctional graphene woven fabrics, *Scientific Reports* **2**, 395 (2012).
301. R.H. Fowler and L.W. Nordheim, Electron emission in intense electric fields, *Proceedings of the Royal Society London, Series A* **19**, 173 (1928).
302. Wanga, W., Zenga, L., Lianga, J., and Lianga, Z. (2012) The synthesis of graphene on carbon nanotube surface, *Technical Digest of the 25th International Vacuum Nanoelectronics Conference, 2012*, pp. 370–371.
303. Liu, J. and Zeng, B. (2012) Vertically aligned graphene sheets grown on carbon nano, Technical Digest of the 25th International Vacuum Nanoelectronics Conference, 2012, pp. 234–235.
304. S.H. Jo, D.Z. Wang, J.Y. Huang, *et al.*, Field emission of carbon nanotubes grown on carbon cloth, *Applied Physics Letters* **85**, 810 (2004).

305. B.Q. Zeng, G.Y. Xiong, S. Chen, *et al.*, Field emission of silicon nanowires grown on carbon cloth, *Applied Physics Letters* **90**, 033112 (2009).
306. S.H. Jo, D. Banerjee, and Z.F. Ren, Field emission of zinc oxide nanowires grown on carbon cloth *Applied Physics Letters* **85**, 1407 (2004).
307. Jiang, N., Wang, H.-X., Sasaoka, H., and Nishimura, K. (2012) Nano-carbon field electron emission films synthesized by a confined-plasma chemical vapour deposition system, *Technical Digest of the 25th International Vacuum Nanoelectronics Conference, 2012*, pp. 326–327.
308. M. Hori and M. Hiramatsu, *Advances in Science and Technology* **48**, 119 (2006).
309. J. Yu, E.G. Wang, and X.D. Bai, Electron field emission from carbon nanoparticles prepared by microwave-plasma chemical-vapor deposition, *Applied Physics Letters* **78**, 2226 (2001).
310. Wang, H.-X., Jiang, N., Sasaoka, H., and Nishimura, K. (2012) Characterization of three-dimensional nano-carbon field emission emitters, *Technical Digest of the 25th International Vacuum Nanoelectronics Conference, 2012*, pp. 328–329.
311. A. Yafyasov, V. Bogevolnov, G. Fursey, *et al.*, Low-threshold field emission from carbon nano-clusters, *Ultramicroscopy* **111**, 409 (2011).
312. Fursey, G.N., Polyakov, M.A., Kontonistov, A.A. *et al.* (2012) Extremely low threshold of field emission from graphene nanoclusters, *Technical Digest of the 25th International Vacuum Nanoelectronics Conference, 2012*, pp. 98–99.
313. Song, Y., Shin, D.H., and Lee, C.J. (2012) Field emission from graphite nanoplatelets, *Technical Digest of the 25th International Vacuum Nanoelectronics Conference, 2012*, pp. 160–161.
314. H.H. Busta, R.J. Espinova, and A. Silzar, Performance of nanocrystalline graphite field emitters, *Solid State Electronics* **45**, 1039 (2001).
315. J.J. Wang, M.Y. Zhu, R.A. Outlaw, and B.C. Holloway, Free-standing subnanometer graphite sheets, *Applied Physics Letters* **85**, 1265 (2004).
316. V.A. Krivchenko, A.A. Pilevsky, A.T. Rakhimov, *et al.*, Nanocrystalline graphite: Promising material for high current field emission cathodes, *Journal of Applied Physics* **107**, 014315 (2010).
317. T. Wei, Z. Fan, G. Luo, *et al.*, Dispersibility and stability improvement of graphite nanoplatelets in organic solvent by assistance of dispersant and resin, *Materials Research Bulletin* **44**, 977 (2009).
318. S.Y. Choi, M. Mamak, E. Cordola, and U. Stadler, Large scale production of high aspect ratio graphite nanoplatelets with tunable oxygen functionality, *Journal of Materials Chemistry* **21**, 5142 (2011).
319. Koike, T., Jyouzuka, A., Nakamura, T. *et al.* (2011) Structural and electrical properties of inflamed graphite at high temperature as a new field emitter, *Technical Digest of the 24th International Vacuum Nanoelectronics Conference, 2011*, pp. 48–49.
320. Iwai, Y., Koike, T., Hayama, Y. *et al.* (2012) Field emission characteristics of inflamed graphite at high temperature and its applications, *Technical Digest of the 25th International Vacuum Nanoelectronics Conference, 2012*, pp. 274–275.
321. Wang, W., Shao, J., and Li, Z. (2012) Vacuum potential barrier and field emission characteristic of graphene edges, *Technical Digest of the 25th International Vacuum Nanoelectronics Conference, 2012*, pp. 294–295.
322. G. Kresse and J. Furthmuller, Efficiency of ab-initio total energy calculations for metals and semiconductors using a plane-wave basis, *Computation Materials Science* **6**, 15 (1996).
323. W.L. Wang, J.W. Shao, and Z.B. Li, The exchange–correlation potential correction to the vacuum potential barrier of graphene edge, *Chemical Physics Letters* **522**, 83 (2012).
324. W.L. Wang, X.Z. Qin, N.S. Xu, and Z.B. Li, Field electron emission characteristic of grapheme, *Journal of Applied Physics* **109**, 044304 (2011).
325. W.L. Wang and Z.B. Li, Potential barrier of graphene edges, *Journal of Applied Physics* **109**, 114308 (2011).
326. N.W. Ashcroft and N.D. Mermin, *Solid State Physics*, Saunders College Publishing, Philadelphia, PA, 1976.

9

Quantum Electron Sources for High Frequency Applications

9.1 Introduction

One of the most important applications of field emission sources is microwave generation. The vacuum tube is used to amplify, switch, or modulate electrical signals. The vacuum device is more robust than solid-state devices in extreme environments involving high temperature and exposure to various radiations. The critical tradeoff is that the vacuum tubes yield higher frequency/power output but consume more energy than the MOSFET. Vacuum is intrinsically superior to the solid as the carrier transport medium since it allows ballistic transport while the carriers suffer from optical and acoustic phonon scattering in semiconductors. The advantages of solid state and vacuum devices can be achieved together if the macroscale vacuum tube is miniaturized to the nanometer scale. The vacuum nanotubes can provide high frequency/power output while satisfying the metrics of lightness, cost, lifetime, and stability at harsh conditions. The field emission cathodes have been conceived as a substitute for the thermionic cathodes of conventional microwave tubes. This application has driven much of the development of field emitter arrays (FEAs), and has spawned the field of vacuum microelectronics, in which microlithography is used to fabricate FEA cathodes.

The decade of frequencies centered on a terahertz represents the "last" frontier for electronics. The problem is that the terahertz band lies in the range where neither conventional electronics nor optical devices function well. Optical sources that depend on maintenance of a population inversion require small temperature, precluding efficient room-temperature operation. Scaled FETs will eventually deliver sufficient frequency, but output powers will be too low for many applications. Devices based on direct manipulation of electron trajectories in free space potentially offer revolutionary advantages.

In this chapter one of the most important applications of quantum electron sources such as high frequency vacuum nanoelectronics devices is considered. The perspectives of resonant tunneling for realization of microwave devices is described and function principles and parameters of field emission resonant tunneling diode (RTD) are analyzed. Some proposals for generation of THz signals both in solid state and in field emission vacuum devices are

Vacuum Nanoelectronic Devices: Novel Electron Sources and Applications, First Edition.
Anatoliy Evtukh, Hans Hartnagel, Oktay Yilmazoglu, Hidenori Mimura and Dimitris Pavlidis.

presented. The possibility of Gunn effect realization at electron field emission is described. The theory of field emission microwave sources including modulation gated FEA, current density is considered in detail. The possibility of the creation of microwave sources based on carbon nanotube (CNT) FEAs and their advantages are also analyzed.

9.2 High Frequency Application of Resonant Tunneling Diode

RTDs can be used for high-frequency applications [1]. Negative differential resistance (NDR) can be exploited in high-frequency oscillators. Oscillations are obtained by biasing the RTD into the NDR region while embedded within a suitable resonant circuit. The frequency response of the RTD depends on several factors: (1) the oscillation frequency of the waveguide circuit, (2) the charge storage delay in the quantum well (QW), and (3) the transit time across the depletion region of the diode.

Originally maximum operation frequency of RTD has been defined in the following way. The maximum frequency f_{max} is given by

$$f_{\max} = \frac{1}{2\pi\tau_{char}}, \tag{9.1}$$

where τ_{char} is a characteristic time governing the resonant tunneling process. If a signal is applied to an RTD with a frequency greater than f_{max}, the carriers within the RTD cannot follow the signal and hence the device cannot respond in a manner similar to its usual *dc* response. At frequencies higher than f_{max}, the NDR of the RTD vanishes.

However it has been recently demonstrated [2], that higher RTD frequencies are possible. The authors reported the production of a chip that emits 1.111-THz radiation with an output power of 0.1 μW, a new frequency record. The resonator was directly connected to a planar horn antenna in order to transmit the signal. Similar results were obtained at using a slot antenna that served as both resonator and antenna.

Calculation of the frequency response of an RTD has been performed using a variety of methods. The most comprehensive studies rely on numerical calculations using quantum transport schemes such as Wigner functions, direct numerical simulation of the temporal evolution of a wavepacket, and Green's functions. The details of these different approaches are too vast and they are presented in Ref. [3]. Let's consider the estimate of the frequency response of an RTD based on an approximate formulation given in Ref. [4]. At the beginning it is important to make some distinction between a fully resonant and a fully sequential tunneling system. As has been discussed earlier tunneling can proceed either sequentially or resonantly. The simplest approach to determine which of these mechanisms dominates the operation of an RTD is to assess whether the scattering time is greater or less than the carrier lifetime in the quasibound state within the RTD. In a RTD the barrier heights are of finite potential height. The electron wave function spreads over the entire device, penetrating the barriers. Therefore, the resonant states are not bound states but quasibound states with an associated finite lifetime. The lifetime of the quasibound state is called τ_{life}. If the scattering time τ_{scat}, defined as the mean time between collisions, is substantially larger than the carrier lifetime τ_{life}, the effects of scatterings on the frequency response of the device can be neglected. This is because the resonant tunneling process occurs more rapidly than scatterings, such that there is insufficient time for an electron to suffer a scattering during the resonant tunneling. If, on the other hand, τ_{scat} is substantially smaller than the carrier lifetime, an electron would necessarily suffer many collisions before it could resonantly tunnel through the structure. Therefore, the tunneling process cannot be characterized as being resonant and it is best described as a sequential tunneling process. The lifetime can be estimated from the full-width at half-maximum of the transmission peak Γ_r as

$$\tau_{life} = \frac{\hbar}{\Gamma_r}. \tag{9.2}$$

The scattering time can be estimated from the mobility of electrons (μ) within a two-dimensional system as

$$\tau_{scat} = \frac{\mu \mathrm{m}}{q}. \tag{9.3}$$

Typically, it is expected that the resonant tunneling characteristic time will decrease due to scatterings. However, in many cases the change in magnitude of the characteristic times within the resonant and sequential models is negligible. It can be assumed that the two processes, sequential and resonant tunneling, can be characterized by the same lifetimes and that the frequency response is the same in either case.

In general, there are several time scales of importance in RTD: (1) the traversal time, the time needed to tunnel through a barrier, (2) the resonant state lifetime, and (3) the escape time. All these factors influence the overall temporal response of the device. In resonant tunneling the main contribution to the characteristic time is from the well region of the device. In resonant tunneling, the electrons become trapped in a quasibound state and persist for some time before they "leak" out of the well through the second barrier. As a result, the resonant state lifetime can be appreciably larger than the barrier traversal time and the escape time. Therefore, the characteristic time can be estimated by calculating the resonant state lifetime of the RTD.

The resonant state lifetime or, equivalently, the lifetime of the quasibound state can be presented as follows (see Chapter 1)

$$\tau_{life} \approx \frac{2(L_w + 1/k_1 + 1/k_2)}{\sqrt{2E_n/m}(T_1 + T_2)}, \tag{9.4}$$

where E_n is the energy level of the quantized state, L_w is the width of the well and k_1 and k_2 are the imaginary wave vectors within the barriers, T_1 and T_2 are the transmissivity of the first and the second boundary.

The actual frequency dependence of an RTD is difficult to establish theoretically without employing a full quantum mechanical calculation. Nevertheless, an estimate of the upper frequency limit of performance can be made which is thought to be accurate within a factor of 2. The simplest picture is that the maximum frequency of oscillation is given as

$$f_{\max} = \frac{1}{2\pi\tau_{life}}. \tag{9.5}$$

From the lifetime, the steady-state resonant tunneling current can be estimated. Assuming that the quasibound resonant state has a relatively long lifetime, charge buildup will occur within the well. In steady state, the charge buildup in the QW σ_{QW} is related to the current density J as

$$\frac{\sigma_{QW}}{\tau_{life}} = J, \tag{9.6}$$

where it is assumed that the lifetime is associated only with carriers exiting through the collector barrier of the RTD. Charge buildup is maximized, when the transmissivity of the collector barrier is significantly less than that of the emitter barrier. In this case, charge leakage out of the well is suppressed, while charge leakage into it is high, resulting in a buildup of charge in the QW region.

RTDs can be used in oscillator circuits. The fact that an RTD shows a NDR enables its use as an oscillator. To make an oscillator using a device exhibiting NDR everything that is required is that the device be connected to a tuned transmission line. The device should terminate one end of the line, leaving the other end with a large discontinuity. The output power is that part of the circulating power that leaks past the discontinuity. A simplified equivalent-circuit model for a RTD is shown in Figure 9.1.

The basic equivalent circuit consists of four elements: the series inductance L_S, the series resistance R_S, the diode capacitance C_j and the negative diode resistance $-R$. The series resistance R_S includes the on-chip interconnects and external wire resistance, the ohmic contacts, and the spreading resistance in the wafer substrate, which is given by $\rho/2d$, where ρ is the resistivity of the semiconductor and d is the diameter of the diode area [5]. The series inductance L_S is due to interconnects, wire bond, and external wires. These parasitic elements establish important limits on the performance of the RTD diode.

The total input impedance Z_{in} of the equivalent circuit of Figure 9.1 is given by

$$Z_{in} = \left[R_S + \frac{-R}{1+\left(\omega RC_j\right)^2} \right] + j\left[\omega L_S + \frac{-\omega C_j R^2}{1+\left(\omega RC_j\right)^2} \right]. \tag{9.7}$$

As can be seen from Equation (9.7) the resistive (real) part of the impedance will be zero at certain frequency, and the reactive (imaginary) part of the impedance will also be zero at another frequency. These frequencies are denoted by the resistive cutoff frequency f_r and the

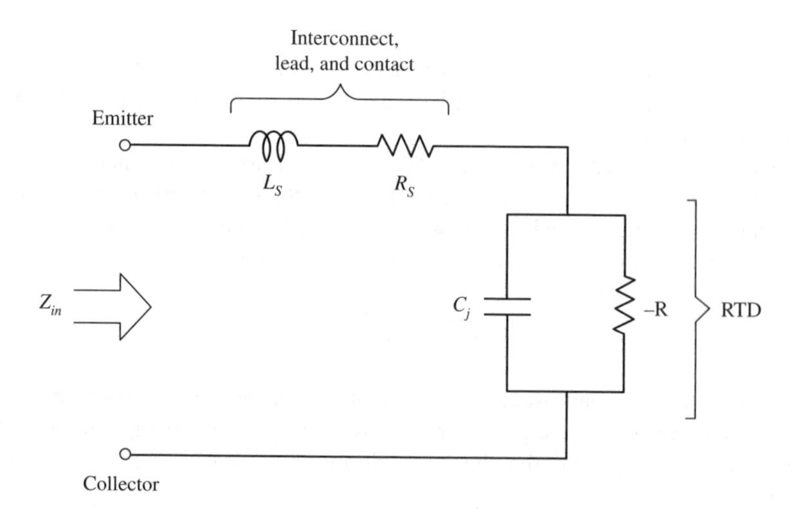

Figure 9.1 Equivalent-circuit model of RTD

reactive cutoff frequency f_x, respectively. These frequencies are given by

$$f_r = \frac{1}{2\pi RC_j}\sqrt{\frac{R}{R_S} - 1}, \tag{9.8}$$

$$f_x = \frac{1}{2\pi}\sqrt{\frac{1}{L_S C_j} - \frac{1}{(RC_j)^2}}. \tag{9.9}$$

Since R is bias dependent, so are the cutoff frequencies. These resistive and reactive cutoff frequencies specified at the bias of R_{min} are

$$f_{r0} \equiv \frac{1}{2\pi R_{\min} C_j}\sqrt{\frac{R_{\min}}{R_S} - 1} \geq f_r, \tag{9.10}$$

$$f_{x0} = \frac{1}{2\pi}\sqrt{\frac{1}{L_S C_j} - \frac{1}{(R_{\min} C_j)^2}} \leq f_x, \tag{9.11}$$

and

$$R_{\min} \approx \frac{2V_P}{I_P}, \tag{9.12}$$

where V_P and I_P are the peak voltage and peak current, respectively.

Since at that bias, the value of R is at its minimum (R_{min}), f_{r0} is the maximum resistive cutoff frequency at which the diode no longer exhibits net negative resistance; and f_{x0} is the minimum reactive cutoff frequency (or the self-resonant frequency) at which the diode reactance is zero. It follows that the diode would oscillate, if $f_{r0} > f_{x0}$. In the most applications where the diode is operated into the negative-resistance region, it is desirable to have $f_{x0} > f_{r0}$ and $f_{r0} \gg f_0$, f_0 being the operating frequency. Equations (9.10) and (9.11) show that to fulfill the requirement that $f_{x0} > f_{r0}$, the series inductance L_S must be low.

The switching speed of a RTD diode is determined by the current available for charging the capacitance and the average RC product. Since R, the negative resistance, is inversely proportional to the peak current, a large tunneling current is required for fast switching. A figure of merit for RTD diodes is the speed index, which is defined as the ratio of the peak current to the capacitance at the valley voltage, I_P/C_j.

Another important quantity associated with the equivalent circuit is the noise figure, which is given as

$$NF = 1 + \frac{q}{2k_B T}|RI|_{\min}, \tag{9.13}$$

where $/RI/_{min}$ is the minimum value of the negative resistance-current product on the current-voltage characteristic.

For triode- and transistor-like three-terminal devices, the maximum power P_m that can be delivered to a load is [6–8]

$$P_m = \frac{E_m^2 v_s^2}{X_o (2\pi f_T)^2}. \tag{9.14}$$

In Equation (9.14), E_m is the critical field at which electrical breakdown occurs, v_s is the electron velocity, X_o is the output impedance level, and f_T is the cutoff frequency. Equation (9.14) can be used to understand the difference between solid-state and vacuum devices. In a solid-state device, f_T can be quite high because device dimensions are small. However, the electron velocity in a solid-state device cannot exceed approximately 10^7 cm/s because of electronic collisions with the semiconductor lattice, whereas for vacuum tubes even relativistic velocities can be attained. The breakdown process in a semiconductor is initiated by valence- to conduction-band transitions, which typically require energies only of the order of 1 eV. In contrast, secondary emission processes, which can be minimized by proper choice of materials and geometry, determine breakdown in tubes. Furthermore, the heat dissipated in the semiconductor is more problematic in a solid-state device and often limits the output power. In general, semiconductors have much lower thermal conductivity than metals, so a properly designed microwave tube can provide better thermal paths to dissipate heat. Consequently, for quite fundamental reasons, the output power provided by a microwave tube can be much higher than that provided by a solid-state device.

9.3 Field Emission Resonant Tunneling Diode

In vacuum micro- and nanoelectronics applications it is very important to employ systems with high emission efficiency at sufficiently low applied voltage. It has become a common practice to use, for this purpose, sharp emitter systems with tiny emitting areas down to the atomic scale. An alternative approach would be to decrease the work function and to change fundamental parameters that determine the tunneling emission using materials with low effective masses, low electron affinity, quantum-size effects, and so on [9–13]. The electron emitters with multilayer structures containing nanoparticles are actively investigated [14–21]. The coherent resonant electron tunneling [20], sequential electron tunneling [15], and single electron tunneling [16] are the main electron transport mechanisms at emission from multilayer quantum emitters. In previous theoretical and experimental investigations the variety of Si-based multilayer quantum emitters among the multilayer cathodes, nanocomposite $Si_xO_yN_z$(Si) films, nanocomposite SiO_2(Si) films, nanocrystalline silicon, metal-oxide-semiconductor structure have been studied [14, 19–22]. But further search and optimization of efficient quantum emitter with narrow energy and spatial electron distribution and for application in high frequency electronics is inevitable. The Si-based multilayer tunneling structures are promising for obtaining the electron sources with enhanced parameters. ***New type of field emission*** RTD ***has been proposed and investigated both*** theoretically and experimentally in Ref. [23].

The diode consists of Si emitter covered with SiO_x layer as an input potential barrier and Si layer as a QW. Vacuum layer is an output potential barrier of a double-barrier quantum structure (DBQS) and then electron transition takes place in the following vacuum transit layer. Energy diagram of the emission RTD is shown in Figure 9.2. The dash line indicates the path of an electron, emitted with a resonant energy E from the emitter conduction band to an anode through nth energy level in the QW. The height of the SiO_x input potential barrier ϕ at its interface with the Si emitter layer for an electron emitting with zero energy in the absence of electric field is equal to the difference of electron affinities χ of Si and SiO_x. The height of the output potential barrier at the interface Si QW with a vacuum barrier layer is determined by the electron affinity χ of Si. The energy diagram has been calculated with an assumption that electric field is constant inside every layer.

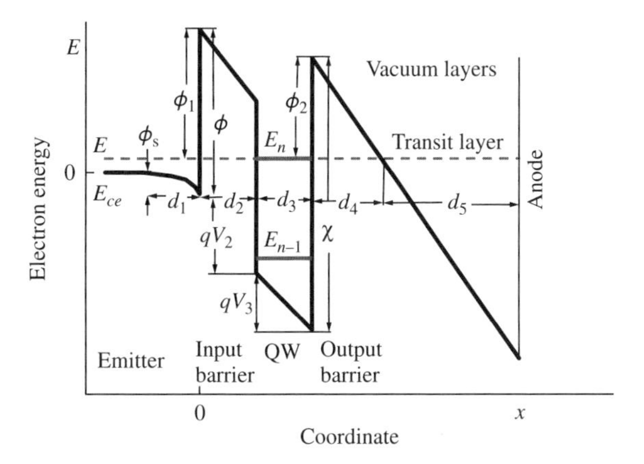

Figure 9.2 Energy band diagram for RTD based on the Si-SiO$_x$-Si cathode. Reproduced with permission from Ref. [23]. Copyright (2012). American Vacuum Society

In order to investigate RTD microwave impedance characteristics in framework of a small-signal model it is assumed that every characteristic of a diode contains the direct and alternating components.

9.3.1 Direct Emission Current

Numerically solving the time-independent Schrödinger's and Poisson's equation system in the same way as in Refs. [20, 24] there was obtained the emission characteristics with negative differential emission conductivity of diode and electron dwelling (life) time in the *n*th subband of the QW. The boundary conditions are the continuity of the electron wave function, its derivative and electric induction at the interfaces of DBQS layers. The proposed model is based on the models of resonant tunneling in a semiconductor DBQS [25] and resonant tunneling emission from multilayer cathode [20, 24]. The differences of electron effective masses and dielectric constants for the adjacent layers of the DBQS and dependencies of the diode potential profile and energy eigen values in the QW on electric field are taken into account. The accumulation layer is considered only in relation to its contribution in band bending and the additional emission from the layer is disregarded. This is the cause of absence of additional diode current resonant peaks, which correspond to emission from accumulation layer energy subbands.

The dependencies of a charge Q_s and potential drop $\varphi_s = \phi_s/q$ on electric field F_s in the accumulation layer are numerically calculated solving once integrated Poisson equation

$$F_s = \sqrt{2 \times [\exp(\beta\varphi_s) - \beta\varphi_s - 1]}/\beta L_D \tag{9.15a}$$

$$Q_s = -\varepsilon_1\varepsilon_0 F_s, \tag{9.15b}$$

with the boundary conditions of electric induction continuity at the emitter-DBQS interface. Here $\beta = q/k_B T$, where q, T, k_B are the electron charge, temperature, and the Boltzmann constant, ε_0 and ε_1 are the dielectric constants of vacuum and emitter, L_D is the electron Debye length in the emitter.

The calculations of emission direct current density are performed in accordance with the following expression [24] where emission from accumulation layer is disregarded

$$J = A \int_0^{\infty} \left\{ \left[D_0 + \sum_{n=1}^{N} D_n \right] \times \ln[1 + \exp[(E_F - E)/k_B T] \right\} dE. \tag{9.16}$$

In Equation (9.16) A is a well-known coefficient $A = 4\pi Mqm_1{}^* k_B T/h^3$, where M is the degeneracy degree of an emitter conduction band, $m_1{}^*$ is the electron effective mass in the emitter layer, h is the Plank constant, E and E_F are the longitudinal and Fermi energies of the emitting electron in emitter, N is the number of resonant levels in the QW, and D_0 and D_n are the transmission coefficients for the nonresonant tunneling and resonant tunneling through the nth energy resonant subband in the QW of the DBQS respectively.

The expression for D_n results from simulating the electron waves interference process in the DBQS like Fabry-Perot resonance [24, 26]. In fact, the wave function of an electron in the QW, as well as an electromagnetic field in a resonator, is a superposition of plane waves, impinging on potential barriers and reflecting from them. A complex coefficient of electron wave transmission through the DBQS equals to a ratio of wave parts transmitted through the DBQS to incident on it.

When taking into account all waves with any reflection multiplicity, the coefficient is presented by the following infinite geometrical progression:

$$\delta = \eta_1 \eta_2 \exp[(jE_n/\hbar - \tau_p^{-1})\tau_0] \times \sum_{l=0}^{\infty} \{\zeta_1 \xi_2 \exp[2(jE_n/\hbar - \tau_p^{-1})\tau_0]\}^l. \tag{9.17}$$

Here η_1, η_2 are the complex coefficients of electron wave transmittance for the input and output potential barriers consequently and ζ_1, ζ_2 are the same coefficients of that wave reflection, l is the number of reflections of the electron wave in the QW, τ_p is the electron momentum relaxation time in the QW layer, $\tau_0 = d_3/\upsilon_n$ is the time, inverse to a rate of electron collisions with the potential barriers in the QW, d_3 is QW width, E_n is the eigen value of electron energy of the nth resonant energy level in the QW and υ_n is the electron velocity corresponding to the energy level, $\hbar$ is the reduced Plank constant, and j is the imaginary unit. After summation of the progression Equation (9.17) and multiplication of the sum by a complex-conjugate expression a real transmission coefficient of an electron wave power for resonant tunneling emission has been obtained. This coefficient and the same coefficient for nonresonant tunneling emission are given by the following expressions [24]:

$$D_n(E - E_n) = T_1 T_2 (\hbar/\Gamma_n \tau_n)^2 \exp(-\tau_n/\tau_p)/\{1 + 4[(E - E_n)/\Gamma_n]^2 \exp(-\tau_n/\tau_p)\}. \tag{9.18a}$$

$$D_0(E) = T_1 T_2 \exp[-t(E)/\tau_p]. \tag{9.18b}$$

In Equations (9.18a) and (9.18b) $\tau_n = 2\tau_0/(T_1 + T_2)$ is an electron dwelling time in the nth energy subband, $t(E) = d_3/\upsilon(E)$ is the time of an electron single transit through the QW with nonresonant energy E, $\upsilon(E) = \{2 \cdot [E + q \cdot (\varphi_s + V_2 + V_3/2)]/m_3{}^*\}^{1/2}$ is the thermal velocity of electron in the QW layer, d_k is the kth diode layer width starting with the accumulation layer, and $m_k{}^*$ and V_k are the electron effective mass and voltage drop in the layer. The energy E is counted out from emitter conduction band bottom level E_{ce} outside of the accumulation layer. $\Gamma_n = (\tau_n{}^{-1} + \tau_p{}^{-1})h/2\pi$ is the total width of the nth resonant energy subband in the QW, T_1 and T_2 are the transparencies of the input and the output potential barriers for electron, emitting with the energy E at electric field in vacuum F. The expressions for $T_1 = \eta_1\eta_1{}^*$ and $T_2 = \eta_2\eta_2{}^*$ are the following:

$$T_1 = \exp\{-[\pi\varepsilon_2\sqrt{m_2^*}/(\hbar q^2 F)] \times [\phi_1^{3/2} - \alpha(\phi_1 - V_2)^{3/2}]\} \tag{9.19a}$$

$$T_2 = \exp\{-[\pi\sqrt{m_0}/(\hbar q^2 F)]\phi_2^{3/2}\}. \tag{9.19b}$$

Here α equals 1 or 0 depending on the trapezoidal or triangular input barrier shape defined by values of the electric field and the height and width of a barrier layer. In Equations (9.19a) and (9.19b) ε_k is the dielectric permittivity of the kth diode layer, m_0 is a free electron mass, $\phi_1 = \phi - q\varphi_s - E$ and $\phi_2 = \chi - q(\varphi_s - V_2 - V_3) - E$ are the heights of the input and output barriers consequently for the electron emitting from emitter conduction band with energy E at electric field in vacuum F (see Figure 9.2).

9.3.2 *Microwave Characteristics*

The small signal analysis of the diode microwave characteristics is fulfilled at the calculated values of the direct current characteristics as parameters. These parameters are the resonant values of electric field and current density, the time of electron dwelling on the resonant level in the QW and cathode differential injection conductivity, σ. The latter is equal to the derivative of the direct current over the direct electric field on dropping interval of the diode current-field characteristics. As it is assumed within the framework of a small-signal analysis of diode impedance [27, 28], the total current alternating in time consists of two alternating components, conduction and induction currents. Equations system for the alternating quantities has a different view in the vacuum transit layer from that in an injection region of the diode. The injection region consists of the accumulation layer and the DBQS. A phase delay of the alternating current concerning the electrical voltage is due to both inertness of the resonant-tunneling injection from DBQS in the transit vacuum layer and finite time of an electron transit in the vacuum layer.

The equation system for an alternating small-signal electric field $e(x, t)$, the total alternating current density $i(t)$, its conduction and induction components, $i_c(x, t)$ and $i_d(x, t)$, at coordinate x of a transit vacuum layer at a point in time t is the following [29]:

$$i(t) = i_c(x, t) + i_d(x, t) \tag{9.20a}$$

$$i_c(x, t) = \gamma \times i(t) \times \exp[-j\omega\tau(x)] \tag{9.20b}$$

$$i_c(x, t) = \sigma \times e(x, t) \times \exp\{-j[\varphi + \omega\tau(x)]\} \tag{9.20c}$$

$$i_d(x, t) = \varepsilon\varepsilon_0 \frac{\partial e(x, t)}{\partial t}. \tag{9.20d}$$

It is assumed that the dependencies of all the small-signal values on the time have harmonic character and in the used approximation the values are proportional to *exp(jωt)*. In Equations (9.20a)–(9.20d) $\omega = 2\pi f$, where f is the alternating signal frequency, $\varphi = \omega\tau_n$ is the phase delay of the alternating current related to alternating voltage due to resonant emission delay, $\tau(x)$ is the time of an electron transit from the injecting interface DBQS-vacuum to a point x in the transit vacuum layer, γ is the injection coefficient defined from boundary conditions of the injection and conduction currents equality at the injecting interface of the DBQS and transit layer.

After solving the equations system (9.20a), (9.20b) at the injecting interface the following expression for injection coefficient has been obtained:

$$\gamma = [1 + j\omega\varepsilon_0 \exp(j\varphi)/\sigma]^{-1}. \tag{9.20e}$$

It is assumed in the small-signal model that permittivity in all the DBQS layers is equal to the one in emitter layer, and the dependency of alternating quantities on coordinate is absent. In the vacuum transit layer $\varepsilon = 1$ and, unlike the total emission alternating current, both of its components depend on coordinate. These components are expressed as the functions of alternating electric field and the total current when using the equation system Equations (9.20a)–(9.20d). The obtained equation leads to the expressions for diode differential impedance at coordinate x in kth layer of an injection region:

$$\partial Z/\partial x = [\sigma \exp(-j\varphi) + j\omega\varepsilon_0\varepsilon_k]^{-1}, \; (k = 1 - 4) \tag{9.21a}$$

and in the transit layer

$$\partial Z/\partial x = \{1 - \sigma \exp[-j(\varphi + \theta(x)] \times [\sigma \exp(-j\varphi) + j\omega\varepsilon_0]^{-1}\}/(j\omega\varepsilon_0), \; (k = 5). \tag{9.21b}$$

When integrating Equations (9.21a) and (9.21b), zero initial velocity and uniformly accelerated motion of an electron in the vacuum transit layer are taken into account. The integration of equation over all the active diode layers leads to analytical expressions for active R and reactive X components of the diode small signal impedance Z

$$Z = \int_{-d_1}^{\sum_{k=2}^{5} d_k} (\partial Z/\partial x)dx = R + jX \tag{9.22a}$$

$$R = B[A_1 \cos\varphi + A_2(\sin\varphi - \sigma/\omega\varepsilon_0)]/P_5 + \sigma d_{inj} \cos\varphi/(\omega^2{\varepsilon_0}^2{\varepsilon_1}^2 P_1) + R_s \tag{9.22b}$$

$$X = B[A_2 \cos\varphi - A_1(\sin\varphi - \sigma/\omega\varepsilon_0)]/P_5 - d_{inj}(1 - \sigma \sin\varphi/\omega\varepsilon_0\varepsilon_1)/(\omega\varepsilon_0\varepsilon_1 P_1) - d_5/\omega\varepsilon_0. \tag{9.22c}$$

Here $A_1 = \cos\theta + \theta\sin\theta - 1$, $A_2 = \theta\cos\theta - \sin\theta$, $P_k = (\sigma\cos\varphi/\omega\varepsilon_0\varepsilon_k)^2 + (1 - \sigma\sin\varphi/(\omega\varepsilon_0\varepsilon_k))^2$, $B = \sigma u/\omega^4{\varepsilon_0}^2$, $\theta = \omega\tau_d$, $d_{inj} = \sum d_k$ is the width of an injection region with summation over layer numbers $k = 1$–4 starting with emitter accumulation layer (see Figure 9.2), $u = qF/m_0$ is an electron acceleration in the transit layer, τ_d and θ are time and angle of an electron transit from the DBQS-transit layer interface to anode, $R_s = \rho_c + \rho_e l_e$ is a diode parasitic series resistance [27], where ρ_e, l_e are the average resistivity and the total width of all the diode epitaxial layers and ρ_c is a specific contact resistance of an emitter. The contact spreading resistance is absent for the vacuum RTD, as distinct from the same semiconductor diode [27].

Carrying out small signal analysis [27] there has been assumed the equality of permittivity in all the layers of an injection region to its value in Si. It is a rough approximation in the view of great difference of permittivity in different layers of the region. However, performed calculations show the negligibly small contribution of the second term in Equations (9.22b) and (9.22c) to diode microwave impedance compared to the contribution of other terms. Therefore, applied approximation does not noticeably change the impedance value.

The direct emission current calculations on the contrary are carried out at the real parameters of every diode layer.

9.3.3 *Calculation of the Direct Emission Current*

The calculations of both direct and microwave diode characteristics are performed for the temperature of 300 °K, doping level 10^{18} cm^{-3} and emitter layer full width of 500 nm respectively. Used small width of layers both the input barrier and QW of 1 nm allows us to minimize the time of electron dwelling on resonant energy level in the QW and to increase the diode operating frequency. Linear dependencies of parameters such as electron effective mass, dielectric constant, and energy bandgap in the SiO_x layer on stoichiometry index x ($x = 0$ for Si and $x = 2$ for SiO_2) in the layer were assumed.

The input barrier height values calculated in this way are 1.28, 1.92, and 2.88 eV for three types of diodes with $x = 0.8$, 1.2, and 1.8 in SiO_x barrier layer. The dependencies of direct emission characteristics of diodes are shown in Figure 9.3. There are three energy levels in the cathode coating QW and the same number of current resonance peaks corresponding to resonant tunneling through the levels at the input potential barrier heights of $\phi = 1.28$ and 1.92 eV, but four levels and the current resonance peaks at $\phi = 2.88$ eV.

Table 9.1 presents calculated time-independent parameters of resonant tunneling emission through two upper energy levels in the QW at the above pointed values of input barrier heights. Here, σ_n is the maximum value of cathode negative differential injection conductivity when resonant tunneling occurs through the nth energy level in the QW at corresponding resonant electric field F_n. The shortest time of electron resonant tunneling and the highest negative differential injection conductivity correspond to the highest diode operating frequency. When the barrier height decreases and electron resonant energy eigen value in the QW increases, the time of resonant tunneling decreases as a result of potential barriers transparency increasing. Therefore, resonant tunneling only through the higher energy levels in the QW is interesting for the high-speed RTD creation. Because of prolonged time of the resonant tunneling process through the lower resonant levels they are not promising for high-speed microwave application.

The analysis of Table 9.1 data demonstrates that resonant tunneling time and negative injection conductivity have the lowest values when resonant tunneling occurs through the fourth highest resonant energy level in the QW with an input barrier height $\phi = 2.88$ eV. When resonant tunneling occurs through the third highest resonant energy level in the QW of the diode with the barrier height $\phi = 1.92$ eV, the tunneling time is among the lowest values and negative injection conductivity is among the highest values for the diodes studied. The

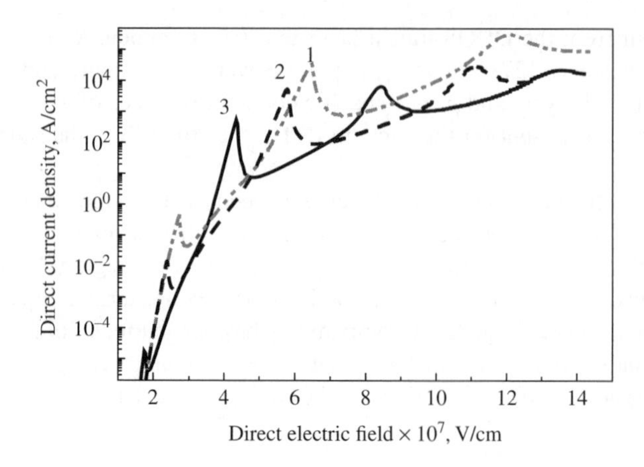

Figure 9.3 Direct current-electric field characteristics of cathode at the input potential barrier height $\phi = 1.28,\ 1.92,\ 2.88$ eV (curves 1–3). Reproduced with permission from Ref. [23]. Copyright (2012). American Vacuum Society

Table 9.1 Time-independent parameters of direct resonant-tunneling emission.

ϕ (eV)	1.28		1.92		2.88	
N	2	3	2	3	3	4
F_n (10^7 V/cm)	6.48	12.15	5.78	11.27	8.47	13.7
τ_n, *(fs)*	1.4 10^2	5.2	2.2 10^2	5.3	7.0	1.8
σ_n *(mS/cm)*	−50.2	−26.0	−7.8	−30.5	−4.9	−0.7

combination of these injection parameters is the best among the parameters obtained for diodes with other input barrier height (Table 9.1). For the following reason, the computed high frequency impedance and admittance characteristics are analyzed below only for the diode with $\phi = 1.92$ eV. The same characteristics obtained for a diode with the other barrier heights were proved to be worse.

9.3.4 *Calculation of Microwave Parameters*

Figure 9.4 presents the dependences of microwave negative conductance on the frequency for the diode studied, with the input barrier height $\phi = 1.92$ eV. Every dependence is obtained at a transit angle $\theta_k = \pi k/20$, where $k = 1, 2, \ldots$ is a curve number when resonant tunneling occurs through the second (Figure 9.4a) and the third (Figure 9.4b) energy levels in the QW. As it appears from the analysis of dependencies, there is only one peak on the frequency dependence, and it is located in the submillimeter wavelength range.

The active component of the diode impedance is determined predominantly by the first term with the higher-order value as compared to the value of the other two terms in Equation (9.22b). The first term presents the impedance active component of a transit layer which depends on

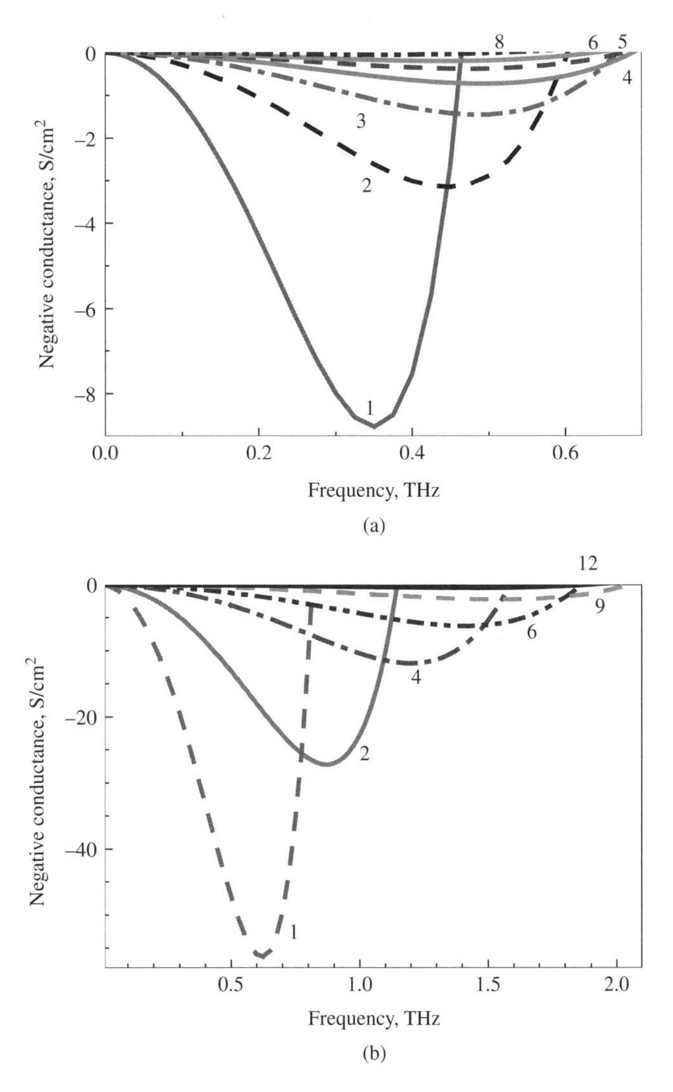

Figure 9.4 Frequency dependencies of negative conductance at transit angle $\theta_k = \pi k/20$ (k is curve number) with resonant tunneling through (a) the second and (b) the third energy level in the QW. Reproduced with permission from Ref. [23]. Copyright (2012). American Vacuum Society

the electron delay in both resonant emission and transit in vacuum. Its value is proportional to the negative differential injection conductivity and the electron acceleration in the vacuum transit layer. The second term in Equation (9.22b) presents the impedance active component of an injection layer and is proportional to the layer width, and so, its value is noticeably less than that of the first term in the whole frequency band of diode negative resistance except for at the edges of the band. The third term is the diode parasitic series resistance, the value of which is frequency independent taking into account the absence of the contact spreading resistance.

The sum of the two last terms reaches the first term value only near the edges of the negative conductance band.

The greatest peaks of both negative conductance and negative resistance are at very small electron transit angle. At increasing the transit angle up to $\theta = \theta_{up}$ the peak values of both negative conductance and negative resistance decrease but the frequencies of the peaks increase. It is also accompanied by the broadening of the negative conductance frequency band due to an increase in the upper boundary frequency. After surpassing θ_{up} the further increase of a transit angle leads to upper boundary frequency lowering and the actual band narrowing. The bandwidth vanishes completely when the transit angle reaches the value θ_{max}, which is maximal for the presence of a negative conductance. The θ_{max} value is at around $2\pi/3$ when the tunneling takes place through the third level in QW. It is a little lower than the value when the resonant tunneling occurs through the second resonant level.

At any transit angle the peak value of the negative conductance is higher, when the resonant tunneling process occurs through the third energy level rather than through the second one. The higher frequency band of the negative conductance and higher maximum peak of the negative resistance corresponding to the resonant emission through the third resonant level than the emission through the second level result from the above-mentioned higher values of injection conductivity and electron acceleration in the first case. The maximum values of the upper frequency of a negative resistance band are around 2 and 0.7 THz and they are reached at transit angle values θ_{up} close to 0.45π and 0.2π accordingly, with resonant tunneling through the third energy level in QW and the second one.

The frequency dependencies of the reactive component of diode admittance at different transit angle values are shown in Figure 9.5. The admittance dependencies of diode in the negative conductance frequency band at electric field values corresponding to resonant tunneling through the second and the third energy levels in QW are shown in Figure 9.6.

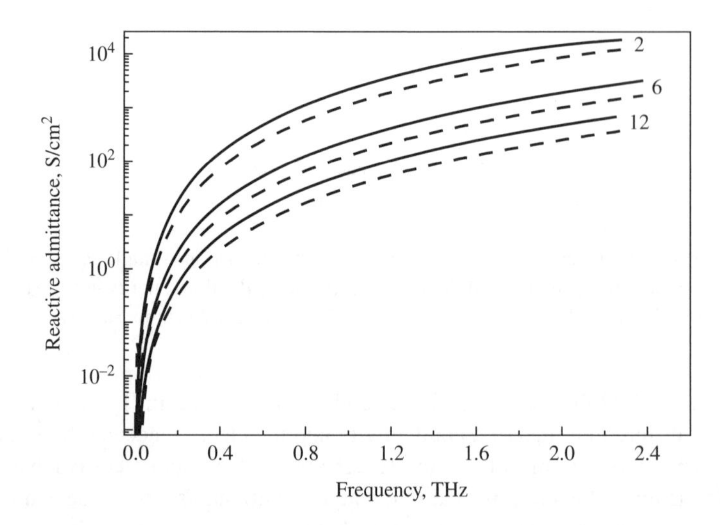

Figure 9.5 Frequency dependencies of reactive components of diode admittance at transit angle $\theta_k = \pi k/20$ (k is the curve number) with resonant tunneling through the second (solid lines) and the third (dash lines) energy level in the QW. Reproduced with permission from Ref. [23]. Copyright (2012). American Vacuum Society

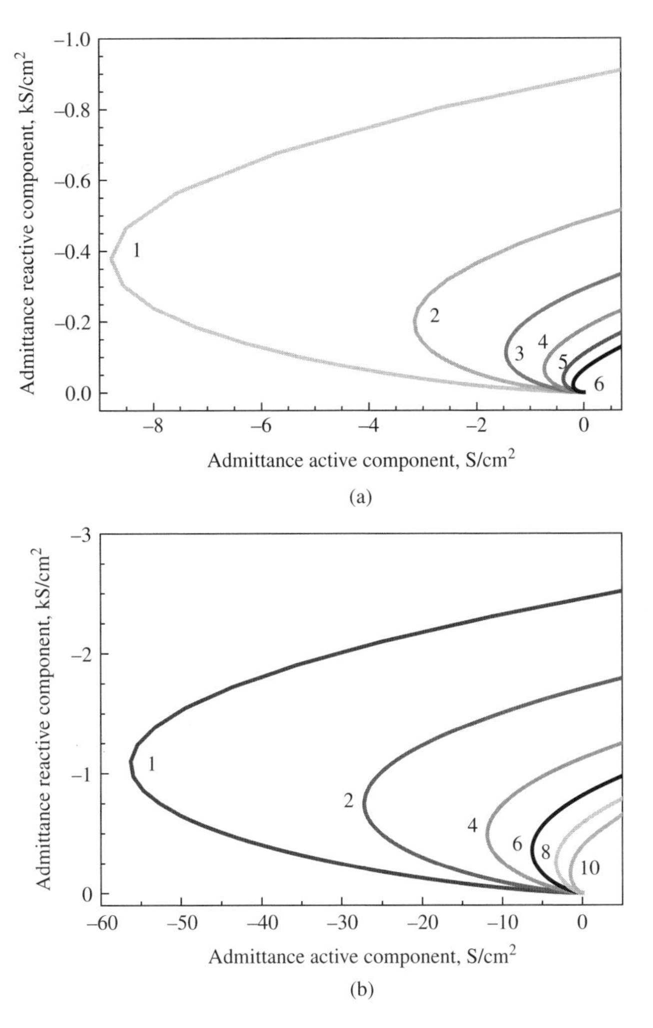

Figure 9.6 Diode in negative resistance region when resonant tunneling occurs through (a) the second and (b) the third energy level in the QW at the transit angle $\theta_k = \pi k/20$ (k is the number of curve). Reproduced with permission from Ref. [23]. Copyright (2012). American Vacuum Society

An analysis of Equation (9.22c) for the reactive component of diode impedance shows that the first term in the expression is inductive. It is determined by the processes of electron delay in the QW and in the vacuum transit layer. The second term in Equation (9.22c) is the capacitive reactance of an injection region. It is smaller than the first one due to noticeably smaller width of an injection region when compared to transit layer width. The ratio of the third term presenting the capacitive reactance of the vacuum transit layer to the first term is around 10 at frequencies close to upper boundary of a diode negative conductance frequency band. It decreases to a value of slightly more than that of a unit with frequency decreasing to the lower boundary value of the negative resistance band. Therefore, diode reactive component of impedance remains capacitive in the negative resistance band of the diode, and frequency dependency of the reactance in high-frequency part of the band is close to the same dependency of capacitance reactance of the transit vacuum layer.

Increase of the transit angle, caused by an increase of width of a transit vacuum layer at a particular fixed frequency, leads to increasing of reactive component of diode impedance and decreasing of reactive component of admittance. In the most part, it is a result of the dependency of a transit vacuum layer capacitance on the width of the layer.

The full vacuum layer of a diode includes the vacuum layer of the output potential barrier in addition to the transit vacuum layer. The width of the output potential barrier layer for an electron emitting with the resonant energy E from the emitter conduction band is given by $d_4 = \phi_2/qF_n$. It is inversely proportional to resonant electric field F_n, which is sufficiently large for the studied vacuum RTD (see Table 9.1) when resonant tunneling occurs through the upper resonant levels in the QW. This is the reason for a considerably lower output barrier layer width in comparison with the same of the transit vacuum layer, which is defined by $d_5 = qF_n\theta^2/2m_0\omega^2$. Therefore, the width of the whole vacuum layer is approximately equals to the width of the transit vacuum layer. It grows as the value of the resonant electric field increases and is nearly twice less in the case of resonant tunneling through the second energy level in the QW as compared with tunneling through the third level at the same transit angle value.

Table 9.2 presents the diode microwave characteristics calculated for diode with the width of a transit vacuum layer d_5, which corresponds to transit angle θ at frequencies f of negative conductance peaks shown in Figure 9.4. Here, G_{max} is the value of negative conductance close to its maximum values, which are reached at frequency f for the diode with the transit layer width d_5. The greatest negative conductance value of 57 S/cm^2 is reached at frequency of 0.63 THz for a diode with a transit layer width of 1.56 μm, when resonant tunneling occurs through the third resonant level in the QW. When resonant tunneling occurs through the second resonant level, the greatest value is around 9 S/cm^2 at a frequency of 0.35 THz when transit layer width is close to 2.59 μm.

The advantages of the proposed emission RTD diode over the semiconductor RTD are in the higher values of electron velocity in a vacuum transit layer, the resonant-tunneling emission electric field, the current density, and lower parasitic series resistance. Owing to those, the maximal values of the negative conductance and operating frequency for the vacuum diode exceed the same values in available semiconductor RTDs. The analysis performed shows that the vacuum RTD based on Si-SiO$_x$-Si cathode operates in the range of up to submillimeter-wave lengths when resonant tunneling occurs through the upper resonant level in the QW of the multilayer coating with layers width of around 1 nm. The experimental

Table 9.2 Diode microwave parameters at frequencies of negative conductance peaks.

n	2		3		
θ	0.05π	0.1π	0.05π	0.1π	0.2π
f (THz)	0.35	0.45	0.63	0.90	1.25
G_{max} (S/cm)	8.8	3.3	57	28	13
d_5 (μm)	2.59	6.27	1.56	3.06	6.35

investigations of the proposed electron emission Si-based resonant-tunneling diode revealed a resonant peak on emission current–voltage characteristics and in such a manner supported the model developed [23].

9.4 Generation of THz Signals in Field Emission Vacuum Devices

More and more applications in imaging science and technology call for the well development of THz wave sources. There are mainly two methods for the generation of electromagnetic (EM) waves. One is a quantum mechanical method, which deals with the use of the conduction electrons in a solid, that is, laser, suitable for optical radiation. The other is a classical process, which deals with the use of streaming electrons in vacuum, that is, maser, suitable for microwave generation. The opposing limitations of the two processes result in the long-recognized millimeter and submillimeter gap in the electromagnetic spectrum where the achievable power falls to low levels from both long- and short-wavelength regions. The well-known electron cyclotron masers (ECMs) provide the solution [29]. The ECM has undergone a remarkably successful evolution from basic research to device implementation, almost filling the gap. However, degradation of power generation in THz band is still present.

The search for efficient terahertz sources based on solid state devices appears to be particularly interesting due to the observation of THz radiation from the ultra-fast transport of charge carriers in semiconductors. Considerable efforts have been made both theoretically and experimentally to understand and enhance the generation of pulsed THz radiation from semiconductor surfaces [30–34] and *p-i-n* structures [35–37] excited by femtosecond laser pulse. It is well known that the two main processes that contribute to the emission of THz signals in semiconductors are the transport of photoexcited carriers in electric field region and instantaneous polarization that arises during optical excitation. THz radiation induced by several other mechanisms in bulk semiconductors and nanometric heterostructures also has been explored, such as coherent polar-phonon oscillations [38], inflection of carriers in the valley in GaN [39], ballistic motion of carriers with negative effective mass [40], ballistic oscillations in parabolically shaped potential well [41] and nanometer transistors [42], and Bloch oscillating electrons in a semiconductor superlattice (SL) [43].

The generation of THz signals by the periodic quasi-ballistic resonant motion of electron on the basis of the contributed action of electron acceleration in a potential well and reflection at the heterointerfaces have been demonstrated by a Monte Carlo simulation in Ref. [44]. The proposed mechanism is distinct from those mentioned above. It is exploited to generate THz signals in semiconductors, namely the elastic reflection of ballistic electrons at a heterojunction interface. The concept is based on the periodic quasi-ballistic motion of electrons caused by the combined action of electron acceleration/deceleration in a submicron potential well and the reflection at the heterointerface. The electrons are accelerated by an external electric field to the opposing barrier. There they are reflected and travel to the opposite energy barrier against the externally applied field. Eventually their velocity is reduced to zero and reversed in the direction of the externally applied electric field, as shown in Figure 9.7. In this way the current resonates as electron bunch backward and forward for a given time. The transit times between reflections are of such short durations that the transport is mainly ballistic. This resonance phenomenon slowly dies down after a given number of periods as the coherency of electron transport is destroyed by scattering. After a certain number of such periods, the amplitude has decayed particularly noticeably and the applied bias has to be reversed in order to repeat

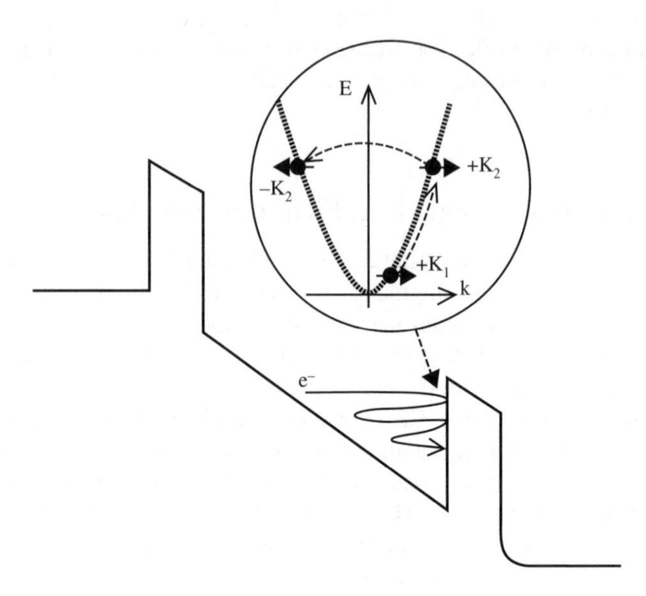

Figure 9.7 Schematic illustration of a quasi-ballistic electron undergoing multiple reflections from the heterojunction interface. In the insert, an electron, k_1, in the conduction band accelerates ballistically to k_2 and then undergoes a reflection at the heterojunction interface and reverses its wavevector to $-k_2$ [44].

this phenomenon at the opposite barrier. This means, such a bias for optimum THz signal generation needs to be switched with repetition frequencies of typically 100 GHz Therefore, such resonance effects in submicron heterojunction wells appear to represent a new way of generation THz signals.

The new material graphene does not have a bandwidth and is semimetal with charge carriers of zero mass. The bandgap can be formed by confining the graphene width in nanoribbon or nanoconstriction structures. For example, the induced bandgap by a 20 nm wide nanoribbon is about 50 meV, while for a nanoconstriction with a 20 nm constriction width this can be about 130 meV. The charge carrier mass then increases, but it is still very small. The energy gap of graphene stripes is similar to that of CNTs, whose value is $E_g(\text{eV}) = 0.8/d(\text{nm})$ ($d = \text{CNT}$ diameter). To obtain a ballistic graphene resonator the structure needs to fabricate consisting of the following structure: narrow width-large width-narrow width (Figure 9.8). It is advantageous not to use monolayer (ML) graphene (where backscattering is possibly problematic) as it would require flipping the pseudospin in contrast to bilayer graphene, where backscattering is however allowed. Multilayer graphene would also be expected to enhance the resulting signal amplitude. Such ballistic resonators has been proposed in Ref. [45].

Amplification and generation of a high frequency electromagnetic wave are the common interest of FEA based devices [46]. The vacuum electronic device based on field emission mechanism for the generation of THz waves has been proposed in Ref. [47]. It is well known that a transition radiation is emitted when an electron passes through an ideally conducting screen in vacuum [48] and the diffraction radiation is emitted when an electron of a constant velocity passes by a metallic structure [49, 50]. The field-emission electrons directly pass through the cathode-anode gap, bunching and synchronizing with the THz wave through the

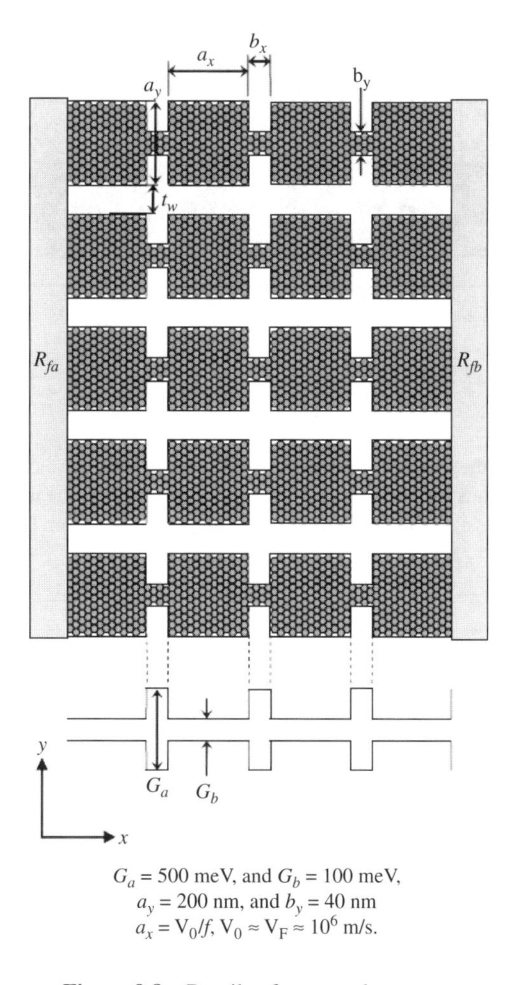

Figure 9.8 Details of proposed structure

well-tailored metallic structures, that is, the oscillating electric field in the y direction coupled through in the x direction. The schematic of the field emission based THz wave generator proposed in Ref. [47] is shown in Figure 9.9. The anode consists of six coupled cavities and the cathode is some kind of FEA. The monitored time evolution of diode voltages for two cases, the field emission and the explosive emission, respectively, is shown in Figure 9.10. The diode voltage corresponding to the field emission starts oscillating earlier than that corresponding to the explosive emission. The average of the data is also shown in the figure.

From the analysis of the simulation results on THz wave generated by the field emission device it was determined that the output power of the THz wave for the field emission case was about 0.074 W/μm^2 and that for the explosive emission case was 0.05 W/μm^2. The estimated values of the corresponding beam power were about 1.53 and 1.29 W/μm^2, respectively. The electronic efficiency of the device, that is, the output power divided by the beam power, can be estimated to be up to 4%.

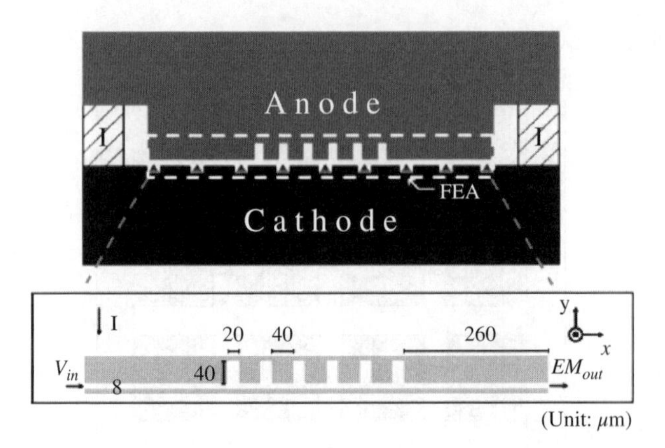

Figure 9.9 Schematic of the field emission based THz generator with the cathode and anode indicated in the figure. Here, FEA stands for "field emission array," and area "*I*" is the insulator. The corresponding MAGIC simulation model is also shown. Reproduced with permission from Ref. [47]. Copyright (2005), American Vacuum Society

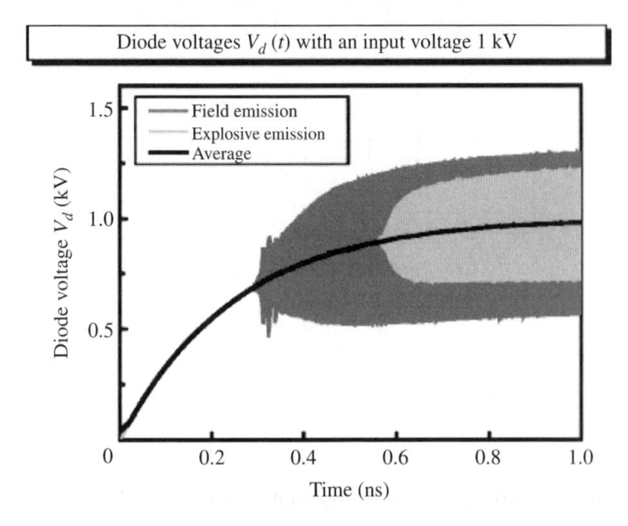

Figure 9.10 Monitored diode voltage $V_d(t)$ curves with an input voltage 1 kV. Reproduced with permission from Ref. [47]. Copyright (2005), American Vacuum Society

The simulations show that there is a resonant interaction of tunneling electrons with a radiation field [51] and this effect may be used to gate field emission current with a laser [52]. These simulations suggest that this mechanism could be used to switch currents in times as short as 10 fs with a pulsed laser and to generate ultrawide band signals from dc to over 100 THz by photomixing [53–55]. The power of the signals generated by resonant laser-assisted field emission to the external load is proportional to the square of the oscillating emitted current. The oscillating current may be increased by (1) increasing the dc current density by using microprotrusions and macro-outgrowths [56] or "supertips" [57] and (2) increasing the optical radiation at the apex by the antenna effect [58] or surface plasmons [59]. By combining these techniques it may be possible to obtain current oscillations of 1 mA or greater.

The heterostructure field emitter of electrons to generate THz signals by laser pulses has been proposed in Ref. [60]. A new THz generator is based on pulsed photo-assisted field emission of electron bunches into vacuum involving space-charge limitation effects. This space-charge limitation of the emitted electrons enables the use of lower cost laser or LED pulses. The applied electric field for electron emission into a vacuum can be kept just below the value for emission to occur. By illumination using optical laser pulses, electrons are then emitted as bunches due to the limitation occurring very rapidly by the space charge of the emitted electrons. Studies of these processes [44] have shown that sufficient electrons are then emitted very rapidly for space-charge limitation to set in. This effect starts after a Femtosecond time scale to limit the emission by space charge of the emitted electron bunch which starts traveling toward the collector. This bunch of electrons generates a THz spectrum.

The laser pulses need to have a pulse duration which can be as long as the main THz period due to the space charge limitation effect, so that the cost of the light source with typically 0.2 ps pulse duration is significantly lower than the Q-switched laser systems with Femtosecond capability employed generally for direct THz generation out of the spectrum of the optical pulses. The time evolution of this space charge limitation effect has been simulated. The case for a particular example is presented by Figure 9.11. The transit length for the electron bunches in vacuum for Figure 9.11 is to be 1 mm.

The semiconductor heterostructure emitter needs to be based on suitable photon energies. As an example the structure shown in Figure 9.12 has been studied. The electron-hole generation is in the InGaAs layer. The holes will be removed via the p^+-GaAs and ohmic contact, the electrons are emitted by the applied electric field in the vacuum by traversing the undoped GaAs overlayer. In order to extract the THz signal the whole structure is designed to operate

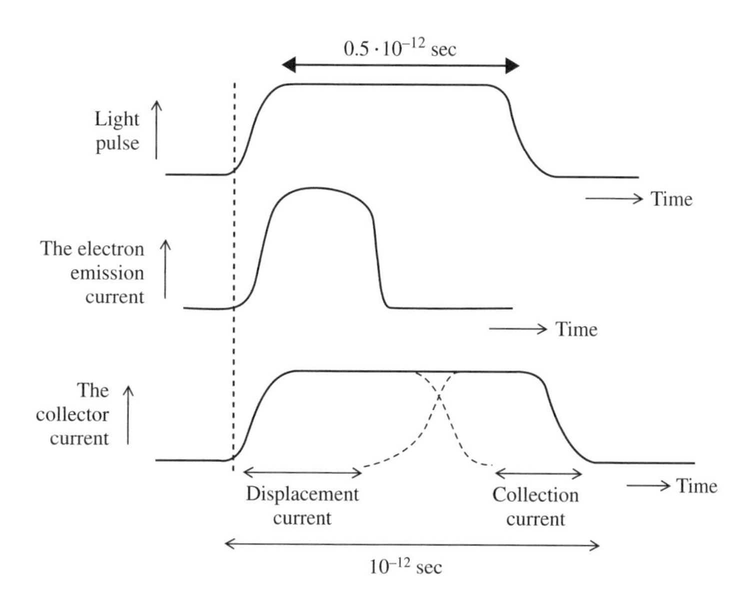

Figure 9.11 Time dependencies. Copyright (2010) IEEE. Reprinted with permission from Ref. [60]

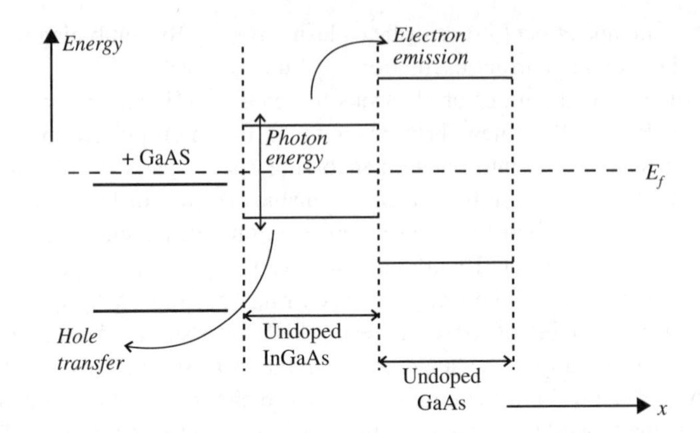

Figure 9.12 Heterostructure emitter for photo-assisted electron emission. Copyright (2010) IEEE. Reprinted with permission from Ref. [60]

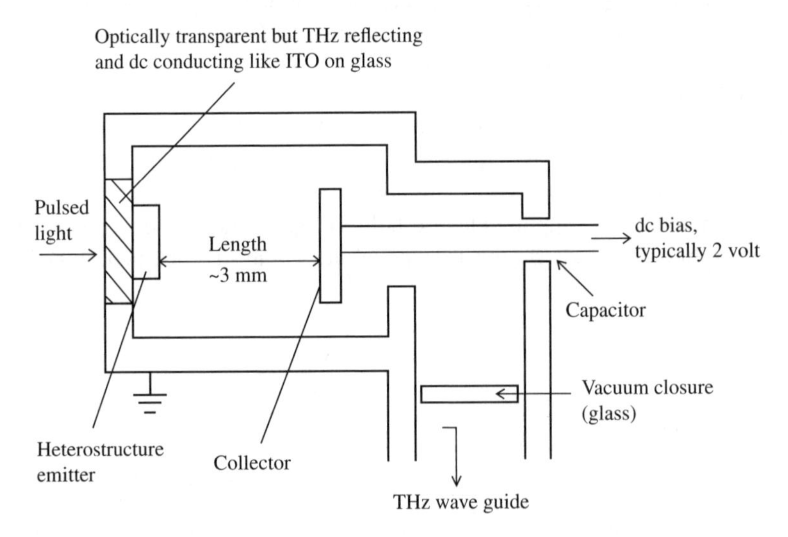

Figure 9.13 Proposed vacuum resonator. Copyright (2010) IEEE. Reprinted with permission from Ref. [60]

inside an evacuated resonator as shown in Figure 9.13. Here the illumination of the InGaAs is generated by a red laser, whose photon energy is active only in the InGaAs and transparent in the GaAs. The light has to be transferred into semiconductor emitter from the back side via an optically transparent, but current carrying and THz reflecting layer such as glass substrate with ITO (Indium Tin Oxide) coating [61]. As laser the Fabry-Perot diode with power level around 20 mW can be used.

Instead of vacuum chamber, the equivalent case of ballistic transport in short semiconductor material was experimentally verified [62] for the output frequency 45 GHz. The transit distance was 144 nm. The experimental system consisted of airbridges connecting pad antenna for waveguide transmission. In this case, no light was used to assist electron bunch transfer but a lower-frequency rf signal of 13 GHz.

CNT cold cathode based THz wave technology was considered in Ref. [63]. A schematic of the proposed novel type of THz radiation source using a CNT cold cathode is illustrated in Figure 9.14. A CNT cold cathode generates a pre-bunched electron beam which can be made by applying a *dc*-biased alternating electron field between the cathode and anode. For the generation of a pre-bunched electron beam, the electron field emission has more advantage over the thermal electron emission.

Because of the inherent high nonlinearity of the electron emission from a CNT cold cathode, a vacuum tube which adopts a CNT cold cathode can be more adequate for the operation scheme using a density modulated electron beam. In principle, a vacuum tube can generate electromagnetic power even with an infinitesimally small current by using the density modulated electron beam. It can be a remarkable merit if the operation frequency is in the THz range. If the velocity modulated operation is employed, in the THz range, a minimum current which is required to start generation of the electromagnetic power generally is easy to exceed a practically available value. A metal grid based on the silicon frame is fabricated by manipulating a silicon wafer. A THz photonic crystal is also fabricated by the similar fabrication method to make a high quality factor resonator.

A novel miniaturized tunable, portable THz source for 0.1–6 THz has been proposed in Ref. [64]. The control voltage with the high frequency is generated using a miniaturized Dynatron oscillator. The oscillator is connected to the anode of the Dynatron triode, which modulates the anode voltage due to self-excited oscillations. The modulated anode voltage is used to control the emission of two miniaturized electron beams, which emit in phase shifted powerful electron pulses in coaxial and opposite directions. As the charges flow in antenna wire, the free flying electron beam pulses emit dipole radiation into a resonator which is adapted in its dimensions to the expected wavelength of the emitted dipole radiation. As in Klystrons up to 47% of the power in the free beams is emitted into the THz radiation. Unlike transmitter antennas the oscillation charges are flying here in free space. This avoids additional power loss by Joules heating of the resistive wire. There is expected a source power of up to 0.1 W for such sources.

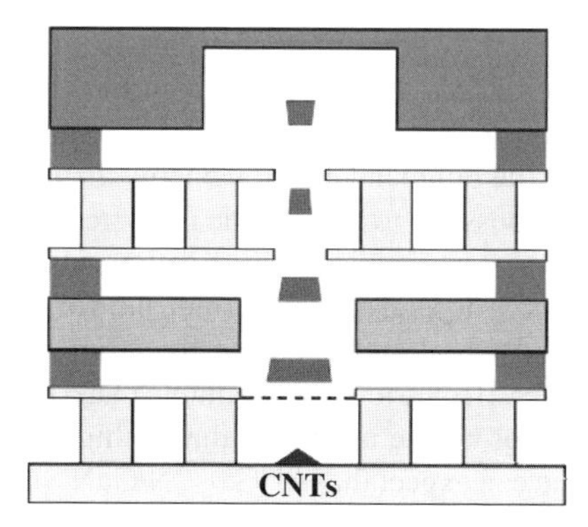

Figure 9.14 Schematic of a proposed THz radiation source using a CNT cold cathode to supply a bunched electron beam. Copyright (2012) IEEE. Reprinted with permission from Ref. [63]

THz technology has become attractive due to the low energy content and nonionizing nature of the signal. This property makes it suitable for imaging and sensing applications [65]. However, the detection and generation of THz signals (300 GHz to 3 THz) is technologically challenging due to the lack of sufficient power sources. In this context, the use of a vacuum THz amplifier, such as a traveling wave tube (TWT) or a klystron could be a possible approach to increase the output power of the existing sources [66].

A critical part of those devices is represented by the electron gun that has to deliver sufficient current in order to allow an amplification of the THz signal. Cold cathodes based guns represent a possible solution, because they can reach high current density, several tens of amps per centimeter square [67] and have many advantages, respect to thermionic cathode, like the room-temperature operation, the possibility of patterning to create the desired beam shape and the reduced dimensions [68].

Micro-electro-mechanical systems (MEMSs) would offer a compact source of submillimeter radiation, but traditional micro-fabrication techniques do not allow the precision and accuracy required to fabricate a device such as a klystron with the required tolerances [69]. The new advances in the MEMS fabrication technique of deep reactive-ion etching (DRIE) have made feasible the possibility of a compact, micro-fabricated submillimeter radiation source. A MEMS-fabricated THz klystron amplifier has been proposed [70]. With the well development of field emission and pseudospark electron sources today [71–73], an electron beam with enough high current density makes realizing a klystron amplifier in the THz regime possible.

9.5 AlGaN/GaN Superlattice for THz Generation

Semiconductor SLs proposed by Esaki-Tsu [74] have been shown to exhibit negative differential conductance (NDC) due to the Bragg reflection of miniband electrons. The NDC of SL induces traveling dipole domains and self-sustained current oscillation.

NDC in semiconductor SLs [74] is at the origin of various proposals for compact submillimeter wave sources. The NDC in dc-biased SLs results in traveling electrical domain formation that has been used in a 147 GHz microwave source made of the InGaAs/GaAs SL [75, 76]. Another type of SL source, the Bloch oscillator [74] is projected to oscillate at the Bloch frequency (terahertz region) and exploits the existence of high-frequency NDC. At the moment, no live example of the Bloch-type source exists because the NDC at zero frequency induces electric-field domains, thus preventing electrons from oscillating at the Bloch frequency [77, 78] Operation of the SL source relies on carrier dynamics specific to a narrow conduction band. High-frequency and zero-frequency NDC may or may not appear simultaneously depending on details of the miniband electron energy dispersion. If the dispersion is not of the simple cosine-type, it is possible to arrange the high-frequency NDC while suppressing the dc instability. This prevents electrical domain formation and allows Bloch oscillations.

The output power of the source depends on the current and voltage swing in the NDC region. Basically, it would be beneficial to high-power device operation if the SL structure was made of a wide band-gap semiconductor. III-Nitrides are a suitable material system for millimeter wave generation due to the benefits of nitride materials, that is, high power, high breakdown voltage, and so on. Therefore, one can expect that $Al_xGa_{1-x}N$/GaN SLs can offer an advantage for generating high power mm-waves. The GaN/AlGaN SL is a possible candidate for high-power submillimeter wave source. GaN-based electronic devices can sustain higher voltage and are less sensitive to the high dislocation density as compared to narrow-gap GaAs–InAs-based devices. $Al_xGa_{1-x}N/GaN$ SLs have in fact been predicted to have the potential for generating THz-range signals [79]. Short-period AlGaN/GaN SLs should provide harmonic electron oscillations at multiples of the fundamental Bloch frequency.

Let's consider the properties of an AlGaN/GaN SL relevant to microwave source feasibility [80]. The wurtzite (0001)AlGaN/GaN SL is the intrinsic Stark SL where the polarization fields shift the energy levels of confined electrons. Proper design of a GaN/AlGaN SL source should account for intrinsic electric fields. Polarization fields in a SL stem from the spontaneous polarization in the bulk and lattice-mismatch-induced piezoelectric component.

The dynamic properties of the SL depend on the conduction-band profile as determined by the conduction band offset. We assume that the band gap in the relaxed $Al_xGa_{1-x}N$ alloy layer is described as follows:

$$E_g^0(x) = xE_g^{AlN} + (1-x)E_g^{GaN} - bx(1-x), \tag{9.23}$$

where $E_g{}^{AlN} = 6.2$ eV, $E_g{}^{GaN} = 3.42$ eV, and the bowing parameter $b = 1$ eV, x is the Al composition of the alloy. We use linear interpolations to calculate the electron effective mass m, (in units of free electron mass) and the lattice parameter a in the basal plane of the alloy:

$$m(x) = xm_{AlN} + (1-x)m_{GaN}; \tag{9.24}$$

$m_{AlN} = 0.27$, $m_{GaN} = 0.2$,

$$a(x) = xa_{AlN} + (1-x)a_{GaN}; \tag{9.25}$$

$a_{AlN} = 3.11$ Å, $a_{GaN} = 3.19$ Å.

In-plane tensile strain in thin (pseudomorphic) AlGaN barriers is given as

$$\xi(x) = \frac{a_{GaN} - a(x)}{a(x)}. \tag{9.26}$$

The band edge positions in the AlGaN layers are given below with the reference energy taken at the valence-band edge of the relaxed layer. The position of the conduction *C(x)* and valence *V(x)* bands are given as

$$C(x) = E_g(x) + 3A_c\xi(x), \tag{9.27}$$

$$V(x) = 2\xi(x)\left[(D_2 + D_4) - (D_1 + D_3)\frac{C_{13}(x)}{C_{33}(x)}\right], \tag{9.28}$$

$$C_{13}(x) = (5x + 103);\ C_{33}(x) = (-32x + 405), \tag{9.29}$$

where the deformation potentials are given in electron volts and the elastic constants in gigapascals [81] $A_c = -4.6$, $D_1 = -1.7$, $D_2 = 6.3$, $D_3 = 8$, and $D_4 = -4$, respectively. We assume that deformation potentials are independent of the Al content x.

Using Equations (9.23)–(9.29), the strained layer band gap, that is, the energy difference between conduction and heavy hole band edges, is given by

$$E_g(x) = C(x) - V(x) = 3.42 + 1.39x + x^2. \tag{9.30}$$

The natural valence-band offset ΔE_v between AlN and GaN is 0.8 eV [82]. Assuming linear interpolation between AlGaN and GaN binary ends, $\Delta E_v(x) = 0.8x$eV, the conduction-band offset is found as follows:

$$\Delta E_c(x) = E_g(x) - E_g^{GaN} - \Delta E_v(x) = 0.603x + 0.99x^2. \tag{9.31}$$

Built-in electric fields in a SL, caused by spontaneous and piezoelectric polarizations, can be calculated using the following equations:

$$F_{well} = \frac{P_{total} d_b}{\varepsilon_0[\varepsilon(x)d_w + \varepsilon(0)d_b]}, \tag{9.32}$$

$$F_{barrier} = -\frac{P_{total} d_w}{\varepsilon_0[\varepsilon(x)d_w + \varepsilon(0)d_b]}, \tag{9.33}$$

where the polarization is given as

$$P_{total}(x) = P_{sp}(0) - P_{sp}(x) - 2\xi(x)\left[e_{31} + e_{33}\frac{C_{13}(x)}{C_{33}(x)}\right], \tag{9.34}$$

where $P_{sp}(x) = (-0.052x - 0.029)$ C/m^2 is the spontaneous polarization in $Al_xGa_{1-x}N$ alloy; $e_{31} = 0.3$, $e_{33} = 1$ are GaN piezoelectric constants (C/m^2), $\varepsilon(x)$ is the dielectric constant of AlGaN, d_b, d_w are thicknesses of barrier and well, respectively, and $d = d_w + d_b$ is the SL period. The resulting conduction band profile including polarization effects is shown in Figure 9.15.

The analytically obtained profile from Equations (9.31)–(9.33), shown in Figure 9.15, was confirmed by numerical calculations with help of the Poisson–Schrödinger solver. Solver-based calculations account for polarization fields by including delta-doped regions that simulate polarization charges at the interfaces.

The first miniband width in InGaN/AlGaN/GaN SLs calculated in Ref. [81] does not account for polarization fields. Here, the miniband energy dispersion in the polarization-distorted band

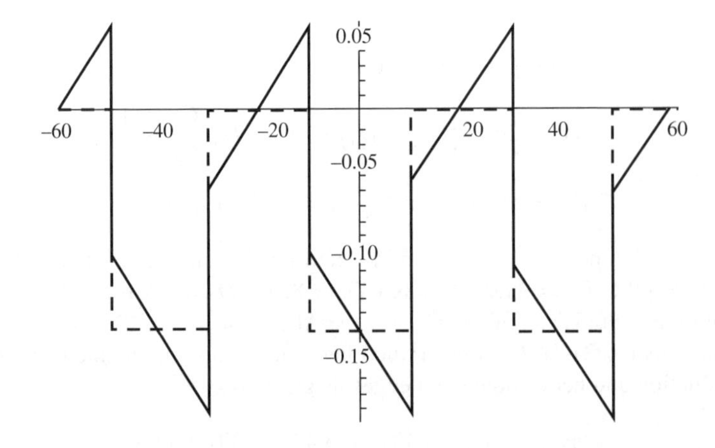

Figure 9.15 Conduction-band profile in AlGaN/GaN SL (eV) in growth direction (Å): solid line – polarization fields included and dashed line – flat-band approximation; $x = 0.18$ and $d_b = d_w = 20$ Å. Reproduced with permission from Ref. [80]. Copyright (2004), AIP Publishing LLC

profile was calculated within the tight-binding approximation [80]:

$$\varepsilon(k) - \varepsilon_0 = \frac{\sum_n h(n)\exp(iknd)}{\sum_n I(n)\exp(iknd)}, \quad (9.35)$$

where

$$h(n) = \int \varphi^*(z)h(z)\varphi(z - nd)dz; \quad (9.36)$$

$$I(n) = \int \varphi^*(z)\varphi(z - nd)dz, \quad (9.37)$$

where $h(n)$ is the hopping integral and $\varphi(z)$ is the wave function of a single SL period shown in Figure 9.16. The wave function in Figure 9.16 was calculated by solving Schrödinger equation with the potential also shown in Figure 9.16. It should be noted that in the range of compositions ($0.18 < x < 0.4$) and SL periods $d < 50$ Å, the calculation of the miniband energy dispersion requires more than first-nearest-neighbor terms in the dispersion law [80]. The calculated miniband width is shown in Figure 9.17.

Theoretical investigations showed that the traveling dipole domain oscillation frequency was found to be in the THz range (Figure 9.18). The short-period condition was satisfied with an aluminum composition x between 18 and 40% and a value for the sum of the barrier and QW thickness d of less than 50 Å. Based on the above mentioned theoretical expectations, $Al_xG_{1-x}N$/GaN SL layers complying with the short-period conditions were designed and grown in house by metal organic chemical vapor deposition (MOCVD) on sapphire substrates. The SL consisted of 50 periods of AlGaN/GaN layers with 15 Å thick AlGaN barrier layer with 34% aluminum composition and a 15 Å thick GaN layer. The SL was embedded between GaN buffer layers with Si-doping (4×10^{18} cm^{-3}), which served for contacting through ohmic contacts. Growth of SL layers was optimized in terms of interface roughness

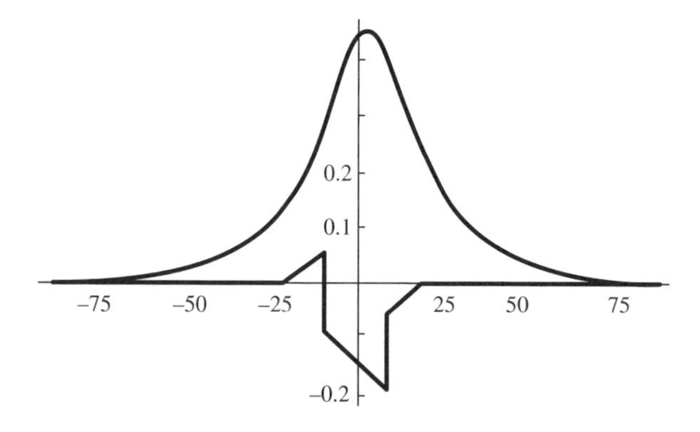

Figure 9.16 The single-period potential (eV) and electron wave function (a.u.) as functions of a distance from the well (Å); $x = 0.18$ and $d_w = 20$ Å. Reproduced with permission from Ref. [80]. Copyright (2004), AIP Publishing LLC

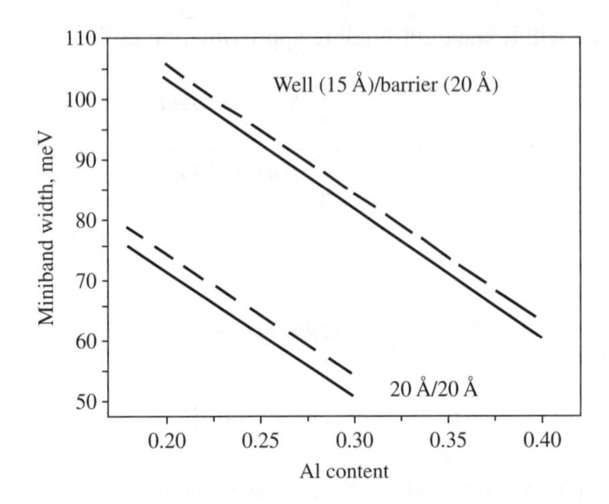

Figure 9.17 First miniband width in AlGaN/GaN SL: dashed line – flat-band approximation and solid line – polarization fields included. Reproduced with permission from Ref. [80]. Copyright (2004), AIP Publishing LLC

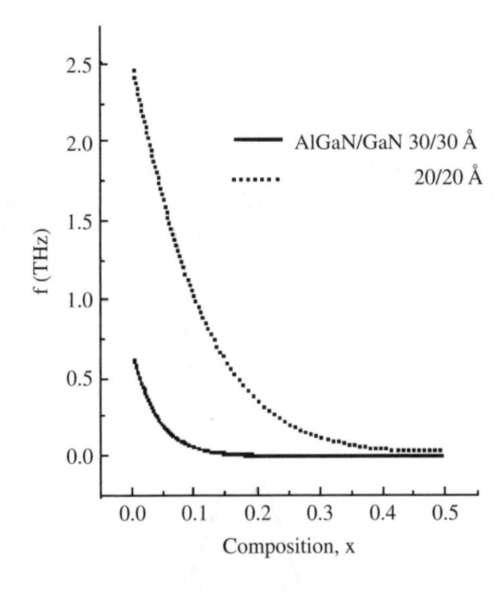

Figure 9.18 Calculated oscillation frequency in AlGaN/GaN SL diodes

and layer thickness control. Good uniformity of layer thickness and interface roughness is observed in both directions of the wafer and the interface roughness is between 0.5 and 1 nm.

Figure 9.19 shows the cross-sectional and top view of the diode designs. Photographs of the fabricated devices are shown in Figure 9.20. Mesa etching of 750–800 μm was achieved by argon plasma reactive ion etching. Then, Ti/Al/Au (600 Å/1800 Å/2500 Å) ohmic contact metallization were deposited by electron beam evaporation. The samples were annealed at 550 °C for 30 seconds and 850 °C for 30 seconds in N_2 atmosphere. Transmission line measurement (TLM) measurements showed an inner ohmic contact

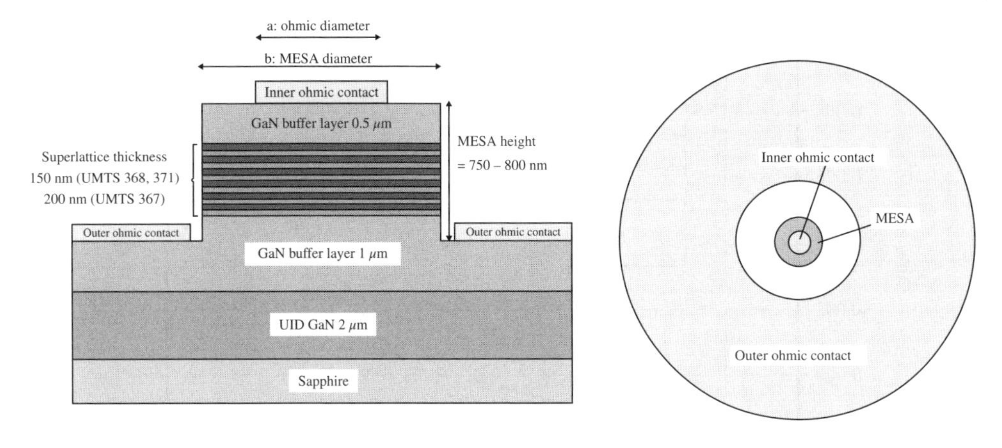

Figure 9.19 Cross-sectional and top view of the AlGaN/GaN superlattice diodes

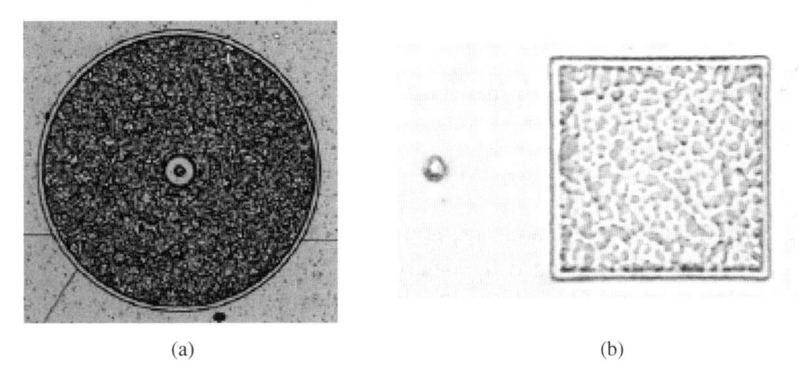

Figure 9.20 Photographs of fabricated AlGaN/GaN superlattice diodes, (a) circular and (b) square designs

resistance $R_c = 0.14\,\Omega$-mm, sheet resistance $R_s = 50\,\Omega$/sq, and specific contact resistance $R_{sc} = 0.42 \times 10^{-5}\,\Omega{\cdot}\text{cm}^2$. The current-voltage characteristics of the SL diodes were measured using a Keithley 2602 source meter at room temperature [83]. All three design of SL were characterized but the following results are referred only to design ampere, since this presented the most profound NDR effects. The voltage was swept from −5 to 5 V with 50 mV step size in the continuous mode. The anode and cathode ohmic contacts were probe contacted with a tip of 10 μm diameter. Figure 9.21 shows the measurement results for the 30 μm diameter diode. Measurements IV#1–7 were done consecutively with a time lapse of 1 minutes. A NDR region was seen from 3 to around 3.9 V and the NDR current decreased during consecutive measurements. The current–voltage measurement results suggest room temperature NDR presence with a peak-to-valley ratio (PVR) of 1.3 and a relative broad voltage range, which could be of interest for practical implementations. DC measurements were also done for 20 μm diameter SL diodes with opposite sweep polarity (Figure 9.22a). However, in this case, no NDR was observed. A similar complex behavior was also observed in GaN/AlN

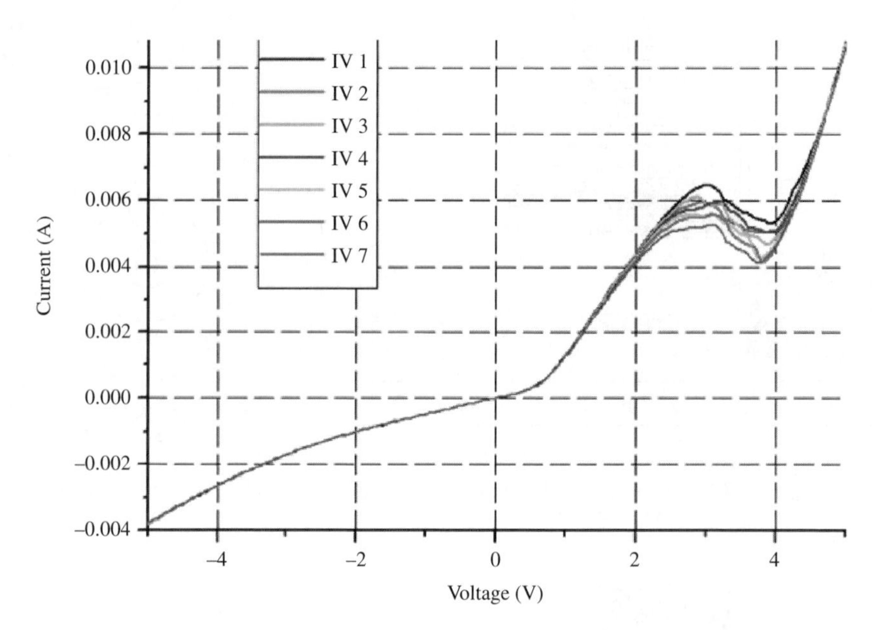

Figure 9.21 Current–voltage characteristics of a 30 μm diameter superlattice diode upon consecutive measurements

RTDs [84, 85]. The behavior of the peaks in the I-V characteristics depend on the previous charge state of the device produced by electrical bias. They suggested that the effects of polarization fields in wurtzite nitride structures and the presence of defects existing in the GaN well and the AlGaN/GaN interface induce the observed complex current behavior. It was shown by measuring the capacitance as a function of frequency, that the electron trapping and de-trapping effects by traps located in the AlN/GaN interfaces are responsible for the observed behavior. A similar explanation may be applied to the AlGaN/GaN SL structures of this work which include traps at the interface.

In order to verify the presence of the above described trapping effect in defects, capacitance-voltage measurements were performed on the same device [86]. As can be seen in Figure 9.22b, the SL diode displayed asymmetric C-V characteristics with negative capacitance features. The capacitance varied from about −22 pF at −5 V to slightly 0 pF as the voltage increased to 0 V and then decreased slowly to −5 pF when the device was biased with increased positive voltage. Negative capacitance effects have been shown in various semiconductor devices, such as p-n junctions, Schottky diode, and GaAs/AlGaAs QW infrared photodetectors [87]. In general, the negative capacitance effect is due to the nonradiative recombination of injected carriers into trap levels [88, 89], or due to the capture-emission of injected carriers between multilevels [87, 90]. For further study of the negative capacitance effect, another 20 μm diameter SL diode was tested (Figure 9.23). A frequency dependence was observed, which tends to disappear (flattening of the curves) when the frequency is increased. It appears that the observed effects are related to time dependent parameters that occur at low frequencies and as the frequency moves from hertz to kilohertz and megahertz they tend to disappear since they can no longer respond to the input signal. The

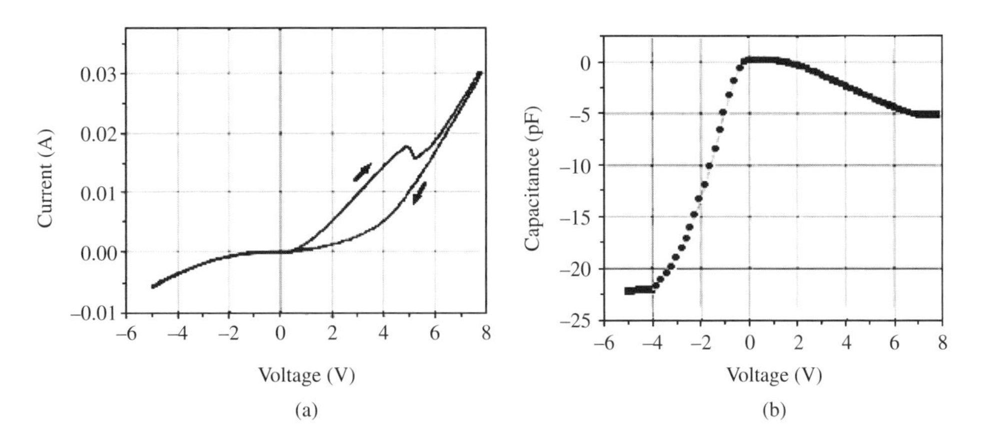

Figure 9.22 (a) Current–voltage characteristics of a 20 μm diameter SL diode with up sweep (solid line) and down sweep (dashed line). (b) Capacitance–voltage measurements (at 1 MHz, room temperature) on the same device with up sweep

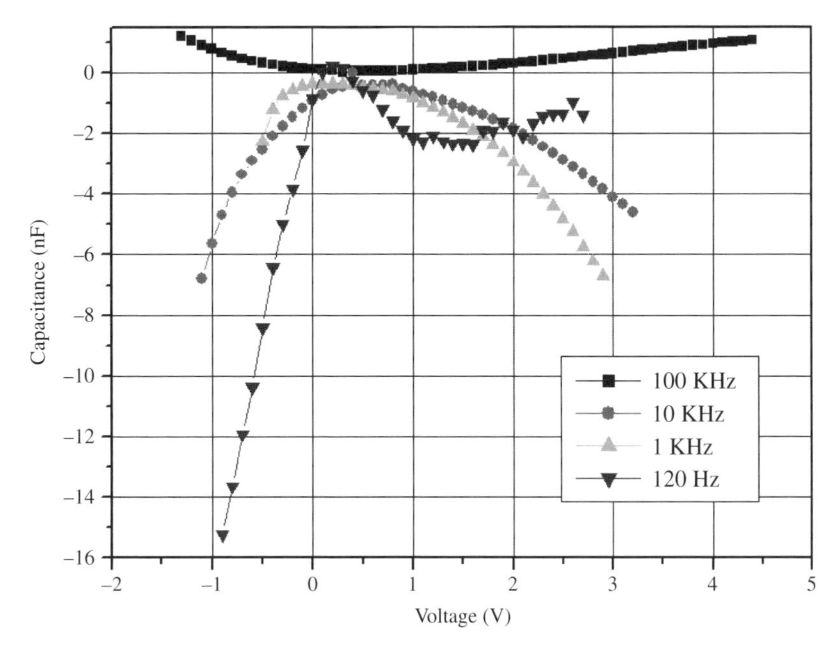

Figure 9.23 C-V characteristics at room temperature of a 20 μm diameter superlattice diode with variable frequency of measurement

observed properties set up the first steps for further fundamental investigations of GaN-based SLs in view of their design and optimization for high frequency applications.

Resonant tunneling (RT) is a major mechanism in many modern electronic but also optical devices such as Quantum Cascade Lasers (QCLs). Figure 9.24 shows the RT mechanism in a QCL by means of a band diagram [91].

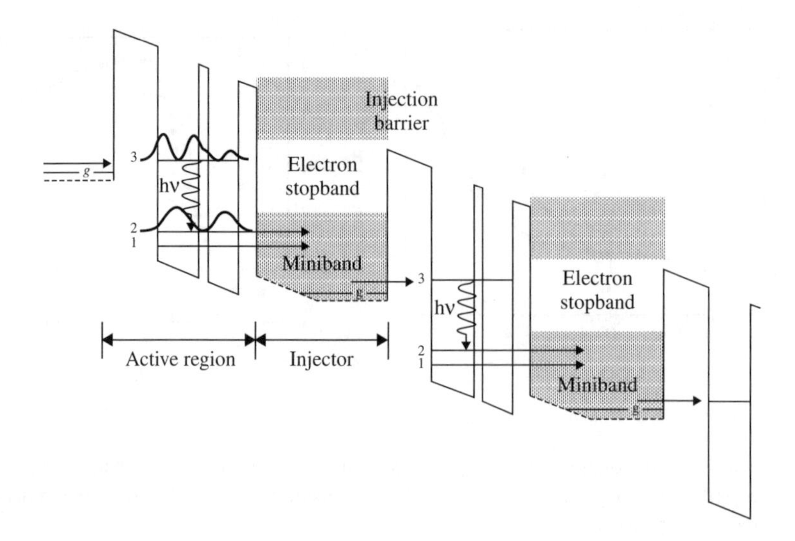

Figure 9.24 Schematic energy diagram of two periods of a QCL with vertical transition showing also the moduli squared of the wavefunctions of the laser transition and the relevant energy levels. The heterostructure is grown using the $Ga_{0:47}In_{0:53}As$–$Al_{0:48}In_{0:52}As$ material system. Copyright (1998) IEEE. Reprinted with permission from Ref. [91]

The QCL operation is based on a vertical transition and the design is made such that the active region with only two coupled QW's and a strong spatial overlap between the wavefunctions of the laser transition (levels 3 and 2), the oscillator strength of which should be maximized. RT between states of the same transverse momentum near the bottom of subbands and near the bottom of subbands g and $n = 3$ takes place upon application of the appropriate bias so that the ground state of the injector becomes degenerate with the excited state of the active region.

The presence of RT in diodes is manifested by measuring the I-V characteristics with a voltage source that has an internal impedance smaller than the magnitude of the NDR. Experiments in $Ga_{0:47}In_{0:53}As$-$Al_{0:48}In_{0:52}As$ QCLs did not allow NDR observation due their control circuitry which acts as a current generator with a very large internal impedance that is greater than $|dV/dI|$. Moreover, NDR in QCLs structures is not expected to be large due to the presence of other subbands at lower energies, such as, for example, $n - 1$, and $n = 2$, that form a parallel channel for scattering of electrons in the ground state of the injector.

The large up to 2.1 eV conduction-band offsets of GaN/Al(Ga)N QWs allow extension of the intersubband transition wavelength to the near-infrared spectral region, while their large longitudinal optical phonon energies (90 meV) allow the development of terahertz sources. QCLs in the near-infrared (1.5–3 μm) and far-infrared (25–70 μm) ranges can take advantage of the properties of this material system.

Tunneling transport in nitride QWs, is limited due to the relatively large densities of dislocations and other structural defects in such wide bandgap semiconductors [92]. Defects may act as strong scattering centers, as well as parallel current paths partially shorting out the QWs in high-resistance vertical-transport devices. As indicated earlier, although NDR-like effects have been observed when measuring the I-V characteristics of RT diodes, these effects become weaker and eventually diminish with time due to filling of tap states. Designs have been explored in Ref. [92] where electrons traverse thick periodic structures based on triple-QW repeat units, similar in principle to typical QCL active regions. The observed characteristics corresponding to vertical transport in such structures were stable and nonlinear corresponding to scattering-assisted sequential tunneling. Interface roughness was thought to be the likely dominant scattering mechanism. Designs explored for these studies included RF plasma-assisted Molecular Beam Epitaxy (MBE) grown 20 repetitions of $GaN/Al_{0.15}Ga_{0.85}N$ triple-QW periods on free-standing GaN substrates. The individual layers in each repeat unit had nominal thickness of the following number in MLs: 12 (barrier), 13 (nominally intrinsic doping (NID) well), 7 (barrier), 9 (NID well), 9 (barrier), and 15 (5×10^{10} cm^{-2} n-doped (Si) well). Bottom and top contact layers were highly doped to a degenerate level using Si to ensure good electrical conductivity. The conduction band lineups of two repeat units are shown in Figure 9.25, under different bias conditions. The horizontal axes in these plots indicate position along the growth direction, increasing toward the cap layer from left to right. The squared envelope functions of the relevant subbands are also shown referenced to their respective energy levels. The operation and observed characteristics are described below.

At 0 V (Figure 9.25a), the electrons supplied by the Si donors in each repeat unit mostly reside in the ground-state subband of the widest well, that is, the 15-ML thick well located to the right of each period. These subbands are energetically separated considerably (40 and 65 meV for the wells immediately above and immediately below, respectively) from the ground states of the neighboring wells. Therefore, the entire structure is in a state of relatively high resistance, especially at low temperatures. As the bias is increased in either direction, the populated subbands become energetically aligned with the ground states of the adjacent wells downstream and vertical transport starts taking place by sequential tunneling (either resonant or scattering-assisted). This leads in its turn to a reduction of the differential resistance.

It should be noted that due to the asymmetric nature of this multiple-QW structure, the voltage drop required to achieve the subband alignment necessary for efficient electronic transport is different depending on whether it is applied from the top or bottom contact. For example, the ground-state subband of each 15-ML well is brought into alignment with that of its overlaying well by an external electric field F of 78 kV/cm, corresponding to a voltage V of about 2.6 V across the 20 repeat units (Figure 9.25b), while a larger electric field of 109 kV/cm corresponding to a voltage of 3.7 V is required to achieve ground-state subband alignment between each 15-ML well and its adjacent well below (Figure 9.25c). As a result, the expected transition to a low-resistance state should occur at a larger value of the applied voltage in the case of "negative" or "reverse" bias.

The measured current–voltage I-V characteristics of the above described device are shown in Figure 9.26a in the presence of positive (i.e., top to bottom) and negative (Figure 9.26b) voltage pulses.

Different heat-sink temperatures (20, 100, 200, and 280 K) were used corresponding to an increased value of current for given voltage in order of increasing current for fixed voltage. The device low-temperature differential resistance $R_{diff} = dV/dI$, obtained from the measured I-V traces at 20 K, is plotted versus voltage in Figure 9.26c, where the solid and dotted lines correspond to negative and positive bias, respectively. The measured I-Vs are nonlinear and indicate a change from high resistance near 0 V to lower resistance for higher voltage values as expected from simulations. The forward and reverse turn-on voltages are also shown in Figure 9.26c by vertical arrows. The temperature dependence of the measured I-Vs is consistent with tunneling

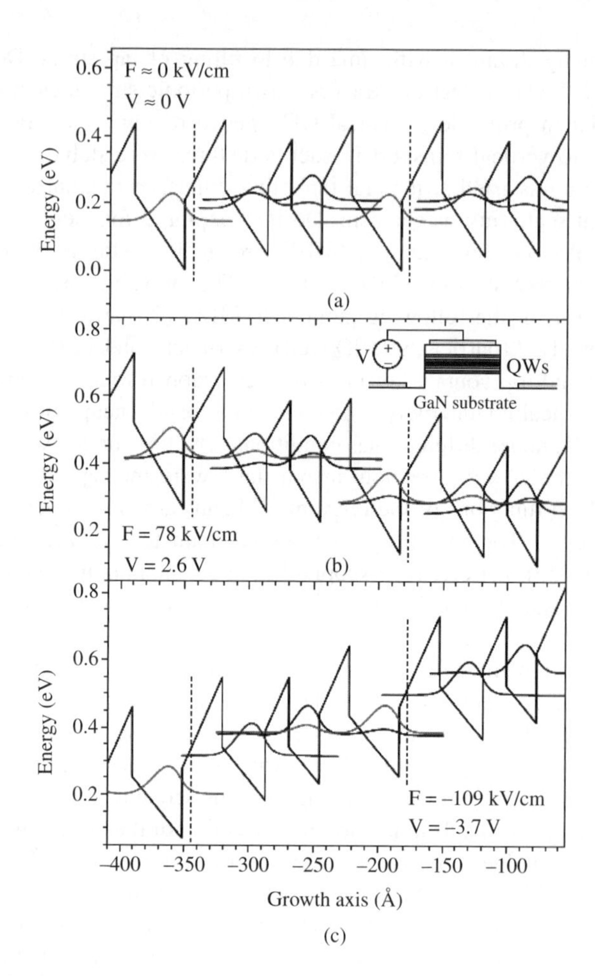

Figure 9.25 Conduction-band lineup of two repeat units and squared envelope functions of the ground-state subbands, under different bias conditions: (a) near zero applied voltage V; (b) for $V = 2.6$ V, as required for efficient tunneling transport in the forward direction; and (c) for $V = -3.7$ V, as required for efficient tunneling transport in the reverse direction. The vertical dashed lines in each plot indicate the boundaries between adjacent repeat units. The inset of Figure 9.25b shows a schematic cross-sectional view of a processed device, and defines the voltage polarity convention used in this work. Reproduced with permission from Ref. [92]. Copyright (2010), AIP Publishing LLC

transport. As the temperature is increased the current turn-on becomes more gradual and shifts to lower values of the applied bias, due to the thermal broadening of the electronic distribution in the initially occupied subbands.

The above studies demonstrate the feasibility of vertical transport by efficient sequential tunneling in complex GaN/AlGaN multiple-QW structures. The possibility of some parallel current paths due to dislocations and other structural defects cannot at this stage be excluded. It should also be noted that no NDR is observed in the obtained results. One notes that the measured differential resistances remain relatively constant as the bias is increased above the

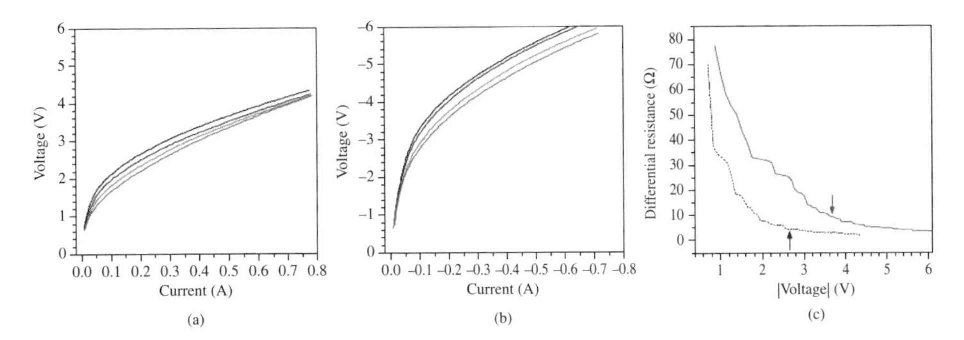

Figure 9.26 Measured I-V characteristics at different heat-sink temperatures, using (a) positive and (b) negative voltage pulses. The four curves in each plot correspond to a temperature of 20, 100, 200, and 280 K in order of increasing current for fixed voltage. (c) Differential resistance obtained from the measured I-V curves at 20 K under reverse solid line and forward dotted line bias, plotted as a function of voltage. The arrow next to each trace denotes the corresponding theoretical turn-on voltage. Reproduced with permission from Ref. [92]. Copyright (2010), AIP Publishing LLC

turn-on value, indicating that scattering plays a major role in the measured transport properties. Thus, efficient elastic tunneling can occur even when the initial subband is at higher energy than the final one, with the required change in plane wave vector provided by the scattering mechanism involved. A likely candidate is scattering by interface roughness, which features a weak dependence on intersubband energy separation over a wide range [93]. Far-infrared light emission was also observed indicating the possibility of photon-assisted tunneling, although the measured spectra are too weak and broad for a conclusive identification of the emission mechanism.

In addition to the sequential tunneling characteristics in multi-QWs described above, results with some reproducibility of NDR have reported on c-plane GaN substrates under special (negative) pre-biasing [94]. However, the majority of repeatable NDR characteristics have been obtained primarily from nonpolar nitride RTDs. These include cubic heterostructures [95], wurtzite nitrides on nonpolar (m-plane) GaN substrates [96], and nitride nanowires [97].

It was shown clear presence of NDR at low-temperature from low Al-composition double-barrier AlGaN/GaN heterostructures grown by plasma-assisted MBE on high-quality free-standing GaN substrates with low (5×10^6 cm^{-2}) dislocation density and thus smaller likelihood of impact of dislocations on vertical transport [98]. Moreover, the Al-composition in the barrier layer was limited to 18% to minimize effects such as relaxation of the heavily-strained AlGaN barrier layers during MBE or subsequent device processing. The low Al-composition in the barrier also limits polarization discontinuities and reduces the possibility of electrical breakdown during operation by reducing the built-in electric fields; the electric field across a barrier depends not only on applied bias but also on internal polarization fields. Details of the layer thicknesses, doping, and alloy composition of the studied RTD structures are shown in Figure 9.27 together with Atomic Force Microscopy (AFM) results from the as-grown RTD surface.

The layers were grown by MBE under Ga-rich conditions at 745 °C. A Si-doped, 400 nm-thick buffer layer served as bottom contact layer, while the active region was composed of a $Al_{0.18}Ga_{0.82}N$/GaN/$Al_{0.18}Ga_{0.82}N$ QW structure and is sandwiched between

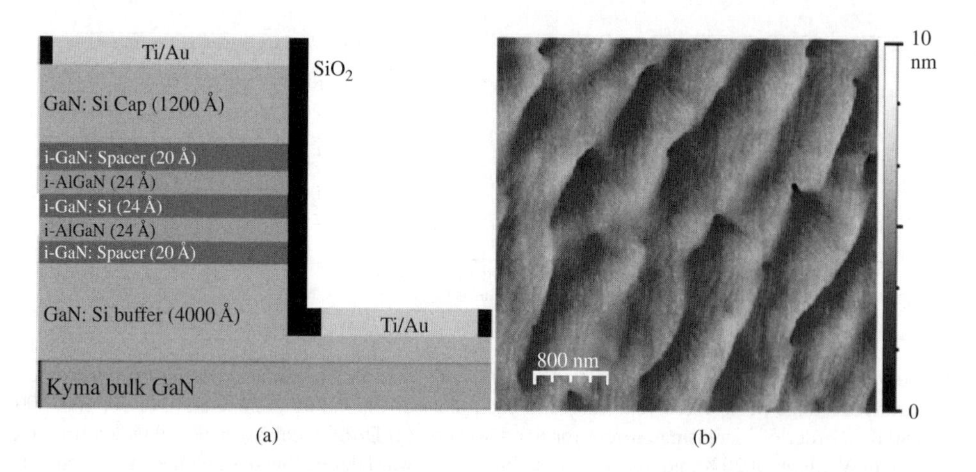

Figure 9.27 (a) A schematic view of the designed RTD structure and (b) AFM image of as-grown RTD surface. The $4 \times 4\,\mu m^2$ region has a RMS roughness of 8 Å. Reproduced with permission from Ref. [98]. Copyright (2012), AIP Publishing LLC

undoped 20 Å GaN spacer layers. The whole structure was capped with a 1200 Å GaN layer with silicon doping at a level of $1 \times 10^{19}\,cm^{-3}$ that also serves as the top contact layer. Devices fabricated on these layers had a minimum size of $4\,\mu m \times 4\,\mu m$, were mounted on Cu heat sinks and wire-bonded to Au pads.

Figure 9.28a shows typical I-V characteristics obtained from $4\,\mu m \times 4\,\mu m$ devices. Small but clear NDR signatures were observed around 1.5 and 1.7 V at 77 K with peak current density of 122 kA/cm^2. It is important to note that the I-Vs are identical on the voltage ramp-up and ramp-down. No significant degradation of the device performance was observed after 20 I-V measurements, in contrast to most previous reports on c-plane GaN. The PVR remained the same after 20 measurements, but the position of the resonances appears to shift to slightly higher voltages after reverse bias to −3 V (shown in Figure 9.28a). This shift of about 0.05 V may be related to charge redistribution along interfaces, possibly even to charge trapping on defects. Figure 9.28b gives a typical I-V curve under reverse bias scanned from 0 to −2 V. A small feature is also visible in reverse bias around 0.75 V. We note that at least one NDR feature is expected in reverse bias, but the reverse current is expected to be smaller due to an effectively thicker potential barrier. The I-V characteristics do not change significantly if the temperature is lowered to 4 K, and the two NDR features are distinguishable up to 110 K. No NDR features were observed from larger size devise. This could be possibly explained by the current leakage through screw dislocations. Since the number of the conductive dislocations is proportional to the area of the mesa, the increase of leakage current with the device area eventually obscures the NDR [99].

Figure 9.29 shows the conduction band diagram in the active region of the RTDs at zero bias as calculated with the self-consistent solution of Schroedinger/Poisson equations [97].

One observes two quantized states at $E_1 = 0.204$ eV and $E_2 = 0.480$ eV. The voltage difference between the two observed NDR features was found to be consistent with the energy difference between the quantized calculated energies (~0.2 eV). Resonant tunneling occurs from the lowest quantum level in the triangular well on the left to the quantum state in the

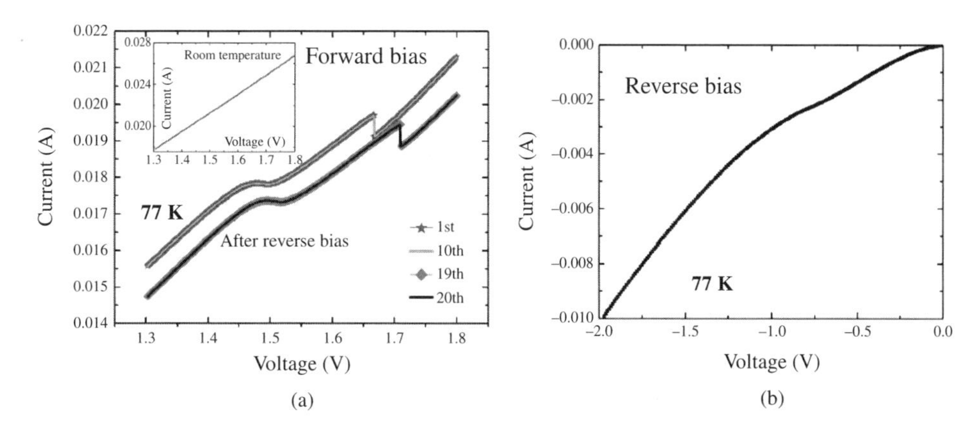

Figure 9.28 I-V characteristics of an RTD at 77 K by sweeping the bias on the mesa top while recording current through mesa. (a) Voltage was scanned from 1.3 to 1.8 V for 20 times and the data is shown for the 1st, 10th, 19th, and 20th measurements. Note that the I-Vs are identical on the voltage ramp-up and ramp-down. The device was reversely biased after the 18th measurement as shown in (b). The inset of (a) shows the room temperature I-V curve for the same device. (b) I-V for reverse bias scanned from 0 to −2V applied to the same device. Reproduced with permission from Ref. [98]. Copyright (2012), AIP Publishing LLC

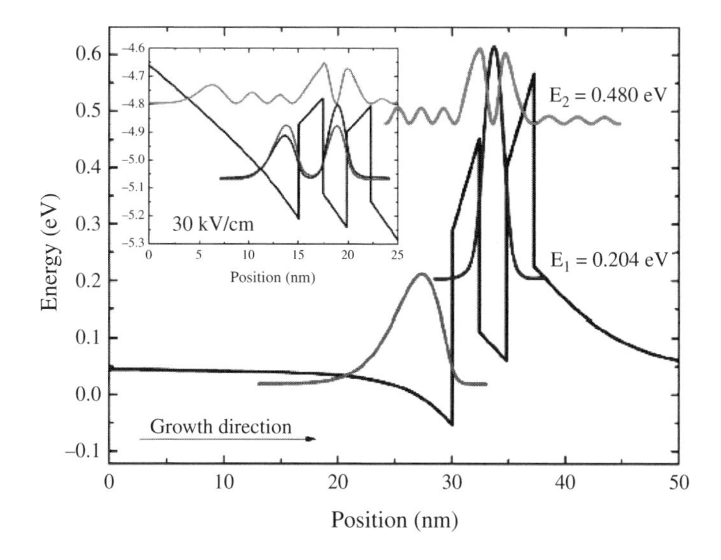

Figure 9.29 Schematic conduction band diagram of the active region at zero bias. The eigenvalues of the confined wave functions in the quantum well are $E_1 = 0.204$ eV and $E_2 = 0.480$ eV at 77 K. The inset shows the structure under a uniform field of 30 kV/cm. Reproduced with permission from Ref. [98]. Copyright (2012), AIP Publishing LLC

well, and then to the continuum (inset of Figure 9.29). Due to the close spacing of the levels in the triangular well at low bias, tunneling into the lowest quantized state should be broadened and as a result the NDR signature should be weak, consistent with the obtained data. At higher bias the quantized levels in the triangular well become better separated and hence the NDR feature is expected to get sharper. Under higher bias, resonant tunneling occurs from the lowest level of the triangular well directly into this quasi-bound state across a single AlGaN barrier (the second barrier is below E_2).

Characterization of the above devices as a function of temperature revealed that the PVR of NDR increases when cooling down the device [99]. The highest PVR at 6 K, was 1.05 at a current density of 383 kA cm^{-2}.This relatively modest value of maximum PVR is likely to be due to large parallel leakage though defects, most likely threading dislocation originating from the substrates. The successive increase of the PVR with decreasing temperatures is consistent with theoretical predictions of resonant tunneling in nitrides and measurements in other material systems. Theoretical considerations suggest that the temperature-dependence of the coherent tunneling is due to increased broadening of the supply function with increasing temperature that enhances the nonresonant component of the tunneling current [100]. The PVR diminishes because the valley current increases faster with temperature than the peak current. Moreover, the temperature effect on tunneling is enhanced by inelastic scattering. Inelastic scattering, mainly phonon and interface scattering, further reduces the peak amplitude and broadens the resonant transmission probability through the double-barrier structure as compared to the coherent case. Overall, it appears that the observed emergence of NDR at low temperature is consistent with the fact that resonant tunneling is present and that charge trapping has no significant effect on the studied RTDs.

Studies of *n*-GaN/AlGaN/n-GaN heterostructures point out that the vertical leakage current is high and associated with percolation due to alloy fluctuations in the ternary AlGaN barrier. By eliminating the random alloy barriers one should therefore be in a position to better control vertical transport [101]. To verify these expectations, the authors studied MBE grown structures consisting of a 15–20 nm degenerately doped n^+-GaN top contact layer, a 30–35 nm thick $Al_xGa_{1-x}N$ barrier on top of a n^+-GaN template, on which samples are grown serves as the bottom *n*-contact as shown in Figure 9.30a.

A series of three samples with Al-composition of the AlGaN barrier equal to 16, 27, and 37% in the AlGaN barrier and a thickness of 30 nm were investigated for this purpose. The simulations in Figure 9.30b shows that a AlGaN barrier with 16% Al leads to a heterojunction barrier of more than 0.5 eV from the Fermi level at equilibrium. The current densities measured experimentally were found to be orders of magnitude higher than the theoretically predicted values that were estimated to correspond to negligible leakage up to a reverse bias of 3 V (Figure 9.30c). Moreover, the leakage was not found to be influenced by the barrier height. Tunneling or thermionic emission are not therefore the mechanisms responsible for current transport since they are theoretically expected to vary by orders of magnitude with such changes in the barrier height. Moreover, the temperature dependent tests of Figure 9.30d did not indicate significant variation of the characteristics and thus leakage in such heterostructures does not appear to be related to thermal activation such as trap-assisted tunneling or thermionic emission over the barrier.

Studies of the I-V characteristics of GaN/AlGaN/GaN structures made on low (10^5 cm^{-2}) and high (10^8 cm^{-2}) screw type Threading Dislocation Density (TDD) GaN templates, showed no major difference in vertical current transport, suggesting therefore that TDDs are not the

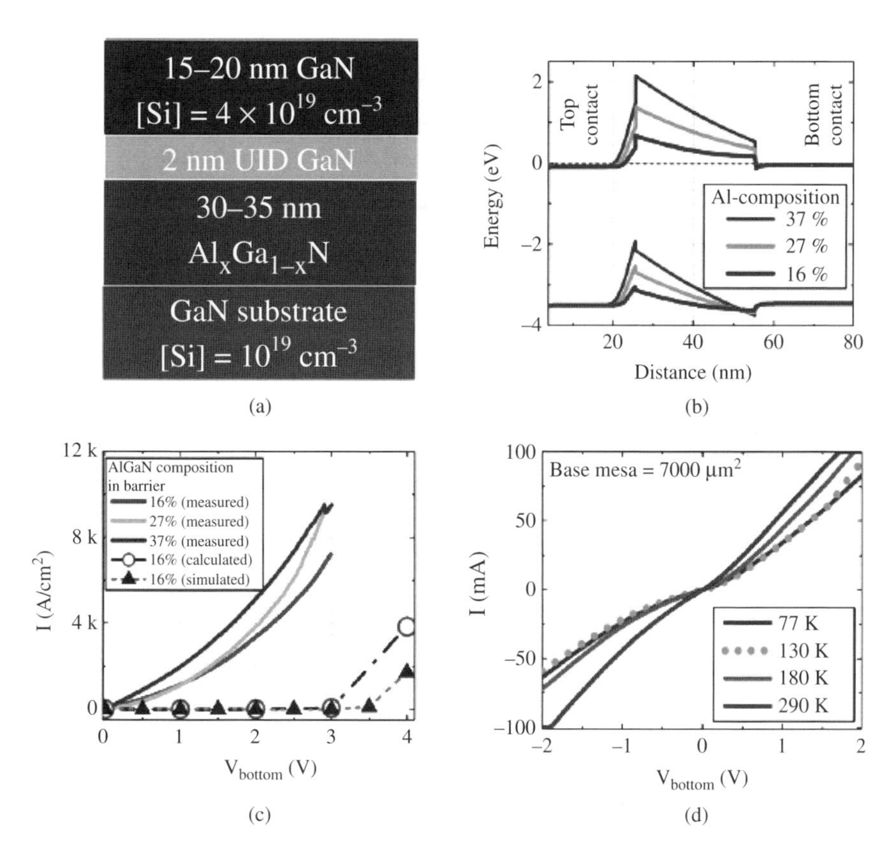

Figure 9.30 Effect of heterojunction barrier height on leakage (a) typical epitaxial stack of devices used in this study. (b) Energy band diagrams of device structures with $Al_xGa_{1-x}N$ barrier with $x = 0.16$, 0.27, and 0.37. (c) Calculated, simulated, and experimentally measured vertical leakage current densities corresponding to an $Al_{0.16}Ga_{0.84}N$ barrier compared to measured leakage for AlGaN barrier with 25 and 37% Al-composition. (d) Temperature-dependent vertical leakage current showing weak temperature dependence. Reproduced with permission from Ref. [101]. Copyright (2013), AIP Publishing LLC

prime mechanism responsible for high currents in such heterostructures. This together with the fact that as mentioned earlier the heterojunction barrier height does not affect the current transport and independent studies show that the current density is almost temperature independent, suggest that electrons see almost zero effective barrier to transport across the heterojunction. It was proposed for this reason that the dominant current mechanism in such heterostructures is linked to percolation-based transport through the Ga-rich regions of AlGaN [101]. This can be caused by composition fluctuations in ternary alloys due to statistical distribution of Ga and Al atoms in group-III sites.

To avoid random alloy fluctuations, two different designs were proposed [101] (1) design D: digital AlGaN barrier with alloy barrier grown by repetition of 2 ML AlN/4 ML GaN digital periods; (2) design B1: nonrandom alloy barrier based on a polarization-engineered binary GaN barrier (Figure 9.31b,c); and compared with (3) design R: random $Al_{0.3}Ga_{0.7}N$ barrier.

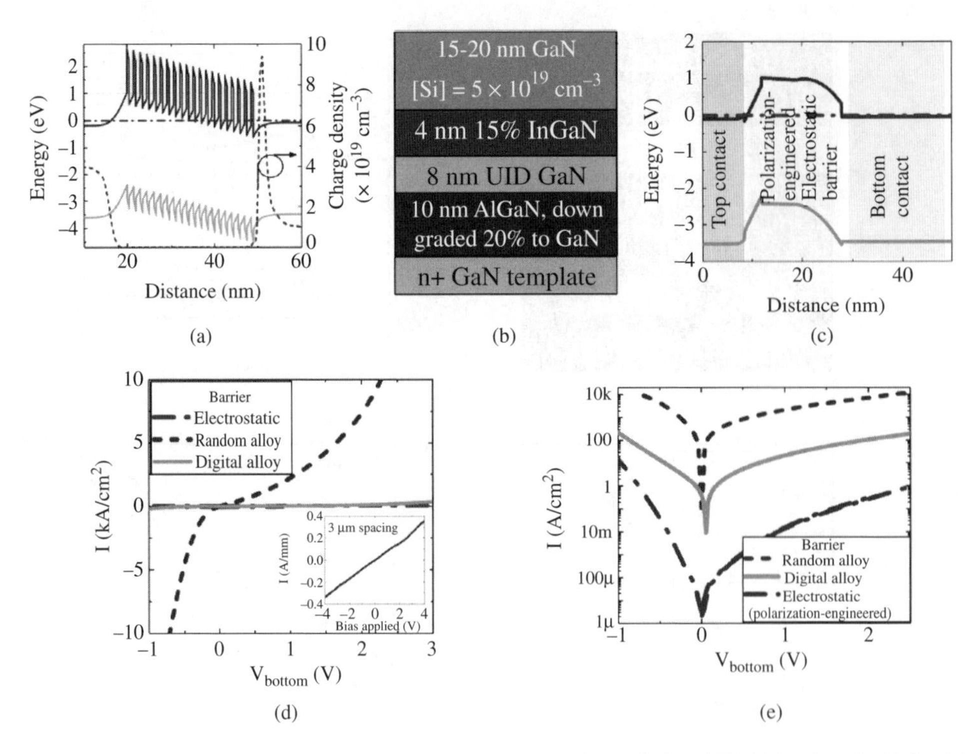

Figure 9.31 (a) Energy band diagram of the sample D1 with a digital AlGaN barrier. (b) Epitaxial stack of the sample with polarization-engineered binary GaN barrier. (c) Energy band diagram of the same. Comparison of measured leakage for samples with random alloy, binary GaN, and digital alloys as barriers in (d) linear (e) log scale. Inset to (d) shows the lateral I-V between two top-contacts of the same device indicating its linear and Ohmic nature (the digital alloy sample). Reproduced with permission from Ref. [101]. Copyright (2013), AIP Publishing LLC

Design D avoids the presence of statistical fluctuations that could lead to Ga-rich regions in the AlGaN. The number of digital periods was such that the barrier corresponded to 25–30 nm of AlGaN with average composition 25% verified by dynamic XRD simulations. Figure 9.31a shows the energy band diagrams corresponding to this design and confirms that the effective barrier height in this case was similar to that for ternary AlGaN. In design B1, a thin InGaN layer provides a polarization-induced dipole that creates an electrostatic barrier that would not be permeable to percolation effects. An AlGaN layer was used to provide a dipole at the uniform intrinsic doping (UID) GaN/AlGaN interface, leads to a flat energy band profile at equilibrium. As can be seen, a thin InGaN layer provides a polarization-induced dipole that creates an electrostatic barrier that would not be permeable to percolation effects.

Figure 9.31d,e show the I-V characteristics of designs R, D1, and B1. Designs with nonrandom digital (D1) and electrostatic binary GaN (B1) barriers had more than 3 orders of magnitude lower leakage current density than those with a random alloy (R) for reverse bias less than 2 V, even though the effective barrier height from the energy band diagram is nominally the same. The low reverse bias leakage characteristic is not top-contact limited since the lateral current between two top-contact pads of a device exhibits linear ohmic characteristics (inset to Figure 9.31d). Overall, the study shows that considerable reduction in vertical leakage is possible using a polarization-engineered electrostatic and a digital alloy ($Al_{0.3}Ga_{0.7}N$) as barriers through elimination of ternary random alloy in the barrier and avoiding percolation-based transport of electrons.

The results presented in this section provide a detailed analysis of transport in Quantum-Cascade and Resonant Tunneling structures based on wide bandgap semiconductors and demonstrate the possibility of NDR presence in them. Control of such characteristics is a key in developing novel, high performance electronic and optoelectronic components.

9.6 Gunn Effect at Electron Field Emission

GaN may be potentially useful for NDR Gunn diode applications. Diodes of this type have been predicted to be very promising for terahertz signal generation [102] with moderate to high power levels up to several hundred milliwatts [103, 104]. The energy band diagram of GaN is similar to that of GaAs having two subbands for light and heavy electrons in the conduction band. This suggests that the conditions necessary for microwave signal generation are satisfied.

To obtain Gunn-effect oscillations in the field-emission current, the studies of the peculiarities of electron field emission for well-studied n-GaAs [105, 106] were performed in some works [107–109]. GaAs contains two conductive band valleys with different energies and light and heavy electrons. A saturation region with instabilities on the current–voltage characteristics has been observed for an n-GaAs field emitter. This characteristic is different from fabricated n^+-GaAs structures. The possible frequency for bunched electron field emission current has been estimated as more than 10 GHz [107, 108].

It is well known that for a stable oscillation generation (Gunn effect) in a solid state diode both contacts have to be ohmic with high conductivity [105, 106]. But in the case of a vacuum diode there is the vacuum barrier on the one side. The peculiarity of a vacuum diode with a tip emitter is the charge accumulation near the cathode, resulting in the sharp growth of the electric field in this region. It often causes the emitter breakdown in measurements. The theoretical calculation showed that for low electron affinity values $\chi \leq 0.1$ eV and vacuum space lengths $L > 0.5 \times 10^{-5}$ cm catastrophic instabilities of the current appeared [110]. In the case of emitters with $\chi = 0$ a stable sharp current pulse was observed. So the necessary condition for a stable microwave generation in vacuum Gunn diodes is zero or even negative electron affinity (NEA) of emitters.

The surface dipoles have a great effect on the electron affinity. It was shown that hydrogen terminated surface of diamond exhibits a NEA while the adsorbate free surface exhibits positive electron affinity (PEA) [111–114]. The NEA was also observed for some other wide band gap semiconductors [115–118]. The ultraviolet photoelectron spectroscopy technique allowed one to determine the electron affinity of III-nitrides and their alloys [118]. The results suggest an electron affinity of 3.4 eV for GaN and NEA for AlN. The extrapolation of the $Al_xGa_{1-x}N$ data suggests that the electron affinity changes from positive to negative at $x = 0.65$. The termination of the semiconductor surface can result in a NEA for an initially PEA semiconductor. For instance, coating the GaN surface with cesium allows a NEA.

Additional necessary conditions for microwave generation in a vacuum Gunn diode are an energy band diagram similar to GaAs with two subbands (light and heavy electrons) in the conduction band and an effective electron field emission.

The energy difference between minima of lower subband (Γ valley, E_Γ, electron affinity χ_Γ) and upper satellite subband (L valley, E_L, χ_L) in GaAs $\Delta E_{\Gamma-L} \approx 0.31$ eV is too small to have a remarkable difference between electron field emission current from the mentioned bands. Much more favorable condition occurs for the case of GaN, where $\Delta E_{\Gamma-X} \approx 1.1$ [119]. The electron affinity χ_X for the GaN upper valley decreases over ~1.1 eV [119].

The peculiarities of electron field emission from nanostructured surface of GaN have been investigated in Ref. [120].

The GaN structure used in the experiments was grown by low-pressure MOCVD on (0001) c-plane sapphire using a modified EMCORE GS-3200 system with trimethylgallium (TMGa) and ammonia (NH_3) as source materials. A buffer layer of about 25-nm-thick GaN was first grown at 510 °C on sapphire. The structure was used is a single 1.5-μm-thick n-GaN layer ($n = 2 \times 10^{17}$cm^{-3}). It was grown at the temperature 1100 °C. A patterned Ti mask served to provide electrical contact to the sample as well an etch mask. The Ti metal contacts were unannealed. Quantum dots were formed by photoelectrochemical etching of GaN [121]. With an additional current and under process-specific conditions it produced high etch selectivity at the dislocations. Photoelectrochemical etching of GaN is the means of great improving of the chemical reactivity of GaN at room temperature. Ultraviolet illumination generates electron–hole pairs at the semiconductor surface, which enhance the oxidation and reduction reactions within an electrochemical cell. The GaN was etched for 6 minutes in a stirred 0.1 M KOH solution using Hg lamp illumination. The scanning electron microscope image of nanometer-size tips on GaN surface is shown in Figure 9.32.

Electron field emission characteristics were measured at 2×10^{-8} mbar. Since the GaN was grown on an insulating sapphire substrate, the cathode electrode was formed on the GaN surface using silver paste. A Si wafer was used as an anode electrode. A 7.5 μm kapton spacer

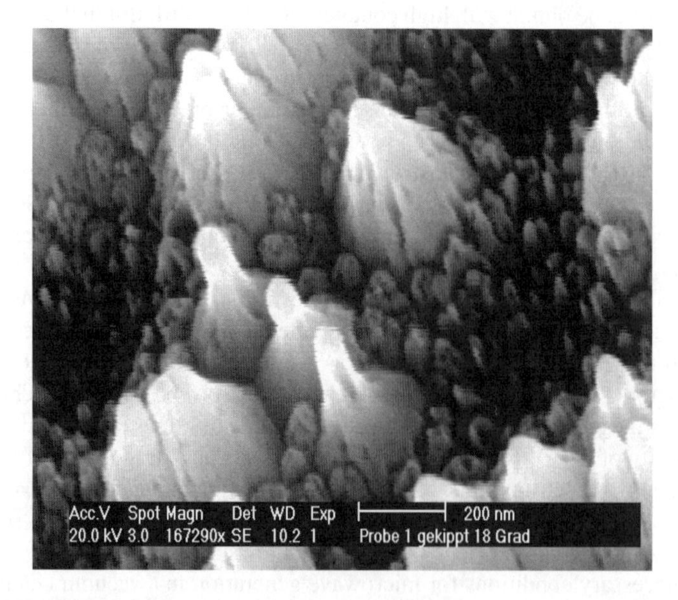

Figure 9.32 SEM image of photoelectrochemical etched GaN sample. Reproduced with permission from Ref. [120]. Copyright (2005), AIP Publishing LLC

defined the distance between the GaN emitter and Si anode. The hole in the spacer foil defined the anode area with $A = \pi \times (350\ \mu m)^2$.

The current–voltage characteristics of emission current and corresponding Fowler–Nordheim plot are shown in Figure 9.33. The sample produces the maximum current density of about $J = 40\ mA/cm^2$. It was suggested that the decreased turn-on voltage is a result of two factors: the small tip radii with increased field enhancement coefficient and lowering of work function due to appearance of a quantum-size effect. The latter causes the increasing of the energy band gap (decreasing electron affinity) due to the appearance of local energy levels in conduction band. The oscillations of the emission current are observed at defined values of applied voltage.

The electron field emission current in the Fowler–Nordheim curve clearly demonstrates two straight line regions with different (about twice) slopes in an unusual way: at low applied fields

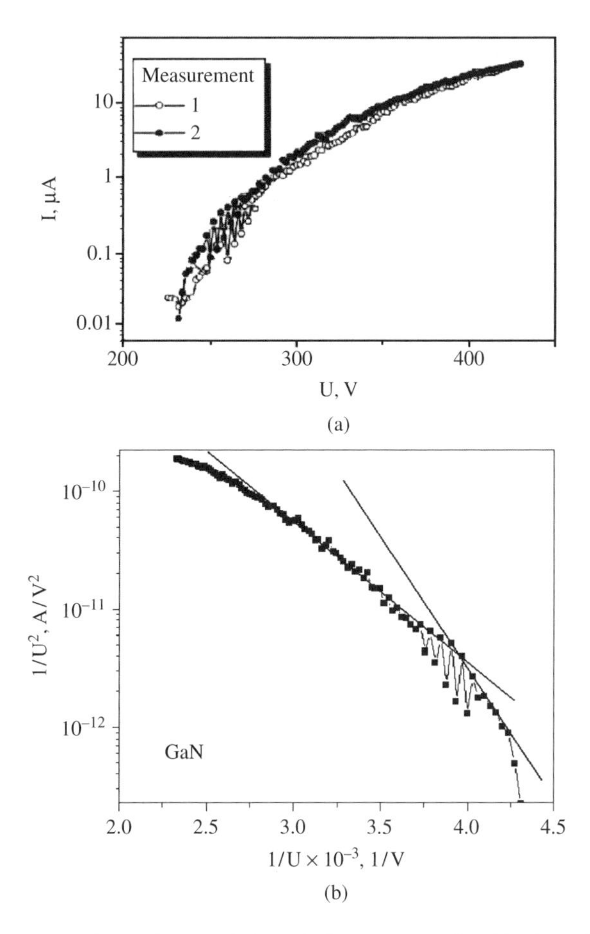

Figure 9.33 (a) Current–voltage characteristic and (b) corresponding Fowler–Nordheim plot of the electron field emission from nanostructured GaN surface. Regions of different slopes as well as oscillating behavior are shown. Reproduced with permission from Ref. [120]. Copyright (2005), AIP Publishing LLC

the slope has been larger than at high voltages. Note, that at very large fields there was observed a saturation tendency due to the lack of electrons on top of the tip cathodes. In the region between the straight lines in the Fowler–Nordheim characteristics there were observed rather regular oscillations, which were connected with the Gunn effect (electron transfer effect), arising due to the redistribution of carriers between Γ-X valleys at critical external field F^*. At large values of F the occupation of the upper X band is very large and main emission occurs from this valley.

It is possible to determine the different electron affinities from the two slopes in the Fowler–Nordheim characteristics (Figure 9.33b). At the beginning the electric field enhancement coefficient due to geometry of tips was estimated (see Figure 9.32) to be $\beta^* \approx 30$. The electric field enhancement coefficient was calculated by using the floating sphere model [122] according to

$$\beta^* \approx h/r + 3, \tag{9.38}$$

where h is the tip height and r is the radius curvature of the tip.

From slopes of the Fowler–Nordheim plot two values of the electron affinity have been obtained: $\chi_1 = 2.3$ eV and $\chi_2 = 1.5$ eV. Taking into account the emission barrier lowering due to image forces, corresponding electron affinity values of $\chi_1 = 3.5$ eV and $\chi_2 = 2.8$ eV are provided. These two values of electron affinity were considered to be related to the electron field emission from the Γ valley and the X valley of GaN, respectively. The energy difference between these valleys is $\Delta\Phi = \chi_1 - \chi_2 = 0.7$–$0.8$ eV. The calculated $\Delta\Phi = 0.7$–0.8 eV from the Fowler–Nordheim characteristics is lower than the value from the literature ($\Delta\Phi = 1.1$ eV) [119]. This discrepancy can be, in particular, caused by quantum-size confinement effect (Figure 9.34). The energy shift of valley positions was calculated according to [123]

$$E = \frac{n^2 h^2}{8m^* d^2}, \tag{9.39}$$

where m^* is the electron effective mass, d is the size of the structure (radius of the tip), h is Plank's constant and $n = 1$.

The shift of energy position of lower (Γ) valley in nanometer size structure is larger ($\Delta E = 0.47$–0.21 eV, for $d = 2$–3 nm) due to the smaller electron effective mass ($m * = 0.2m_0$ light electrons [124]) in comparison with the shift of the upper (X) valley ($\Delta E = 0.11$–0.05 eV, for $d = 2$–3 nm, $m * = 0.88m_0$, heavy electrons [124]). Taking into account this effect there was obtained the following value of the energy difference $\Delta\Phi = 1.0 - 1.2$ eV. This is in good agreement with the literature data [119]. The existence of nanowires on the tip (Figure 9.32) with a size of 2–3 nm can be supposed. The dependencies of energy gaps $E_g(\Gamma)$ and $E_g(X)$ on the GaN nanowire diameter are shown in Figure 9.35. As can be seen, there is a decrease of $\Delta\Phi = E_g(X) - E_g(\Gamma)$ with smaller nanowire diameters.

It is also possible to determine the energy difference between the Γ valley and the X valley in another way and to compare it with the value obtained from the Fowler–Nordheim plot. The electric field in the semiconductor can be calculated as $F_0 = V/L, F_0{}^* = \beta * F_0, F_s = F_0{}^*/\varepsilon_s$, where F_0 is the macroscopic electric field strength in vacuum, V is the applied bias, L is the cathode–anode distance, $\beta*$ is the field enhancement factor, $F_0{}^*$ is the microscopic (enhanced) electric field in vacuum near the cathode surface, ε_s is the relative dielectric constant in the semiconductor, and F_s is the electrical field in the semiconductor near the cathode surface.

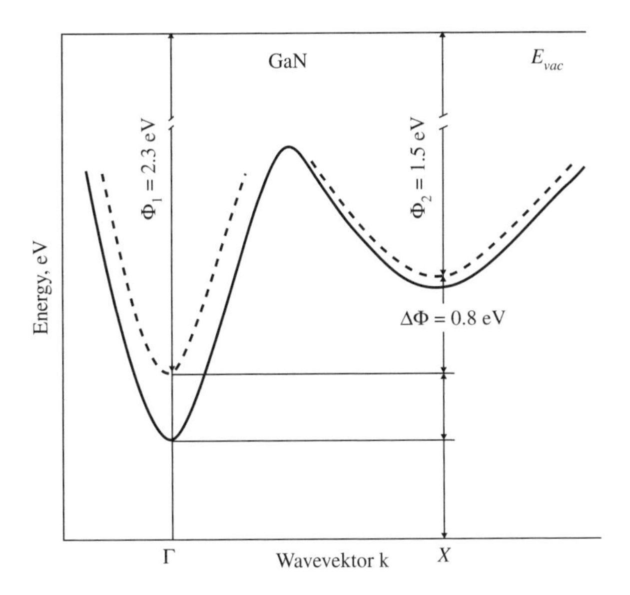

Figure 9.34 Energy shift of valley positions due to the quantum-size confinement effect in GaN nanostructures. Reproduced with permission from Ref. [120]. Copyright (2005), AIP Publishing LLC

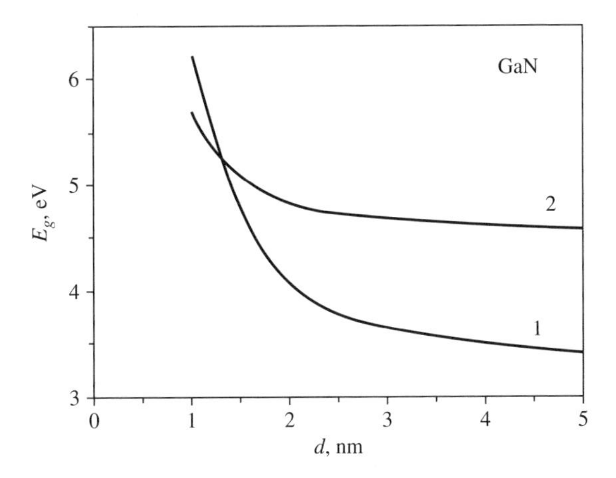

Figure 9.35 Energy gaps $E_g(\Gamma)$ (1) and $E_g(X)$ (2) on GaN as a function of nanowire diameter. Reproduced with permission from Ref. [120]. Copyright (2005), AIP Publishing LLC

With these relations the electrical field F_s close to the surface in GaN is 1.05×10^6 V/cm. The necessary threshold field of $F_{th} = 150$ kV/cm for GaN Gunn diodes can be achieved in the material. According to proposed in Ref. [120] model at small fields the electron field emission takes place from the lower subband (higher slope in the Fowler–Nordheim plot (Figure 9.33)). With increasing of the fields the electrons are heated up and transited to the upper subband from which the emission also takes place (lower slope in the Fowler–Nordheim plot (Figure 9.33)). To transfer electrons from the lower to the upper subband the carriers have to be heated and provided with the energy $\Delta\Phi = 0.8$ eV. With growing electric field in the semiconductor the drift velocity (v_D) of electrons and their kinetic energy (ΔE_k) is increased. When ΔE_k equals to $\Delta\Phi$ the electrons transfer into the upper subband. Simple relations give the possibility to calculate ΔE_k in a strong electric field within the semiconductor:

$$\Delta E_k = m^* v_D^2/2,\ v_D = \mu_D F_{th}, \tag{9.40}$$

where m^* is the electron effective mass, μ_D is the electron mobility, and F_{th} is the threshold field for GaN Gunn diodes.

Using the value of $F_{th} = 150$ kV/cm and $\mu = 900$ cm^2/V s, and $m_e^* = 0.2m_0$ for the lower subband from Ref. [125], one obtains $\Delta E_k = 1.04$ eV. Taking into account the field dependence of the mobility at high electric fields it gives lower value for ΔE_k. It is in good agreement with $\Delta\Phi$ obtained from the measured Fowler–Nordheim characteristics (Figure 9.33).

The current from the X valley becomes dominant already at a rather small concentration n_X, because the barrier is much lower, than for the Γ valley: $\Delta\Phi = (\Phi_\Gamma - \Phi_X) \approx 0.8$ eV (for GaN). Namely in this region of $F_s = F_0^*/\varepsilon_s$ the Gunn effect (electron transfer effect) can occur. The strong decrease of the scattering in quantum wires due to a small electron–phonon interaction promotes this process [126].

9.7 Field Emission Microwave Sources

Most long range telecommunication systems are based upon microwave links. In order to satisfy the present power (tens of Watts) and bandwidth requirements (30 GHz), satellites employ TWT's based on thermionic cathodes. However, TWTs are bulky and consume large proportions of the valuable space and weight budget in a satellite. Thus, any miniaturization of the TWTs would lead to rather significant cost savings in a satellite launch, and indeed aid the implementation of micro-satellites. Solid state devices cannot be used in this high frequency, high bandwidth regime because the maximum power attained by solid state devices today at 30 GHz is ~1 W. Therefore the most effective way to reduce the size of a TWT is *via* direct temporal modulation of the e-beam, for example, in a triode configuration. Cold cathodes with the ability of being modulated at 30 GHz do not exist presently. The approach to replace the thermionic emitters by electron field emitters in microwave applications has been developed in many works [127–134]. FEAs stand to strongly impact device performance when physical size, weight, power consumption, beam current, and/or high pulse repetition frequencies are an issue. For RF power amplifiers, lower input capacitance results in a higher operating frequency [135]. The primary goals for rf power amplification based on FEAs, therefore, are to maximize the transconductance and minimize the input capacitance of the FEA. At the same time, reasonable input impedance must be maintained in order to avoid the inefficiencies and complexities of complicated impedance matching schemes often required for high-frequency, high-power, solid-state devices [129].

In order to use the resulting electron beam from possibly a whole matrix of such field emitters for the generation of electromagnetic signals at millimeter- or submillimeter wave frequencies, it is necessary to adopt concepts of microwave tubes such as klystrons, TWTs, magnetrons, carcintrons, and so on [7, 136]. The basic requirement for tubes with separate input and output electrodes such as klystrons is that space-charge waves or velocity modulation of the beam electrons are induced into the beam by either suitable capacitive or inductive electrodes. Another possibility in the emission system might be the input emission velocity modulation via microwave modulation of the photon flow [137]. The slower electrons are caught up by the faster ones so that the space charge bunching occurs. Those bunches represent space charge accumulations which can be used to induce an enhanced microwave signal into the second electrode of a capacitive or inductive pick-up with attached resonator structure. Input velocity modulators or output space charge bunch pick-ups are similar to antennas.

Another microwave-tube structure is represented by TWTs in that the helix or any other related slow-wave structure induces at the input side velocity modulation which goes into space charge density modulation at the end of the wave structure, where the signal is also induced as an amplified output. The electromagnetic wave component needs to be efficiently slowed down to slightly less than the electron velocity. The slow-wave structure is basically of broad-band nature so that the in and out connections are usually coaxial and not hollow conductors. Also here, matching components need to be inserted at in- and output terminals. The cathode emits an unmodulated electron beam. In the first section of the circuit, the *rf* input signal imposes a small velocity modulation on the electron beam, which launches longitudinal space-charge waves. As the electrons drift through the microwave tube, the beam modulation cycles between kinetic-energy modulation, that is, velocity modulation, and potential-energy modulation, that is, density modulation. If the initial modulation is small, it is increased by passing the beam trough intermediate interaction regions, where *rf* modulation of the beam current exchanges energy with an electromagnetic wave. In the last interaction region (the output region), the relative phase of plasma wave and the electromagnetic wave are adjusted to maximize energy transfer from the beam to the electromagnetic wave. The electron beam requires a sufficiently high electron density so that for the velocity modulation of an equivalent acceleration potential of say 10% an equivalent current modulation yields the power level of the order of the input signal power. For 1 μW and an acceleration voltage difference of 100 V (for beam potentials of 1 kV) a beam current requirement of at least 10 nA for 100% coupling would result. The amplified output power (say power amplification of 100) would then require more than 1 μA to ensure satisfactory operation. This can, however, only be possible with a matrix of filed-emission points. In order to avoid breakdown effects with 1 kV the dimensional details are derived. It was reported about a 100 W Cold-Cathode TWT [138]. The operation was at 5 GHz, and the cathode current was 100 mA, and its voltage was 3500 V. The signal gain was 32 dB and the circuit length was 16 cm. For the frequency of 10 GHz the 13.5 dB gain and 10 W output power at cathode current of 50 mA have been obtained [139].

There have been suggestions to employ electron beams from field-emission structures as means to generate electromagnetic power. This can only be achieved, if the various interacting components are designed appropriately for matching and optimal space charge wave pick up. Unfortunately, the normal current levels are not easily made sufficiently strong so that satisfactory space charge waves result. But recently the new type of electron field emission cathode based on nanogranular materials that produce high current level has been proposed [140]. An electron beam of high power density is employed to generate nanostructures with dimensions >20 nm, being composed of amorphous to nanogranular materials, for example, with gold or platinum crystals of 2–5 nm diameter embedded in a Fullerene matrix. Those compounds are generated in general by secondary or low energy electrons in layers of inorganic, organic, organometallic compounds absorbed to the sample. Those are converted into nanogranular materials by the electron beam. Nanogranular composites like Au/C or Pt/C with metal nanocrystals embedded in a Fullerene matrix have hopping conduction with zero-dimensional Eigen-value characteristics and can carry "Giant Current Densities" and show "Anomalous Electron Transport" with values from >1 MA/cm^2 to 0.1 GA/cm^2 without destruction of the materials [141, 142, 57].

Microelectronic FEAs can modulate the beam density at high frequency and with good spatial localization, extending the frequency range of density-modulated amplifiers by orders of magnitude. In FEA structures, the grid (or gate) is fabricated in nearly the same plane as the emitting surface, dramatically reducing interception current and increasing transconductance. The microtriode using a FEA cathode is illustrated in Figure 9.36. The use of field emission also eliminates the need to dispense a continuous supply of low work function material as it is often done in thermionic cathodes. This material, which is vaporized in the tube, can coat the grids and grid-cathode insulators, resulting in secondary emission and shorts.

The detailed consideration of electron field emission microwave devices has been performed in Ref. [8]. Here we shortly describe the analysis of some main parameter of such devices.

9.7.1 Modulation of Gated FEAs

The key to the performance advantages of inductive output amplifiers (IOAs) is the emission gating of the electron beam at the cathode surface before acceleration to anode potential. The cathode assembly that performs this modulation is usually an old technology pushed to its fundamental limitations (i.e., gridded thermionic cathodes) or a new technology pushed to its present limits of performance (i.e., FEAs or laser-driven photo-cathodes). The critical measures of the performance of any emission gated cathode are low transit time, high transconductance, and low capacitance. The current density must be sufficiently high for good performance, but not too high for good beam optics.

To achieve acceptable gain in an IOA, it must be possible to modulate the emission from a gated cathode with a low power input signal. Let's consider a number of important factors that influence the suitability of a gated cathode for high frequency modulation [8].

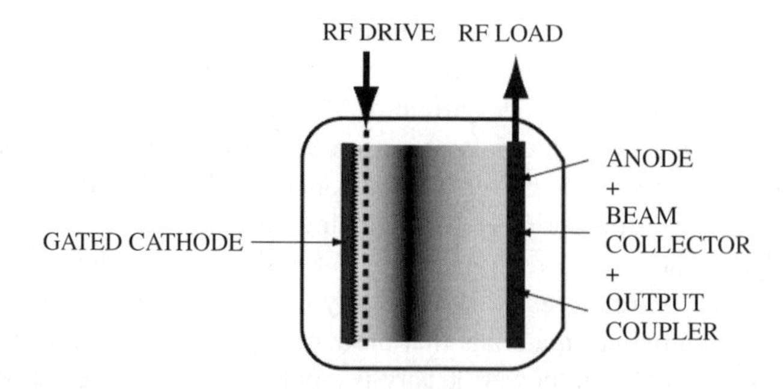

Figure 9.36 Vacuum microtriode. *Rf* signal is applied to the gate electrode of the FEA, providing a density-modulated beam. The anode both collects the beam and delivers amplified current to the *rf* load [8].

9.7.1.1 Transit Time

The performance of a gated cathode will degrade, if the *rf* fields experience by an electron change appreciably during its transit from the emitting surface to the gate–anode region. In this context, "transit time" refers to the time that an electron spends under the influence of the electric field between the cathode and the gating structure. In gridded thermionic cathodes, it is the time for an electron to reach the plane of the grid, while in FEAs it is the time for an electron to reach the gate potential. In a gridded thermionic cathode, the *dc* bias voltage on the grid is usually negative with respect to the cathode in order to suppress the extraction of thermally emitted electrons by the anode electric field. The grid polarity is rarely positive because emitted current will then be intercepted by the grid, which would unacceptably load the input circuit and damage the grid at high power densities. Consequently, the electric field that accelerates electrons away from a thermionic cathode is relatively small, and in fact it must be negative for part of each *rf* cycle in Class C operation. In contrast, the strong electric fields at the emitting surface of an FEA accelerate emitted electrons to high velocity immediately upon emission. In addition, the gate electrode is approximately co-planar with the emitting tip, so that the electron passes from the influence of the oscillating gate potential into that of the anode static field in a short distance.

Emission-gated cathodes offer the most dramatic performance advantages in Class C operation. Under these conditions the accurate determination of the limitations imposed by transit-time effects requires simulations of 2-D electron trajectories that include time-varying space charge and electrons returning to the emitting surface. However, transit-time effects in thermionic and field emission cathodes can be roughly compared by focusing on the gross distinctions between the ţwo structures. The gate voltage of a field emitter, V_g, modulates the current by causing electron emission, while the grid voltage of a thermionic emitter, V_{gr}, modulates the current by suppressing electron extraction from the thermally emitted cloud on the cathode surface. For a space-charge-limited thermionic cathode with an ideal grid (an ideal grid is a thin, perfectly conducting sheet that intercepts no current), the limiting current density J_{CL} is determined by the Child–Langmuir law [143, 144]:

$$J_{CL} = \frac{4\varepsilon_0}{9}\left(\frac{2e}{m}\right)^{1/2}\frac{V_{gr}^{3/2}}{d^2}. \tag{9.41}$$

In Equation (9.41), d is the cathode-to-grid separation. Consequently, the ratio of the full-on voltage $V^+{}_{gr}$ to cut-off voltage $V^-{}_{gr}$ required for a ratio of full-on current I^+ to cut-off current I^- of $I^+/I^- = 1000$ is

$$\frac{V_{gr}^+}{V_{gr}^-} = \left(\frac{I^+}{I^-}\right)^{2/3} = 100. \tag{9.42}$$

Thus, the electrons emitted near cutoff depart the cathode surface with only 1% of the acceleration of electrons emitted near full-on conditions.

In contrast, for a field emitter in which the current is given by

$$I_b = AV_g^2 \exp(-B/V_g); \tag{9.43}$$

the ratio of the currents is

$$\frac{I^+}{I^-} = \left(\frac{V_g^+}{V_g^-}\right)^2 \exp\left[-\frac{B}{V_g^+}\left(1 - \frac{V_g^+}{V_g^-}\right)\right]. \tag{9.44}$$

For $B = 750$ V and $V^+{}_g = 75$ V, the reasonable values for today's field emitters, Equation (9.44) yields $V^+{}_g/V^-{}_g = 1.6$. The field that accelerates electrons that are emitted near cutoff is over 60% of that at full-on conditions. This simple example shows why field emission is inherently better adapted to Class C amplifiers than thermionic emission; no field emitted electron can linger in the time-varying electric field of the gate.

In addition to the differing cut-off conditions of thermionic and field emission cathodes, the transit time under full-on conditions differs substantially as well. The field between the gate (grid) and the emitting surface is approximately constant and equal to the potential change divided by the gate (grid)-cathode distance. For the thermionic cathode, the electric field in the cathode-to-grid region is sufficient to extract the required current density, as determined by the Child-Langmuir law of Equation (9.41). Using $E = V_{gr}/d$ in Equation (9.41) and solving for E gives

$$E = \left[\left(\frac{9}{4}\frac{J_{CL}}{\varepsilon_0}\right)^2 \frac{dm}{2e}\right]^{1/3}. \tag{9.45}$$

For $J_{CL} = 2$ A/cm^2 and $d = 250$ μm, Equation (9.45) predicts that $E = 264$ kV/cm. The resulting transit time τ is

$$\tau = \int_0^d \frac{dz}{v(z)} = \sqrt{\frac{m}{2e}} \int_0^d \frac{dz}{\sqrt{\phi(z)}} = \sqrt{\frac{2dm}{eE}} = 10ps. \tag{9.46}$$

This corresponds to the cutoff frequency $f_c = 1/2\pi\tau = 1.6$ GHz. As the gate voltage declines toward cut-off, the transit time approaches infinity, resulting in the return of some electrons to the cathode.

The potential on the axis of symmetry for a gated field emitter with an anode has been derived in Ref. [145]:

$$V(z) = V_g \frac{E_{tip}z}{V_g + E_{tip}z}\left(1 + \frac{E_0 z}{V_g}\right). \tag{9.47}$$

In Equation (9.47), E_{tip} is the field at the emitter tip on its center axis, and E_0 is the background field due to the anode. An emitted electron can be significantly influenced by the gate when $V(z) < V_g$. Solving Equation (9.47) for $V(z) = V_g$ yields an upper bound to the extent of the control region, $z_g = V_g/\sqrt{E_0 E_{tip}}$. Since collisions can be neglected, the electron velocity $v(z)$ is determined by the electrostatic potential $\phi(z)$ as

$$\frac{1}{2}mv^2(z) = e\phi(z) = eV_g \frac{E_{tip}z}{V_g + E_{tip}z}\left(1 + \frac{E_0 z}{V_g}\right). \tag{9.48}$$

The electron velocity is

$$v(z) = \sqrt{\frac{2eV_g}{m}}\sqrt{\left(\frac{E_{tip}z}{V_g + E_{tip}z}\right)\left(1 + \frac{E_0 z}{V_g}\right)}. \tag{9.49}$$

Then, the transit time is

$$\tau = \int_0^{z_g} \frac{dz}{v(z)} = \sqrt{\frac{mV_g}{2eE_{tip}E_0}}\int_0^1 \sqrt{\frac{u + \sqrt{E_0/E_{tip}}}{u(1 + \sqrt{E_0/E_{tip}}u}}du \cong \sqrt{\frac{mV_g}{2eE_{tip}E_0}}. \tag{9.50}$$

In Equation (9.50), $\sqrt{E_0/E_{tip}} \ll 1$ has been used in approximating the integral. In a typical FEA, a gate voltage of 75 V produces a tip field of 0.5 V/A. The anode field must be large enough to draw the entire field emitted current away from the grid, yet small enough to avoid arc breakdown. The value of 20 kV/cm is reasonable for moderate emission currents; this is much higher than for thermionic emission because of the very high local current densities obtained from FEAs. The transit time is then $\tau = 0.15$ ps, which is nearly 3 orders of magnitude shorter than the thermionic case and corresponds to $f_c \cong 1000$ GHz.

9.7.1.2 Input Impedance

Although the close spacing of the gate and cathode diminishes the transit time, it increases the grid-cathode capacitance. Further, the gate-cathode region constitutes a distributed transmission line, as depicted in Figure 9.37. A detailed analysis of the voltage distribution within the FEA and the input impedance presented by the FEA has been provided in Ref. [135]. The simplified versions are given here.

The array is assumed to be composed of cells that repeat with periodicity a. The gate capacitance of each repeat cell arises from the capacitance through the gate insulator, C_{pc}, and the gate-tip capacitance C_{tc}, as shown in Figure 9.37a. If the extent of the array in the direction of propagation (z direction, hereafter called the length) is l and the array width is w, the capacitance per unit length, C, is

$$C = \frac{w(C_{tc} + C_{pc})}{a^2}. \tag{9.51}$$

If the effects of the gate-tip holes are neglected [135], the resistance per unit length is $R = \rho_g/wt$, where ρ_g and t are the resistivity and thickness of the gate metal, respectively. Using a TEM transmission-line approximation [146], the inductance per unit length is $L = \mu_0 h/w$, where μ_0 is the permeability of free space and h is the gate-insulator thickness.

Consider the gate to be excited by the superposition of a dc voltage V_g^{dc} and a sinusoidal rf gate voltage V_g^{rf}. Using the equivalent transmission line of Figure 9.37b, the rf gate current $I_g^{rf}(z,t) = Re[\widetilde{I}_g^{rf}(z)e^{j\omega t}]$ and the rf gate voltage $V_g^{rf}(z,t) = Re[\widetilde{V}_g^{rf}(z)e^{j\omega t}]$ are determined by

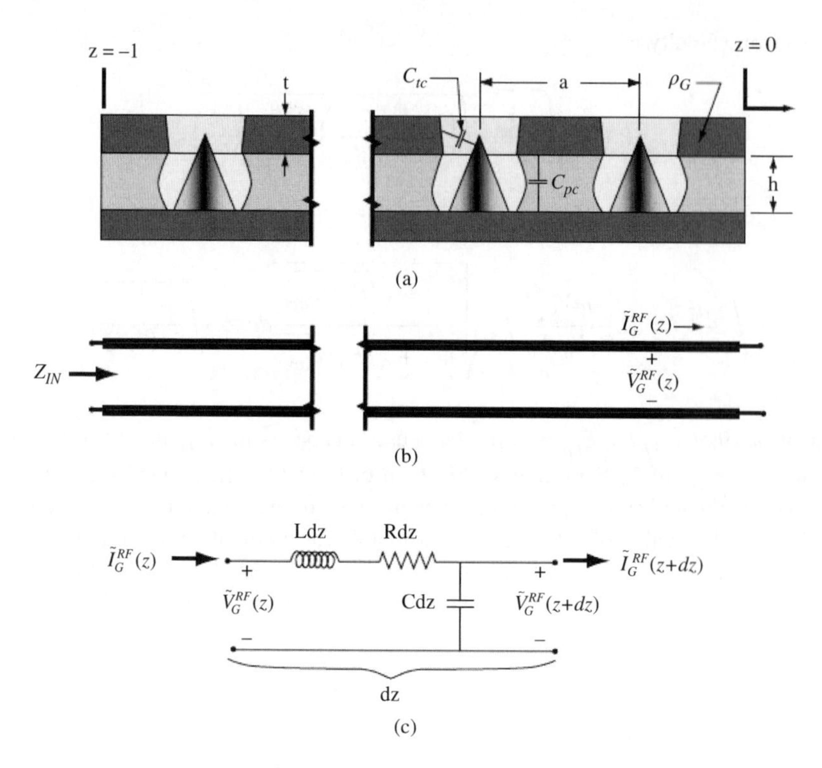

Figure 9.37 Transmission-line effects for a gated FEA. (a) The relevant parameters of a gated-FEA input circuit; (b) the equivalent transmission line; and (c) the incremental transmission line used to calculate the gate-voltage distribution [8]. Copyright (2001) by John Wiley & Sons, Inc. This material is reproduced with permission of John Wiley & Sons, Inc.

the transmission-line equations:

$$\frac{\partial \widetilde{I}_g^{rf}(z)}{\partial z} = -j\omega C \widetilde{V}_g^{rf}(z)$$

$$\frac{\partial \widetilde{V}_g^{rf}(z)}{\partial z} = -(R + j\omega L)\widetilde{I}_g^{rf}(z). \tag{9.52}$$

Solving Equation (9.51) subject to the boundary condition that the open circuit exists at $z = 0[\widetilde{I}_g^{rf}(0) = 0]$ gives

$$\widetilde{V}_g^{rf}(z) = V_0 \cos(\beta z)$$

$$\widetilde{I}_g^{rf}(z) = \frac{V_0}{jZ_0} \sin(\beta z), \tag{9.53}$$

where

$$\beta = \omega\sqrt{LC}\left(1 - j\frac{R}{\omega L}\right)^{1/2} = \frac{\omega}{a}\sqrt{\mu_0 h(C_{pc} + C_{tc})}\left(1 - \frac{j}{\omega}\frac{\rho_g}{\mu_0 h}\right)^{1/2}$$

$$Z_0 = \omega\sqrt{\frac{L}{C}}\left(1 - j\frac{R}{\omega L}\right)^{1/2} = \frac{a}{w}\sqrt{\frac{\mu_0 h}{C_{pc} + C_{tc}}}\left(1 - \frac{j}{\omega}\frac{\rho_g}{\mu_0 h}\right)^{1/2}.$$

Thus, the *rf* input impedance of the gate Z_{in}, is

$$Z_{in} = \frac{\widetilde{V}_g^{rf}(-l)}{\widetilde{I}_g^{rf}(-l)} = -jZ_0\cot(\beta l). \tag{9.54}$$

For $|\beta l| \ll 1$, the cotangent function can be expanded, and

$$Z_{in} = \frac{Rl}{3} + \frac{1}{j\omega(Cl)} + j\omega\frac{Ll}{3} = \left(\frac{\rho_g}{3t}\right)\left(\frac{l}{w}\right) + \frac{1}{j\omega N_{tip}(C_{tc}+C_{pc})} + j\omega\left(\frac{\mu_0 h}{3}\right)\left(\frac{l}{w}\right). \tag{9.55}$$

Here, the quantity N_{tip} is the total number of tips in the array.

Equation (9.53) shows that each tip does not experience the same gate–tip voltage. To examine the effects of the gate-voltage distribution on the emission current, let's consider the beam current to be composed of a *dc* component $I_b{}^{dc}$ and a small *rf* component $I_b{}^{rf}(t)$, so that $I_b(t) = I_b{}^{dc} + I_b{}^{rf}(t)$. If an individual tip emits current $I_{tip}(V_g)$ at a gate voltage V_g, the current emitted per unit length, $K_b(z, t)$, is given by

$$K_b(z,t) = \frac{w}{a^2}I_{tip}[V_g^{dc} + V_g^{rf}(z,t)] \cong \frac{w}{a^2}[I_{tip}(V_g^{dc}) + g_{mtip}V_g^{rf}(z,t)]. \tag{9.56}$$

On the right-hand side of Equation (9.56), the approximation is valid only in small-signal conditions, and the transconductance per tip, $g_{m\,tip}$, is given by

$$g_{mtip} = \left.\frac{\partial I_{tip}}{\partial V_g}\right|_{V_g^{dc}}. \tag{9.57}$$

For small-signal sinusoidal excitation, the *rf* beam current is

$$I_b^{rf}(t) = \int_{-l}^{0} K_b^{rf}(z,t)dz = \frac{w g_{mtip}}{a^2}\mathrm{Re}\left[e^{jwt}\int_{-l}^{0}\widetilde{V}_g^{rf}(z)\,dz\right]. \tag{9.58}$$

Using Equation (9.53) and writing $I_b{}^{rf}(t) = Re(\sim I_b{}^{rf} e^{jwt})$,

$$I_b^{rf}(t) = \frac{w g_{mtip}}{a^2}\int_{-l}^{0}\widetilde{V}_g^{rf}(z)dz] = N_{tip}g_{mtip}\frac{\sin(\beta l)}{\beta l}. \tag{9.59}$$

Thus, the reduction of the transconductance by the nonuniform gate voltage is expressed by the term $\sin(\beta l)/\beta l$, implying that the entire array will not be effectively modulated unless $|\beta l| \ll 1$.

Small dimensions of FEA, together with the high emission current required by a microwave tube, often result in input impedance much lower than 50 Ω. Within a factor of 2, the total capacitance per cell, $C = C_{pc} + C_{tc}$, may be estimated as the parallel-plate capacitance $C = \varepsilon a^2/h$. Suppose that the total emission current of 100 mA is required from FEA, for which $a = 1\,\mu\text{m}$, $h = 1\,\mu\text{m}$, and $I_{tip} = 1\,\mu\text{A}$. Neglecting resistive losses, $\beta = 4.2\ \text{cm}^{-1}$, which implies that l must be less than 250 μm for $\beta l < 0.1$. In order to emit 100 mA at 1 μA/tip, 10^5 tips ($N_{tip} = 10^5$) must be used. For $l = 250\ \mu\text{m}$, the array width $w = N_{tip}a^2/l$ must be at least 400 μm. Using Equation (9.54), the input reactance is then approximately 5 Ω at 10 GHz. Consequently, an impedance-matching network must be inserted between the power source and FEA to efficiently couple the power to the FEA, as shown in Figure 9.38a.

Impedance-matching considerations are important because they can affect the FEA design and packaging techniques. In the equivalent circuit of Figure 9.38b, the source is represented by conductance $g_s = 1/r_s$, and the FEA is represented by a series connection of resistor r_L and capacitor c_L. By Poynting's theorem [147], the input admittance of the matching circuit, Y_l, is

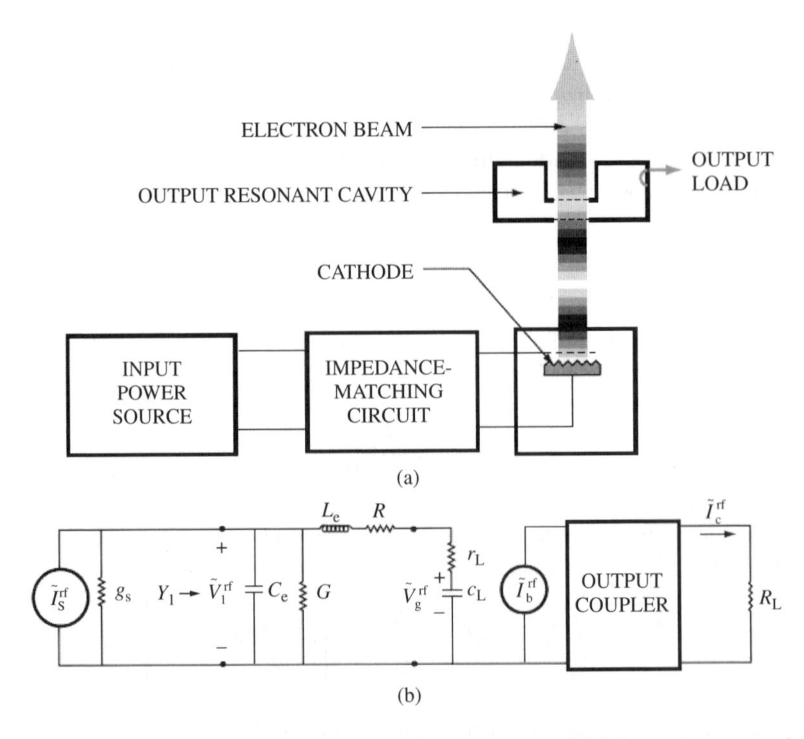

Figure 9.38 Input circuit of FEA. (a) A power source is coupled to the gate of an FEA by an impedance-matching network and (b) an equivalent circuit for the input networks. The impedance-matching network is a resonant circuit that is resonant at the operating frequency. The resonant circuit is characterized by circuit elements C_e, G, L_e, and R [8]. Copyright (2001) by John Wiley & Sons, Inc. This material is reproduced with permission of John Wiley & Sons, Inc.

given by

$$Y_1 = \frac{2P_d + 4j\omega(\langle W_e\rangle - \langle W_m\rangle)}{|\widetilde{V}_1^{rf}|^2}. \tag{9.60}$$

In Equation (9.60), P_d is the power dissipation, and $\langle W_m\rangle$ and $\langle W_e\rangle$ are the average magnetic and electric energies, respectively. For optimum power transfer, $Y_1 = g_s$, so that $\langle W_m\rangle = \langle W_e\rangle$, that is, the circuit is resonant. Near the resonance, the circuit can be approximated by an effective inductance L_e and capacitance C_e, as shown in Figure 9.38b. The values of L_e and C_e are chosen so that resonance occurs at the design frequency ω, presenting a parallel resonance at port 1 and a series resonance at port 2. Parallel conductance G and a series resistance R are added to L_e and C_e, respectively, to represent losses in the matching circuit. For simplicity, it will be assumed that

$$\frac{\omega L_e}{R} = \frac{\omega C_e}{G} = Q. \tag{9.61}$$

The quantity Q is the quality factor of the matching circuit. Circuit analysis yields the ratio of the *rf* output power P_O^{rf} to the *rf* power available from the source, P_A^{rf}

$$\frac{P_O^{rf}}{P_A^{rf}} = \frac{4r_L g_s}{[1 - \chi_c + (g_s + G)]^2 + \chi_c\left[\frac{(r_L + R)}{\eta} + \eta(g_s + G)\right]^2}. \tag{9.62}$$

In Equation (9.62),

$$\chi_c = \omega^2 L_e C_e\left(1 - \frac{1}{\omega^2 L_e c_L}\right)$$

$$\eta = \sqrt{\frac{L_e}{C_e}\left(1 - \frac{1}{\omega^2 L_e c_L}\right)}. \tag{9.63}$$

The maximum value of P_O^{rf} as a function of ωL_e is

$$\left(\frac{P_O^{rf}}{P_A^{rf}}\right)_{max} = \left(\frac{r_L}{r_L + R}\right)\left(\frac{g_s}{g_s + G}\right) \tag{9.64}$$

and occurs when

$$\omega L_e = \sqrt{\frac{r_L + R}{g_s + G}}\sqrt{1 - (r_L + R)(g_s + G)} + \frac{1}{\omega c_L} \cong \sqrt{\frac{r_L}{g_s}} + \frac{1}{\omega c_L}$$

$$\omega C_e = \sqrt{\frac{g_s + G}{r_L + R}}\sqrt{1 - (r_L + R)(g_s + G)} \cong \sqrt{\frac{g_s}{r_L}}. \tag{9.65}$$

It is clear from Equation (9.64) that r_L must be much larger than R to avoid power loss by the matching circuit. An estimate of R and G can be obtained by using Equations (9.61) and

(9.65), which gives

$$R = \frac{\omega L_e}{Q} \cong \frac{1}{Q}\left(\sqrt{\frac{r_L}{g_s}} + \frac{1}{\omega c_L}\right)$$

$$G = \frac{\omega C_e}{Q} \cong \frac{1}{Q}\sqrt{\frac{g_s}{r_L}}. \tag{9.66}$$

Defining the load quality factor $Q_L = 1/\omega r_L c_L$, Equation (9.65) gives a condition for efficient impedance matching

$$r_L \gg R \cong \frac{1}{Q}\left(\sqrt{r_L r_s} + \frac{1}{\omega c_L}\right) \text{ or } Q \gg \sqrt{\frac{r_s}{r_L}} + Q_L. \tag{9.67}$$

The matching circuit must be designed so that Equation (9.67) is not violated. Alternately, the FEA designer, faced with unavoidable circuit losses, must design both the FEA and the FEA packaging with Equation (9.67) in mind. The matching circuit can be realized in the variety of ways. Stub transmission lines near the emitting area can add the shunt inductance needed to match the capacitance of the FEA. For narrowband operation, a quarter-wave impedance transformer [146] can be used. Lumped-element circuits are often more compact, but suffer from low quality factors.

The emission current is modulated by the *rf* voltage that is applied to the gate $\widetilde{V}_g^{rf}$, as shown in Figure 9.38b. This voltage is given by

$$|\widetilde{V}_g^{rf}|^2 = \frac{2P_O^{rf}}{r_L(\omega c_L)^2} = \frac{2g_s P_A^{rf}}{(\omega c_L)^2(g_s + G)(r_L + R)} \cong \frac{2Q_L}{\omega c_L} P_A^{rf}. \tag{9.68}$$

As an example, at 10 GHz each quadrant of the FEA cathode [148] has input impedance $Z_{FEA} = 2.5 - j12\Omega$ and requires a peak-to-peak voltage of approximately 20 V ($|\widetilde{V}_g^{rf}| = 10$ V) to modulate the emission current. The required power, assuming lossless matching, is

$$P_A^{rf} = \frac{\omega c_L |\widetilde{V}_g^{rf}|^2}{2Q_L} = \frac{100}{2 \times 12.5 \times (12.5/2)} = 0.6 \text{ W}.$$

9.7.1.3 Beam-Current Modulation

In the absence of transit time delays, the modulated beam current $I_b(t)$ of a gated cathode is given by substituting the gate voltage $V_g(t)$ into the current–voltage relation of the cathode. In the case of a gated FEA, the voltage modulation is usually sinusoidal, and the current–voltage relation of an FEA is taken to be the F-N relation of Equation (9.43). Because the characteristic curve is nonlinear, the resulting beam-current waveform will include harmonic frequencies. Computer simulations must be used to exactly obtain the emission current modulation that results from a given gate voltage modulation. However, an approximate analysis, coupled with Equation (9.68), can be used to estimate the beam-current modulation produced by the FEA. The gate voltage is assumed to be

$$V_g(t) = V_g^{dc} - V_g^{rf}\cos(\omega t). \tag{9.69}$$

Then, defining $\chi = V_g^{rf}/V_g^{dc}$ and using Equation (9.43), the emission current is

$$I_b = A[V_g^{dc} - V_g^{rf}\cos(\omega t)]^2 \exp\left[-\frac{B}{V_g^{dc} - V_g^{rf}\cos(\omega t)}\right]$$

$$\cong A(V_g^{dc})^2 \exp\left[-\frac{B}{V_g^{dc}[1-\chi\cos(\omega t)}\right]. \quad (9.70)$$

By Fourier analysis,

$$\frac{1}{1-\chi\cos(\omega t)} \cong 1 + \frac{2\chi}{1+\sqrt{1-\chi^2}}\cos(\omega t). \quad (9.71)$$

If Equation (9.71) is inserted into Equation (9.70), the emission current can be expressed in terms of the fundamental and harmonic frequencies as

$$I_b(t) \cong A(V_g^{dc})^2[I_0(\delta) + 2\sum_{k=1}^{\infty} I_k(\delta)\cos(\omega t)]. \quad (9.72)$$

In Equation (9.72),

$$\delta = \frac{2B}{V_g^{dc}\sqrt{1-\chi^2}}\left(\frac{\chi}{1+\sqrt{1-\chi^2}}\right),$$

and $I_k(z)$ is the modified Bessel function of the first kind. The identity

$$e^{z\cos(\theta)} = I_0(z) + 2\sum_{k=1}^{\infty} I_k(z)\cos(k\theta)]$$

has been used [149]. The appropriate modulation will depend on the application: the frequency multiplier will require a more strongly modulated beam than a linear amplifier. Equation (9.72) is a good indicator of the fraction of the beam energy that can be converted to electromagnetic energy in the fundamental frequency, provided that the inductive output circuit only extracts power from the beam to the circuit. If the output circuit is lengthened to increase the modulation of the beam before extraction begins, space-charge effects and nonlinear interactions between the beam and the inductive output can result in conversion of power between the harmonics [150–153].

Writing $I_b(t) = I_b^{dc} + Re[\sim I_b^{rf} e^{j\omega t}]$, the *dc* and *rf* components of the beam current are given by

$$I_b^{dc} = A(V_g^{dc})^2 \exp\left[-\frac{B}{V_g^{dc}\sqrt{1-\chi^2}}\right] I_0(\delta)$$

$$|\widetilde{I}_b^{rf}| = 2A(V_g^{dc})^2 \exp\left[-\frac{B}{V_g^{dc}\sqrt{1-\chi^2}}\right] I_1(\delta) = 2I_b^{dc}\frac{I_1(\delta)}{I_0(\delta)}. \quad (9.73)$$

The values of χ and δ can be estimated for optimal impedance matching, using Equation (9.68).

The transconductance of a voltage-controlled current source is another indicator of the efficiency by which gate-voltage *rf* modulation is converted to emission-current *rf* modulation [153]. It is defined as the incremental change in beam current divided by the incremental change in gate potential, $g_m = \partial I_b / \partial V_g$. In the absence of transit time effects, the transconductance is the slope of the characteristic curve $I_b(V_g)$; if the characteristic curve is nonlinear, the transconductance will depend upon V_g. The transconductance of a gated FEA is thus

$$g_m = \frac{\partial I_b}{\partial V_g} = AV_g\left(2 + \frac{B}{V_g}\right)\exp\left(-\frac{B}{V_g}\right) = \left(2 + \frac{B}{V_g}\right)\frac{I_b}{V_g}. \tag{9.74}$$

This transconductance, like the current, is exponentially sensitive to the F-N B parameter. To relate cathode performance to the gain of an IOA, a generalized transconductance α may be defined as the incremental *rf* current that results for an increment in *rf* gate-drive power, that is, $\alpha = \partial|\widetilde{I}_b^{rf}|/\partial P_g{}^{rf}$. Since the application of an oscillating potential to the gate performs the emission gating, α is related to g_m as

$$\alpha = \frac{\partial |I_b^{rf}|}{\partial P_g^{rf}} = \frac{\partial |I_b^{rf}|}{\partial V_g} \times \frac{\partial V_g}{\partial V_g^{rf}} \times \frac{\partial V_g^{rf}}{\partial P_g^{rf}} = g_m \times 1 \times \frac{\partial V_g^{rf}}{\partial P_g^{rf}} = \frac{2Q_L}{\omega c_L} P_A^{rf} g_m. \tag{9.75}$$

In Equation (9.75), Equation (9.68) has been used to relate $V_g{}^{rf}$ to $P_A{}^{rf}$. This, of course, assumes optimal impedance matching. More generally, the relation between the drive power and the *rf* voltage at the gate depends upon the input circuit as discussed earlier.

9.7.2 Current Density

In most cases, both emission current and current density are limited by reliability considerations. These are discussed below. However, fundamental limits apply to the current that can be obtained from field emitter cathodes.

9.7.2.1 Space Charge

As emission current from the cathode increases, the reduction of the field near the cathode by the space charge of the emitted electrons can no longer be neglected. For an FEA diode, the space-charge-limited current density is described by the Langmuir-Child law of Equation (9.41) and is determined by the reduction of the extraction field at the tip by the space charge of the emitted electrons [154, 155]. However, for a gated field emitter, space charge does not greatly diminish the tip field but rather, gives rise to large gate current. Because of the high current density that is required, such space-charge effects must be considered in any microwave tube design.

The one-dimensional analysis in Ref. [156] can be extended to provide some insight into the nature of these effects. A gated FEA with *dc* voltages V_g and V_a applied to the gate and anode, respectively, is depicted in Figure 9.39. To minimize confusion, the polarities of the

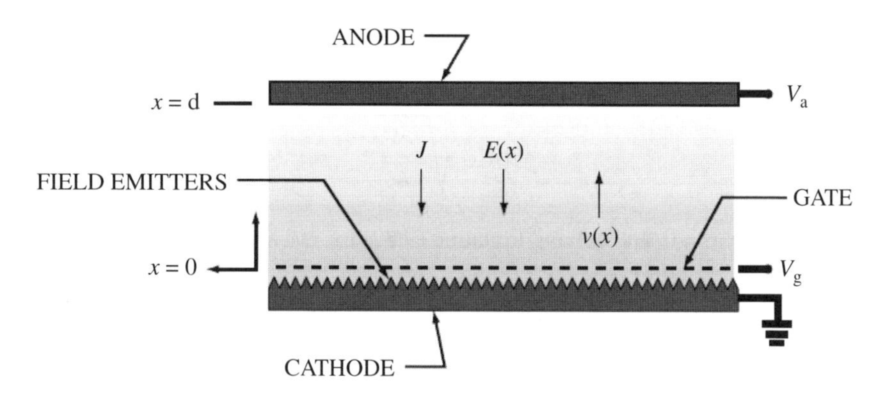

Figure 9.39 Simple model used to estimate space-charge effects in a gated-FEA cathode [8]. Copyright (2001) by John Wiley & Sons, Inc. This material is reproduced with permission of John Wiley & Sons, Inc.

current density J, electric field $E(x)$, and electron velocity $v(x)$ are defined to be positive for electron flow from the cathode to the anode. In the gate-anode region ($0 < x < d$), Poisson's equation relates the electrostatic potential $\phi(x)$ to the charge density $\rho(x)$:

$$\frac{\partial^2 \phi(x)}{\partial x^2} = \frac{\rho(x)}{\varepsilon_0}. \tag{9.76}$$

For static conditions, the current density J is independent of x and is given by

$$J = \rho(x)v(x). \tag{9.77}$$

Because field emitters operate in UHV, any electron collisions with gaseous molecules can be neglected, so that $v(x)$ is given by

$$\frac{m}{2}v^2(x) = e\phi(x). \tag{9.78}$$

Using the polarity definitions of Figure 9.39, the electric field $E(x)$ is given as

$$E(x) = \frac{d\phi(x)}{dx} = \frac{m}{e}\frac{dv}{dt}. \tag{9.79}$$

Differentiating Equation (9.79),

$$\frac{d^2\phi}{dx^2} = \frac{m}{e}\frac{d}{dx}\left(\frac{dv}{dt}\right) = \frac{m}{e}\frac{d}{dt}\left(\frac{dv}{dt}\right) = \frac{m}{ev}\frac{d^2v}{dt^2}. \tag{9.80}$$

In view of Equations (9.76) and (9.77), Equation (9.80) becomes

$$\frac{d^2v}{dt^2} = \frac{eJ}{\varepsilon_0 m}. \tag{9.81}$$

The emitted electrons are assumed incident upon the gate-anode region with a velocity derived from the gate voltage, that is,

$$v(0) = \sqrt{\frac{2eV_g}{m}}. \tag{9.82}$$

Defining $t = 0$ at $x = 0$, and solving Equation (9.82) for $v(t)$ and $x(t)$,

$$v(t) = \frac{eJ}{2m\varepsilon_0}t^2 + \frac{eE_s}{m}t + \sqrt{\frac{2eV_g}{m}}$$

$$x(t) = \frac{eJ}{6m\varepsilon_0}t^3 + \frac{eE_s}{m}t^2 + \sqrt{\frac{2eV_g}{m}}t. \tag{9.83}$$

In Equation (9.83), E_s is the electric field at $x = 0$. If the emitted electrons reach the anode at time T, then at $t = T$ Equation (9.83) becomes

$$\sqrt{\frac{2eV_a}{m}} = \frac{eJ}{2m\varepsilon_0}T^2 + \frac{eE_s}{m}T + \sqrt{\frac{2eV_g}{m}}$$

$$d(t) = \frac{eJ}{6m\varepsilon_0}T^3 + \frac{eE_s}{m}T^2 + \sqrt{\frac{2eV_g}{m}}T. \tag{9.84}$$

In a gated FEA, J is determined by V_g through Equation (9.43) and is thus a given quantity. As a result, Equation (9.84) determines E_s and T as a function of V_g, V_a, J, and d. This allows $x(t)$ and $\phi[x(t)]$ to be determined from Equations (9.83) and (9.78). Figure 9.40 displays the dependence of ϕ upon x for several values of emission current, using the FEA parameters of Figure 9.41.

As J increases, $E_s = d\varphi/dx$ diminishes until, analogous to the Child–Langmuir law, $E_s = 0$ at current density J_L given by

$$J_L = J_{CL}\left(1 - \sqrt{\frac{V_g}{V_a}}\right)\left(1 + 2\sqrt{\frac{V_g}{V_a}}\right)^2. \tag{9.85}$$

The quantity J_L only roughly estimates the upper limit of the current density. A number of important factors have been neglected in this simple analysis, including the two-dimensional (2-D) geometry of the FEA and the spreading of the emitted-electron beam. In most cases, numerical simulations [157] must be used to accurately determine these effects. More importantly, the redirection of the emission current from the anode to the gate electrode occurs at current densities lower than J_L. The resulting high gate current degrades FEA performance and enhances failure probability. Figure 9.41 shows experimental data from a 6100-tip array that has been tested in a UHV probing apparatus [134]. The FEA was approximately 25×25 μm in area and the probe anode was spaced about 0.46 mm from the FEA. Figure 9.41a shows how anode current saturates as a result of space-charge effects.

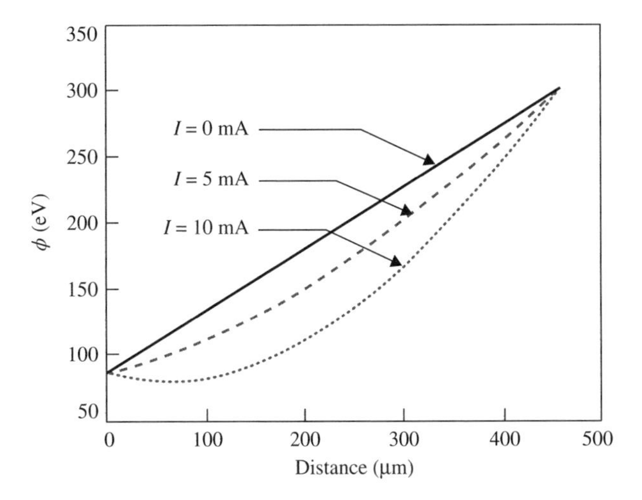

Figure 9.40 Calculated dependence of the electrostatic potential in the gate–anode region for several values of emitted current. As the emission current increases, the electric field near the cathode surface diminishes. At sufficiently high emission current, increased gate current results [8]. Copyright (2001) by John Wiley & Sons, Inc. This material is reproduced with permission of John Wiley & Sons, Inc.

As the anode voltage is increased, higher values of emitted current can be achieved, as it is indicated by Equation (9.85). Figure 9.41b shows F-N plots of the same data and includes the gate current. The departure of the anode current from F-N dependence and the accompanying gate-current increase are evident.

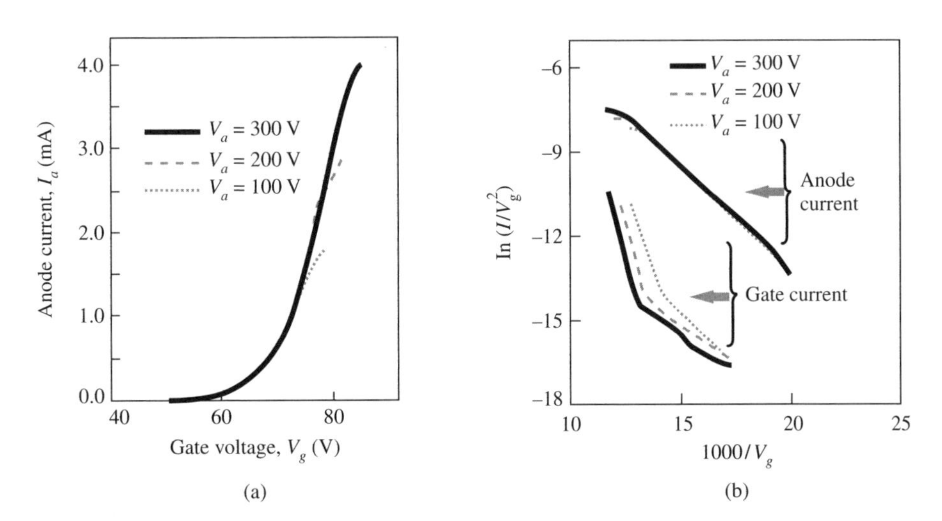

Figure 9.41 Experimental manifestation of space-charge effects. (a) Anode current vs. gate voltage, showing the saturation of anode current at high emission levels because of the space charge. (b) F-N plots of gate and anode current, showing the increase of gate current that accompanies anode-current saturation [8]. Copyright (2001) by John Wiley & Sons, Inc. This material is reproduced with permission of John Wiley & Sons, Inc.

Since gain depends so strongly upon minimizing the gate-to-cathode capacitance, a small-area source that operates near peak intensity will generally provide the best simultaneous gain and efficiency. Since efficiency is also improved when the bulk of the beam passes close to the output circuit electrodes, the optimum electron beam geometry is a thin annulus. Because such compact annular beams improve the performance of rf output couplers, it is advantageous to draw the maximum current density consistent with a reasonable cathode lifetime. This raises issues in electron gun design, including initial velocity effects, beam spreading, axial demodulation, and beam stability. All these are of concern in a design context. Electron guns for IOAs should be designed to exploit cathodes such as FEAs that are capable of emitting hundreds of amperes per square centimeter.

9.7.3 CNT FEAs

Current cold cathode technology suffers because it is based upon integrated grid systems where the spacing between the cathode and the grid is filled with an insulator (typically SiO_2 or Al_2O_x), which leads to unavoidably high capacitance. Hence, they cannot be directly modulated at gigahertz frequencies and can only provide the primary (*dc*) beam [158].

In the case of CNT arrays, this modulating grid is not integrated and can be located at an extended distance of 10–100 μm from the emitters, where the space between emitter and grid is evacuated (10^{-6} mbar) with a consequent reduction in grid/cathode capacitance of the order of 20–50 times, therefore allowing for high frequency operation. The schematic of such an amplifier is shown in Figure 9.42.

In order to be commercially viable and truly competitive with thermionic cathodes for this application, CNT cathodes must deliver current densities in the range of 1–2 A/cm^2. This can only be achieved by controlling the uniformity in height and diameter of the CNTs by understanding their detailed deposition characteristics. Indeed, various CNT types can be grown. By way of an example, Figure 9.43 shows an emitter formed from nominally metallic multiwalled CNTs.

There has been demonstrated the operation of this system at 1.5 GHz using various radio frequency-input powers to generate different macroscopic electric fields at the apex

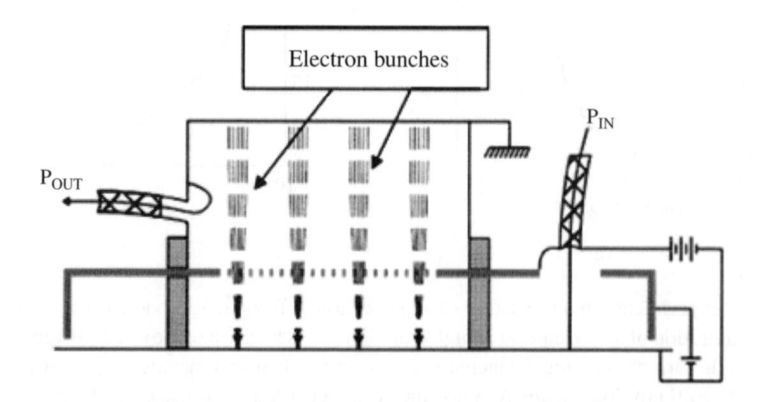

Figure 9.42 Schematic of CNT based vacuum microwave amplifier. Reproduced from Ref. [159] with permission from the Royal Society of Chemistry. Copyright (2004) Royal Society of Chemistry

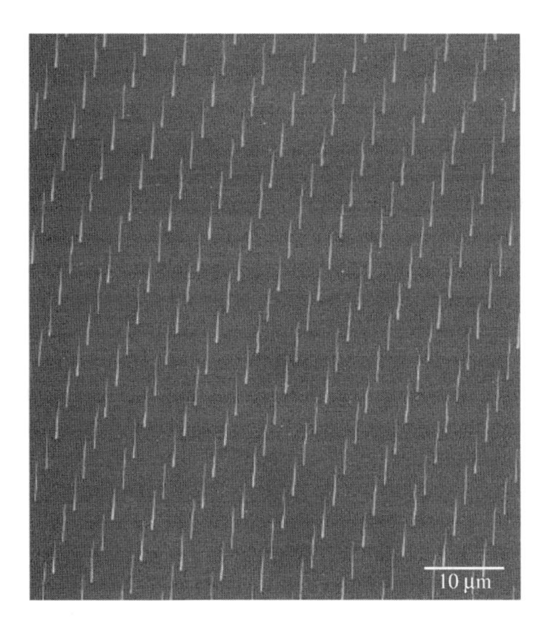

Figure 9.43 Array of MWCNTs with σ tip diameter (~4.1%) and σ height (~6.3%). Copyright (2012) IEEE. Reprinted with permission from Ref. [158]

of carbon-nanotube emitters [158]. The spectrum analyzer attached to the output antenna confirmed the presence of the fundamental 1.5-GHz peak in the cavity. The cathodes were operated at 1 mA and 1.5 GHz for 40 hours without degradation or a decrease in current output (within the measurement error of 5%). With the applied radiofrequency electric field of 29 MV/m, the output at the anode reaches 3.2 mA, with an average current density of 1.3 A/cm^2. This corresponds to peak current of 30 mA and a current density of 12 A/cm^2 in the output waveform [160]. There has been demonstrated the modulation at 32 GHz with 1.4 A/cm^2 peak current density from 82% modulation [160]. In this work, a high cathode–grid spacing (100 μm) was used and the applied field was modulated using a resonant cavity. With this configuration, the cathode–grid capacitance was extremely low making it compatible with very high frequency operation. However, the consequence of using a resonant cavity is that the modulated electron source can only be used in narrow bandwidth amplifiers. In order to circumvent this there has been developed a CNT-based optically controlled field emission cathode [160, 161]. In this case, as opposed to the cathodes described above, the applied electric field is constant. The modulation of the emitted current is obtained through optical excitation. Such devices are thus compatible with high frequency and very large bandwidth operation is possible [161]. Using this new photocathode concept, the first CNT-based photocathode using Si *p–i–n* photodiodes and MWCNT bundles has been fabricated.

This photocathode consists of an array of MWCNTs, in which each MWCNT is electrically connected to an n^+ doped area defined in an intrinsic layer grown on a p^+ doped semiconducting substrate (Figure 9.44). Due to the localization of the n^+ areas, the *p–i–n* diodes are independent of each other. Under an applied field, the MWCNTs emit and the diodes are reversely biased. The voltage drop (ΔV) appears across the diodes, mainly in depletion regions that extend under each n^+ area. Upon illumination by photons, with energy exceeding the band gap of the semiconductor, electron–hole pairs are generated in or near to the depletion region. The carriers are then swept across the depletion region by the electric field and electrons are subsequently emitted by the MWCNTs. When the optical excitation is turned off, the diode current becomes very small (equal to the leakage current of the diode). The electrons previously accumulated at the MWCNT apex are still field emitted though the available electron population to be emitted is substantially reduced. Thus, the emitted current decreases and adjusts to the diode leakage current. During this process, as the emission current has been higher than the diode current, the MWCNTs have become positively biased compared to the substrate. The voltage drop within the diode (ΔV) has attained its maximum value. It is of prime importance to estimate this value and to design the photocathode in order to avoid diode breakdown. Using such device, for the first time, optical modulation of the emission current from CNTs by pulsing the laser source as shown in Figures 9.44 and 9.45 has been demonstrated.

As shown in Figure 9.45a, the device was tested under ultra-high vacuum (10^{-9} mbar) in a triode configuration with cathode–grid distance of 100 μm. The anode consists of a glass plate coated with a thin ITO layer. The photocathode is then illuminated through this transparent and conductive anode, using the optical source. To create free carriers in the intrinsic Si, photons of energy roughly equal to or greater than the band gap (1.12 eV for silicon, corresponding to a wavelength of less than 1100 nm) have to be absorbed. A 532 nm green laser was consequently employed as the optical source. Using a controlled laser-pulse at this frequency this photocathode delivers 0.5 mA with an internal quantum efficiency of 10% and an I_{ON}/I_{OFF} ratio of 30.

In Figure 9.45b, the emitted current of the photocathode is plotted as a function of the cathode–grid applied voltage and for absorbed optical powers P_{OPT} ranging from 0 to 11.8 mW (left axis). The black curve represents the non-illuminated photocathode emission current, and the remaining colored curves represent the photocathode emission current for various absorbed optical powers (P_{OPT}). The absorbed optical powers were determined taking into account the

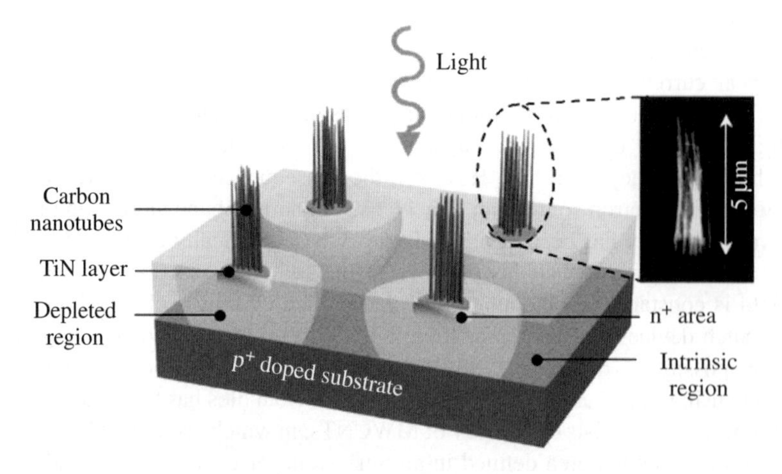

Figure 9.44 Schematic view of the new MWCNT based photocathode [161]. Copyright (2008) IOP Publishing. Reproduced with permission. All rights reserved

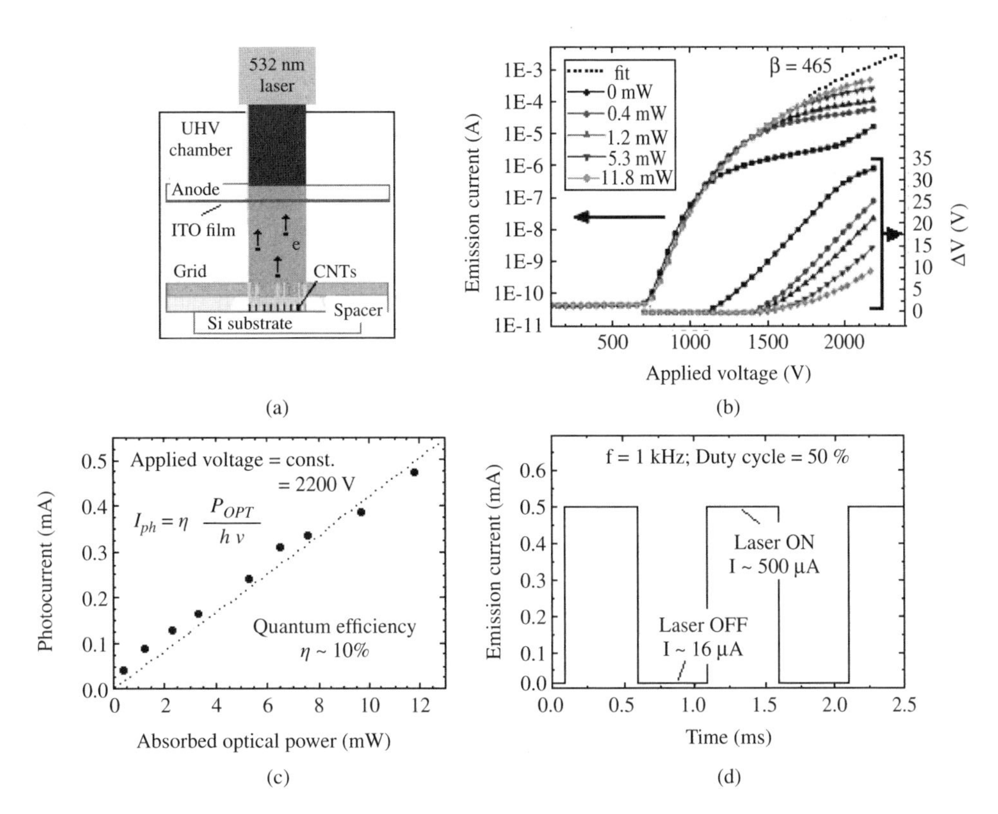

Figure 9.45 (a) Schematics of the experimental setup with the cathode–grid assembly (spacer thickness = 100 μm), the transparent and conductive anode, and the 532 nm laser; (b) the emitted current of the photocathode as a function of the cathode–grid applied voltage; (c) photocurrent on absorbed optical power dependence; and (d) switch on–switch off test [161]. Copyright (2008) IOP Publishing. Reproduced with permission. All rights reserved

transparency of the ITO/glass, the grid and the reflection from the Si surface. The dashed curve is the Fowler-Nordheim (F-N) fit for the experimental curves in the low emission current region, that is, where the effect of the limiting current of the *p–i–n* diode is negligible. At higher voltages, the curves deviate from the F-N law and exhibit different saturations, corresponding to the different absorbed optical powers. In this saturation regime, a voltage drop appears across the *p–i–n* diode and limits the emission current because the *p–i–n* photocurrent is insufficient to feed the emission, even under high illumination.

The first demonstration was performed at relatively low frequency as shown in Figure 9.45d. The device was operated for 2 months at 0.5 mA without any observable degradation in current. This is due to the optical stability of *p–i–n* diodes and excellent emission stability of the MWCNTs. By incorporating high frequency *p–i–n* photodiodes (e.g., GaInAs photodiodes) into the structure and driving them with 1.55 μm telecommunication lasers, photocathodes operating in the 10–30 GHz range can be envisaged.

9.8 Conclusion

The consideration and analysis of possible application of the quantum electron sources in high frequency vacuum nanoelectronics devices showed that some new approaches based on realization of resonant tunneling effect and Gunn effect are very promising. Some proposals for generation of THz signals both in solid state and in field emission vacuum devices have been presented. The theory of field emission microwave sources including modulation gated FEA, current density has been considered in detail. Gated FEA cathodes have operational characteristics that should enable superior performance in microwave inductively output amplifiers. Further, the modulation of FEAs at microwave frequencies is possible, and sufficiently high emission currents have been experimentally demonstrated in clean and well-controlled environments.

The further downscaling of cathodes in conjunction with low work function materials may decrease the turn-on gate voltage to less than 1 V, thus enabling these devices to be competitive with modern semiconductor technology. These benefits can be attained by the use of matured IC technology to fabricate nanoscale vacuum tubes and facilitate circuit integration.

References

1. K.F. Brennan and A.S. Brown, *Theory of Modern Electronic Semiconductor Devices* (John Wiley & Sons, Inc., 2002).
2. M. Feiginov, C. Sydlo, O. Cojocari, and P. Meissner, Resonant-tunnelling-diode oscillators operating at frequencies above 1.1 THz, *Applied Physics Letters* **99**, 233506 (2011).
3. K.F. Brennan, and P.P. Ruden, *Topics in High Field Transport in Semiconductors,* (Singapore: World Scientific, 2001).
4. H.C. Liu, and T.C.L.G. Sollner, High-frequency resonant-tunneling devices, in *Semiconductors and Semimetals*, ed. R.K. Willardson and A.C. Beer (San Diego, CA: Academic Press, 1994), pp. 359–419.
5. S.M. Sze and K.K. Ng, *Physics of Semiconductor Devices*, (John Wiley & Sons, Inc. New York, 2007).
6. J.M. Early, Maximum rapidly-switchable power density in junction triodes, *IRE Transactions on Electron Devices* **6**, 322–325 (1959).
7. E.O. Johnson, Physical limitations of frequency and power parameters of transistors, *RCA Review* **26**, 163–177 (1965).
8. R.A. Murphy, M.A. Kodis, Cold cathode microwave devices, in *Vacuum Microelectronics*, ed. W. Zhu (John Wiley & Sons, Inc., New York, 2001) pp. 349–391.
9. K.L. Jensen, Theory of field emission in *Vacuum Microelectronics*, ed. W. Zhu (New York, John Wiley & Sons, Inc., 2001), pp. 33–104.
10. A.A. Evtukh, V.G. Litovchenko, R.I. Marchenko, *et al.*, Parameters of the tip arrays covered by low work function layers, *Journal of Vacuum Science and Technology B* **14**, 2130 (1996).
11. K.L. Jensen and F.A. Buot, The methodology of simulating particle trajectories through tunneling structures using a Winger distribution approach, *IEEE Transactions on Electron Devices* **ED-38**, 2337 (1991).
12. V.G. Litovchenko and Yu. V. Kryuchenko, Field emission from structures with quantum wells, *Journal of Vacuum Science and Technology B* **11**, 362 (1993).
13. A.A. Evtukh, V.G. Litovchenko, R.I. Marchenko, *et al.*, The enhanced field emission from microtips covered by ultrathin layers, *Journal of Vacuum Science and Technology B* **15**, 439 (1997).
14. A.A. Evtukh, V.G. Litovchenko, and M.O. Semenenko, Electrical and emission properties of nano-composite SiO_x(Si) and SiO_2(Si) films, *Journal of Vacuum Science and Technology B* **24**, 945 (2006).
15. L.D. Filip, M. Palumbo, J.D. Carey, and S.R.P. Silva, Two-step electron tunneling from confined electronic states in a nanoparticle, *Physical Review B* **79**. 245429 (2009).
16. O.E. Raichev, Coulomb blockade of field emission from nanoscale conductors, *Physical Review B* **73**, 195328 (2006).
17. C. Kim, H.S. Kim, H. Qin, and R.H. Blick, Coulomb-controlled single electron field emission via a freely suspended metallic island, *Nano Letters* **10**, 615 (2010).

18. W.M. Tsang, V. Stolojan, B.J. Sealy, *et al.*, Electron field emission properties of Co quantum dots in SiO_2 matrix synthesised by ion implantation, *Ultramicroscopy* **107**, 819 (2007).
19. H. Shimawaki, Y. Neo, H. Mimura, *et al.* (2010) Photo-assisted electrom emission from MOS-type cathode based on nanocrystalline silicon. Proceedings of the 23rd International Vaccum Nanoelectronic Conference, Palo Alto, CA, July 26–30, 2010, pp. 74–75.
20. V. Litovchenko, A. Evtukh, Yu. Kryuchenko, *et al.*, Quantum size resonance tunneling in the field emission phenomenon, *Journal of Applied Physics* **96**, 867 (2004).
21. V. Litovchenko and A. Evtukh, Vacuum nanoelectronics, In *Handbook of Semiconductor Nanostructures and Nanodevices*, Vol. **3**. Spintronics and Nanoelectronics, ed. A.A. Balandin and K.L. Wang (American Scientific Publishers, Los Angeles, CA. 2006), pp. 153–234.
22. H. Mimura, Y. Abe, J. Ikeda, *et al.*, Resonant Fowler–Nordheim tunneling emission from metal-oxide-semiconductor cathodes, *Journal of Vacuum Science and Technology B* **16**, 803 (1998).
23. A. Evtukh, V. Litovchenko, N. Goncharuk, and H. Mimura, Electron emission Si-based resonant-tunneling dio, *Journal of Vacuum Science and Technology B* **30**, 022207 (2012).
24. N. M. Goncharuk, The influence of an emitter accumulation layer on field emission from a multilayer cathode, *Materials Science and Engineering* **A353**, 36, (2003).
25. A. Matiss, A. Poloczek, W. Brockerhoff, *et al.* (2007) Large-signal analysis and AC modelling of sub micron resonant tunneling diodes. European Microwave Conference, pp. 207–210.
26. A.S. Tager, Low-dimensional quantum effects in submicrometer semiconductor structures and prospects of their application in microwave electronics. I. Physical fundamentals, *Electronic techniques*, Electronics of Microwave Frequencies, **9**, 21 (1987) (in Russian).
27. R. Tsu and L. Esaki, Tunneling in a finite superlattice, *Applied Physics Letters* **22**, 562 (1973).
28. S. M. Sze, *Physics of Semiconductor Devices*, Chapter 10, vol. **2**, pp. 207–212 (Ed. New York: John Wiley & Sons, Inc., 1981).
29. K.R. Chu, The electron cyclotron maser, *Reviews of Modern Physics* **76**, 1 (2004).
30. X.C. Zhang, B. Xu, J. Darraw, and D. Auston, Generation of femtosecond electromagnetic pulses from semiconductor surfaces, *Applied Physics Letters* **56**, 1011 (1990).
31. V.L. Malevich, Monte Carlo simulation of THz-pulse generation from semiconductor surface, *Semiconductor Science and Technology* **17**, 551 (2002).
32. M. Nakajima and M. Hangyo, Study of THz radiation from semiconductor surfaces excited by femtosecond laser pulses under laser illumination, *Semiconductor Science and Technology* **19**, S264 (2004).
33. J.S. Hwang, H.C. Lin, K.I. Lin, and X.C. Zhang, Terahertz radiation from InAlAs and GaAs surface intrinsic-N^+ structures and the critical electric fields of semiconductors, *Applied Physics Letters* **87**, 121107 (2005).
34. R. Ascazubi, I. Wilke, K.J. Kim, and P. Dutta, Terahertz emission from $Ga_{1-x}In_xSb$, *Physical Review B* **74**, 075323 (2006).
35. W. Sha, A.L. Smirl, and W.F. Tseng, Coherent plasma oscillations in bulk semiconductors, *Physical Review Letters* **74**, 4273 (1995).
36. R. Kersting, K. Unterrainer, G. Strasser, *et al.*, Few-cycle THz emission from cold plasma oscillations, *Physical Review Letters* **79**, 3038 (1997).
37. A. Reklaitis, Monte Carlo analysis of terahertz oscillations of photoexcited carriers in GaAs *p-i-n* structures, *Physical Review B* **74**, 165305 (2006).
38. D.N. Thao, S. Katayama, T.D. Khoa, and M. Iida, Calculation of THz radiation due to coherent polar-phonon oscillations in *p–i–n* diode structure at high electric field, *Semiconductor Science and Technology* **19**, S304 (2004).
39. J.T. Lu and J.C. Cao, Terahertz generation and chaotic dynamics in GaN NDR diode, *Semiconductor Science and Technology* **19**, 451 (2004).
40. Z.S. Gribnikov, A.N. Korshak, and N.Z. Vagidov, Terahertz ballistic current oscillations for carriers with negative effective mass, *Journal of Applied Physics* **80**, 5799 (1996).
41. M. Eckardt, A. Schwanhauber, L. Robledo, *et al.*, Exotic transport regime in GaAs: absence of intervalley scattering leading to quasi-ballistic, real-space THz oscillations, *Semiconductor Science and Technology* **19**, S195 (2004).
42. J. Lusakowskia, J.W. Knap, N. Gyakonova, *et al.*, Voltage tuneable terahertz emission from a ballistic nanometer InGaAs/InAlAstransistor, *Journal of Applied Physics* **97**, 064307 (2005).
43. C. Waschke, H.G. Roskos, R. Schwedler, *et al.*, Coherent submillimeter-wave emission from Bloch oscillations in a semiconductor superlattice, *Physical Review Letters* **70**, 3319 (1993).

44. D.S. Ong and H.L. Hartnagel, Generation of THz signals based on quasi-ballistic electron reflections in double-heterojunction structures, *Semiconductor Science and Technology* **22**, 981 (2007).
45. H. Hartnagel, D.S. Ong, and S. Al-Daffaie (2013) Proposal of a THz signal generator by ballistic electron resonance device based on graphene, *Conference WOCSDICE 2013, Warnemuende, Germany, 2013.*
46. K. Yokoo and T. Ishihara, (1995) Field emission monotron for THz emission, *Conference Digest of IVMC'95, Oregon, 1995*, pp. 123–127.
47. M.-C. Lin, K.-H. Huang, P.-S. Lu, *et al.*, Field-emission based vacuum device for the generation of terahertz waves, *Journal of Vacuum Science and Technology B* **23**, 849 (2005).
48. F.G. Bass and V.M. Yakovenko, Theory of radiation from a charge passing through an electrically inhomogeneous medium, *Soviet Physics Uspekhi* **8**, 420 (1965).
49. B.M. Bolotovskii and G.M. Voskresenskii, Diffraction radiation, *Soviet Physics Uspekhi* **9**, 73 (1966).
50. P.M. Van den Berg and A.J.A. Nicia, Diffraction radiation from a charge moving past an obstacle, *Journal of Physics A* **9**, 1133 (1976).
51. M.J. Hagmann, Mechanism for resonance in the interaction of tunneling particles with modulation quanta, *Journal of Applied Physics* **78**, 25 (1995).
52. M.J. Hagmann, Simulations of the interaction of tunneling electrons with optical fields in laser-illuminated field emission, *Journal of Vacuum Science and Technology B* **13**, 1348 (1995).
53. M.J. Hagmann, Simulations of the generation of broadband signals from DC to 100 THz by photomixing in laser-assisted field emission, *Ultramicroscopy* **73**, 89 (1998).
54. M.J. Hagmann, M. Brugat, M. Rodriguez, *et al.*, Prototype optoelectronic device for generating signals from dc to 10 GHz by resonant laser-assisted field emission, *Journal of Vacuum Science and Technology B* **19**, 72 (2001).
55. K. Alonso and M.J. Hagmann, Comparison of three different methods for coupling of microwave and terahertz signals generated by resonant laser-assisted field emission, *Journal of Vacuum Science and Technology B* **19**, 68 (2001).
56. L. Kontorovich, M. V. Loginov, and V. N. Shrednik, Atom probe determination of the multicomponent material thermo-field microprotrusion parameters, *Journal of Vacuum Science and Technology B* **15**, 495 (1997).
57. H.W.P. Koops, C. Schossler, A. Kaya, and M. Weber, Conductive dots, wires, and supertips for field electron emitters produced by electron-beam induced deposition on samples having increased temperature, *Journal of Vacuum Science and Technology B* **14**, 4105 (1996).
58. M.J. Hagmann, Intensification of optical electric fields caused by the interaction with a metal tip in photofield emission and laser-assisted scanning tunneling microscopy, *Journal of Vacuum Science and Technology B* **15**, 597 (1997).
59. A. Wokaun, Surface enhancement of optical fields. Mechanism and applications, *Molecular Physics* **56**, 1 (1985).
60. H.L. Hartnagel and M. Dragoman (2010) Heterostructure field emitter of electrons to generate THz signals by laser pulses, *Technical Digest of IVNC 2010, Davis, CA*, pp. 42–43.
61. H.L. Hartnagel A.L. Dawar, A.K. Jain *et al.*, *Semiconductor Transparent Thin Films* (Institute of Physics Publishing, Bristol and Philadelphia, PA, 1995).
62. H.L. Hartnagel, D.S. Ong, and I. Oprea, Balistic electron wave swing (BEWAS) to generate THz-signal power, *Frequenz* **63**, 3 (2009).
63. S.-G. Jeon, J.-I. Kim, G.-J. Kim *et al.* (2012) CNT cold cathode based THz wave technology, Technical Digest of IVNC 2012, Jeju, Korea, 2012, pp. 104–105.
64. H.W.P. Koops, S.A. Daffaie, H.L. Hartnagel, and A. Rudzinski (2012) Development of a miniaturized dynatron THz-oscillator with a FEBIP system, *Technical Digest of IVNC 2012, Jeju, Korea, 2012*, pp. 70–73.
65. P. Mukherjee and B. Gupta, Terahertz (THz) frequency sources and antennas – A brief review, *International Journal of Infrared Milli Waves* **29**, 1091 (2008).
66. G. Ulisse, F. Brunetti, and A. Di Carlo (2009) Design of an electron gun with FEA cathode for THz devices, *Technical Digest of IVNC 2009, Hamamatsu, Japan, 2009*, pp. 225–226.
67. X. Li, C. Yang, G. Bai, *et al.*, Investigation of FEAs applied in vacuum electron gun, *Applied Surface Science* **215**, 249 (2003).
68. D.R. Whaley, B.M. Gannon, C.R. Smith, *et al.*, Application of field emitter arrays to microwave power amplifiers, *Plasma Science* **28**, 727 (2000).
69. P.H. Siegel, A. Fung, H. Manohara *et al.* (2001) Nanoklystron: a monolithic tube approach to THz power generation, *Proceedings of the 2nd International Conference on Space THz Technolohy, San Diego, CA, February 2001*, pp. 14–16.

70. M.C. Lin, D.N. Smithe, P.H. Stoltz, *et al.* (2009) A microfabricated klystron amplifier for Thz waves, *Technical Digest of IVNC 2009, Hamamatsu, Japan, 2009*, pp. 189–190.
71. J.A. Nation, L. Schächter, F.M. Mako, *et al.*, Advances in cold cathode physics and technology, *Proceedings of the IEEE* **87**, 885–889 (1999).
72. J. Christiansen and Ch. Schultheiss, Production of high current particle beams by low pressure spark discharges, *Zeitschrift fur Physik A* **290**, 35–41 (1979).
73. A. W. Cross, H. Yin, W. He, *et al.*, Generation and application of pseudospark-sourced electron beams, *Journal of Physics D: Applied Physics* **40**, 1953–1956 (2007).
74. L. Esaki and R. Tsu, Superlattice and Negative Differential Conductivity in Semiconductors, *IBM Journal of Research and Development* **14**, 61 (1970).
75. E. Schomburg, R. Scheuerer, S. Brandl, *et al.*, InGaAs/InAlAs superlattice oscillator at 147 GHz, *Electronics Letters* **35**, 1491 (1999).
76. S. Schomburg, K. Brandl, T. Hofbeck, *et al.*, Generation of millimeter waves with a GaAs/AlAs superlattice oscillator, *Applied Physics Letters* **72**, 1498 (1998).
77. H. Kroemer, (2000) Large-amplitude oscillation dynamics and domain suppression in a superlattice Bloch oscillator, preprint, cond-mat/0009311.
78. (a) Y. Romanov and Y.Y. Romanova, On a superlattice bloch oscillator, *Physics of the Solid State* **46**, 164 (2003); (b) Y. Romanov, L.G. Murokh, and N.J.M. Horing, Negative high-frequency differential conductivity in semiconductor superlattices, *Journal of Applied Physics* **93**, 4696 (2003).
79. V.I. Litvinov, V.A. Manasson, and L. Sadovnik (2000) GaN-based teraherz source. *Proceedings of the SPIE, 4111, Terahertz and Gigahertz Electronics and Photonics II,* pp. 116–123.
80. V. Litvinov, V. Manasson and D. Pavlidis, Short-period intrinsic Stark GaN/AlGaN superlattice as a Bloch oscillator, *Applied Physics Letters* **85**, 600 (2004).
81. M. Kumagai, S.L. Chuang, and H. Ando, Analytical solutions of the block-diagonalized Hamiltonian for strained wurtzite semiconductors, *Physical Review B* **57**, 15303 (1998).
82. H. Wei and A. Zunger, Calculated natural band offsets of all II-VI and III-V semiconductors: Chemical trends and the role of cation *d* orbitals, *Applied Physics Letters* **72**, 2011 (1998).
83. S. Seo, D. Pavlidis, T. Karaduman, and O. Yilmazoglu (2006) Fabrication and DC characterization of AlGaN/GaN SL diode for millimeter wave generation. Proceedings 30th Workshop on Compound Semiconductor Devices and Integrated Circuits (WOCSDICE 2006), May 2006, pp. 51–53.
84. A.E. Belyaev, O. Makarovsky, D.J. Walker, *et al.*, Resonance and current instabilities in AlN/GaN resonant tunnelling diodes, *Physica E* **21,** 752 (2004).
85. C.T. Foxon, S.V. Novikov, A.E. Belyaev, *et al.*, Current–voltage instabilities in GaN/AlGaN resonant tunnelling structures, *Physica Status Solidi C*, 2389 (2003).
86. S. Seo, D. Pavlidis, and O. Yilmazoglu (2007) DC and negative capacitance properties of AlxGa1-xN/GaN superlattice diodes. Proceedings of the 31st Workshop on Compound Semiconductor Devices and Integrated Circuits (WOCSDICE 2007), May 2007, pp. 61–63.
87. A.G.U. Perera, W.A. Shen, M. Ershov, *et al.*, Negative capacitance of GaAs homojunction far-infrared detectors, *Applied Physics Letters* **74**, 3167, (1999).
88. X. Wu, S. Yang, and H.L. Evans, Negative capacitance at metal-semiconductor interfaces, *Journal of Applied Physics* **68**, 2485 (1990).
89. N.C. Chen, P.Y. Wang, and J.F. Chen, Low frequency negative capacitance behavior of molecular beam epitaxial GaAs *n*-low temperature-*i*-*p* structure with low temperature layer grown at a low temperature, *Applied Physics Letters* **72**, 1081 (1998).
90. M. Ershov, H.C. Liu, L. Li, *et al.*, Unusual capacitance behavior of quantum well infrared photodetectors, *Applied Physics Letters* **70**, 1828 (1997).
91. C. Sirtori, F. Capasso, J. Faist, *et al.*, Resonant tunneling in quantum cascade lasers, *IEEE Journal of Quantum Electronics* **34**, 1722, (1998).
92. F. Sudradjat, W. Zhang, K. Driscoll, *et al.*, Sequential tunneling transport characteristics of GaN/AlGaN coupled-quantum-well structures, *Journal of Applied Physics* **108**, 103704 (2010).
93. T. Unuma, M. Yoshita, T. Noda, *et al.*, Intersubband absorption linewidth in GaAs quantum wells due to scattering by interface roughness, phonons, alloy disorder, and impurities, *Journal of Applied Physics* **93**, 1586 (2003).
94. M. Boucherit, A. Soltani, E. Monroy, *et al.*, Investigation of the negative differential resistance reproducibility in AlN/GaN double-barrier resonant tunnelling diodes, *Applied Physics Letters* **99**, 182109 (2011).

95. N. Zainal, S.V. Novikov, C.J. Mellor, *et al.*, Current-voltage characteristics of zinc-blende (cubic) $Al_{0.3}Ga_{0.7}N$/GaN double barrier resonant tunneling diodes, *Applied Physics Letters* **97**, 112102 (2010).
96. C. Bayram, Z. Vashaei, and M. Razeghi, Reliability in room-temperature negative differential resistance characteristics of low-aluminum content AlGaN/GaN double-barrier resonant tunneling diodes, *Applied Physics Letters* **97**, 181109 (2010).
97. L. Rigutti, G. Jacopin, A.D. Bugallo, *et al.*, Investigation of the electronic transport in GaN nanowires containing GaN/AlN quantum discs, *Nanotechnology* **21**, 425206 (2010).
98. D. Li, L. Tang, C. Edmunds, *et al.*, Repeatable low-temperature negative-differential resistance from $Al_{0.18}Ga_{0.82}N$/GaN resonant tunneling diodes grown by molecular-beam epitaxy on free-standing GaN substrates, *Applied Physics Letters* **100**, 252105 (2012).
99. D Li, J Shao, L Tang, *et al.*, Temperature-dependence of negative differential resistance in GaN/AlGaN resonant tunneling structures, *Semiconductor Science and Technology* **28**, 074024–074028 (2013).
100. O. Vanbesien, R. Bouregba, P. Mounaix, and D. Lippens, Temperature dependence of peak to valley ratio in resonant tunneling double barriers in *Resonant Tunneling in Semiconductors*, ed. L.L. Chang, E.E. Mendez, C. Tejedor (NATO ASI Series B, Vol. **277**) (New York: Plenum), 1991, pp. 107–116.
101. D.N. Nath, Z.C. Yang, C.-Y. Lee, *et al.*, Unipolar vertical transport in GaN/AlGaN/GaN heterostructures, *Applied Physics Letters* **103**, 022102 (2013).
102. V.N. Sokolov, K.W. Kim, V.A. Kochelap, and D.L. Woolard, Terahertz generation in submicron GaN diodes within the limited space-charge accumulation regime. *Journal of Applied Physics* **98**, 064507 (2005).
103. E. Alekseev and D. Pavlidis, Large-signal microwave performance of GaN-based NDR diode oscillators, *Solid State Electronics* **44**, 941 (2000).
104. O. Yilmazoglu, K. Mutamba, D. Pavlidis, and T. Karaduman, Measured negative differential resistivity for GaN Gunn diodes on GaN substrate, *Electronics Letters* **43**, 480 (2007).
105. J.E. Carroll, *Hot Electron Microwave Generators* (Edward Arnold Publishers, London, 1970).
106. M.E. Levinshtein, Yu. K. Pozela, and M.S. Shur, *Gunn Effect* (Soviet Radio, Moscow, 1975) (in Russian).
107. K. Yokoo, (1999) New approach for high frequency electronics based on field emission array, in Proceedings of the Second Workshop on Vacuum Microelectronics, *Proceedings of the Second Workshop on Vacuum Microelectronics, Wroclaw, Poland, 1999*, pp. 30–40.
108. O. Yilmazoglu, H. Hartnagel, H. Mimura, *et al.* (2000) Gunn effect based modulation beam generation with GaAs lateral field emitter structures, *Proceedings of the Second European Field Emission Workshop, Spain, 2000.*
109. A. Evtukh, H. Hartnagel, V. Litovchenko, and O. Yilmazoglu, Two mechanisms of negative dynamic conductivity and generation of oscillations in field-emissions structures, *Materials Science and Engineering A* **353**, 27 (2003).
110. V.E. Chayka, N.F. Zhovnir, and D.V. Mironov, Comparative characteristics of diode and triode structures of vacuum microelectronics based on ip cathodes, *Electronics and Communications* **17**, 89 (2002) (in Russian).
111. F.J. Himpsel, J.A. Knapp, J.A. Van Vechten, and D.E. Eastman, Quantum photoyield of diamond(111) – A stable negative-affinity emitter, *Physical Review B* **20**, 624 (1979).
112. J. Weide, Z. Zhang, P.K. Baumann, *et al.*, Negative-electron-affinity effects on the diamond (100) surface, *Physical Review B* **50**, 5803 (1994).
113. P.K. Baumann and R.J. Nemanich, Negative electron affinity effects on H plasma exposed diamond (100) surfaces, *Diamond and Related Materials* **4**, 802 (1995).
114. R.E. Thomas, T.P. Humphreys, C. Pettenkofer, *et al.*, Influence of surface terminating species on electron emission from diamond surfaces, *Materials Research Society Symposium Proceedings* **416**, 263 (1996).
115. R.J. Nemanich, P.K. Baumann, M.C. Benjamin, *et al.*, Negative electron affinity surfaces of aluminum nitride and diamond, *Diamond and Related Materials* **5**, 790 (1996).
116. M.J. Powers, M.C. Benjamin, L.M. Poster, *et al.*, Observation of a negative electron affinity for boron nitride, *Applied Physics Letters* **67**, 3912 (1995).
117. M.C. Benjamin, C. Wang, R.F. Davis, and R.J. Nemanich, Observation of a negative electron affinity for heteroepitaxial AlN on α(6H)-SiC(0001), *Applied Physics Letters* **64**, 3288 (1994).
118. R.J. Nemanich, Electron affinity of AlN, GaN and AlGaN alloys. In *Properties, Processing and Applications of Gallium Nitride and Related Semiconductors*, ed. J. H. Edgar, S. Strite, I. Akasaki, *et al.*, EMIS Datareview Series, vol. **23** (INSPEC, IEE, 1998), pp. 98–103.
119. M.E. Levinshtein, S.L. Rumyantsev, and M.S. Shur, *Properties of Advanced Semiconductor Materials: GaN, AlN, InN, BN, SiC, SiGe* (John Wiley & Sons, Inc., New York, 2001).

120. V. Litovchenko, A. Evtukh, O. Yilmazoglu, *et al.*, Gunn effect in the field-emission phenomena, *Journal of Applied Physics* **97**, 044911 (2005).
121. C. Youtsey, L.T. Romano, R.J. Molnar, and I. Adesiva, Rapid evaluation of dislocation densities in n-type GaN films using photoenhanced wet etching, *Applied Physics Letters* **74**, 3537 (1999).
122. T. Utsumi, Vacuum microelectronics: What's new and exciting, *IEEE Transactions on Electron Devices* **38**, 2276 (1991).
123. A.F. Kravchenko and V.N. Ovsyuk, *Electronics Processes in Solid-State Low-Dimensional Systems* (Novosybirsk University Publisher 2000) (in Russian).
124. C. Bulutay, B.K. Ridley, and N.A. Zakhleniuk, Full-band polar optical phonon scattering analysis and negative differential conductivity in wurtzite GaN, *Physical Review B* **62**, 15754 (2000).
125. C.-K. Sun, Y.-L. Huang, S. Keller, *et al.*, Ultrafast electron dynamics study of GaN, *Physical Review B* **59**, 13535 (1999).
126. R.D. Fedorovich, A.G. Naumovets, and P.M. Tomchuk, Electron and light emission from island metal films and generation of hot electrons in nanoparticles, *Physics Reports* **328**, 73 (2000).
127. K.R. Shoulders, Microelectronics using electron-beam-activated machining techniques, *Advances in Computers* **2**, 135 (1961).
128. C.A. Spindt, A thin-film field-emission cathode, *Journal of Applied Physics* **39**, 3504 (1968).
129. D. Palmer, H.F. Gray, J. Mancusi, *et al.*, Silicon field emitter arrays with low capacitance and improved transconductance for microwave amplifier applications, *Journal of Vacuum Science and Technology B* **13**, 576 (1995).
130. F. Charbonnier, Developing and using the field emitter as a high intensity electron source, *Applied Surface Science* **94/95**, 26 (1996).
131. C.A. Spindt, C.E. Holland, P.R. Schwoebel, and I. Brodie, Field-emitter-array development for microwave applications, *Journal of Vacuum Science and Technology B* **14**, 1986 (1996).
132. H. Takemura, Y. Tomihari, N. Furutake *et al.* (1997) A novel vertical current limiter fabricated with a deep trench forming technology for highly reliable field emitter arrays, *Technical Digest of the 1997 IEEE-IEDM, IEEE, Piscataway, NJ, 1997*, p. 709.
133. C.A. Spindt, C.E. Holland, P.R. Schwoebel, and I. Brodie, Field emitter array development for microwave applications. II, *Journal of Vacuum Science and Technology B* **16**, 758 (1998).
134. K.L. Jensen, Field emitter arrays for plasma and microwave source applications, *Physics of Plasmas* **6**, 2241 (1999).
135. J.P. Calame, H.F. Gray, and J.L. Shaw, Analysis and design of microwave amplifiers employing field-emitter arrays, *Journal of Applied Physics* **73**, 1485 (1993).
136. J.A. Eichmeier and M.K. Thumm *Vacuum Electronics*, Springer-Verlag, Berlin, 2008.
137. H.L. Hartnagel, O. Yilmazoglu, V. Litovchenko *et al.* (2009) Field emission from quantum structures and concepts for suitable microwave circuit structures, *Technical Digest of IVNC 2009, Hamamatsu, Japan, 2009*, pp. 77–78.
138. D.R. Whaley, R. Duggal, C.M. Armstrong, *et al.*, 100 W operation of a cold cathode TWT, *IEEE Transactions on Electron Devices* **56**, 895 (2009).
139. D. Whaley, R. Duggal, C. Armstrong, *et al.* (2013) High average power field emitter cathode and testbed for X/Ku-band cold cathode TWT, *Abstract Book of IVNC 2013, Roanoke VA, 2013*, pp. 25–26.
140. H.W.P. Koops and H. Fukuda, (2013) "Giant current density" and "anomalous electron transport" observed at room temperature with nanogranular materials, *Abstract Book of IVNC 2013, Roanoke VA, 2013*, pp. 81–82.
141. J. Kretz, M. Rudolph, M. Weber, and H.W.P. Koops, Three-dimensional structurization by additive lithography, analysis of deposits using TEM and EDX, and application to field-emitter tips, *Microelectronic Engineering* **23**, 477 (1994).
142. F. Floreani, H.W. Koops, and W. Elsäßer, Concept of a miniaturised free-electron laser with field emission source, *Nuclear Instruments and Methods in Physics Research Section A* **483**, 488–492 (2002).
143. C.D. Child, Discharge from hot CaO, *Physics Reviews* **32** (Series 1), 492 (1911).
144. I. Langmuir, The effect of space charge and initial velocities on the potential distribution and thermionic current between parallel plane electrodes, *Physics Reviews* **21** (Series 2), 419 (1923).
145. K.L. Jensen, M.A. Kodis, R.A. Murphy, and E.G. Zaidman, Space charge effects on the current-voltage characteristics of gated field emitter arrays, *Journal of Applied Physics* **82**, 845 (1997).
146. S. Ramo, J.R. Whinnery, and T. van Duzer, *Fields and Waves in Communication Electronics*, John Wiley & Sons, Inc.: New York, 1965.
147. R.E. Collin, *Foundations for Microwave Engineering*, New York: McGraw-Hill, 1966.

148. R.A. Murphy, C.T. Harris, R.H. Mathews *et al.* (1997) Fabrication of field-emitter arrays for inductive output amplifiers. Proceedings of the IEEE International Conference on Plasma Science, San Diego, CA, 1997, p. 127.
149. M. Abramowitz and I.A. Stegun, *Handbook of Mathematical Functions*, Dover: New York, 1965.
150. J.E. Rowe, A large-signal analysis of the traveling-wave amplifier : Theory and general results, *IRE Transactions on Electron Devices* **3**, 39 (1956).
151. Y.Y. Lau and D. Chernin, A review of the ac space-charge effect in electron–circuit interactions, *Physics of Fluids B: Plasma Physics* **4**, 3473 (1992).
152. H.P. Freund, E.G. Zaidman, and T.M. Antonsen, Theory of helix traveling wave tubes with dielectric and vane loading, *Physics of Plasmas* **3**, 3145 (1996).
153. D.N. Smithe, B. Goplen, M.A. Kodis, and N.R. Vanderplaats (1995) Predicting twystrode output performance. Proceedings of the IEEE International Conference on Plasma Science, Madison, WI, 1995.
154. J.P. Barbour, W.W. Dolan, J.K. Trolan, *et al.*, Space-charge effects in field emission, *Physics Reviews* **92**, 45 (1953).
155. W.A. Anderson, Role of space charge in field emission cathodes, *Journal of Vacuum Science and Technology B* **11**, 383 (1993).
156. Y.Y. Lau, Y. Liu, and R.K. Parker, Electron emission: From the Fowler–Nordheim relation to the Child–Langmuir law, *Physics of Plasmas* **1**, 2082 (1994).
157. L. Yun-Peng and L. Enze, Space charge of field emission triode, *Applied Surface Science* **76/77**, 7 (1994).
158. W.I. Milne and M.T. Cole, (2012) Carbon nanotubes for field emission applications, *Technical Digest of IVNC 2012, Jeju, Korea, 2012*, pp. 8–11.
159. W.I. Milne, K.B.K. Teo, G.A.J. Amaratunga, *et al.*, Microwave devices: Carbon nanotubes as field emission sources, *Journal of Materials Chemistry* **14**, 933–943 (2004).
160. K.B.K. Teo, E. Minoux, L. Hudanski, *et al.*, *Nature* **437**, 968 (2005).
161. L. Hudanski, E. Minoux, L. Gangloff, *et al.*, Carbon nanotube based photocathodes, *Nanotechnology* **19**, 105201 (2008).

Index

Vacuum Nanoelectronic Devices: Novel Electron Sources and Applications, First Edition.
Anatoliy Evtukh, Hans Hartnagel, Oktay Yilmazoglu, Hidenori Mimura and Dimitris Pavlidis.